SAFETY OF SEA TRANSPORTATION

PROCEEDINGS OF THE INTERNATIONAL CONFERENCE ON MARINE NAVIGATION AND SAFETY OF SEA TRANSPORTATION (TRANSNAV 2017), GDYNIA, POLAND, 21–23 JUNE 2017

Safety of Sea Transportation

Marine Navigation and Safety of Sea Transportation

Editors

Adam Weintrit & Tomasz Neumann
Gdynia Maritime University, Gdynia, Poland

CRC Press
Taylor & Francis Group
Boca Raton London New York

CRC Press is an imprint of the
Taylor & Francis Group, an **informa** business
A BALKEMA BOOK

Published by:
CRC Press/Balkema
P.O. Box 447, 2300 AK Leiden, The Netherlands
e-mail: Pub.NL@taylorandfrancis.com
www.crcpress.com – www.taylorandfrancis.com

First issued in paperback 2020

Typeset by V Publishing Solutions Pvt Ltd., Chennai, India

ISBN 13: 978-0-367-73614-9 (pbk)
ISBN 13: 978-1-138-29768-5 (hbk)

**Visit the Taylor & Francis Web site at
http://www.taylorandfrancis.com**

**and the CRC Press Web site at
http://www.crcpress.com**

Contents

General Chair of TransNav Conference
Prof. Dr. Adam **Weitntrit**, FRIN, FNI, Master Mariner

Executive Chair of TransNav Conference
Dr. Tomasz **Neumann**

Honorary Committee Members
President of the Nautical Institute: Captain David (Duke) **Snider** FNI
Rector of Gdynia Maritime University: Prof. Dr. Janusz **Zarębski**

Host of TransNav Conference
Prof. Dr. Leszek **Smolarek**, Dean of Faculty of Navigation of Gdynia Maritime University

List of Scientific Program Committee Members of the 12th International Conference on Marine Navigation and Safety of Sea Transportation – TransNav 2017 21-23 June 2017, Gdynia, Poland

*Prof. Agnar **Aamodt**, Norwegian University of Science and Technology, Trondheim, Norway*
*Prof. Ajith **Abraham**, Scientific Network for Innovation and Research Excellence, Auburn, Washington, USA*
*Prof. Teresa **Abramowicz-Gerigk**, Gdynia Maritime University, Gdynia, Poland*
*Prof. Michele **Acciaro**, Kühne Logistics University, Hamburg, Germany*
*Prof. Sauli **Ahvenjarvi**, Satakunta University of Applied Sciences, Rauma, Finland*
*Prof. Paolo **Alfredini**, University of São Paulo, Polytechnic School, São Paulo, Brazil*
*Prof. Daniel N. **Aloi**, Oakland University, Rochester, Michigan, The United States*
*Prof. Anatoli **Alop**, Estonian Maritime Academy, Tallin, Estonia*
*Prof. Karin **Andersson**, Chalmers University of Technology, Gothenburg, Sweden*
*Prof. Yasuo **Arai**, Marine Technical College, Ashiya, Hyogo, Japan*
*Prof. Terje **Aven**, University of Stavanger (UiS), Norway*
*Prof. Michael **Baldauf**, Word Maritime University, Malmö, Sweden*
*Prof. Andrzej **Banachowicz**, West Pomeranian University of Technology, Szczecin, Poland*
*Prof. Marek **Banaszkiewicz**, Space Research Center, Polish Academy of Sciences, Warsaw, Poland*
*Prof. Marcin **Barlik**, Warsaw University of Technology, Poland*
*Prof. Eugen **Barsan**, Constanta Maritime University, Romania*
*Prof. Milan **Batista**, University of Ljubljana, Ljubljana, Slovenia*
*Prof. Ghiorghe **Batrinca**, Constantza Maritime University, Romania*
*Prof. Raphael **Baumler** , World Maritime University, Malmö, Sweden*
*Prof. Angelica **Baylon**, Maritime Academy of Asia & the Pacific, Philippines*
*Prof. Knud **Benedict**, University of Wismar, University of Technology, Business and Design, Germany*
*Prof. Christophe **Berenguer**, Grenoble Institute of Technology, Saint Martin d'Hères, France*
*Prof. Heinz Peter **Berg**, Bundesamt für Strahlenschutz, Salzgitter, Germany*
*Prof. Tor Einar **Berg**, Norwegian Marine Technology Research Institute, Trondheim, Norway*
*Prof. Carmine Giuseppe **Biancardi**, The University of Naples „Parthenope", Naples, Italy*
*Prof. Vitaly **Bondarev**, Baltic Fishing Fleet State Academy, Kaliningrad, Russia*
*Prof. Neil **Bose**, Australian Maritime College, University of Tasmania, Launceston, Australia*
*Prof. Jarosław **Bosy**, Wroclaw University of Environmental and Life Sciences, Wroclaw, Poland*
*Prof. Alexey **Boykov**, Moscow State Academy of Water Transport, Moscow, Russia*
*Prof. Andrzej **Bujak**, WSB Schools of Banking, Wrocław, Poland*
*Prof. Zbigniew **Burciu**, Gdynia Maritime University, Gdynia, Poland*
*Sr. Jesus **Carbajosa Menendez**, President of Spanish Institute of Navigation, Spain*
*Prof. Doina **Carp**, Constanta Maritime University, Romania*
*Prof. Ayşe Güldem **Cerit**, Dokuz Eylül University, Izmir, Turkey*
*Prof. Shyy Woei **Chang**, National Kaohsiung Marine University, Taiwan*
*Prof. Adam **Charchalis**, Gdynia Maritime University, Gdynia, Poland*
*Prof. Wu **Chen**, Hong Kong Polytechnic University, Hong Kong*
*Prof. Andrzej **Chudzikiewicz**, Warsaw University of Technology, Poland*
*Prof. Frank **Coolen**, Durham University, The United Kingdom*
*Prof. Thomas **Cottier**, University of Bern, Switzerland*
*Prof. Kevin **Cullinane**, University of Newcastle upon Tyne, The United Kingdom*
*Prof. Jerzy **Czajkowski**, Gdynia Maritime University, Gdynia, Poland*
*Prof. Krzysztof **Czaplewski**, Gdynia Maritime University, Gdynia, Poland*
*Prof. Ireneusz **Czarnowski**, Gdynia Maritime University, Gdynia, Poland*
*Prof. Mirosław **Czechowski**, Gdynia Maritime University, Gdynia, Poland*
*Prof. German **de Melo Rodriguez**, Polytechnic University of Catalonia, Barcelona, Spain*

Prof. Robert **De Souza**, National University of Singapore NUS, Singapore
Prof. Decio Crisol **Donha**, Polytechnic School, University of São Paulo, Brazil
Prof. Patrick **Donner**, World Maritime University, Malmö, Sweden
Prof. Eamonn **Doyle**, Irish Institute of Master Mariners, Cork, Ireland
Prof. Branislav **Dragović**, University of Montenegro, Kotor, Montenegro
Prof. Daniel **Duda**, Polish Naval Academy, Polish Nautological Society, Poland
Prof. Czesław **Dyrcz**, Polish Naval Academy, Gdynia, Poland
Prof. Marek **Dzida**, Gdańsk University of Technology, Gdańsk, Poland
Prof. Milan **Džunda**, Technical University of Košice, Slovakia
Prof. Billy **Edge**, North Carolina State University, US
Prof. Bernd **Eissfeller**, Institute of Space Technology and Space Applications, Universitaet der Bundeswehr Munich, Germany
Prof. Ahmed **El-Rabbany**, University of New Brunswick; Ryerson University in Toronto, Ontario, Canada
Prof. Naser **El-Sheimy**, The University of Calgary, Canada
Prof. Akram **Elentably**, King Abdulaziz University (KAU), Jeddah, Saudi Arabia
Prof. Tarek A. **Elsayed**, Arab Academy for Science, Technology & Maritime Transport, Alexandria, Egypt
Prof. William **Emery**, Colorado University, Boulder, The United States
Prof. Sophia **Everett**, Victoria University, Melbourne, Australia
Prof. Odd M. **Faltinsen**, Norwegian University of Science and Technology, Trondheim, Norway
Prof. Jeffrey **Falzarano**, Texas A&M University, Corpus Christi, US
Prof. Alfonso **Farina**, SELEX-Sistemi Integrati, Rome, Italy
Prof. Andrzej **Fellner**, Silesian University of Technology, Katowice, Poland
Prof. Andrzej **Felski**, Polish Naval Academy, Gdynia, Poland
Prof. Yanming **Feng**, Queensland University of Technology, Brisbane, Queensland, Australia
Prof. Włodzimierz **Filipowicz**, Gdynia Maritime University, Gdynia, Poland
Prof. Renato **Filjar**, University College of Applied Sciences, Bjelovar, Croatia
Prof. Börje **Forssell**, Norwegian University of Science and Technology, Trondheim, Norway
Prof. Alberto **Francescutto**, University of Trieste, Trieste, Italy
Prof. Erik **Franckx**, Vrije Universiteit Brussel, Brussels, Belgium
Prof. Jens **Froese**, Jacobs University Bremen, Germany
Prof. Masao **Furusho**, Kobe University, Japan
Prof. Wiesław **Galor**, Maritime University of Szczecin, Poland
Prof. Yang **Gao**, University of Calgary, Canada
Prof. Aleksandrs **Gasparjans**, Latvian Maritime Academy, Riga, Latvia
Prof. Jerzy **Gaździcki**, President of the Polish Association for Spatial Information; Warsaw, Poland
Prof. Avtandil **Gegenava**, Georgian Maritime Transport Agency, Maritime Rescue Coordination Center, Georgia
Prof. Mirosław **Gerigk**, Gdańsk University of Technology, Gdańsk, Poland
Prof. Hassan **Ghassemi**, Amirkabir University of Technology (AUT), Teheran, Iran
Prof. Witold **Gierusz**, Gdynia Maritime University, Gdynia, Poland
Prof. Dariusz **Gotlib**, Warsaw University of Technology, Warsaw, Poland
Prof. Martha R. **Grabowski**, Le Moyne College; Rensselaer Polytechnic Institute, US
Prof. Dorota **Grejner-Brzezinska**, The Ohio State University, United States of America
Prof. Norbert **Gruenwald**, University of Applied Sciences Technology, Business and Design, Wismar, Germany
Prof. Marek **Grzegorzewski**, Polish Air Force Academy, Deblin, Poland
Prof. Andrzej **Grzelakowski**, Gdynia Maritime University, Gdynia, Poland
Prof. Marek **Grzybowski**, Gdynia Maritime University, Gdynia, Poland
Prof. Lucjan **Gucma**, Maritime University of Szczecin, Poland
Prof. Stanisław **Gucma**, Maritime University of Szczecin, Poland
Prof. Carlos **Guedes Soares**, Instituto Superior Técnico, Universidade Técnica de Lisboa, Portugal
Prof. Seung-Gi **Gug**, Korea Maritime and Ocean University, Pusan, Korea
Prof. Hans-Dietrich **Haasis**, University of Bremen, Germany
Prof. Jerzy **Hajduk**, Maritime University of Szczecin, Poland
Prof. Esa **Hämäläinen**, University of Turku, Finland
Prof. Jong-Khil **Han**, Sungkyul University, Anyang, Gyeonggi-do, South Korea
Prof. Kazuhiko **Hasegawa**, Osaka University, Osaka, Japan
Prof. Peter J. **Hayes**, California Maritime Academy, California State University, Vallejo, California, The United States
Prof. Bernhard **Hofmann-Wellenhof**, Graz University of Technology, Graz, Austria
Prof. Serge Paul **Hoogendoorn**, Delft University of Technology, Delft, The Netherlands
Prof. Mohammed **Hossam-E-Haider**, Military Institute of Science and Technology, Dhaka, Bangladesh
Prof. Qinyou **Hu**, Shanghai Maritime University, Shanghai, China
Prof Carl **Hult**, Kalmar Maritime Academy, Linnaeus University, Kalmar, Sweden
Prof. Marek **Idzior**, Poznań University of Technology, Poznań, Poland
Prof. Stojce Dimov **Ilcev**, Durban University of Technology, South Africa
Prof. Akio **Imai**, Kobe University, Japan
Prof. Toshio **Iseki**, Tokyo University of Marine Science and Technology, Tokyo, Japan,
Prof. Marianna **Jacyna**, Warsaw University of Technology, Poland
Prof. Jacek **Jania**, University of Silesia in Katowice, Poland
Prof. Ales **Janota**, University of Žilina, Slovakia
Prof. Maurice **Jansen**, Erasmus University Rotterdam, The Netherlands
Prof. Jacek **Januszewski**, Gdynia Maritime University, Gdynia, Poland
Prof. Piotr **Jędrzejowicz**, Gdynia Maritime University, Gdynia, Poland

*Prof. Wacław **Morgaś**, Polish Naval Academy, Gdynia, Poland*
*Prof. Junmin **Mou**, Wuhan University of Technology, Wuhan, China*
*Prof. Reinhard **Mueller-Demuth**, Hochschule Wismar, Germany*
*Prof. Takeshi **Nakazawa**, World Maritime University, Malmoe, Sweden*
*Prof. Janusz **Narkiewicz**, Warsaw University of Technology, Poland*
*Prof. Rudy R. **Negenborn**, Delft University of Technology, Delft, The Netherlands*
*Prof. John **Niedzwecki**, Texas A&M University, College Station, Texas, The United States*
*Prof. Nikitas **Nikitakos**, University of the Aegean, Chios, Greece*
*Prof. Andy **Norris**, The Royal Institute of Navigation, University of Nottingham, The United Kingdom*
*Prof. Gabriel **Nowacki**, Military University of Technology, Warsaw, Poland*
*Prof. Tomasz **Nowakowski**, Wrocław University of Technology, Poland*
*Prof. Oliver Chinagorom **Ojinnaka**, University of Nigeria, U.N.E.C, Nigeria*
*Prof. Aykut I. **Ölcer**, World Maritime University, Malmö, Sweden*
*Prof. Stanisław **Oszczak**, University of Warmia and Mazury in Olsztyn, Poland*
*Prof. Zbigniew **Otremba**, Gdynia Maritime University, Gdynia, Poland*
*Prof. Kjell Ivar **Øvergård**, University College of Southeast Norway*
*Prof. Photis **Panayides**, Cyprus University of Technology, Nicosia, Cyprus*
*Prof. Dimos **Pantazis**, Technological Educational Institute of Athens, Greece*
*Prof. Gyei-Kark **Park**, Mokpo National Maritime University, Mokpo, Korea*
*Prof. Jin-Soo **Park**, Korea Maritime and Ocean University, Pusan, Korea*
*Prof. António Manuel dos Santos **Pascoal**, Instituto Superior Tecnico, Lisboa, Portugal*
*Mr. David **Patraiko**, The Nautical Institute, London, The United Kingdom*
*Prof. Vytautas **Paulauskas**, Maritime Institute College, Klaipeda University, Lithuania*
*Prof. Jan **Pawelski**, Gdynia Maritime University, Gdynia, Poland*
*Prof. Thomas **Pawlik**, Bremen University of Applied Sciences, Germany*
*Prof. Egil **Pedersen**, Norwegian University of Science and Technology, Trondhein, Norway*
*Prof. Michael Robert **Phillips** , University of Wales Trinity Saint David, Swansea, Wales, The United Kingdom*
*Prof. Zbigniew **Pietrzykowski**, Maritime University of Szczecin, Poland*
*Prof. Francisco **Piniella**, University of Cadiz, Spain*
*Prof. Marzenna **Popek**, Gdynia Maritime University, Gdynia, Poland*
*Prof. Thomas **Porathe**, Norwegian University of Science and Technology, Trondheim, Norway*
*Prof. Malek **Pourzanjani**, South African International Maritime Institute (SAIMI), Port Elizabeth, South Africa*
*Prof. Tomasz **Praczyk**, Polish Naval Academy, Gdynia, Poland*
*Prof. Boris **Pritchard**, University of Rijeka, Croatia*
*Prof. Adam **Przybyłowski**, Gdynia Maritime University, Gdynia, Poland*
*Prof. Dorota **Pyć**, University of Gdańsk, Poland*
*Prof. Refaat **Rashad**, Arab Academy for Science and Technology and Maritime Transport in Alexandria, Egypt*
*Prof. Martin **Renilson**, Australian Maritime College, University of Tasmania, Launceston, Australia*
*Prof. Jonas **Ringsberg**, Chalmers University of Technology, Gothenburg, Sweden*
*Prof. Jerzy B. **Rogowski**, Gdynia Maritime University, Gdynia, Poland*
*Prof Hermann **Rohling**, Hamburg University of Technology, Hamburg, Germany*
*Prof. María Carolina **Romero Lares**, World Maritime University, Malmö, Sweden*
*Prof. Paulo Rosa **Santos** , University of Porto (FEUP), Porto, Portugal*
*Prof. Heinz **Ruther**, University of Cape Town, Rondebosch, South Africa*
*Prof. Abdul Hamid **Saharuddin**, Universiti Malaysia Terengganu (UMT), Terengganu, Malaysia*
*Prof. Helen **Sampson**, Cardiff University, Wales, The United Kingdom*
*Prof. Jens-Uwe **Schröder-Hinrichs**, World Maritime University, Malmö, Sweden*
*Prof. Viktoras **Sencila**, Lithuanian Maritime Academy, Klaipeda, Lithuania*
*Prof. Zahid Ziyadkhan **Sharifov**, Azerbaijan State Marine Academy, Baku, Azerbaijan*
*Prof. Chaojian **Shi**, Shanghai Maritime University, China*
*Prof. Mirosław **Siergiejczyk**, Warsaw University of Technology, Poland*
*Prof. Jacek **Skorupski**, Warsaw University of Technology, Poland*
*Prof. Wojciech **Ślączka**, Maritime University of Szczecin, Poland*
*Prof. Roman **Śmierzchalski**, Gdańsk University of Technology, Poland*
*Prof. Leszek **Smolarek**, Gdynia Maritime University, Gdynia, Poland*
*Prof. Henryk **Śniegocki**, Gdynia Maritime University, Gdynia, Poland*
*Prof. Joanna **Soszyńska-Budny**, Gdynia Maritime University, Gdynia, Poland*
*Prof. John Dalsgaard **Sørensen**, Aalborg University, Denmark*
*Prof. Jac **Spaans**, Netherlands Institute of Navigation, The Netherlands*
*Prof. Cezary **Specht**, Gdynia Maritime University, Gdynia, Poland*
*Prof. Vernon **Squire**, University of Otago, New Zealand*
*Prof. Andrzej **Stateczny**, Gdańsk University of Technology, Poland*
*Prof. Andrzej **Stepnowski**, Gdańsk University of Technology, Poland*
*Prof. Manivannan **Subramaniam**, Malaysian Maritime Academy (ALAM), Kuala Sungai Baru, Melaka, Malaysia*
*Prof. Jan **Szantyr**, Gdańsk University of Technology, Poland*
*Prof. Janusz **Szpytko**, AGH University of Science and Technology, Kraków, Poland*
*Prof. Tomasz **Szubrycht**, Polish Naval Academy, Gdynia, Poland*
*Prof. Elżbieta **Szychta**, Lodz University of Technology, Lodz, Poland*
*Prof. Marek **Szymoński**, Polish Naval Academy, Gdynia, Poland*
*Prof. Hai Tung **Ta**, University of Science and Technology, Hanoi, Vietnam*

List of reviewers

Prof. Teresa **Abramowicz-Gerigk**, Gdynia Maritime University, Gdynia, Poland
Prof. Michele **Acciaro**, Kühne Logistics University, Hamburg, Germany
Prof. Sauli **Ahvenjarvi**, Satakunta University of Applied Sciences, Rauma, Finland
Prof. Paolo **Alfredini**, University of São Paulo, Polytechnic School, São Paulo, Brazil
Prof. Anatoli **Alop**, Estonian Maritime Academy, Tallin, Estonia
Prof. Karin **Andersson**, Chalmers University of Technology, Gothenburg, Sweden
Prof. Ghiorghe **Batrinca**, Constantza Maritime University, Romania
Prof. Angelica **Baylon**, Maritime Academy of Asia & the Pacific, Philippines
Prof. Carmine Giuseppe **Biancardi**, The University of Naples „Parthenope", Naples, Italy
Prof. Doina **Carp**, Constanta Maritime University, Romania
Prof. Shyy Woei **Chang**, National Kaohsiung Marine University, Taiwan
Prof. Adam **Charchalis**, Gdynia Maritime University, Gdynia, Poland
Prof. Andrzej **Chudzikiewicz**, Warsaw University of Technology, Poland
Dr. Dimitrios **Dalaklis**, World Maritime University, Malmö, Sweden
Prof. German de Melo **Rodriguez**, Polytechnic University of Catalonia, Barcelona, Spain
Prof. Sophia **Everett**, Victoria University, Melbourne, Australia
Prof. Andrzej **Fellner**, Silesian University of Technology, Katowice, Poland
Prof. Erik **Franckx**, Vrije Universiteit Brussel, Brussels, Belgium
Prof. Aleksandrs **Gasparjans**, Latvian Maritime Academy, Riga, Latvia
Prof. Avtandil **Gegenava**, Georgian Maritime Transport Agency, Maritime Rescue Coordination Center, Georgia
Prof. Hassan **Ghassemi**, Amirkabir University of Technology (AUT), Teheran, Iran
Prof. Marek **Grzegorzewski**, Polish Air Force Academy, Deblin, Poland
Prof. Andrzej **Grzelakowski**, Gdynia Maritime University, Gdynia, Poland
Prof. Marek **Grzybowski**, Gdynia Maritime University, Gdynia, Poland
Prof. Lucjan **Gucma**, Maritime University of Szczecin, Poland
Prof. Esa **Hämäläinen**, University of Turku, Finland
Prof. Peter J. **Hayes**, California Maritime Academy, California State University, Vallejo, California, The United States
Prof Carl **Hult**, Kalmar Maritime Academy, Linnaeus University, Kalmar, Sweden
Prof. Marianna **Jacyna**, Warsaw University of Technology, Poland
Prof. Ales **Janota**, University of Žilina, Slovakia
Prof. Maurice **Jansen**, Erasmus University Rotterdam, The Netherlands
Prof. Lech **Kobyliński**, Polish Academy of Sciences, Ship Handling Research and Training Centre, Ilawa, Poland
Prof. Uday **Kumar**, Luleå University of Technology, Luleå, Sweden
Prof. Kezhong **Liu**, Wuhan University of Technology, Wuhan, Hubei, China
Prof. Dieter **Lompe**, Hochschule Bremerhaven, Germany
Prof. Chin-Shan **Lu**, Hong Kong Polytechnic University, Hong Kong
Prof. Cezary **Łuczywek**, Gdynia Maritime University, Gdynia, Poland
Prof. Mirosław **Luft**, University of Technology and Humanities in Radom, Poland
Prof. Zbigniew **Łukasik**, University of Technology and Humanities in Radom, Poland
Prof. Tihomir **Luković**, University of Dubrovnik, Croatia
Prof. Margareta **Lützhöft**, Australian Maritime College, University of Tasmania, Launceston, Australia
Prof. Michael Ekow **Manuel**, World Maritime University, Malmoe, Sweden
Prof. Francesc Xavier **Martinez de Oses**, Polytechnical University of Catalonia, Barcelona, Spain
Prof. Jerzy **Matusiak**, Helsinki University of Technology, Helsinki, Finland
Prof. Boyan **Mednikarov**, Nikola Y. Vaptsarov Naval Academy, Varna, Bulgaria
Prof. Max **Mejia**, World Maritime University, Malmö, Sweden
Prof. Jerzy **Merkisz**, Poznań University of Technology, Poznań, Poland
Prof. Jerzy **Mikulski**, University of Economics in Katowice, Poland
Prof. Waldemar **Mironiuk**, Polish Naval Academy, Gdynia, Poland
Prof. Nikitas **Nikitakos**, University of the Aegean, Chios, Greece
Prof. Gabriel **Nowacki**, Military University of Technology, Warsaw, Poland
Prof. Tomasz **Nowakowski**, Wroclaw University of Technology, Poland
Prof. Zbigniew **Otremba**, Gdynia Maritime University, Gdynia, Poland
Prof. Kjell Ivar **Øvergård**, University College of Southeast Norway
Mr. David **Patraiko**, The Nautical Institute, London, The United Kingdom
Prof. Jan **Pawelski**, Gdynia Maritime University, Gdynia, Poland
Prof. Thomas **Pawlik**, Bremen University of Applied Sciences, Germany
Prof. Marzenna **Popek**, Gdynia Maritime University, Gdynia, Poland
Prof. Adam **Przybyłowski**, Gdynia Maritime University, Gdynia, Poland
Prof. Dorota **Pyć**, University of Gdańsk, Poland
Prof. María Carolina **Romero Lares**, World Maritime University, Malmö, Sweden
Prof. Helen **Sampson**, Cardiff University, Wales, The United Kingdom
Prof. Jens-Uwe **Schröder-Hinrichs**, World Maritime University, Malmö, Sweden
Prof. Viktoras **Sencila**, Lithuanian Maritime Academy, Klaipeda, Lithuania
Prof. Zahid Ziyadkhan **Sharifov**, Azerbaijan State Marine Academy, Baku, Azerbaijan
Prof. Mirosław **Siergiejczyk**, Warsaw University of Technology, Poland
Prof. Leszek **Smolarek**, Gdynia Maritime University, Gdynia, Poland

Safety of Sea Transportation
Introduction

A. Weintrit & T. Neumann
Gdynia Maritime University, Gdynia, Poland
Chairman and Secretary of TransNav Conference

I am pleased to present another volume of the proceedings related to the TransNav International Conference on Marine Navigation and Safety of Sea Transportation, published by CRC Press/Balkema, Taylor and Francis Group.

The 12th International Conference TransNav 2017 was held from 21 to 23 June 2017, organized jointly by the Faculty of Navigation of the Gdynia Maritime University and The Nautical Institute (http://transnav.am.gdynia.pl).

The Conference Proceedings are addressed to scientists and professionals in order to share their expert knowledge, experience, and research results concerning all aspects of navigation, safety of navigation and sea transportation. The book contains original papers contributing to the science of broadly defined navigation: from the highly technical to the descriptive and historical, over land through the sea, air and space, to promote young researchers, new field of maritime sciences and technologies.

The Twelfth Edition of the most innovative World conference on maritime transport research is designed to find solutions to challenges in waterborne transport, navigation and shipping, mobility of people and goods with respect to energy, infrastructure, environment, safety and security as well as to the economic issues.

The focus of TransNav Conference is high-quality, scholarly research that addresses development, application and implications, in the field of maritime education, maritime safety management, maritime policy sciences, maritime industries, marine environment and energy technology. Subjects of papers include nautical science, electronics, automation, robotics, geodesy, astronomy, mathematics, cartography, hydrography, communication and computer sciences, command and control engineering, psychology, operational research, risk analysis, theoretical physics, ecology, operation in hostile environments, instrumentation, administration, ergonomics, economics, financial planning and law. Also of interest are logistics,

transport, mobility and ocean technology. The TransNav Conference provides a forum for transportation researchers, scientists, engineers, navigators, ergonomists, and policy-makers with an interest in maritime researches.From contemporary issues to the scientific, technological, political, economic, cultural and social aspects of maritime shipping, transportation and navigation, the TransNav publishes innovative, interdisciplinary and multidisciplinary research on marine navigation subjects and is set to become the leading international scholarly journal specialising in debate and discussion on maritime subjects. The TransNav is especially concerned to set maritime studies in a broad international and comparative context.

The content of the **TransNav 2017** Conference Proceedings, **Part 2**, with subtitle *Safety of Sea Transportation*, is partitioned into sixteen separate chapters (sections) titled: *Sustainability, Intermodal and Multimodal Transportation* (covering four papers), *Safety and Hydrodynamic Study of Hydrotechnical Structures* (covering three papers), *Bunkering and Fuel Consumption* (covering three papers), *Gases Emission, Water Pollution and Environmental Protection* (covering eight papers), *Occupational Accidents* (covering two papers), *Supply Chain of Blocks and Spare Parts* (covering three papers), *Electrotechnical Problems* (covering six papers), *Ships Stability and Loading Strength* (covering two papers), *Cargo Loading and Port Operations* (covering three papers), *Maritime Education and Training* (covering three papers), *Human Factor, Crew Manning and Seafarers Problems* (covering seven papers), *Economic Analysis* (covering two papers), *Mathematical Models, Methods and Algorithms* (covering five papers), *Fishery* (covering two papers), *Legal Aspects* (covering four papers), and *Aviation* (covering two papers).

In each of them readers can find a few papers. Papers collected in the first chapter, titled *Sustainability, Intermodal and Multimodal*

Transportation, concern sustainability (towards a new consciousness), informational provision of risk management in maritime and multimodal transportation, port cities smart and sustainable development challenges (Gdynia case study), and analysis of infrastructure ports and access road and rail to Tri-City seaport.

In the second chapter there are described problems related to safety of hydrotechnical structures: hydrodynamic study of nautical and shore protection structures in Santos Bay, safety of ships moored to quay walls, and tools for evaluation quay toe scouring induced by vessel propellers in harbour basins during the docking and undocking manoeuvring.

The third chapter deals with bunkering and fuel consumption. In this section there is described ship fuel consumption prediction under various weather condition based on DBN (Deep Belief Networks), analysis of electric powertrain application to drive an inland waterway barges, and bunkering and bunkering decisions (the literature review).

The fourth chapter deals with gases emission, water pollution and environmental protection. In this section there are described the following issues: coastal dynamics and danger of chemical pollution of southeast sector of the Azov Sea, a new vision to monitoring tank cleaning, oil spill modelling with Pisces II around Bay of Izmir, evaluating air emission inventories and indicators from ferry vessels at ports, the structural types of oil spill response organizations (the comparisons of countries on oil spill response operations), the concept of "green ship": new developments and technologies, noise reduction in railway traffic as an element of greening of transport, and noise in road transport as a problem in European dimension.

The fifth chapter deals with occupational accidents. The content of the fifth chapter concerns investigation of occupational accidents on board with fuzzy AHP method, and a research on occupational accidents aboard merchant ships.

The sixth chapter, titled *Supply Chain of Blocks and Spare Parts,* covers the following issues: simulation-based modeling of block assembly area at shipyards, exploring the potential of 3D printing of the spare parts supply chain in the maritime industry, and modelling of short sea shipping tanker in Black Sea.

In the seventh chapter there are described the following electrotechnical problems: the impact of electromagnetic interferences on transport security system of certain reliability structure, increasing energy efficiency of commercial vessels (by using led lighting technology), electromagnetic compatibility of the radio devices in maritime shipping, universal recuperation system of electricity from the exhaust system of an internal combustion engine as the engine of small capacity, a new innovative turbocharger concept numerically tested and optimized with CFD, and green shipping hybrid diesel-electric new generation marine propulsion technologies.

In the eighth chapter there are described problems related to hips stability and loading strength: assessment of the realistic range of variation of ship equivalent metacentric height governing synchronous roll frequency, and buckling strength of rectangular plates with elastically restrained edges subjected to in-plane impact loading.

The ninth chapter deals with cargo loading and port operations. The contents of the ninth chapter cover: a comparison of loading conditions effects on the vertical motions of turret-moored FPSO, the analysis of container vessel service efficiency in the aspect of berth and handling equipment usage in Polish ports, and development investments at container terminals in the case of cargo congestion.

In the tenth chapter there are described problems related to Maritime Education and Training (MET): innovation methods of assessment and examination system for universities engaged in Bologna process, effects of deck cadets' working conditions on quantity and perceived quality of sleep among marine science students, and online learning technology in the academic educational process.

The eleventh chapter deals with human factor, crew manning and seafarers problems. The contents of the eleventh chapter concern the following issues: underlying causes of and potential measures to reduce long-term sick leave among employees in the service department on board Swedish passenger vessels, human reliability analysis of a complex pilotage operation, supporting seafarer and family well-being in the face of traumatic events: a before, during and after model, grey list danger of Turkish flagged vessels, companies and new seafarers generation needs and expectations (finding a balance), the connection between teamwork and political correctness competence provision for the seafarers, and evaluation of occupational risk factors for cardiovascular disease in Romanian seafarers.

In the twelfth chapter title *Economic Analysis* there are presented the following issues: economic analysis of introducing free public transport, and economic analysis of a vessel in service equipped with an LNG fueled ship engine.

The thirteenth chapter deals with mathematical models, methods and algorithms. In this section there are described the following issues: safety analysis of a new and innovative transhipping concept (a comparison of two Bayesian network models), extensions of the Cayley-Hamilton theorem to transfer matrices of linear systems, a method of assessing the safety of technical systems of the ship, reliability assessment of vessel's main engine by combining Markov analysis integrated with time dependent failures and method of vehicle routing problem.

In the fourteenth chapter there are described fishery problems: logistical approach to a fishing-industrial complex functioning, and the effects of burnout level on job satisfaction: an application on fishermen.

The fifteenth chapter deals with legal aspects. The content of the fifteenth chapter concerns the following issues: legal issues concerning the UN Convention on the Conditions for Registration of Ships (1986), development of competencies when taking into consideration the nowadays challenges inside organizations, investigation of the piracy causes (an quantitative research), and analysis of the Lusitania tragedy: crime or conspiracy?.

The last chapter deals with the problems related to airports and aviation. The content of the sixteenth chapter covers the current challenges within security systems at international airports, and ecological aspects associated with an operation of aviation electronic support systems.

Each paper was reviewed at least by three independent reviewers. As Editor, I would like to express my gratitude to distinguished authors and reviewers of submitted papers for their great contribution for expected success of the publication. Let me congratulate the authors and reviewers for their excellent work.

Sustainability, Intermodal and Multimodal Transportation

Sustainability – Towards a New Consciousness

M. Denc
Gdynia Maritime University, Gdynia, Poland

ABSTRACT: The idea of sustainable development makes the basis for a new thinking about civilization and strikes fundamental aspects of to-date human activity. It assumes the rejection of the present model of development chiefly focused on pursuing infinite economic growth; therefore it requires a transformation of consciousness in the direction of perceiving the relationship and harmony between economic, human and social values and their interdependence with nature. The article is to analyze sustainable shipping in the context of the wider concept of sustainable development. I verify the hypothesis about the necessity of its implementation through a process of multidimensional and long-term education covering the aspects of responsibility for the use of commodities, responsibility for present and future generations and human responsibility for nature as the three aspects of the proper functioning of sustainable shipping. The implementation of the sustainability intentions is still at an early stage and thus, it deserves a closer look and a thorough discussion within the circles of people related to maritime industry, maritime trade and economy if shipping is to become truly sustainable.

"Our problems stem from our acceptance of this filthy, rotten, system."

S. Stallone, actor

"It doesn't matter how green you make your lifestyle. Capitalism will never be sustainable."

BONO, U2 singer

"Attacking the rich is not envy, it is self-defence. The hoarding of the wealth is the cause of poverty. The rich are not indifferent to poverty; they create it and maintain it."

Jodie Foster, actress

"Sustainable development is like teenage sex – everybody claims they are doing it but most people aren't, and those that are, are doing that very badly."

Chris Spray, Northumbrian Water

1 THE IDEA OF SUSTAINABLE DEVELOPMENT

Developing countries claim that the global contamination of the environment, the breach of stability of eco-systems and climate change correspond to the model adopted by Western countries. Harald Welzer shares with us his observations that since the beginning of the human activity that is some 40 thousand years, the recent 250 years have destroyed the foundations of survival more than the previous 39 750. Western civilizations not only exploited resources without any restrictions, but also have produced environmentally harmful substances, which already hinder and will hinder survival and development in the future [1]. To make the matter worse, they also "visited" many other countries, felt at home and helped themselves with their resources and lavishly bestowed them with heaps of detritus of any kind. The founder of sociobiology, Edward Wilson noted: "if we assumed that the whole world would be at the consumer level of the USA, and the population of the globe would amount to 6 billion people, it would require development of space in the form of four additional planets earth. No one has ever questioned these numbers. Meanwhile the demagogues, supporters and promoters of the so called *"American way of life"* proclaim: "We want the rest of the world to become like us." This is insane. There is only one solution: we need to improve the economic situation of the majority and at the same time attempt to save such a large part of the planet as possible " [2]. Declaration only and failing to take these steps will have further disastrous consequences for the ecosystem of the whole globe.

"An outline of the new concept of development, taking into account the negative pressure on the anthropological dimension of nature, was put forward at the UN conference in Stockholm in 1972.

The first official document formulating quite vague goals of this idea was the Stockholm Declaration on the natural environment. Its main assumption was to emphasize that the subjugation of nature by human economic activity has led to such a depletion of natural resources that further unrestricted continuation of this practice will undoubtedly result in the destruction of the life on Earth. Hence, an alternative to unsustainable growth and overexploitation was to become the concept of development, which consists of harmonizing activities corresponding to the capabilities of the environment. Since then, the nature has become axiologically highlighted element of human development with its economy based only on environmentally friendly technology. This concept was raised under the name of *sustainability*.

Sustainability in the broadest sense is a set of postulated and recommended principles for practical use:

1 The principle of respect for sustainable development, which is also called the principle of green economy and its development based on rational and long term optimization of biosphere resources;

2 The principle of the integrity of the environment (ecosystem integrity); its essence is a recommendation to "think globally (as a whole, holistically), but act locally."

3 The principle of development through rational use of natural resources by means of appropriate technology and system of production which take into account and provide for the conservation of nature;

4 The principle of prevention, also known as the principle of elimination of pollution at source.

5 The principle of response to existing environmental threats, also called - not always justifiably - the principle of passive policy;

6 The principle of cooperation (partnership) and social (public) participation, also called the principle of community participation for solving environmental problems or the principle of socialization;

7 The principle of regionalization understood as a demand to adapt to the requirements of the protection of regional and local conditions and to enable local and regional authorities to choose the tools for implementing the idea of sustainable development;

8 The principle of the rule of law, which in the context of specific countries indicates the need for the reconstruction of the system of environmental law and the manner of its implementation, to make sure that every rule is strictly observed and there is no way to replace or amend the regulations in the name of so called the " higher necessity" or "public interest";

9 The principle of intergenerational observance of ecological justice. This principle of sustainable development deserves special attention. While the above mentioned rules focus mainly around environmental protection issues, the last principle of emphasizes that the protection, in addition to nature, deserves the human kind, especially the future generations BERNER (2006). This concept reformulated the concept of sustainable development, so that it covered not only shift in the economy onto the eco-friendly tracks, but also highlighted the necessity of changes in the overall conditions in which human kind is now, will be soon, and in the future. [5]

A widely accepted and frequently quoted definition of sustainable development comes from Report of the World Commission on Environment and Development: *Our Common Future*, also known as *Brundtland Report*: "Sustainable development is development that meets the needs of the present without compromising the ability of future generations to meet their own needs. It contains within it two key concepts: the concept of 'needs', in particular the essential needs of the world's poor, to which overriding priority should be given; and the idea of limitations imposed by the state of technology and social organization on the environment's ability to meet present and future needs." [6]

The idea of sustainable development seems to be reasonable alternative to the Washington consensus based on the monetarist economics and the principle of trickle down ("The *Trickle-Down* theory is the principle that the poor, who must subsist on table scraps dropped by the rich, can be best served by giving the rich bigger meals" – CHER, a singer) to the poorer stratum. The new paradigm of social development abandons the US models implemented during the current globalization, both in terms of the environment, as well as in economic strategy (Gawor, 2006: 99-120). However, I have serious concerns whether the elites will ever give up the established order, no matter what they say on the air, and further assume that the vast majority of developed civilization brought up in the spirit of consumerism with the current state of consciousness, habits and desires is not ready to make sacrifices in the form of reducing consumption and changing their lifestyles.

Rio de Janeiro held two important conferences on the Environment and Sustainable Development. The first one was held in 1992 under the mandate of the UN General Assembly in its resolution 64/236, on new ideas that allow for sustainable development, green economy and poverty eradication and the creation of an institutional framework for the development of these projects. The second conference was held in June 2012, two decades after the first. This shows how the shift of the paradigm to the environmentally friendly one is difficult to achieve and time-consuming under the current global order. The Kyoto Protocol of 1997, which obliges signatories to progressively reduce atmospheric emissions of greenhouse gases has not been ratified by the US and China because they pollute the most of the atmosphere, as well as India. Lack of the will on the part of the US also caused the withdrawal of Russia, Japan and Canada. It brings us to the conclusion that summits devoted to this issue have been more soap than substance. We are continuously witnessing stalling or a strange pursuit – evasion game making pretence of taking crucial measures.

"The specificity of the concept of sustainable development and its practical value are of the utmost importance for the construction of a new human civilization. All implemented local, national and transnational environmental programs, the recommendations of Agenda 21, the UN decade on education on sustainable development, are new forms of civilization quality. This in turn, the whole concept of sustainable development becomes a practical opposition to the globalization reaching its apogee. [7]

This requires reframing of the current vision and the purpose of the society functioning. First of all, abandonment of the principle of maximizing profits at all costs and its replacement with the principle of profit optimization. The former principle absolutizes one dimension of social existence of the human kind – the economic activity. On the contrary, the latter allows updating other aspects of human existence. It brings the attention that there are certain economic limits that must be met for the development of the human spirit as the most important aspect. In this way the shift from the unlimited quantitative approach to the qualitative approach is possible. (The world circulated with photos showing Bill Gates – a retired chairman of Microsoft, playing the guitar as his idea for retirement. For this you do not need a fortune of a global corporation head. Vide: M. Minta-Kobus, *Ostatni dzień w pracy*, „Dziennik", 28-29.06.2008, p. 32.)

Current trends tell us to measure social development with mathematical indicators which limit the social good only to the growth of production and services in a given time, expressed through monetary transactions. These are quantitative changes. "If a natural disaster happens such as a hurricane or an earthquake, the immediate consequence is the growth of its GNP (Gross National Product), what reflects an increased activity in order to repair the damage. If a society is affected by an epidemic, GNP increases due to the construction of new hospitals and the hiring of new employees in the health service. If the crime rate rises, the GNP also increases with the number of police officers and the construction of new prisons"

[8] - highlights James Goldsmith revealing the controversial nature of quantitative indicators and showing that this line of reasoning is beyond comprehension from the humanistic point of view.

An interesting proposal of a new meter was put forward by Grzegorz Kołodko. His Integrated Prosperity Index (IPI) consists of: a) the level of Gross Domestic Product per capita (GDP per capita); b) the level of subjective well-being based on the analysis of the level of his life, health condition, the state of social and political life; c) the state of the environment and its impact on man; d) assessment of leisure time and cultural values, which one can then pursue [9]. Only then, the praiseworthy goals of sustainability can truly and globally materialize. Otherwise, they will be nothing but another lofty project that failed in the Golden Age of pretence, hypocrisy and many a time falsity and unfairness.

The whole concept and yearning for sustainability and harmony is not new. Many humanists have brought our attention to the legitimacy of the immeasurable in its nature qualitative criterion for ages. Thomas More, an English social philosopher and a noted Renaissance humanist, created an interesting utopia - a vision of a community prevailing in the equality and justice with no private property. His novel's principal message recommends that the nation should be "polite and cheerful, active, fond of relax, but when there is work to be done - sufficiently resistant to handle physical hardships. It means pursuit for social order and discipline rather than liberty [10]. A notable Polish historian and philosopher Bogdan Suchodolski criticizing the criteria for material comfort, places countries on the progress ladder and asks how many people from those countries live in tension and stress and how many are truly happy and smile honestly. Smile as a criterion of progress - what an oddity - scoff economists and technocrats respecting only numbers and graphs. But is not it actually an important and authentic measure? And should it not be defended? [11] ".

2 EDUCATION AND SUSTAINABLE SHIPPING

2.1 *Sustainable education*

At the core of the above mentioned problems lies the lack of education and upbringing that take into account three pillars of sustainable development: economic, social and environmental simultaneously, which failed both in socialism and capitalism. People under socialism were not taught responsibility for the entrusted goods and resources, respect for the common good and attention to nature. Under capitalism, especially with neoliberal face, people stayed too focused on generating infinite profits and surplus value, competition and calculations utterly ignoring the harm done to nature and not bothering to control the exploitation of the resources they were blessed with. Both systems, however, have something in common – neither of them has instilled the responsibility for another man and future generations. Although the moral maxim and principle of altruism also known as the Golden Rule or ethic of reciprocity *What you wish upon others, you wish upon yourself* takes back to ancient Egypt and is found in many human cultures, ideologies and religions [12], there was no room for thorough education in this field or that remained in contradiction with the established order and mentality. There is a need to create conditions in which human beings could grow with a sense of responsibility for their activity, nature and other people - the ones who live next to us, here and now and the future generations. Psychologically, it involves a person empathizing with others. Philosophically, it involves a person perceiving their neighbour also as "I" or "self". Sociologically, 'love your neighbour as yourself' is applicable between individuals, between groups, and also between individuals and groups. In economics, "without some kind of reciprocity, society would no longer be able to exist. [13]

What does sustainable education mean then? Sustainable education is primarily concerned with shaping values, particularly the value of respect, respect for others (including current and future generations), respect for diversity, respect for the environment and resources of our planet. It takes into account the concern about the quality of education meeting the following criteria: (i) it is holistic and interdisciplinary education, (ii) shapes the values, (iii) develops critical thinking and ability to assess values, (iv) uses a wide range of teaching methods and learning, (v) requires the active participation, (vi) refers directly to everyday life both personal and professional, (vii) responds to local challenges. Sustainability in education is shaped from the perspective of all areas of human development and includes challenges for our civilization. It is addressed to all and at every stage of life and at every level of education - from early childhood to adulthood [14]. The role of education is to awake reflective thinking about global inequities and share the necessary know-how, experience, technology, skills with the ones who lack the knowledge to equal opportunities. In the other words, where there is imbalance, there has to flow selfless support and primarily the consent for "the weaker" to become partners on equal conditions with a reasonable threshold of entry to make them fit for global competition. Neither should there any longer be tolerance for child labour or modern slavery which remain largely unknown *and hidden from view and cloaked in social customs, this being*

convenient for economic exploitation purposes [15] and these need to be brought out of the shadows.

Hence there is a necessity of global socialization which should result in an internationalization of values related to *sustainability*. People need to soak with these values until they become a part of their daily life, a conscious strategy of survival controlled by mentality and personality, an element of the life - protection mechanism - just as drivers and pedestrians for instance, are aware of the fragility of life and lurking dangers and naturally avoid them. In the other words, if something is followed by a majority of people, then it gains a positive recognition and response. Such educational policy could be a giant leap towards a fairer, safer and more sustainable world. These are perspectives and concerns, I suggest, that they need to be thoroughly considered and on no account neglected if sustainability is to be achieved.

2.2 *Sustainable Shipping*

The development and activities of shipping and ports operations significantly influence the condition of global waters, coasts and infrastructure related to them. There is a number of concerns how effectively and efficiently preserve and protect biodiversity, biosecurity and prevent further climate change and questions how deteriorated areas can be revived and brought to their pristine condition to reverse negative effect of the maritime industry, navigation, ports operations and coastal works related to the business as well as the area expansion for further development.

The Sustainable Shipping Initiative (SSI) aims at:
- improvement of shipyard operation (more sustainable building and disposal of ships),
- implementation of new low energy technology for ships' operation,
- implementation of good shipping practice,
- financing ships promoting sustainable shipping,
- supporting countries which lack the requisite resources, experience or skills to implement IMO treaties through an Integrated Technical Co-operation Programme (ITCP) [16].

If the above mentioned concepts are to come to pass, the shipping sector needs to be regulated with a transparent law and thoroughly supervised the way there would be no room for breaking, bending or evading the law and thoroughly supervised to eliminate any breach of the ecosystem protection regime. I see it as a qualitative shift from "green shipping, which is mostly about limiting the negative and replacing it with a positive (eliminating the negative). This can be best achieved through acting at the roots i.e. education and training relevant to sustainable education to develop values and practices remaining in compliance with sustainability. In the other words, if we manage to convince the majority of people to abandon the fierce competition and follow the practices without indulging in excessive resource consumption, through a greater concern for others, this could be a giant leap towards improving sustainability. Thus, we should dare to re-evaluate the economic aspect to optimize profit and provide a greater equity in the global distribution of wealth and resources through a reduction in the economic growth and consumption habits of the developed world in favour of underprivileged and developing countries. Such policies could in the longer term produce a fairer and safer world, and might lead to a greater sense of well being of people as humans.

3 CONCLUSIONS

The process of popularization and implementation of sustainable development and the transformations required to materialize the concept are multidimensionally conditioned. The current destructive (environment), wasteful and excessive (resources), exploitive (the man) and unfair and competitive (selective partnerships and regionalization of interests) human activity remains in conflict with the idea of sustainability. Its success is largely dependent on the human factor, which obviously requires more time to gestate and transform consciousness. A number of phenomena and processes related to culture, environment, politics, finance and economic sphere must be re-evaluated and perceived from an utterly different perspective, so far unknown. Change, as much else, begins at home. Without changing the consciousness of the dominant cultures the realization of the postulates will not be feasible. The originators cannot engage credibly in the process, and expect others to implement sustainability, unless they begin with themselves. Moreover, change will not come if we endlessly look around and procrastinate, waiting for better opportunities, some extraordinary circumstances, people or time.

REFERENCES

[1] G.Cimek , Utopia polityki wobec wyzwań współczesności (w:) Utopia- wczoraj i dziś, pod red. T, Sieczkowskiego, D. Misztala, Wydawnictwo Adam Marszałek, Toruń 2010, pp. 343 -361 .

[2] Socjalizm jest dobry dla mrówek, „Dziennik" – „Europa", nr 175, 11.08.2007.

[3] Declaration of the United Nations Conference on the Human Environment; http://www.unep.org/documents.multilingual/default.asp?documentid=97&articleid=1503

[4] The Governing Programmes of the United Nations

[5] L. Gawor L., Diametros no. 9 Idea Zrównoważonego Rozwoju jako projekt nowej ogolnoludzkiej cywilizacji (September 2006) p.87

[6] http://www.un-documents.net/our-common-future.pdf (WCED, 1997) Chapter II "Towards sustainable development", p.41

[7] L.Gawor, Diametros no.9, p. 101

[8] J. Goldsmith, dz. cyt., p. 4 -5. Za: G. Cimek , Utopia polityki…

[9] Vide. G. W. Kołodko, Wędrujący świat, Warszawa 2008, p. 270 G. Cimek , Utopia polityki…

[10] T. More, Utopia, translation. K. Abgarowicz, De Agostini, Altana 2001, s. 166.

[11] Suchodolski, Dwie cywilizacje uniwersalne, [w:] Strategia obrony i rozwoju cywilizacji humanistycznej. pod red. B. Suchodolskiego, Warszawa 1997, pp. 25 -26.

[12] A. Flew, "golden rule". A Dictionary of Philosophy. London: Pan Books 1979. p. 134

[13] R. Swift, Pathways & possibilities. New Internationalist - Issue 484 (July 2015).

[14] ESD in the UK in 2008: A Survey of Action. The UN Decade of Education for Sustainable Development 2005-2014, Report by Education for sustainable Development Indicators Advisory Group, published by the UK National Commission for UNESCO, May 2008, p.11 UNESCO - 2006, Framework for the UN DESD International Implementation Scheme, UNESCO, Paris, pp. 3-4

[15] Chapter 2. Child labour and slavery in modern society, June, 2014, pp. 13-55

[16] http://www.climateactionprogramme.org/climate-leader-papers/

Proceedings of 12ᵗʰ International Conference on Marine Navigation and Safety of Sea Transportation, TransNav 2017
21-23 June 2017, Gdynia, Poland

Informational Provision of Risk Management in Maritime and Multimodal Transportation

S.S. Moyseenko & L.E. Meyler
Baltic Fishing Fleet State Academy of the Kaliningrad State Technical University, Kaliningrad, Russia

ABSTRACT: The paper discusses the actual issue of the provision of the risk management system with an essential information resource. Since the quantitative risk assessments are calculated using the statistical information, there is a need to systematize the formation of databases. An analysis of the accidents reports allows to conclude that, in many cases, there is no enough causal relationship of accidents. It reduces the importance of the information in the context of the identification and analysis of risks as well as the use of statistical data for the probabilistic risk assessment calculations. The paper describes an approximate composition of the necessary information for the risk management tasks. A typical logistical scheme of the information provision of the risk management system in the maritime and multimodal transportation is suggested.

1 INTRODUCTION

1.1 *The statement of the problem*

The risk assessment in the maritime and multimodal cargo transportation is based primarily on the data analysis of accidents, emergencies in the past (Abchuk 1983, Bogalecka & Popek 2007, Topalov & Torskiy 2007, Perez-Labajos 2008, Kristiansen 2010, Moyseenko & Meyler 2011, Moyseenko et al. 2013, Karahalios et al. 2015). The quality and completeness of the analysis of this information depends, ceteris paribus, on the reliability and sufficiency of occurred accidents and the correct outcomes for the organization performing the investigation of accidents. An analysis of risks of accidents for different types of vessels allows to develop measures to prevent such events in the future (Katsman & Ershov 2004). The availability of information for shipping companies is an important condition. Torskiy et al. (2016) underline that "The "prescriptive" approach to the navigation safety, which is currently used in the world maritime field, is based on long-term experience and ship accidents investigation results". Skorupski (2013) noted that for different transport modes "Accidents investigation is usually conducted in terms of searching for the reasons of these events and to make preventive recommendations aimed at elimination of these causes". The information is

needed considering that the accidents are caused by factors associated with a vessel (failure, design defect), man (human error, workload), and environment (Smolarek 2010). In particular, Ziarati et al. (2011) presented scenarios of accidents that were developed by carefully studying maritime accident reports in the past focusing on the most critical/dangerous emergency.

According to International Labour Organization: code of practice (1996): "reports of accidents and near accidents should be discussed at safety and health committee meetings on board ship and steps should be taken to minimize the possibility of recurrences. The reports should also be discussed by shore management, and, if necessary, the ship owner's safety and health policy should be amended to take account of the conclusions of the investigation". In particular, reports (2004 – 2016) of the Marine Accident Investigation Branch (UK) contain many descriptions of accidents of merchant vessels under 100 gross tons or over and fishing vessels. Reports include the factual information on the place and conditions of an accident, its analysis, conclusions, recommendations to prevent such or similar events in the future. Similar information can be picked up, for example, from the Collection of the Russian sea/river transport inspectorate (2016), where also causes of accidents (structural, operational, organizational and human factors) are given. However, the available information on accidents very often is presented generally (e.g. the type and the number of accidents). It is not informative enough for the use in calculating the probabilistic risk assessments. For example, cause-and-effect relationships of accidents of fishing vessels were identified (Moiseenko & Meyler 2015) on the base of an analysis of emergencies and main types of risks in fishery were defined. The method of calculating the quantitative estimating the predicted risk was elaborated using the method of fuzzy sets/expert evaluations and the theory of probability.

Various methods and models for analyzing consequences of accidents on transport are used at present. In particular, Chauvina et al. (2013) presented the Human Factors Analysis and Classification System (HFACS), which has been adapted to the maritime context and used to analyze human and organizational factors (environmental and personnel) in collisions. Akyuz (2017) presented the hybrid accident analysis model that integrates an Analytical Network Process (ANP) method with HFACS. The model is established to enhance safety and prevent loss of life or injury in maritime transportation industry.

Nevertheless, it should be noted that problems of monitoring information on the accidents with vessels at sea and with land vehicles are solved partially in many cases. In particular, the analysis of the information on the accident rate of maritime and fishing vessels shows:
- in many cases, the inquiry materials had no cause-effect relationship of happened accidents;
- internal and external factors and their influence on the development of emergency situations did not adequately take into account in the investigations of the conditions of accidents occurrences;
- information on emergencies was not adequately provided;
- the main attention in the investigation of accidents was paid to the human factor, while the system "operator-machine-environment" should be considered.

Thus, the lacks of information on accident rates in the past significantly reduce its value and possibilities to use for the risk management purposes. In this regard, the task of organizing the system of monitoring and analytical data processing accidents is relevant. It allows to form a database on the level of the industry/regions and large shipping/transport companies.

2 STRUCTURE OF INFORMATION FOR TASKS OF RISK MANAGEMENT SOLUTION

2.1 Scenario of an emergency

The risk assessment and its management are based on the use of information concerned to the following kinds of risks:
- natural, i.e. risks related to the elemental forces of nature;
- transport, i.e. risks related to cargo transportation with various modes of transport, in particular arising when loading and unloading, storage and movement of goods; damage and loss of cargoes;
- navigational, i.e. risks of the vessels collision, grounding, damages of the propeller-steering group, elements of the navigation aids on fairways, wharves and vessels when mooring, etc.;
- anthropogenic impacts, i.e. risks generated by technical and technological and industrial human activities, for example risks of failures/accidents of the technical equipment, production lines, fire, etc.;
- environmental, i.e. a probability of a damage to the environment, as well as the life and health of third parties, in particular fuel spill during bunkering, during cargo operations with harmful substances, etc.;
- fishing risks, i.e. risks associated with catching in ocean fisheries, for example, catching ground, winding on the screw propeller, damage or loss of fishing gear, collisions of vessels, etc. (Moyseenko et al. 2014a, b). Also criminogenic, military, piracy and other kinds of risks can be considered (Duda & Wardin 2012, Nincic 2013).

The calculation of the probabilistic estimates of an emergency occurrence and the risk implementation is based on the use of expert evaluations and statistical data on accidents, loss of catch, cargo, any damage. To use the statistics on accidents, the occurrence of another kind of an emergency it is necessary to reflect in the investigation: the type and function of the vessel; time and place of accident/emergency; the season and conditions of the environment. Also it is very important to know what internal and external factors could influence the occurrence and development of the situation; what actions have been taken to prevent and/or reduce the level of the damage.

As a result, it is necessary to clearly define the chain: "conditions - causes – consequences". It is advisable to describe the scenario of an emergency. The analysis of the scenario helps to obtain information that will be useful in the future. On the one hand it is "learning from others' mistakes" and for the analysis of possible scenarios for the development of a specific situation, for development and further evaluation of the effectiveness of measures to prevent negative outcomes. According to the opinion by Szymanski (2007): "there is an urgent need to create simple, understandable and workable risk management process suitable for shipping companies". An example of the emergency scenario during loading/unloading operations at the port is presented in Figure 1. There are two possible outputs (negative or positive) according to this scenario.

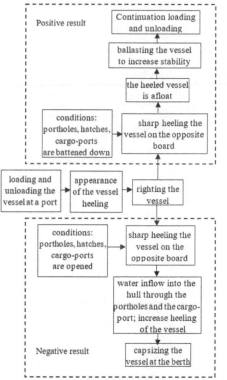

Figure 1. A possible scenario of emergency situation: the loss of stability and capsizing the vessel at the port.

2.2 *Formation of the database on the transport accident rate*

The information on accidents and other emergency situations will be of the greatest practical importance in the case of the availability of a representative sample. The data should be differentiated according to areas of navigation, types and function of vessels, a season. It is necessary to have data on the number of vessels at the navigating area in order to calculate the probabilistic evaluations of the risk. It allows to calculate the expectation of the risk of emergency situation. As far as the calculation of probabilistic estimates based on retrospective statistical data the sample should be representative, there is a need to establish integrated databases. Such databases can be created at the level of the large shipping

companies, regional information and analytical centers, local authorities and ministries of transport. Herewith databases of the lower levels should be integrated into higher-level databases. Real opportunities to create representative databases and the use of the previous experience to solve practical problems are offered in this way. Statistics can also be used to calculate the probabilistic assessments of emergency situations and identification of the risk, the study of cause-and-effect relations, etc. As an example of the logistical scheme of information provision with the risk management system on maritime transport in Russia is shown in Figure 2.

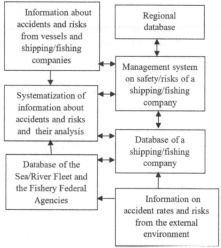

Figure 2. Logistical scheme of the information provision of the risk management system

The information received from vessels contains data about the threats of arising accidents or emergency situation.

Shipping/fishing companies put the information about the accidents to the block of the systematization and analysis and further to the database. The information on accidents and risks from the external environment is data about:
- accident rates of vessels and other transport modes;
- approaching hurricanes, tropical cyclones, tsunamis, etc;
- changes of conditions of navigation in straits, channels, etc;
- threats of pirate attacks, areas of military operations, a criminal situation in ports and at areas of the traffic, etc.

The block "Systematics of risk information and analysis" performs an analytical work on the systematization of data from different sources, a data analysis and a risk assessment. Then results are send to the block "Safety management system/risks of the shipping company" and in an accident rates database. Measures to reduce the risk level and formation of management solutions are elaborated in this block.

3 CONCLUSIONS

The results of the data analysis of accidents on the transport fleet show that there is a lack of information on accidents in the past. It reduces significantly its value and possibilities to use such information for risk management purposes. In consequences of this situation the problems of improving the organization of monitoring and analytical processing of data on accidents have an actuality. Formation of a database on the level of the industry/region and possibly large shipping/fishing companies will take efficient actions to prevent emergencies in the future.

Determining the structure of information provision of the system of the assessment and risk management, systematization of data on accidents will significantly expand the possibilities of using statistics and experience to ensure safety of navigation and fishing.

Designing the system of a logistical interaction of information sources on the accident rates, an occurrence of the threat of emergency situations/accidents, activation of the piracy and a criminogenic environment will allow to define more full the qualitative and quantitative risk assessment and to elaborate the most effective management decisions to improve the safety of human activities at sea.

REFERENCES

Abchuk, V. 1983. *Risk theory in the marine practice*. Leningrad: Sudostroenie (in Russian)

Akyuz, E. 2017. A marine accident analyzing model to evaluate potential operational causes in cargo ships. *Safety Science*. 92, February 2017: 17–25.

Bogalecka, M. & Popek, M. 2007. Analysis of sea accidents in 2006. In: A. Weintrit & T. Neumann (eds), *Proc. of the Int. Symp. TransNav - Advances in Navigation and Safety of Sea Transportation*, Gdynia, Poland, June 20-22, 2007: 639 - 642.

Collection of character accidents at sea occurred with vessels flying the flag of the Russian Federation 2016. www.sea.rostransnadzor.ru

Chauvina, C., Lardjaneb, S., Morela, G., Clostermannc, J-P & Langarda, B. 2013. Human and organizational factors in maritime accidents: Analysis of collisions at sea using the HFACS. *Accident Analysis & Prevention*. 59, October 2013: 26–37.

Duda, D. & Wardin, K. 2012. Influence of pirates' activities on maritime transport in the Gulf of Aden region. *TransNav, the International Journal on Marine Navigation and Safety of Sea Transportation*. 6(2), June 2012: 195-202.

ILO code of practice. Accident prevention on board ship at sea and in port. Geneva, International Labour Office, 2nd edition, 1996.

Karahalios, H., Yang, Z. & Jin, W. 2015. A risk appraisal system regarding the implementation of maritime regulations by a ship operator. *Maritime Policy & Management*. 42(4): 389-413.

Katsman, F., & Ershov A. 2004. Accident rate of maritime fleet and security of navigation. Transport of the Russian Federation. 5: 82 – 84. (in Russian)

Kristiansen, S. 2010. *Maritime Transportation: Safety Management and Risk Analysis*. Elsevier

MAIB (2004-2016). Accident Investigation Reports – www.maib.gov.uk

Moyseenko, S. & Meyler, L. 2011. *Safety of marine cargo transportation*. Kaliningrad: BFFSA Publ. House (in Russian).

Moyseenko, S., Meyler, L. & Faustova, O. 2013. Formation of the integrated risk assessment of a disaster in multimodal cargo transportation. *Proc. X Baltic intern. maritime forum*, Swetlogorsk, May, 28–31. Kaliningrad: BFFSA Publ. House: 265 – 270. (in Russian)

Moyseenko, S., Meyler, L., Bondarev, V. & Faustova, O. 2014a. Analysis of the problem of risk assessment in commercial fishing. *Proc. XI Baltic intern. maritime forum*. Swetlogorsk, May 26–30. Kaliningrad: BFFSA Publ. House: 76-82. (in Russian)

Moyseenko, S., Skrypnik, V. & Faustova, O. 2014b. Differential - integral approach to modeling the processes of development of emergencies in navigation and ocean fishing. *Bull. Adm. Makarov' State University of sea and river fleet*. 4(26): 47-53. (in Russian)

Moyseenko, S. & Meyler, L. 2015. A method of risks assessment for the fishing vessels accidence. *Bull. Adm. Makarov' State University of sea and river fleet*. 5(33). December 2015: 47-55. (in Russian)

Moyseenko, S., Meyler, L. & Bondarev, V. 2015. Risk assessment for fishing vessels at fishing grounds. *TransNav, the International Journal on Marine Navigation and Safety of Sea Transportation*. 9(3), September 2015: 351 – 355.

Nincic, D. Maritime security: future trends and challenges 2013. *Proc. 14-th Ann. Gen. Assembly Int. Assoc. Mar. Univ. (IAMU) AGA14. – New Technological Alternatives for Enhancing Economic Efficiency*, Constanta, Romania, October 26-28, 2013: 169 – 176.

Perez-Labajos, C. 2008. Fishing safety policy and research. *Marine Policy*. 32: 40-45.

Skorupski, J. 2013. Method of serious traffic incidents analysis with the use of stochastic timed Petri nets. *TransNav, the International Journal on Marine Navigation and Safety of Sea Transportation*. 7(4): 599-606.

Smolarek, L. 2010. Finite discrete Markov model of ship safety. *TransNav, the International Journal on Marine Navigation and Safety of Sea Transportation*. June 2010. 4(2): 223-226.

Szymanski, K. 2007. Risk management – do we really need it in shipping industry? In: A. Weintrit & T. Neumann (eds), *Proc. of the Int. Symp. TransNav - Advances in Navigation and Safety of Sea Transportation*, Gdynia, Poland, June 20-22, 2007: 611 - 612.

Topalov, V. & Torskiy, V. 2007. *Risks in navigation*. Odessa: Astroprint (in Russian)

Torskiy, V., Topalov, V. & Chesnokova, M. 2016. Conceptual Grounds of Navigation Safety. *TransNav, the International Journal on Marine Navigation and Safety of Sea Transportation*. 10(1), March 2016: 79-82.

Ziarati, R., Ziarati, M. & Acar, U. 2011. Developing scenarios based on real emergency situations. In M. Pourzanjani & D. Martinovi (eds), *Proc. intern. conf. IMLA 19. September 28– October 1 Opatija, Croatia*: 363 – 372.

Port Cities Smart&Sustainable Development Challenges – Gdynia Case Study

A. Przybyłowski
Gdynia Maritime University, Gdynia, Poland

ABSTRACT: The sustainable development of the cities, including port ones, represents one of the major challenges for the future of the planet in the 21st century, relatively to the contribution and adaptation to climate change, natural resources consumption, energetic transition, population mobility, welfare and security, pollution, the global economic growth. A smart sustainable development can start from the port areas' new circular metabolism which should be extended to the whole city/region, thus modifying the land and space use through new efficient planning and design able to increase the comprehensive quality of the urban landscape. The aim of the paper is to present the urban sustainable development challenges, taking as a case study the Sustainable Urban Mobility Plan (SUMP) elaboration process in the city of Gdynia. The research has been based on a literature review and Author's practical experience as an expert within the EU CIVITAS DYN@MO project. For study justification, reports relating to smart and sustainable seaport areas development and city internal documents have been studied. Current patterns of transportation development are not sustainable and may compound both environmental and health problems for this port city inhabitants. However, an integrated smart development model, supported by sustainable urban planning, may contribute to urban ecological resilience of such port areas.

1 INTRODUCTION

Port cities are key places where economic strength, competitiveness, human capital and global appeal, population and migration processes are increasingly concentrated. Since ports throughout history handled an important kind of transportation, they may dominate the local economy of a coastal city. The present day challenge, is the inability of most coastal cities, to absorb rapidly expanding port developments and population growth. However, port cities and areas have a particular development potential (Girard L. F., 2013). They may take on a key role in launching a smart and sustainable development model, starting from local cultural resources for the activation of the creative processes of a circular economy through a synergistic approach, combining the port's economic, logistic and industrial activities with a cultural heritage regeneration, with the creativity of inhabitants.

The aim of the paper is to present the urban smart and sustainable development challenges, taking as a case study the Sustainable Urban Mobility Plan (SUMP) elaboration process in the city of Gdynia. The research has been based on a literature review and Author's practical experience as an expert within the CIVITAS DYN@MO project funded from the 7th Framework Programme of the European Commission. For study justification, reports relating to smart and sustainable seaport areas development and city internal documents have been studied.

2 SMART & SUSTAINABLE URBAN DEVELOPMENT CHALLENGES

Urbanisation is believed to be an area of significant civilizational changes, as the future of humankind is linked with cities. No more than 5% of the world population lived in the cities in 18th century, while today it is more than 50%, and United Nations forecasts that more than 80% of the population will live in urban area by the end of this century (UNITED NATIONS, 2014). The data confirms the assumptions that contemporary cities are facing a challenge related to keeping the balance of resources and strengthening development which would improve security and the quality of life for numerous citizens. Within the United Nations Human

Settlements Programme works on a new development model under the slogan "The city we need" are being carried out. Urban centres should: integrate their community, close the material and energy flow, stimulate the economy and involve citizens in it, strengthen the identity and sense of belonging, provide security, health, be inexpensive in everyday use and provide a just access to its resources, and be governed at metropolitan level (World Urban Forum, 2014).

The city development/re-development management relates to use of scares resources and transformation of the existing state to desired one (Wojewnik-Filipkowska A., 2017). The management relates to economic, social, technological, and natural systems. The city development and investment management means space management as city is both a physical place of paths and buildings, and also a space of values, beliefs, and relations. The general aim of the city development management is to ensure sustainable development which is manifested by an increase in national income, qualitative changes in the structure of the economy, availability of goods and services for citizens, better standard of living. But translating this concept into action is a challenge. On the strategic level the management means outlining goals at future requirements. It is master planning on the tactical level, and finally, planning, implementation, and evaluation of urban development project on the operational level (Girard, Nijkamp, 1997).

Cities need to invest in improving their efficient organization in order to increase wealth production towards a sustainable development. New unconventional forms of value creation and production are based on circular processes. The smart city image should be linked first of all to technologies able to produce a less dissipative and more efficient city organization. Circularization and synergies are the general principles for smart city sustainable development (Ravetz, J. & all., 2012).

A more efficient and less dissipative (Ravetz, J., 2011) organization of city processes is required in order to face not only the globalized economic competition, but also the energy challenge, the social challenge, and the ecological challenge. They are all based on circular and synergistic processes.

This development strategy should integrate hard and soft values/objectives (that is, both economic and social/environmental, as well as landscape preservation) in a win-win game. It may contribute to city resilience, which is the capacity of the city to react and manage change, maintaining its comprehensive organization and structure.

It is however clear that the creativity does not relate only to technological innovations but also to city organisation and management connected with social-business responsibility and all stakeholders'

participation. These all make urban development projects planning and evaluation to be a challenge in the process of implementing the strategy via investment project management on the operational level. Figure 1 one illustrates the content of new paradigm on the strategic level.

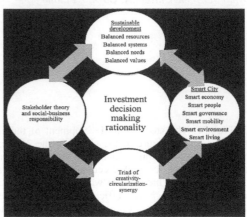

Figure 1. New paradigm of the investment decision making rationality in urban development – strategic level.
Source: Wojewnik-Filipkowska A., 2017.

A new approach to management of city investments is needed. It has been established including the principles of sustainable smart development, triad of creativity – circularisation – synergy, stakeholder theory, and the idea of social-business responsibility. The approach primarily elevates the concept of the public governance which is characterised by a de-centralisation, participation, constructivist and a win-win approach. These attributes increasingly refer to the decision-making process and subsequent implementation.

Port city, which serves as a link between the local economy and the global economy, is an interaction of both urban and port systems, giving rise to its complex and dynamic nature (Hein, C., 2011).. While the development of a port city is an aspect that requires continual research and monitoring, the current literature addressing the issue of sustainable development in port cities is rather limited. In addition, empirical studies often analyse the port system and the urban system separately, with little research attempting to integrate the two systems.

A port areas' smart development should be based on circular processes at three main levels (Girard L. F., 2013):

1 Economic: symbiosis between companies for value creation, allowing synergies and also stimulating circuits between company and community prosperity; between companies and the city; between all actors and the outside territory;

2 Social: able to regenerate interpersonal relationships—often weakened in cities—through relationships with "places";

3 Ecological: all living systems are characterized by circular processes and are able to conserve and reproduce themselves. Circular processes emerge through re-use, recycling and regeneration of materials and energy, with a reduction of negative externalities. The resilience of systems and creativity are stimulated by circular processes, which break down linear metabolism

Harbor areas become vital, when they act as dynamic, complex systems, capable of transforming and adapting to the continuous pressure of change from the outside and when they are able to modify their physical structure regarding space, organization and functions, by combining infrastructures, facilities, installations, *etc.*, while maintaining their own identity (Schubert, D., 2011). Here, green industrial activities, able to decouple economic wealth production from ecological losses, should be grounded while considering the model of living systems (Boulos J., 2016).

The sustainable development of the biggest cities of the world represents a major challenge for the future of the planet in the 21st century, relatively to the contribution and adaptation to climate change, natural resources consumption, energetic transition (the "after oil" transition), population mobility, welfare and security, pollution, the global economic growth (Ducruet, C., 2011). For historical reasons, a great part of these cities, and especially those of emerging countries like Brazil or China, are located on a coast or on a river, thus including a port and playing a special and major role in the national and the global economy (nodes of logistic chain, concentration of population, touristic attractiveness). The integrated planning for cities development has to consider the passengers' mobility and the transport of goods as priorities to reach a sustainable growth (Martell-Flores et al., 2009).

Today, the growing number of usage disputes over increasingly coveted coastal areas is prompting local managers to incorporate urban and port-related issues in overarching planning programs (Zengqi X., Jasmine Siu Lee L., 2016). In particular, planning of the sea front and the buffer zone between the port and the city must contribute decisively to the deployment of more effective, cleaner transport services for the port city as a whole. In general, one of the key global challenges for planners and decision-makers consists in integrating sustainable development goals (environmental and social components, as well as the stimulation of industrial competitiveness) into urban planning.

Mobility planning is a complicated task because of complex and contradictory factors and needs in this process. Political and financial issues pose additional difficulty. The Sustainable Urban Mobility Plan is a concept which contributes to achieving EU objectives on climate and energy. The new European Commission's planning concept proposals treats challenges related to transport in a more sustainable, integrate and comprehensive way (Guidelines Developing and Implementing a Sustainable Urban Mobility Plan, 2013). It focuses on involving stakeholders, coordinating the vision between social and economic sectors (land management, transport, social policy, safety, health, etc.) as well as between authorities. It requires a sustainable, long-term vision of urban area, taking into account broader costs and social benefits. Its aim is to meet the mobility needs of people and companies. It proposes actions improving the quality of life. Such a plan has been adopted in the Polish port city Gdynia.

3 SUSTAINABLE URBAN MOBILITY PLAN FOR GDYNIA PORT CITY

Gdynia is an important European transport node with maritime connections to ports in Europe and worldwide. It is an important component of the Gdańsk-Gdynia-Sopot Metropolitan Area (GGS MA) being the most important metropolitan in North Poland and simultaneously in the South Baltic Sea with a regional yet international influence (Studzieniecki T., 2016). Gdynia has unique natural values, due to its seaside location, in particular, a long, diverse and attractive foreshore zone (Fig. 1).

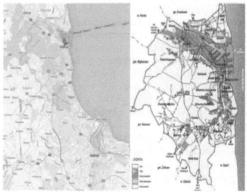

Figure 1. Gdynia city and seaport location
Source: Zarząd Dróg i Zieleni in Gdynia, September 2016.

Gdynia is a city with a population of almost 250,000 and the area of 135 km^2 in the Pomorskie Voivodeship. Some objectives of sustainable transport development cannot be achieved without clearly setting Gdynia in the metropolitan area and in the Pomorskie Voivodeship. This is related to the numerous metropolitan functions and the role of the port and the shipbuilding industry located in the city.

In the urban area, complex relations are formed at the interface of environment, society and economy. A dynamic development of seaports (in particular container terminals) has become an important factor stimulating the economic development. The dynamic development of Gdynia and the accompanying changes in spatial development, connected with the intensification of suburbanisation processes create transport problems that have an impact on the public space, inhabitants quality of life and environment. The most important problems include increasing congestion, resulting in a reduced travel speed, domination of passenger cars in trips, negative transport patterns and behaviours among the inhabitants, strengthened by uncontrolled suburbanisation processes and settlement development in neighbouring communes.

Under the European project entitled CIVITAS DYN@MO (DYNamic citizens @ctive for sustainable MObility) within the CIVITAS II PLUS initiative (funded from the 7th Framework Programme of the European Commission), the Sustainable Urban Mobility Plan (SUMP) has been elaborated and adopted by the city authorities. The plan takes into account the principles of participation, integration and assessment as well as be based on current practices of urban mobility planning. It is based on the idea of an integrated approach, i.e. promoting the sustained growth of available transport means. Due to the depleting natural resources of cheap fossil fuels, it seems natural to seek alternative, non-motorised means of transport across urban areas.

The document covers the urban and suburban areas (the functional area), defines the passenger public transport, non-motorised transport, intermodality, road transport, mobility management, use of Intelligent Transportation Systems (ITS), urban logistics, city road traffic safety, implementation of new use patterns or promotion of clean and energy-saving vehicles (clean fuels and vehicles), considering the related needs identified in the given area.

The purposes of the project are the implementation of modern mobility solutions and the exchange of knowledge and experience between the participant cities, including:
– the development of "web 2.0" systems and services;
– the introduction of city- and inhabitant-friendly electrical vehicles;
– the involvement of inhabitants in the process of mobility planning and service quality improvement.

Based on the strategic framework for the development of Gdynia, this document states the vision of the transport system and mobility improvement in Gdynia in 2016-2025. Four strategic objectives were set for Gdynia, which are actually strategic activity directions (Sustainable Urban Mobility Plan for Gdynia, 2016):
– No. 1 - Attractive and safe urban space;
– No. 2 - Safe and effective transport system;
– No. 3 - Rational transport choices;
– No. 4 - Effective cargo transport in the city.

The expected effects of implementing the integrated action plan for Objective 4 (Tab.1) are as follows: reduced goods traffic in the centre of Gdynia and an increased role of railway transport in the service of container terminals.

Table 1. Integrated action plan for strategic objective no. 4: "Effective cargo transport in the city"

4.1. Better transport accessibility of the seaport	4.2. Establishment of an effective and sustainable urban distribution system	4.3. Support for modern technologies and organisation solutions for goods transport
Development of the road infrastructure servicing the seaport and reducing the pressure on the city road system. Development of the railway infrastructure serving the seaport. Development of the lorry parking infrastructure integrated with the TRISTAR system.	Partnership for sustainable city supplies (city logistics). Organisation of the supply system in the city centre. Regulation of heavy goods transport access to selected city areas.	Use of ITS and ICT in goods transport to optimise goods traffic. Implementation of low emission transport means in the city goods transport and municipal services.

Source: Sustainable Urban Mobility Plan for Gdynia, 2016.

The defined specific objectives and corresponding activity packages respond to the following challenges (Sustainable Urban Mobility Plan for Gdynia, 2016):
– increased turnover in the seaport will cause a further traffic increase in the main city road system;
– railway transport will play an increasing role in the service of container and full-train cargo.

The primary component of the road system in cargo transport in Gdynia is the express road S6 and Estakada Kwiatkowskiego. It is the only access road to the sea port from the south, because the voivodeship road no. 468 in Sopot is closed for heavy traffic. The national road no. 6 to the north is also relatively important. The smallest cargo traffic in car transport takes place on the national road no. 20 and other voivodeship and district roads.

An important role in cargo transport belongs to the route of road S6 and motorway A1, which is part of the national road no. 1, connecting the Tricity seaports with the border city of Cieszyn, of which the length is 419.9 km. It is one of the main longitudinal routes in Poland. It is the Polish part of the international transport route E75 Helsinki-Gdańsk-Łódź-Budapest-Athens. Due to the high lorry traffic intensity and its predicted increase, pilot implementations of modern road surface protection solutions were introduced in the form of weight pre-selection of overloaded lorries in traffic with a vision system and automatic transfer of standard violation notifications to the appropriate law enforcement authorities.

The first pilot weight under the CIVITAS DYN@MO project was installed in ul. J. Wiśniewskiego in December 2015 and in January 2016 data acquisition was initiated. In April 2016, when weather conditions allowed it, the system was first calibrated and finally commissioned for use with the B+(7) category.

Goods train traffic via Gdynia is related mainly with the service of the seaport and other entities around the port. The traffic has been increasing for years and the handling forecasts of the Port of Gdynia Authority S.A. indicate its continued, uninterrupted growth. The reduction of traffic in the railway line no. 202 Gdańsk Główny - Gdynia Główna is related with the modernisation of the corridor E65. Some trains also run along the railway line no. 201. The line is used primarily during difficulties along the main line via Gdańsk and Tczew. The number of intermodal trains on the railway system sections in Gdynia is increasing dynamically. Generally, the traffic is concentrated on the railway line no. 202. In 2015 intermodal trains accounted for as much as 38% of all goods trains on the subject segment of the line 202. The port location and service of loads generated within creates a potential hazard for inhabitants in the case of dangerous goods transport. To minimise those hazards, it is necessary to move a major part of traffic to the line 201, which requires modernisation for that purpose.

A consequence of the economic development based in part on the transport-forwarding-logistics sector is the increase in goods transported by road transport (Europe-wide trend). In case of Gdynia the central location of the seaport is a source of additional, constant load on the city road system. Increased turnover in the port (especially container and some bulk cargo, e.g. grain) has an impact on the road and railway infrastructure throughput. Thus, it is necessary to undertake investment and organisation activities in strict co-ordination with road and railway system managers, as well as other stakeholders of the transport market (terminal operators, railway and road operators, port management, the Maritime Office etc.).

Figure 3 shows road traffic intensity changes 'business as usual' pessimistic scenario by 2025 in particular road segments during the afternoon rush hour on a business day. In all segments subject to modelling, the traffic intensity is increased by from 4% to 161% in the centre of Gdynia and by 31% in Trasa Kwiatkowskiego. In west districts, the

increase in intensity will be equally significant (e.g. in ul. Wielkopolska: from 14.6% to 22.4%).

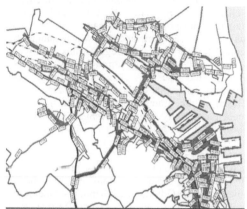

Figure 3. Changes in road traffic intensity for scenario "Unsustainable mobility with an increasing inhabitant mobility" (for the afternoon rush hour).
Source: Zarząd Dróg i Zieleni in Gdynia, September 2016.

Lorry traffic in Gdynia is determined by the service of port terminals and storage and manufacturing zone located near ul. Hutnicza. There is a gradual increase in the lorry traffic intensity, of which the source is economic development and activity of the entities operation in the port of Gdynia. The core of the road system for cargo transport in Gdynia is the express road S6 and Estakada Kwiatkowskiego, which due to the poor technical condition of the oldest segment requires regular repair works, which reduce the throughput of the main transport system.

The universal nature of the port in Gdynia, reinforced by investment projects and terminal activity, implies a high share of road transport in the service of the primary cargo groups (especially containers, other general cargo and grains). The planned increase of handling in the port (in the maximum variant: up to 44 million tonnes in 2027) is a challenge especially for the port hinterland infrastructure. This can also generate a significant burden for the City. Quantitatively and qualitatively, the road infrastructure will be a key factor determining the capacity of the port terminals to acquire and retain the strategic cargo groups defined in the development strategy for the port in Gdynia. Share (%) of railway transport in the service of the seaport for monitoring cargo transported by railway in relation to total cargo will be a key indicator to enable assessing the extent to which the strategic objective no. 4 'Effective cargo transport in the city' have been achieved.

4 CONCLUSIONS

Smart and sustainable development is a particularly big challenge for port cities. However, such a smart sustainable development can start from the port areas' new circular metabolism which should be extended to the whole city/region, thus modifying the land and space use through new efficient planning and design able to increase the comprehensive quality of the urban landscape.

Based on the strategic framework for the development of port city Gdynia, the Sustainable Urban Mobility Plan (SUMP) has been elaborated stating the vision of the transport system and mobility improvement in 2016-2025. Current patterns of transportation development are not sustainable and may compound both environmental and health problems for the Gdynia city inhabitants. One of the key objectives of this document is an effective cargo transport in the city with the support of modern technologies and organisation solutions for goods transport.

Thus, an integrated smart development model, supported by sustainable urban planning, may contribute to urban ecological resilience of such port areas. As it has been mentioned, a holistic approach is needed to face such challenges as climate change, natural resources consumption, energetic transition (the "after oil" transition), population mobility, welfare and security, pollution, etc. in order to prevent pessimistic scenarios and bring desirable long-term effects.

REFERENCES

Boulos J., 2016, Sustainable Development of Coastal Cities-Proposal of a Modelling Framework to Achieve Sustainable City-Port Connectivity, Elsevier, Procedia - Social and Behavioral Sciences 216 (2016) 974 – 985.

Ducruet, C., 2011, Economic development paths of port-cities: specialization vs. diversification, OECD Seminar, CNRS & UMR Géographie - Cités: Paris, pp.1-14.

Giffinger R., Fertner C., Kramar H., Kalasek R., Pichler-Milanovic N., 2007, MEIJERS E., Smart Cities – Ranking of European Medium-Sized Cities, Research Re-port, Vienna University of Technology, Vienna.

Girard L. F., 2010, Creative Evaluations for a Human Sustainable Planning, in: Making Strategies in Spatial Planning, Cerreta M., Concilio G., Monno V. (eds.), 2010, Springer, Dordrecht, Heidelberg, London, New York, p. 305-328.

Girard L. F., 2013, Toward a smart sustainable development of port cities/areas: the role of the 'historic urban landscape approach', in: Sustainability, vol. 5, no 10, p. 4329-4348.

Guidelines Developing and Implementing a Sustainable Urban Mobility Plan, 2013, Rupprecht Consult, European Union.

Hein, C., 2011, Port Cities: Dynamic Landscapes and Global Networks; Routledge: London, UK, New York, NY, USA,.

Morel G., Lima Fernando R., Martell-Flores H., Hissel F., 2013, Tools for an integrated systems approach to sustainable port city planning, urbe. Revista Brasileira de Gestão Urbana (Brazilian Journal of Urban Management), v. 5, n. 2, p. 39-49, jul./dez.

Ravetz, J., 2011, Exploring Creative Cities for Sustainability with Deliberative Visualization. In *Sustainable City and Creativity. Promoting Creative Urban Initiatives*; Fusco Girard, L., Baycan, T., Nijkamp, P., Eds.; Ashgate: Aldershot, UK.

Ravetz, J.; Fusco Girard, L.; Bornstein, L., 2012, A research and policy development agenda: Fostering creative equitable and sustainable port cities. *Bollettino del Dipartimento di Conservazione dei Beni Architettonici ed Ambientali, 12,* 67-69.

Schubert, D., 2011, Seaport Cities: Phases of Spatial Restructuring and Types and Dimensions of Redevelopment. In Port Cities: Dynamic Landscapes and Global Networks, 1st ed.; Hein, C., Ed.; Routledge: London, UK, New York, NY, USA,; pp. 54–69.

Sustainable Urban Mobility Plan for Gdynia, 2016, Wolek M. (ed), CIVITAS DYN@MO, Gdynia.

Studzieniecki T., 2016, The development of cross-border cooperation in an EU macroregion – a case study of the Baltic Sea Region, Elsevier, Procedia Economics and Finance 39 (2016) 235 – 241.

UNITED NATIONS, 2014, Department of Economic and Social Affairs, Population Division, World Urbanization Prospects: The 2014 Revision, Highlights, New York.

World Urban Forum, 2014.

Wojewnik-Filipkowska A., 2017, Rationalisation of Investment Decisions in the Sustainable Management of Urban Development – is a New Paradigm Needed?, Problemy Ekorozwoju – Problems of Sustainable Development, vol. 12, no 1, 79-90.

Zarząd Dróg i Zieleni in Gdynia, September 2016.

Zengqi X., Jasmine Siu Lee L., 2016, A systems framework for sustainable development of Port City: Case study of Singapore's policies, http://www.sciencedirect.com/science/article/.

Proceedings of 12th International Conference on Marine Navigation and Safety of Sea Transportation, TransNav 2017
21-23 June 2017, Gdynia, Poland

Analysis of Infrastructure Ports and Access Road and Rail to Tri-City Seaport

M. Ziemska & P. Szumacher
Gdynia Maritime University, Gdynia, Poland

ABSTRACT: The paper presents the current state of the access road and rail Port of Gdynia and the Port of Gdańsk. The focus is also on port infrastructure. For the analysis shows the most recent projects implemented by the ports with the support of the European Union, which have significantly improved the features mentioned above ports. Additionally, described prospects for development of ports in the coming years.

1 INTRODUCTION

Polska Street sea ports infrastructure was outdated, and its technical parameters do not meet the requirements of modern technologies, handling and used vehicles. NIK report drawn up in 2011. Confirmed reprehensible state of communication connections ports. The problem was the access to the port from the land, as well as from the sea. The quality of Polska Street communication network significantly surpass the standards preserved in Western Europe. Polska Street sea ports are characterized by low to integrate in a sea-land transport networks. This process is a consequence of the location of sea ports in the field of international transport corridors. To correct the problem, actions related to investment in particular the construction of new and reconstruction and the improvement of the existing road and rail networks included in the TEN-T together with adjacent objects, eg. parking, as well as modernized structures intersecting each way, or bridges and overpasses. This is a very important operation, as the rail and road infrastructure provides a direct land access to most areas of the port and merges them with the elements of the main routes.

As a result of economic growth due to a thriving trade in the Baltic and Nordic areas, the development of Morska Street transport to the economy of the country was a necessity, and an opportunity to achieve a better position on the background of competitiveness on the international market. The main objective adopted by the "Programme for the development of Polska Street sea ports by 2020 (With the prospect of 2030) "was so increase the competitiveness of Polska Street seaports, and increase their participation in social and economic aspects of the country. Care was also for the upgrading of ports in the international transport system. Investing activities led to improve the situation led to a significant increase in the quality and condition of the roads and their surface even by over 22% compared to previous years. Increased permissible load of national roads up to 115 kN / axle and significantly increased the number of national roads which access to ports. Seaports develop their potential in the bulk cargo, containerized cargo and supported. Dredging of fairways and swimming pools, as well as the modernization of turntables have significantly improved access to the terminals for larger cargo ships and passenger. Containerization develops and brings big profits, therefore, very important is the development of container terminals, which seek to ensure that meet the demands posed by the ship-owners currently sailing ships.

2 GDYNIA PORT

The International Port of Gdynia is a port of fundamental importance for the Polska Street and is part of the Corridor VI Trans-European Transport Network TEN-T. In the port takes place handling of general cargo and bulk carried in containers and ro-ro. To port can dynamically grow and function properly must fulfill certain conditions, thus providing direct, and at the same time collision-free port connection with the system of national roads is one that is a priority.

The history of the construction of Route name of the engineer Eugeniusz Kawiatkowskiego - the originator and builder of the port of Gdynia - its origins date back to the seventies. In 1974, the government allocated funds for the construction and connection to 1992, was made part of the road named wharf Kawiatkowskiego, which were two separate segments of the route. A few years later, in cooperation with the World Bank resumed construction of the wharf and supplemented by further episodes, and exactly call the streets and Morska Street and Dąbka Street a population of 2.316 kilometers. Due to the high cost of the project was interrupted and was possible only in the years 2004 - 2006 thanks to the acquisition financing from EU funds under the Sectoral Operational Programme Transport. The project "Construction of Kawiatkowskiego - Phase III" envisaged the creation of the connection port of Gdynia with a modern system of roads of national and international by joining the existing flyover of the Tri-City ring road. Implementation of the third stage led to the creation of two two-lane roads with a width of 7m. Also built three road viaducts, two lengths of 55 and 80m at a height of 10.2 to 30m. Episode Wharf - Bypass consists of 2.736 km, and the Wharf- Center 2,728km. The total cost of the project is 231 243 709 PLN (including 75% of EU contribution). With the realized construction of Kawiatkowskiego port of Gdynia got a chance of rapid development, because the connection automatically increased the amount of cargo delivered to him. Thus it gained the country's economy, as the road led to increased integration with the international economy. Total cargo handling in the port of Gdynia is also growing, and port areas are a very attractive aspect for investors.

Another important project that improved access to the port is an investment under the name "Infrastructure road and railway access to the eastern part of the Port of Gdynia" under the Operational Programme Infrastructure and Environment, 2007 - 2013. The location for the project are areas Street Polska Street, Marine Station and zone II Quay Polska Street. Infrastructure was established along the Port Basin IV. The project consists of two stages - development of road and rail access to the elements of the road. Implementation of the first phase consisted mainly of rebuilding Polska Street, Chrzanowskiego Street and Wendy Street. Built entirely new route road with a length of 1.49 km, formed four intersections, as well as tracks on the section 1,49km. To avoid possible collisions built a new railway turnouts, and existing reconstructed. The second phase of the project involved the reconstruction of Polska Street section at the roundabout Chrzanowskiego Street - Gombrowicza Square, whose length was 0,51km. What's more, the upgraded 1.48-kilometer railway line extending

from the flyover in Solidarity Avenue to the roundabout with the Polska Street Chrzanowskiego Street. Fixed the condition of the surface road - rail, to improve intermodal transport near the ferry terminal located at the Quay Polska Street. This was the 0.92 km route. The cost of the investment amounted to 112 555 612 PLN (75% of the amount was financed from EU funds). As a result of the project has been modernized access to the eastern part of the Port of Gdynia through the highway A1, streets and layout of routes such as Poland -Janka Wisniewski - Route Kawiatkowskiego - Tri-Ring Road - Highway A1.

Continuation improve access to the Port of Gdynia required quite expensive, the next upgrade Kawiatkowskiego Route, which also block the movement around the reconstruction indefinitely, and the port lead to paralysis. It would be very inconvenient, because the Port of Gdynia is crucial for the national economy, which is why it was decided to build the so-called. "Red Way". This is one of the major investment to improve the communication port. Construction of the route is designed to solve the problem of fluid connection port on the Tri-City Bypass and the network of national roads. Estimate the most expensive investment so far is estimated at about 900mln Polska Street Zloty, and the lead time is 6 years. After this time Sea will be connected to the port of Gdynia. Beginning of "Red Road" to exit the bypass at the district Chylonia and the end of the Port of Gdynia. The final effect is assumed the two-lane road, which will be implemented in two stages of the project. The first stage involves the creation of single-lane road in each direction, and the cost of their construction is about 666mln zł, of which approx. 100 million Polska Street Zloty from the state treasury. Construction of "Red Road" is also hope for the city, linked to the OPAT-I, which will bypass Rumia and Reda. The road which creates access to the port would be in fact the most expensive in the construction segment, which constitutes the beginning of the Northern Ring Road Tri-City Agglomeration.

We can not forget the western part of the Port of Gdynia. "Reconstruction of rail access to the western part of the Port of Gdynia" is another of the most important investments aimed at increasing the competitiveness of the port. The project is set out in "Description of the planned infrastructure investments of the Port of Gdynia SA" for the period 2014-2020. Rail transport is rapidly increasing, and intermodal transport significantly expanding. Reloading intermodal increase by 10-20% each year. Therefore, it is necessary modernization and reconstruction of existing rail transport links, as well as handling. The investment is linked to the program of TEN-T and is a priority in the strategy for transport development. Location of the project is the western part of the port of Gdynia from the south and north. The southern part includes the reconstruction of the tracks and the construction of a technological road commuting. The result of the above operations intermodal terminal will be ready to handle trains with a length of up to 700m. The planned operation of these trains comprises two variants. The first is the service "4x half the composition of the train," and other one-time support for all cars. The northern part of the port is a project whose goal is to rebuild the station 2,5km from the intermodal terminal BCT Sp z.o.o. to the west, as well as its electrification. Port authorities also plan to implement a remote railway traffic control system and NDS thirteen turnouts will be equipped with heating turnouts (EOR). Part of the work at the terminal BCT has already been completed, but the complete investment will be completed by the end of 2020.

3 ACCESS TO THE PORT OF GDYNIA FROM THE SEA

Stan fairways and the entire infrastructure associated with them, as well as the equipment and installations to assist in the functioning of the port arouses discontent among the authorities of the port of Gdynia. Access to the port is limited primarily by the depth and breadth of approach channels and too small turntables representing certain limit for large commercial vessels and passenger. In connection with the fairly significant problem Board of the Port of Gdynia SA undertook further investments in order to improve access to the port from the sea.

"Reconstruction of the Port Channel in the Port of Gdynia" for the years 2008-2011 was the investment necessary for effective development of the port. The main idea of the project was to increase the permitted draft of ships in port and improve the conditions of navigation within it. There have been dredging work thereby increasing the depth of the Port Channel from 11.5 m to 13.5 m on the stretch of 5.3km from the entrance to the port in the west side (waterfront Bulgarian and Helskie I) and eastern (Pool III). From this point to the port may arriving ships with a draft of at 13m. They rebuilt the Spurs head Pilot increasing the width of the entrance of 84m to 98m, allowing the service vessels of greater width. What's more, it was reconstructed until 6 berths: Dutch, Belgian, Norwegian, Hel I, Slovak and French modernizing it and improving the condition of the bottom using the strengthening of the length of 2474m, and the waterfront Dutch, Hel I, Slovak, French and Norwegian additionally deepened bottom.[8] At that time, the depth of quays posed a barrier to use the dominant size of the ships, which are increasingly popping up in the terminals. Consequently, there was no possibility of their full

load, which of course is not a desirable phenomenon. The project also included the modernization of the port turntables, and more specifically to increase their diameters and deepening. Each of the turntables deepened 2m, from 11.5 to 13.5 m, and further increased the diameter of the turntables 2 and 3. Turntable 2 was increased from 300m to 385m in diameter, and the turntable 3 increased from 270 to 400m in diameter. Made operations allow for more freedom and security during the maneuvers of ships arriving at the port, and also created the possibility arriving vessels of larger dimensions than before. The project is 90 660 788,00 PLN, of which 51 161 570.04 PLN is co-financing from the European Union from the Cohesion Fund under the Programme Infrastructure and Environment.

In recent years much has moved on transport and handling of goods in ro-ro technologies. In this situation, it was signed the next investment project of the Port of Gdynia SA Involving port development through infrastructure development - "Development of port infrastructure to handle ro-ro ships with road and railway access to the Port of Gdynia" for the years 2007 -2013. Total cost of the project amounted to 100 276 130.06 PLN, of Which nearly half the cost of subsidized European Union - 44 858 186.58 PLN. The guiding idea of the project was to Increase the Attractiveness of Polska Street investment, as well as development associated with handling system, ro-ro, and THUS Their improvement and remain 'competitive on the international market. The project included the reconstruction of the road and rail systems, as well as getting rid of some located in the eastern part of the port of unnecessary objects. A very important aspect was also the improvement and dissemination of intermodal transport in the transport of goods through adjustment of Ro-Ro Terminal for rail transport. The project consisted of the fourth phase, which made a number of changes in the inner harbor. The project included reconstruction of 1.2 km of roads, plus construction of an additional technological road of 0.78 km, Which are a combination between the two squares maneuver and components at the terminal. Actions contributed to the expansion of the destroyed street. Romanian is 7m with underground networks. Moreover, the new road was built Which constitute a direct connection squares Street. Polska Street and Janka Wiśniewskiego Street terminal Ro-Ro. The then railway lines were in very bad condition, so it was a reconstruction of about 2km stretch and additions further Top 600m of new tracks to load KR5, Which gave the opportunity to drive through different cars. The investment demolished obsolete objects. The magazine 20, R and bunkers, and THUS spared new surfaces for maneuvering and-storage (22.70 thousand m2). Together with the squares after the reconstruction - previously utilized - port Became available 82,28tys m2. An important aspect of the project was the reconstruction of the lower and upper construction of ro-ro ramp, and further Top modernization of the mooring system.

In the Following years, namely 2014-2020, the Port of Gdynia working with the Maritime Office in Gdynia are planning another investment Fri. "Deepening the approach fairway and waters inside the Port of Gdynia" to the Implementation of the third stage. The project is a continuation of the previous projects for Improving access to the port from the Baltic Sea. General objective is to Increase accessibility to the port ships with a draft of up to 15m, safety and improvement of navigation. All These processes lead to Increased competitiveness on the market. Phase I involves the construction of a new turntable 2 with a length of 240m. Phase II covers only dredging the approach fairway to the port to 17m. Phase III, in turn, takes into account the deepening of the Port Canal and the port waters to a depth of 16m. Probably this is also the reconstruction of the neighboring wharves, Mainly the Strengthening of Their slopes. It is also Necessary to protect underwater installation. Estimated sludge until it approx. 5,700 m3. Phase I was completed in the years 2014-2015, the Implementation of phase II is currently in progress, and for the time of its completion is 2018 years. At the same time we are working on phase III.

4 ACCESS BY ROAD AND RAIL TO THE PORT OF GDAŃSK

4.1 Information about the port and its development strategies

Port of Gdańsk, as a link in the Trans-European Transport Corridor No. VI (Baltic - Adriatic), consists of two main parts: the internal port (located along the Dead Vistula and the port channel) and external port having direct access to the Gulf of Gdansk.

Within the internal port quay are of a universal nature, as well as specialized units: the passenger terminal and a base for ro-ro ships, a base for handling cars and citrus fruits, terminal handling bulk cargo
and a phosphorene. The external port are located on the other hand, among others, Base liquid fuels, coal and liquefied gas. There is also the largest Polish container terminal - DCT. The port area are located a total of 38 different kinds of wharves, piers and terminals. Reserves field in Gdansk are located mainly in the outer port. The dynamic development of this part of the port (including the construction of the terminal mass Sea Invest, location PERN fuel base, the development of DCT container terminal,

construction of the grain terminal) tends to seek further directions of spatial development of the port. Strategies for more intensive use of the port area are:
– Expansion into new areas of the external port,
– Revitalization of the external port associated with the failed investments of the 90s.,
– "Out to sea".

The main objectives of the strategic port of Gdańsk implemented in 2014. In accordance with the maritime policy of the state is the development of a universal port that supports the largest ships that can enter the Baltic, performing the functions of distribution in the Baltic Sea area and in the hinterland, with particular emphasis on handling containerized cargo and intermodal transport. The main priorities and objectives of the development strategy of the Port of Gdańsk include: development of deep-water reloading, development and industrial distribution function, ensuring the availability of the sea and land, revitalization of infrastructure in the Inner Port, development and implementation of the concept of spatial development of the Deepwater Port.

4.2 Road access to the Port of Gdańsk

The most significant impact on the access road to the Port of Gdańsk is called. frame communication Gdansk. For its part the important port consists primarily Western Bypass Tri Gdansk Southern Bypass, Route Sucharskiego with a tunnel under the Dead Vistula and the route of the Slovak (from still missing section of New Kosciuszko.

4.2.1 Sucharskiego route with a tunnel under the Dead Vistula

The route runs from the node Sucharskiego Gdańsk port to the street. Elbląskiej, then through the cable-stayed bridge. Pope John Paul II and the Node Wosia Budzysza along. Sucharskiego to Node Ku Ujsciu. Then of Node Ku Ujściu to the Westerplatte Ferry Terminal. Sucharski Route is a road investment recorded in the construction and modernization of roads in Gdansk in 2007-2013 called Gdańsk Wide Way. According to the schedule, the investment was to be realized in 2009-2011. The project "The combination of national roads - Route Sucharskiego" consists of three investment projects with a total length of about 8.3 km. The aforementioned first two sections of this main road accelerated motion of the two roads two lanes in each direction, the third section leading to the Northern Port

This collective way of a roadway, one lane in each direction. Route Sucharskiego as part of the national road No. 89 is the main road connection DCT container terminal and the Northern Port with the network of other national roads, including the Southern Bypass Gdansk and continue on highway

A1. With this route not only improved road accessibility and increased competitiveness of the Port of Gdansk, but also there has been a reduction of car traffic in the city center. Extension of Route Sucharskiego is a tunnel under the Dead Vistula. Building which for use on April 24 2016. It has 1.4 kilometers long and at its lowest point goes under water at a depth of almost 35m. This is the first underwater road tunnel in Poland. Its construction was one of the biggest and most expensive road projects in the history of Gdansk. The total cost was around 885 million zł. The opening of the tunnel under the Dead Vistula River will cause changes in the organization directions to located on the left bank of the Vistula part of the port, shipyards and other businesses, primarily restricting the movement of lorries over 24 tons in the town center. The tunnel is also important for all who enter to Gdansk from Warsaw, Łódź and Kaliningrad.

The tunnel was constructed using drilling machines Tunnel Boring Machine, which was manufactured in Schwanau in Germany. Construction of the tunnel began in 2011, the same drill in mid-2013. The speed limit is 70 km / hr., So far limited to 50km / h. The tunnel under the Dead Vistula River with a total length of 1,377.5 meter complex is two parallel hollow roads. In each of the strands of the tunnel it was established roadway with two lanes and a hard shoulder. In the tunnel to traffic of cyclists and pedestrians. Traffic in the tunnel will be monitored around the clock.

4.2.2 *Slowackiego Route*

Slowackiego Routeis a project to link the Airport and the Port of Gdańsk. Its total value is 1 420 million Polish Zloty (cost of financing 1 154 million Polish Zloty). The investment project implemented in 2011-2014, planned to build a route length of 10km, but ultimately not completed a 2-kilometer stretch. The element missing in the Slovak Route, and thus the frame communication Gdansk, located in the area of transport to the Port of Gdańsk, the street New Kosciuszko. The cost of this investment is estimated at 100 million, but not reserved any money for this purpose and does not plan to implement it in the coming years. It is quite controversial due to the fact that 100 million does not seem so huge expense to complete the massive investment that is the frame communication Gdansk.

4.2.3 *Projects Sectoral Operational Programme - Transport*

Besides big projects, which undoubtedly is to improve transport framework, in order to improve road access to the Port completed no less important projects on a smaller scale. The projects were implemented on the basis of the contract covering the joint implementation of three investment projects submitted by the PGA SA the Ministry of Infrastructure Operational Programme SOP-Transport, which were to serve simultaneously improving port infrastructure, which is the responsibility of the Port, as well as improved road accessibility to the staging port infrastructure, which is the responsibility of the Municipality of Gdańsk.

*Project 1 "Access Infrastructure to the Industrial Quay".*The aim of the project was to streamlining the communication system connecting the waterfront industrial development areas and the port with the main communication eastern part of the port Route Sucharskiego and increase the cargo handling capacity of Industrial Quay.

The project was reconstructed waterfront, built a new viaduct with a capacity of 50 tons and modernized Ku Ujściu Street. The result was provided by direct access road for heavy truck traffic to the area of the waterfront while the elimination of the residential districts. In addition, conditions were created for the development of approx. 120 hectares of port development. The project was completed in 2007.

Project 2. The project "Improving access to Free Zone in the Port of Gdańsk" - including the redevelopment of the waterfront WOC II and the reconstruction of the road junction Oliwska / Industrial. The scope of tasks included the reconstruction of the road the main access road to the WOC (national road), along with reconstruction of the junction with the railway crossing and tram. Mounted been additional lanes, improved geometry intersections listed sections of roads and reconstructed set of underground and overground technical infrastructure, as well as a network of electric traction. Achieved a marked improvement in the safety and fluidity of movement and provided parking space in front of the customs territory.

*Project 3. The project "Development of the quay and road infrastructure at the Westerplatte Ferry Terminal."*Part of the investment relating to the access road is associated with enlargement of maneuvering and parking and modernization of road junction in the foreland of the Westerplatte Ferry Terminal. The project created the conditions for the transfer of part of the road with a land border crossings, and also had a positive impact on productivity and quality of port services by the introduction of multimodal transport in a wider dimension.

The investment has improved road safety on the foreground and the access roads to the base. The project contributed to the activation of economic areas located in the vicinity of the east side of the port.

4.3 *Rail access to the Port of Gdańsk*

In recent years there has been a percentage increase in the role of railways in operation Tri-city terminals. For comparison, in 2009. Share of rail-to-use DCT Gdańsk was 26% in 2011. - 36% and in 2014. - 35%. The number of containers handled by the railway in 2013 and 2014 totaled approx. 415 thousand. TEU.

4.3.1 *Railway line 226 + bridge over the Dead Vistula*

Improve the capacity on the railway line to the port was required, inter alia due to the investment plans of DCT - soon will open a second terminal, which will increase bandwidth DCT from 1.5 to 4 million TEU per year, as well as due to the recently completed or planned large investments within the port, such as. the construction of transshipment terminal of loose bulk goods, construction logistics park Goodman or the construction of the Universal Deepwater Terminal.PKP Polish Railway Lines SA preparing further investments that will increase the quality of railway infrastructure leading to the ports. Still in the first quarter of this year, Polish Railways are planning to call for tender for project documentation concerning investments in the area of the Port of Gdańsk and strictly dedicated to freight transport station: Gdańsk North Port, Gdańsk Channel Kaszubski, Gdańsk Zaspa Towarowa with lines connecting these stations and the lead-out trains inland. The project is saved in the National Programme for Railway and applying for EU funding in the second call for the "Connecting Europe Facility"- "Connecting Europe". In particular, the technical condition of the station North Port is inadmissible. This station also has historical reasons much more targeted tracks the movement of the inner port than on the external port, which now has greater importance for handling the individual entities operating in the Port.

From the point of view of the largest container terminal in the Tri-DCT operation associated with the improvement of the availability of the terminal by rail it is also to improve the throughput segment from the North Port to the terminal, as well as own infrastructure lane at the same terminal. This is expanded in 2013. railway siding: 4 tracks with a total length of 2.5 km, each of which is able to handle the full composition of the train. Its annual capacity is 780 thousand. TEU.

5 CONCLUSION

In the above study we focus particularly on improving access to the port from the water in case of Gdynia center and developed more access from the hinterland to the Port of Gdańsk. For this analysis it was decided due to the conditions of the two ports - the author took into account the areas where the shortage of sufficient quality infrastructure was an obstacle to development and aspects that improvement was a high priority date.

In conclusion, we would like to explain why it is important to develop access to the port. The quality of the motorway and express road connections of seaports with their hinterland depends on the quality, safety, time and cost of transportation to or from the port. At present, the efficiency of transport links the port with its hinterland and foreland is a decisive factor in choosing the port by cargo and largely determines the profitability of the entire land and maritime transport chain. Therefore, without a proper focus on this factor, the competitiveness of the port drastically reduced.

In line with its strategy, the Port of Gdańsk for directing action to strengthen the position of the port and eventually turn it into a so-called. port V generation, which is the port that is universal intermodal logistics hub that allows you to handle the largest vessels entering the Baltic Sea, an important region-wide industrial center with comprehensive services for intermodal transport, wholesale and distribution center and the key on the southern Baltic hub and the resort to generate added value. For this project it is necessary to continually improve the access road and rail to the port.

Port of Gdynia in the vision is seen as a universal port multimodal logistics hub transport corridor North - South, capable of creating market advantages. Port of Gdynia has also made the necessary steps to become an ocean port, also supports marine transit. The high level of quality of functioning of the Port of Gdynia is to strengthen its market position and cause the well-established, strong brand port will differentiator in attracting users. This vision of the future port also involves focusing on the realization of investment plans for both improved access from the land and from the sea.

REFERENCES

[1] Pluciński M., Polskie porty morskie w zmieniającym się otoczeniu zewnętrznym, Wydawnictwo CeDeWu, Warszawa 2013,
[2] Antonowicz M., Kolejowy transport towarowy, w: Euro Logistics, 02/2013,
[3] Urząd Transportu Kolejowego, Rynek kolejowych przewozów intermodalnych, Gdańsk, 2014,
[4] http://utk.gov.pl/pl/analizy-i-monitoring/statystyka-kwartalna/przewozy-towarowe/
[5] https://www.portgdansk.pl/o-porcie/spot
[6] http://bip.transport.gov.pl/pl/bip/zamowienia_publiczne/oce na_program_porty_2020/px_program_rozwoju_portow_m orskich___zalacznik_do_siwz.pdf
[7] http://www.rynekinfrastruktury.pl/
[8] Smolarek L., Kaizer A., The analysis of dredging project's effectiveness in the Port of Gdynia, based on the interference with vessel traffic. Safety of Marine Transport 2015

Safety and Hydrodynamic Study of Hydrotechnical Structures

Hydrodynamic Study of Nautical and Shore Protection Structures in Santos Bay, Brazil

P. Alfredini, E. Arasaki & J.C.M. Bernardino
Polytechnic School of Sao Paulo University, Sao Paulo, São Paulo

H.L. Puia
Santos Pilots, Santos, São Paulo

G. Silva
Fundação Centro Tecnológico de Hidráulica

R.C. Prats
Engecorps

ABSTRACT: The enlargement of the nautical dimensions of Santos Port Outer Access Channel (Brazil), with the purpose of receiving vessels between Post Panamax Plus and New Panamax, training walls crossing the Offshore Bar are needed. The training walls choice to reduce dredging rates also induces to consider a coupling planning between nautical purposes and shore protection measures. Indeed, the beaches of Santos Municipality have serious erosion problems nowadays due to the urban growth in the backshore, sea level rise and the increasing effect of extreme events of storm surges. For Port and Municipality Authorities decision sup-port, the Hydraulic Laboratory of Polytechnic School of São Paulo University was commissioned to study in a composite mathematical and scale model, evaluations about the morphological impacts of the training walls, considering the nautical purposes and a compatible solution for the shore protection structures.

1 INTRODUCTION

Santos Port, in São Paulo State Coastline (Fig.1), throughputs approximately 15% of Brazilian maritime cargo, more than 110 million tons per year and is the most important maritime cargo transfer terminal in the Southern Hemisphere. The requirement to enlarge and deepen the Santos Port Outer Access Channel (depth, width and radius) to receive Post Panamax Plus and New Panamax, typically vessels of 12,000 TEUs (L_{OA} = 398 m, B = 56.4 m, $T_{full\ load}$ = 15.0 m) training walls crossing the Offshore Port are the best cost-benefit solution to avoid huge dredging rates. This concept was already presented in the first Master Plan of Santos Port, proposed in the sixties, and it is the Phase 1 of the Seaward Conceptual Planning for the Offshore Port (Alfredini, Arasaki & Moreira, 2015).

Furthermore, the sea level rise occurred in the last century (IPCC, 2014) and the increasing rates of extreme events of storm surges require different alternatives concerning sea defenses structures in Santos beaches, mainly in Ponta da Praia location.

According to Alfredini, Arasaki & Moreira (2015), an engineering solution for the Santos Port dredging problem is to construct a rubble mound structure, such as two training walls with a total length of 9 km. It would provide a solution, significantly reducing the OPEX maintenance dredging requirements, and re-designing the adequate channel dimensions suitable for the larger vessels.

Otherwise, due to the storm surge attack in Ponta da Praia area (Fig.1), it is also important to provide a shore protection structure for that area.

For Port and Municipality Authorities decision support, the Hydraulic Laboratory of Polytechnic School of São Paulo University was commissioned to study in a composite mathematical and scale model, evaluations about the morphological impacts of the training walls, considering the nautical purposes and a compatible solution for the shore protection structures.

The goal in this paper is to present the first results of two conceptual projects solution, considering the nautical purposes as mandatory, but also trying to find a compatible solution for the shore protection structures.

2 MATERIAL AND METHODS

2.1 *Study Area*

The study area location, including Port area and Ponta da Praia beach, are presented in Figure 1. Ponta da Praia is the most critical area of Santos Municipality, with erosion problems due to the increasing effects of storm surges and sea level rise (Alfredini et al., 2013).

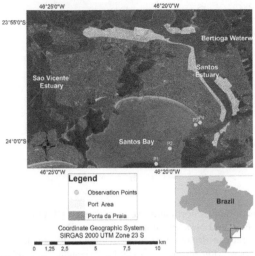

Figure 1. Location map of study area and observation points P1, P2, P3 and P4 (dots).

The data set was obtained with an ADCP gauge from Santos Pilot (Santos Pilots, 2016), located in point P2 (see Fig. 1).

Two conceptual project solutions were studied: the construction of training walls and a segmented breakwater structure.

2.2 Numerical Modelling Description

The effectiveness and efficiency of these structures were analyzed using numerical modeling. Delft3D numerical model (DELTARES), Flow and Wave modules, was used in the present study with the application of complete formulations for shallow water equation finite-difference calculation, the hydrostatic hypothesis and the Boussinesq approximation. The Boussinesq approximation states that if density variations are small the density may be assumed constant in all terms except the gravitational term (Broomans, 2003).

According to Chatzirodou & Karunarathna (2014), Delft3D is a finite difference code that solves the Navier-Stokes equations under the Boussinesq and shallow water assumptions, in 2D or 3D dimensions. For a 3D flow simulation, the system of equations then reads:

$$\frac{\partial(\zeta)}{\partial t} + \frac{\partial\left[(\zeta+d)\bar{U}\right]}{\partial x} + \frac{\partial\left[(\zeta+d)\bar{V}\right]}{\partial y} = S \quad (1)$$

$$\frac{\partial U}{\partial t} + U\frac{\partial U}{\partial x} + V\frac{\partial U}{\partial y} + \frac{\omega}{h}\frac{\partial U}{\partial \sigma} - fV = -\frac{1}{\rho_0}P_x + F_x + M_x + \frac{1}{h^2}\frac{\partial}{\partial \sigma}\left(v_V\frac{\partial u}{\partial \sigma}\right) \quad (2)$$

$$\frac{\partial V}{\partial t} + U\frac{\partial V}{\partial x} + V\frac{\partial V}{\partial y} + \frac{\omega}{h}\frac{\partial V}{\partial \sigma} - fU = -\frac{1}{\rho_0}P_y + F_y + M_y + \frac{1}{h^2}\frac{\partial}{\partial \sigma}\left(v_V\frac{\partial v}{\partial \sigma}\right) \quad (3)$$

where f is the Coriolis parameter; U and V are the horizontal velocities in x and y directions; ω is the vertical velocity in relation to σ coordinates; F_x, F_y are the horizontal Reynold's stresses; v_V is the vertical eddy viscosity; P_x, P_y are the horizontal pressure terms approximated by the Boussinesq assumptions; M_x, M_y are external forces added as source or sink terms in the momentum equations (2), (3); ρ_0 is the reference density; S represents the contributions per unit area due to the discharge or withdrawal of water, evaporation and precipitation; ζ is the water level; d is the water depth in relation to a reference level and h is the total water depth ($h = \zeta + d$).

Delft3D-Flow module grid includes São Vicente Estuary, Santos Estuary and Bertioga Waterway, with 59096 elements. The grid resolution in the study area has 15x15 m and it was used one layer in the vertical direction.

Based on water level and water current field data was possible to calibrate and validate the mathematical model. Boundary conditions as wave height and direction, wind velocity and direction were obtained with the aid of the WAVEWATCH III® model developed by National Oceanic and Atmospheric Administration, National Weather Service, National Centers for Environmental Prediction and Marine Modeling and Analysis Branch (NOAA/NWS/NCEP/MMAB, 2016).

Data analysis was based on the variation due to the presence of the structures in wave significant height and direction and current intensity and direction. The period simulated started on June 17th 2012 and finished on July 15th 2012. It includes the storm event of June 20th, which generated waves with significant higher than 2,5 m in point P1 (Fig. 1). The scenarios simulated were defined as:
- S1 - No Structure: current situation, without any structure;
- S2 - Training walls: future situation in long-term, with nautical purposes;
- S3 – Segmented breakwater: future situation in short-term.

It is important to note that the segmented breakwater will not interfere with the training walls purposes. Thus in a future long term situation, it would not be a problem.

2.3 Scale Model

The Spectral Wave Simulator for Port and Coastal Studies is constructed in a wave basin reproducing in a fixed bed a portion of Santos Bay and Ponta da Praia Beach, at Hydraulic Laboratory of Polytechnic School, University of São Paulo.

This facility (Fig. 2), with dimensions of 37 m x 17 m x 1.5 m, has a water reservoir capacity of 650 m³, 10 independent piston wave generators with 1KW for 500 mm piston translation. Irregular short-crested spectral waves can be reproduced with individual wave period above 0.6 s (f = 1.66 Hz) and wave heights between 8 and 250 mm.

3 RESULTS AND DISCUSSION

3.1 *Wave Height and Direction Effects*

Table 1 shows significant wave height in three of the observation points (P2, P3 e P4, see Fig. 1) to each of the scenarios presented. Simulation results obtained showed that both structures were effective in the interested area. They have blocked wave height, although the segmented breakwaters blocked just at point P4, and training walls. Figure 3 presents the significant wave height in the storm surge event (July 20th at 15h:00).

Comparing Figure 3 (A and B) it is possible to verify the differences between wave significant height in the channel area. Figure 3 (C and D) shows the effect of the segmented breakwater in blocking the wave action in the beach area. It is possible to observe the diffraction effect in breakwater gaps, in spite of this numerical model has limitations to reproduce it.

Wave direction was also compared. Although wave incident direction is parallel to training walls alignment, wave height reduced. Figure 4 zooms the entrance area of training walls and it is possible to verify the influence of the structure in reducing the wave height and also a small direction change inside the channel.

Figure 2. General view of the Spectral Wave Simulator for Port and Coastal Studies installed at Hydraulic Laboratory of Polytechnic School, University of São Paulo.

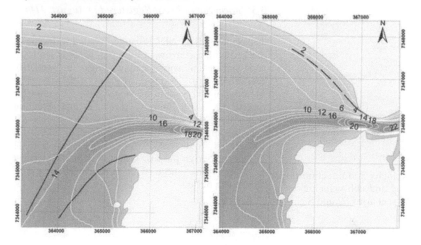

Figure 3. Schematic location of the proposed training walls and segmented breakwater.

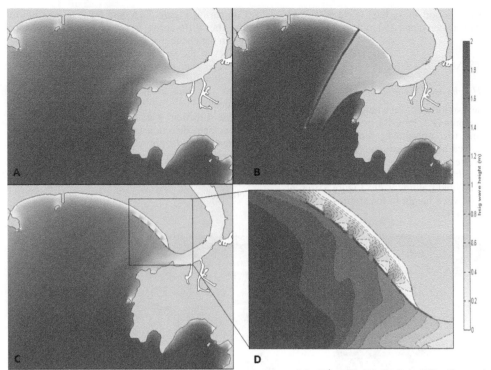

Figure 4. Significant wave height (m) during the storm surge event on July 20th at 15h:00), for three different scenarios: A –S1 - Current situation; B – S2 – Training walls; C – S3- Segmented breakwater; D – Zoom S3.

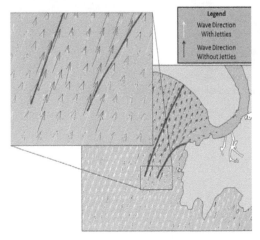

Figure 5. Wave peak direction to S1 (black vector) and S2 (white vector).

Figure 6 shows current velocity and direction for the instant of maximum flood and ebb velocities in the spring tide conditions for scenarios S2 and S3.

3.2 Time Series for Waves Height

Figure 7 presents time-series from June 17th 2012 and finished on July 15th 2012 for wave height in P4. The significant reduction in wave height shows the effectiveness of the segmented breakwater (S3) in protect Ponta da Praia Beach.

3.3 Storm Surge Event Registered in August, 21th 2016

The Santos Pilots ADCP gauge, located in the point marked with the arrow in Figure 8, records sea level and significant wave height (Figs. 9 and 10), according to Santos Pilots (2016). The Meteorological Station of Santos Pilot anemometer records the wind intensity (Fig. 11) and direction (Fig.12).

Confirming that the environmental premises considered in this study are adequate as extreme events for design purposes, in this item is described a recent storm surge with at least 10 years of recurrence period.

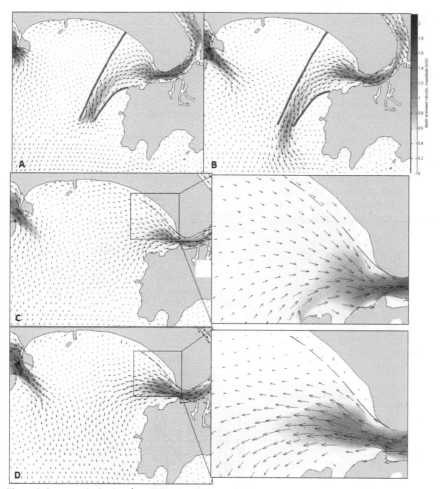

Figure 6. Current velocity (m.s⁻¹) and direction for Scenario 2: A - flood; B – ebb. Scenario C – flood; D – ebb.

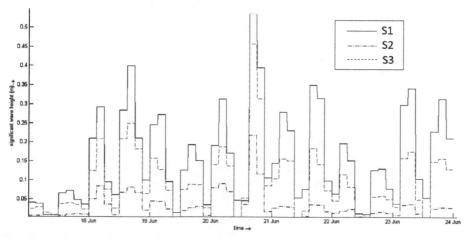

Figure 7. Significant wave height at point P4 to S1 (continuous line), S2 (dashed line) and S3 (dotted line).

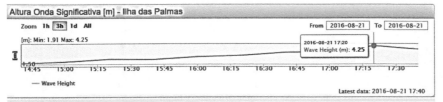

Figure 10. Significant wave height, located in Ilha das Palmas, according to Santos Pilots (2016).

In August, 21th 2016, an extreme event was recorded in the ADCP (Figs. 9 and 10) and in the Meteorological Station (Figs. 11 and 12). The storm surge was induced by wind blowing more than 12 h from SW fetch (Fig. 12), with maximum velocities from 20 to 58 knots (Fig. 11), rising more than 1 m the predicted astronomic sea level in Santos Bay and induced waves (see Fig. 9).

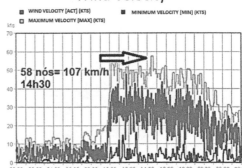

Figure 11. Wind intensity in August, 21th 2016 in Santos Port (Santos Pilots, 2016).

Figure 8. Map with wind direction during the storm surge in Santos Port (Santos Pilot, 2016).

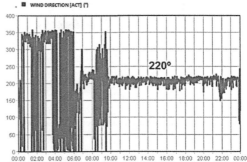

Figure 12. Graphic of wind direction in 21/08/2016 in Santos Port (Santos Pilots, 2016)

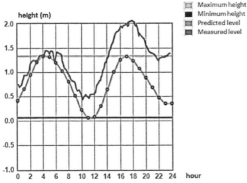

Figure 9. Measured sea level in Santos Port: maximum height (astronomical high water); minimum height (Chart Datum); predicted level (dotted); measured level (continuous line).

3.4 Scale Model Results for the Segmented Breakwater

The scale model runs of scenario S3 with a storm surge similar to the simulated in the numerical model can be illustrated by Figure 13. It is possible to observe the similarity of the diffraction pattern comparing with Figure 4D.

Figure 13. Santos Pilotsal picture of the segmented breakwater in the scale model (Hydraulic Laboratory).

4 CONCLUSIONS

This paper showed the studies that have been made in Hydraulic Laboratory of Polytechnic School of São Paulo University. The goal was to analyze the best solution for Santos Bay problems that involves nautical purposes and shoreline protection.

Two structures were simulated at this area showing alternatives to improve the navigation demands and also the shoreline protection with classical structures like training walls and segmented breakwater.

Model simulations with segmented breakwater showed the effectiveness in reducing wave height more than 50% in Ponta da Praia Beach, which means wave energy reduction of 75% in a storm surge event.

For the training walls model simulations, the current velocities in the Access Channel increased, showing the real possibility to reduce the dredging rates here. Moreover, results obtained also indicated that the structures reduce wave height in Ponta da Praia Beach, which would enhance the protection against shore erosion in this critical area.

Therefore, both structures showed to be efficient in protect Ponta da Praia Beach against wave erosion.

REFERENCES

Alfredini, P.; Arasaki. E.; Pezzoli, A.; Fournier, C. P. 2013. Impact of climate changes on Santos Harbor, São Paulo State (Brazil). *TransNav, the International Journal on Marine Navigation and Safety of Sea Transportation*, 7(4): 609-617.

Alfredini, P; Arasaki, E.; Moreira, A.S. 2015. Design Tide and Wave for Santos Offshore Port (Brazil) Considering Extreme Events in a Climate Changing Scenario. In Adam Weintrit & Tomasz Neumann (Eds.), *Information, Communication and Environment - Marine Navigation and Safety of Sea Transportation* 147-151. CRC Press/ Balkema, Taylor & Francis Group, London.

Broomans, P. 2003. *Numerical accuracy in solutions of the shallow-water equations.* Master Thesis. Delft.

Chatzirodou, A. & Karunarathna, H. 2014. Impacts of tidal energy extraction on sea bed morphology. *Proceedings of 34th Conference on Coastal Engineering*, Seoul, Korea,

DELTARES. 2014. *Delft 3D-FLOW- Simulation of multi-dimensional hydrodynamic flows and transport phenomena, including sediments.* User Manual. 712p. Delft.

IPCC. 2014. Climate Change 2014: Synthesis Report. Contribution of Working Groups I, II and III to the Fifth Assessment *Report of the Intergovernmental Panel on Climate Change.* IPCC, Geneva, Switzerland, 151 pp.

NOAA/NWS/NCEP/MMAB. 2016. User manual and system documentation of WAWEWATCH III ® version 5.16.

Santos Pilots. 2016. http://www.sppilots.com.br/

Proceedings of 12th International Conference on Marine Navigation and Safety of Sea Transportation, TransNav 2017
21-23 June 2017, Gdynia, Poland

Safety of Ships Moored to Quay Walls

V. Paulauskas, D. Paulauskas, B. Placiene & R. Barzdziukas
Klaipeda University, Klaipėda, Lithuania

ABSTRACT: In many ports terminals for big ships are located close to the port entrance and ships are moored very close to the navigational channels by which big ships are passing. Big ships passing close to the ships moored to quay walls create hydrodynamic interaction forces. Hydrodynamic interaction forces together with other external forces, created by wind, waves and current has negative influence on moored ships and quay walls. This article presents theoretical basis for the evaluation of external forces, created by hydrodynamic interaction by passing and moored to quay wall ships, wind, waves as well as currents acting on moored ships. The article also presents simulation and experimental results with real ships in real similar conditions. Klaipeda port oil terminal, which is located close to the port entrance and all ships enter and depart from the port through the port pass close to the ships moored to oil terminal quay walls, is taken as a case study. Calculation methods of the external forces on ships moored to a quay wall and practical recommendations, which can in-crease navigational safety of the ships moored to quay walls, are presented on the basis of theoretical calculations, simulation and experimental result analysis.

1 INTRODUCTION

In many ports terminals for big ships are located close to the port entrance, because in such situation less investment is needed to serve big ships in the port. Terminals of dangerous cargo like oil, gas and chemicals are located close to the port entrance from safety point of view, because it is easier for the ships with dangerous cargo to leave the port in emergency situations [1, 2, 3].

However, big ships which enter the port or leave it through the navigational channels close to the port entrance pass close to the moored ships and create additional hydrodynamic interaction forces between the passing and moored to quay walls ships [4, 6].

Comprehensive theoretical and experimental study of the mentioned situations is very important to find solutions to minimize the impact of passing ships on ships moored to the quay wall, to increase the safety of the moored ships, especially while loading and unloading dangerous cargoes [6, 7, 9].

The theoretical basis for the evaluation of the forces created by passing ships on the ships moored to a quay wall (hydrodynamic interaction), as well as wind, current and waves is presented in many articles [4, 7, 10], however, they do not relate these

factors with other external forces and are not tested experimentally by real ship study.

In this article theoretical calculations and simulation results made by calibrate simulators "Simflex Navigator" and "SimFlex 4" (producer FORCE TECHNOLOGY, Denmark) [8] and experimental results made by the authors in real port conditions are presented.

Klaipeda port oil terminal, located close to the port entrance and the navigational channel used by other big ships sailing in and out the port, was taken as a case study.

2 SITUATION ANALYSIS OF THE PORTS, WHICH HAVE QUAY WALLS CLOSE TO PORT ENTRANCE AND NAVIGATIONAL CHANNELS

In many ports some terminals are located close to the port entrance and navigational channels by which passing big ships create hydrodynamic interaction forces; wind, current and waves also create additional external forces on the ships moored to the quay wall. Klaipeda port oil terminal (Lithuania), Goteborg port oil terminals (Sweden), Milford port (oil and gas terminals) (UK), Sant

Nzare port (France) and many other ports can be mentioned as examples of such ports.

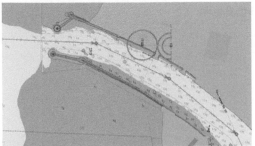

Figure 1. Oil terminal in Klaipeda port is located close to the navigational channel, by which big ships pass

Figure 2. Oil and gas terminals in Milford port are located close to the navigational channel.

Figrue 3. Oil and gas terminals in Sant Nazare port (France) located close to the navigational channel.

Increasing ships parameters and intensity of the port operations, and influence on ships moored close to the navigational channels, especially close to the port entrance request additional attention to ensure safety of moored ships. Study of the forces, created by big ships passing near moored ships, is very important for finding methods and possibilities to decrease the negative impact on ships and quay walls.

3 THE THEORETICAL BASIS OF THE ASSESSMENT OF EXTERNAL FORCES AFFECTING MOORED SHIPS

The theoretical basis of the evaluation of forces acting on ships moored to a quay wall, located close to the navigational channels, created by passing ships such as hydrodynamic interactions between ships and other forces that influence ships moored to

quay walls could be calculated and taken into account for channel and quay wall design as well to ensure safety of ships moored to quay walls. The hydrodynamic interaction forces, which act on moored ships, depend on ship dimensions, speed, distance between the ships and the depths of the channel, and could be evaluated as follows:

$$F_{int} = f(D, v, S, H, ...),\qquad(1)$$

where: D – passing ship displacement; v – passing ship speed; S – distance between the ship moored to a quay wall and the passing ship; H – the depth in the channel.

Hydrodynamic interactions between ships are based on pressure distribution around the sailing ship [9]. Revising the case of impact of ships passing near a ship moored to a quay wall reveals that the passing ship moves and turns the moored ship [7, 9].

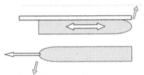

Figure 4. A ship passing by close to a moored ship moves and turns the moored ship.

The forces of hydrodynamic interactions between ships depending on pressure distribution can be expressed in the following way [7]:

$$F_{int} = k \cdot D \cdot v^2 \frac{3B}{S^2} \cdot \frac{H_{kr}}{H},\qquad(2)$$

where: k – coefficient, for calculations could be taken as 0,175; B – width of a passing ship; H_{kr} – critical depth in the channel that is calculated as follows:

$$H_{kr} = \frac{v}{\sqrt{gT}},\qquad(3)$$

where: T - passing ship draft; g – gravitational acceleration.

Wind and current acting on a ship moored to a quay wall can be calculated by the established ship theory methods [6, 9]. Aerodynamic (wind) force could be calculated as follows:

$$R_{ay} = C_{ay} \cdot \frac{\rho_1}{2} \cdot S_x \cdot v_a^2 \cdot \sin q_a,\qquad(4)$$

where: C_{ay} - aerodynamic coefficient, which in average is about 1,07 – 1,30 (specific data could be taken from aerodynamic tube testing); ρ_1 - air density, for the calculations could be taken as 1,25 kg/m³; S_x - ship air projection on middle square; v_a - wind velocity; q_a - wind course angle (the angle to a quay wall).

Current created force, which could be calculated as follows:

$$R_{cy} = C_y^{'} \cdot \sin q_c \cdot \frac{\rho}{2} \cdot L_1 \cdot T_1 \cdot v_c^2 \ , \qquad (5)$$

where: v_c - current velocity; q_c - current course angle (angle to quay wall; L_1 - moored ship length between perpendiculars; T_1 - moored ship average draft; $C_y^{'}$ - hydrodynamic coefficient of the ship hull and its dependence on the current angle to the quay wall can be calculated as follows;

$$C_y^{'} = 0,8 \cdot \sin q_c \ , \qquad (6)$$

Finally, total forces, acting on a ship moored to a quay wall could be calculated as follows:

$$\sum \vec{F} = \vec{F}_{\text{int}} + \vec{F}_{\text{ay}} + \vec{F}_{\text{cy}} \qquad (7)$$

The total force must be absorbed by quay wall fenders and bollards. In case of correct mooring scheme, about 50 % of the total force is absorbed by quay wall fenders and about 50 % should be absorbed by ship's mooring ropes and quay wall bollards.

In case if there is not enough ship mooring rope pretension, the moored ship starts moving along the quay wall and this creates inertia forces ( ), which sometimes could be higher as any other forces. The inertia forces could be calculated as follows;

$$R_{in} = m^{'} \cdot a \ , \qquad (8)$$

where: m - the mass of the moored ship; a - acceleration, which could be calculated as follows:

$$a = \frac{v_1^{'}}{2 \cdot T_P} , \qquad (9)$$

where: $v_1^{'}$ - maximum possible ship movement speed near the quay wall which could be calculated on the basis of ship movement distance near the quay wall and the movement period. The movement distance near the quay wall in the study case could be taken up to 25 – 30 % of the shortest mooring rope depending on pretension of the mooring ropes. Ship movement period (s) near a quay wall could be equal to $L_1/2 - L_1/3$. Finally, inertia forces could be calculated as follows:

$$R_{in} = k_P \cdot l_r \cdot \frac{m \cdot}{2 \cdot T_P^2} \ , \qquad (10)$$

where: k_P - mooring rope pretension coefficient, which could be between 0,05 and 0,25 (0,25 no pretension in all).

The theoretical basis presented above could be used for the calculation of the real forces, acting on a ship moored to a quay wall close to the navigational channels by which other ships are passing; and for design of the quay walls and its elements, like fenders, mooring bollards.

It is very important to take into consideration that ship mooring ropes pretension depends of the ship parameters and it should be equal to the periodical forces, which are created by periodical influence, such as wind, waves, or hydrodynamic interaction of a passing ship. In case if pretension forces of mooring ropes are equal to the periodic forces, the increase of the length of mooring ropes could be minimized and inertia forces will be decreased.

4 EVALUATION OF HYDRODYNAMIC INTERACTIONS AND OTHER EXTERNAL AND INERTIA FORCES

For practical testing of hydrodynamic interactions and other forces acting on ships moored to a quay wall HANDISIZE and POST PANAMAX tankers moored to quay walls in Klaipeda oil terminal were studied. LNG (length 290 m) and POST PANAMAX oil tankers in ballast (length 250 m) were passing a loaded tanker at the distance from 60 m up to 100 m. The speed of the passing ships was 6,0 – 6,4 knots, the wind speed was from 10 m/s up to 15 m/s from different directions, the current direction was parallel to the quay wall and its velocity was from 0,5 up to 2,0 knots. The pretension of ropes of the moored ship was adjusted depending of the possible external forces. Mooring scheme calculations and distribution of forces were evaluated by OPTIMOOR simulation model [5, 8] and by theoretical calculations.

The resulting force acting on the vessel was used for the calculation and preparation of mooring schemes for POST PANAMAX tankers and other ships, as well as for a quay wall design.

Figure 6. POST PANAMAX tanker passing near a tanker moored to a quay wall.

The parameters (distances, periods) of the moored tanker movements near the quay wall were measured from some points from the shore as well as by RTK (Real time kinematic) system on the moored tanker.

Figure 7.Visual measures of the movement of the moored tankers

On the basis of theoretical calculations by the method presented in this article, for the POST PANAMAX moored tanker in tones (T) the external forces are presented in the figures below.

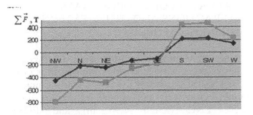

Figure 8. External forces created by the moored POST PANAMAX tanker in ballast (wind 15 m / s (blue) and 20 m / s (red), current - 2 knots and passing a similar size tanker at 80 m distance, including inertia forces)

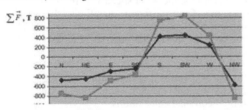

Figure 9. External forces created by the moored POST PANAMAX tanker in ballast (wind 15 m / s (blue) and 20 m / s (red), current - 2 knots and the passing similar size tanker at 80 m distance, including inertia forces).

The resultant maximum horizontal forces acting on the berth, when the vessel is in ballast (red line), loaded to a draft of 13 meters (yellow line), are presented in Figure 10.

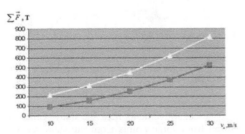

Figure 10. The forces created on the quay by a POST PANAMAX tanker, depending on the wind speed, wind direction from NW to SW) (the loaded ship - yellow line), (ship in ballast - red line) (current - 2 knots) and the passing similar size tanker at a distance of 80 m (including inertia forces).

POST PANAMAX moored tanker movement from the neutral position in both sides when pretention of the mooring ropes were equal to external forces shown that in case passing the LNG tanker, moored POST PANAMAX tanker movements near quay wall were up to +/- 0.3 m, and in case passing POST PANAMAX tanker – were up to +/- 0.25 m.

The received results of calculations of the specified mooring scheme and the primary mooring rope tension of not less than 180 kN were similar.

In this way, it can be said that movement of ships moored to the quay walls with an initial tension of the mooring ropes referred to in calculations should not exceed 0,5 m from the initial moored position while the same size or smaller vessels are passing at a speed greater than 6 knots, which is useful for the tanker loading operations.

Locate tables close to the first reference to them in the text and number them consecutively. Avoid abbreviations in column headings. Indicate units in the line immediately below the heading. Explanations should be given at the foot of the table, not within the table itself. Use only horizontal rules: One above and one below the column headings and one at the foot of the table (Table rule tag: Use the Shift-minus key to actually type the rule exactly where you want it). For simple tables use the tab key and not the table option. Type all text in tables in small type: 10 on 11 points (Table text tag). Align all headings to the left of their column and start these headings with an initial capital. Type the caption above the table to the same width as the table (Table caption tag). See for example Table 1.

5 CONCLUSIONS

1 Ships moored to quay walls near navigational channels are very much affected by external forces and especially by hydrodynamic interaction of passing ships.
2 The methodology presented in this article can be used for calculation of forces acting on ships moored to a quay wall.
3 The pretension of mooring ropes up to the external periodical forces, which could be calculated by the methodology presented in this article, can decrease movement of ships near the quay wall and the inertia forces, which are created by the moored ship, and decrease the impact on the quay wall.
4 Correct calculations of the external forces, pretension of ship mooring lines of the moored ship increase safety of the ship moored close to the navigational channels navigated by big ships.

REFERENCES

1. Choudhury D.; Ahmad S.M. (2008). Stability of waterfront retaining wall subjected to pseudo-dynamic earthquake forces. Journal of Waterway, Port, Coastal and Ocean Engineering, 134(4), 252-260.
2. Danish Hydraulic Institute Staff. MIKE 21 Coastal Hydraulics & Oceanography. User Guide and Reference Manual, Denmark: Danish Hydraulic Institute, 2003. 193 p.
3. EAU 2012 : Recommendations of the Committee for Waterfront Structures – Harbours and Waterways (Ernst & Sohn, 2012) (EAU 2012).
4. José Azcona, , Xabier Munduate, , Leo González, , Tor A. Nygaard. Experimental validation of a dynamic mooring lines code with tension and motion measurements of a submerged chain. Ocean Engineering, Volume 129, 1 2017, Pages 415-427.
5. OPTIMOOR simulation program (OPTIMOOR, 2010).
6. Paulauskas V. 2013. Ships entering the Port. N.I.M.S., Riga, 240 p.
7. Paulauskas V.; Paulauskas D.; Maksimavicius R.; Jonkus M. 2014. Hydrodynamic interactions between ships in narrow channels. Transport, Vilnius. Technika, 29(2). p. 212 - 216.
8. SimFlex Navigator Simulator. Force Technology, Denmark, 2014 (SimFlex 2014).
9. Ships maneuvering theory. Force Technology (Denmark), 4th Editon, 2003. 186 p.
10. Wijffels J.; Paulauskas V. 2010. Ships with big freeboard safety in ports. PIANC Conference, Liverpool, 2010, presentation No. 55.

Tools for Evaluation Quay Toe Scouring Induced by Vessel Propellers in Harbour Basins During the Docking and Undocking Manoeuvring

M. Castells, F.X. Martínez De Osés, A. Martín, A. Mujal-Colilles & X. Gironella
Polytechnic University of Catalonia, Barcelona, Spain

ABSTRACT: The evolution of the shipping industry (increased capacity and size of ships, power and self-propulsion), the increased productivity and rearrangement of spaces (port calls increased in number and frequency, changes in use of docks), the intensification of the use not only of stern main propellers (conventional or azimuthal) but also lateral bow and stern ones, are the leading causes of injury to the toe of the docks. Scouring processes due to manoeuvring actions can produce big consequences on the stability of harbour structures such as docks and protecting dikes. As a consequence, the sedimentation of the eroded sediment reduces the total depth of the harbour basin and navigation channel. At the same time, contaminants settled at the bed of the harbour basins may be resuspended by the effect of vessel's propellers and produce an important environmental problem to harbour authorities.

One of the main problems is the interpretation of parameters related to propellers, but also the fact of not considering important aspects such as different types of propeller, manoeuvring practices for docking and undocking, propulsion system orientation or the confinement effect generated by the dock itself and the shelter of the vessel. This contribution aims to assess docking and undocking manoeuvres which can produce erosion generated by the propulsion of ships. This methodology will analyse manoeuvres patterns and will allow understanding the effects of the sedimentation of the eroded sediment on docking and undocking manoeuvres.

1 INTRODUCTION

Motion Vessels and their propulsion system can produce hydrodynamic. In shallow areas, these effects may have an influence on the seabed. Over the last 20 years, the rise in shipping activities with bigger vessels with an increase of draft and power of engines have led to growing structural problems: scouring effects in the vicinity of the structures affecting the stability and sedimentation of the scoured material in the harbour basin, among others. The sedimentation of the eroded sediment reduces the average depth of the harbour basin and navigation channel and can produce undesired and sometimes uncontrollable motions of the vessel.

Morphodynamic changes inside harbours due to docking and undocking manoeuvring represent an increasing problem for harbour authorities. In particular, old marinas designed to host ships with lower depths and engine powers have to either fill the scouring holes or dredge the sedimentation areas quite often, or alternatively implement bed protection measures in the harbour basins. Both

factors decrease the efficiency and operability of the harbour, causing significant economic losses. This problem affects several harbours around the world with different configurations, morphologies and tidal ranges (Lam et al. 2012). Moreover the eroded sediment is deposited along the harbour reducing water depth level and operative zones for several vessels manoeuvring.

Nowadays, the present propulsion systems are closer to the soil of the docks causing sediment erosion close to toe of the docks which, in turn, may cause severe problems to the docking platforms. Hence these vessels dock at the same position and navigate the same route when approaching the harbour.

The docking and undocking manoeuvres are the most effective in terms of erosion, in particular for vessels without the help of a tugboat or pilot (Mujal-Colilles et al. 2015).

This paper aims to monitor the docking and undocking manoeuvring and its consequences on the sea bed and quay structures in order to design, in the future, the possibility of new docking and undocking manoeuvres to minimize the erosion caused by the propulsion system. Moreover, results will allow understanding the effects sedimentation of the eroded sediment on docking and undocking manoeuvres. Some research of a long set of bathymetries are used to evaluate the effects produced by vessels during docking and undocking manoeuvring with real data of the manoeuvring frequencies and duration.

We focus our analysis on a set of Automatic Identification System (AIS) data at a particular basin obtained during the period of June - July 2016. AIS provides a continuous stream of information of the all AIS enabled vessels in range. The system provides position and speed updates on predefined intervals depending on vessel speed and manoeuvre situation with a sample-rat from 3 seconds for high speed or turning vessels to 15 minutes for ships at anchor (Aarsæther & Moan 2010). The use of AIS data permits to understand the effect of changes to the fairway and to propose improvements to harbour areas.

Morphodynamic changes are identified after periodic bathymetries of the particular basin, dominated by a single vessel. The real location of the harbour basin (Fig. 1) used in this research is kept confidential at the request of the harbour authorities.

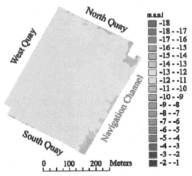

Figure 1. Overview of the area of the harbour location and quays direction.

2 METHODOLOGY

2.1 Bathymetric surveys

Periodic bathymetric surveys were carried out with a multibeam system SeaBeam1185, Elac-Nautik, Germany. The blanking distance from the floating line was 0.65 m and data were recorded at 180 kHz with a boat speed ranging from 3 to 5. The data acquisition average error was around 0.1 m due to an upper layer of mud within the harbour basin of an estimated thickness of 0.5m (Mujal-Colilles et al. 2017)

Geological studies performed by the harbour authorities yield sediment characteristics below the mud layer of $d_{50} = 0.3$ mm and $d_{90} = 1.0$ mm, normal sizes for a harbour located in a deltaic zone. According to the geological studies, the sediment layer with these characteristics reaches up to the -26 m above sea level (asl) level. Figure 2 plots bathymetric data of a real harbour basin with a mean depth of -12m asl where the evolution of berthing depth is represented for eighteen months.

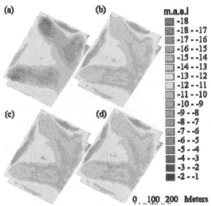

Figure 2. Evolution of the bathymetry along the harbour. (a) June 2014; (b) June 2015; (c) September 2015; (d) December 2015

As seen in Figure 2, the profiles in the NW and SW docks show holes of up to 5 m compared to the mean depth of -12m asl. Parallel to the scouring action, a sedimentation process occurred between June 2014 and December 2015, with the sediment deposition located parallel to the west and north docks and in the middle of the harbour area. The sedimentation is of the order of 2 m. In June 2015, harbour authority decided to dredge the areas with a lower depth in the harbour basin.

2.2 Automatic Identification System (AIS) data

The Automatic Identification System (AIS) is an automatic tracking system for identification and location of vessels by exchanging data via VHF communication to other nearby ships, AIS base stations and satellites. It has become mandatory and was implemented through the International Maritime Organization (IMO) as per requirement of SOLAS (1974) for commercial vessels over 300 GT since 2004. Additional legislation from the EU and the US extended the requirements for having an AIS transmitter on board also to smaller crafts such as fishing boats. The sample rate of the AIS data depends on the ships speed and state of the vessel (IMO 1974).

The original purpose of AIS was meant as an aid to collision avoidance but many other applications have since been developed such as vessel traffic services (VTS), maritime security, fleet and cargo tracking, search and rescue, among others, and is an important source of information for studying maritime traffic and associated critical situation (Mestl et al. 2016). Moreover, it represents an opportunity to study the navigational patterns in coastal and restricted areas, as a quay or harbour.

The Port Authority provided AIS data in form of reports for the period of June and July 2016 for the presented area (Fig. 1) for research purposes. The AIS information received contained mainly, time, latitude, longitude, speed over ground (SOG), and course over ground (COG). The presented data will be therefore anonymized as much as possible.

In this research, the AIS data will be applied to analyse manoeuvres in the constrained area to derive the exhibited manoeuvre patterns, being the first task to transform the collected data into a form that eases the final analysis.

2.3 Case study

To investigate the effect on navigation decisions and external effects on ship manoeuvring it is convenient to test a specific scenario as a demonstration. A detailed study of the docking and undocking manoeuvres by a RoPax vessel type is conducted. Manoeuvring refers to the slow speed movement of the vessel between the ports breakwater (entry/exit) and point berth.

The docking location of the RoPax vessel with a daily frequency is at the SW corner (west quay) and the draft of the vessel is 7 m. The vessel has two controllable pitch propellers and two bow thrusters. The engine power of four main engines is 55440 kW and the power of each bow thruster is 1850 kW. Figure 3 shows the position of the propellers and bow thrusters from the AIS position. This information is useful in order to know the effect of propellers in the harbour basin.

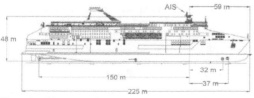

Figure 3. Distances from RoPax AIS position to the controllable pitch propellers and bow thrusters.

The effect of other vessels docking at the same areas could also be considered but with the aim of being conservative, only the RoPax vessel has been taken into account in the current research.

Figures 4 and 5 show vessel traces of three different docking and undocking manoeuvrings (2016, 8th of June; 2016, 1st of July and 2016 6th of July) extracted from AIS data from this area.

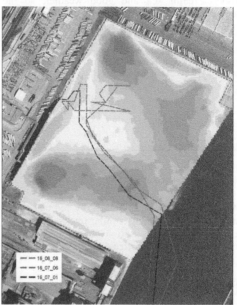

Figure 4. West quay docking manoeuvres scenario (2016, 8th of June; 2016, 1st of July and 2016 6th of July). Colours of the bathymetry along the harbour are defined in Figure 2.

Figure 5. West quay undocking manoeuvres scenario (2016, 8th of June; 2016, 1st of July and 2016 6th of July). Colours of the bathymetry along the harbour are defined in Figure 2.

In order to analyse manoeuvres, expected weather conditions must be known. The weather during these manoeuvrings is presented in Table 1. According to Beaufort Wind and Douglas Sea Scales, it can be considered calm sea and good weather, so the influence of weather in these cases will not affect the manoeuvrability of the vessel.

Table 1. Weather information of docking and undocking maneuverings obtained from Puertos del Estado website. (http://www.puertos.es/en-us/oceanografia/)

		Wind speed (knots)	Wind direction	Significant wave height (m)
16_06_08	Docking (19:05)	6.55	205	0.3
	Undocking (23:32)	2.10	280	0.19
16_07_01	Docking (18:30)	6.02	212	0.43
	Undocking (23:52)	4.12	296	0.36
16_07_06	Docking (19:30)	6.01	222	0.44
	Undocking (23:30)	2.64	249	0.26

As can be observed, both docking and undocking manoeuvres are following similar tracks. In the next section, a detailed description of how the vessel is manoeuvred at port is carried out.

3 RESULTS AND DISCUSSION

3.1 *Docking and undocking maneuvers analysis*

The regular call of the RoPax vessel at the West quay is daily. The AIS vessel manoeuvring data comprises hour, position, course keeping, course changing and speed changing, but does not contain information such as the instantaneous position of the rudder. However, data from AIS has a potential to reveal the preferred navigational patterns and manoeuvre parameters position in use in a specific area and can calculate the curvature of the track line of the vessel. (Aarsæther & Moan 2007)

This section shows the main parameters of a manoeuvre using positioning data obtained from the AIS in order to identify manoeuvring processes.

Main parameters of docking manoeuvring (Course over Ground and Speed over Ground) can be observed in Figure 6. The first rule of berthing is to approach at a low speed and controlled speed. The average results of manoeuvres studied is a manoeuvre to enter the harbour from the middle of the navigational channel starting at a course of 28°-30° at speed of 5.9 knots. The starting position for the vessel has no obstructions on the initial heading and will be in a position to initiate the turning manoeuvre into the harbour with in a relatively short period of time (3 minutes). During the entrance manoeuvring, the vessel will change the course and, as can observe in Figure, the speed will decrease at 3 knots until the vessel is on a course suitable for the final approach. The average entrance manoeuvring

period takes about 19 minutes. The approach angle number of turns is accurately identified in Figure 6.

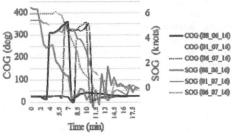

Figure 6. Course over Ground (COG) and Speed over Ground (SOG) for three arrivals manoeuvrings considered.

The same procedure is used to identify departure manoeuvres (Fig. 7).

As can be observed, the turning circle described in all manoeuvres follow similar angle and the speed of the vessel increases being around 3 knots inside the undocking area reaching 6 knots at the navigational channel. The average departure manoeuvring period takes around 13 minutes.

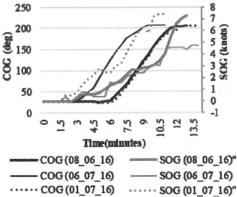

Figure 7. Course over Ground (COG) and Speed over Ground (SOG) for three departure manoeuvrings considered.

From Figures 6 and 7 we can observe similar patterns in both berthing and unberthing manoeuvres in calm sea and good weather. The reason of morphodynamic changes observed in the evolution of periodic bathymetric surveys inside the harbour could be due to the same daily manoeuvres performed by a single RoPax vessel.

3.2 *Influence of shallow water*

Manoeuvring properties are of particular importance in shallow water. Due to squat caused by increased velocities under the vessel, a combination of a mean bodily sinkage plus a trimming effect can produce unstable motion and decrease of course keeping ability.

According to (Quadvlieg and Coevorden 2003) the vessel should have enough course stability and turning ability to fulfil the IMO requirements (IMO 1993) with respect to overshoot angles and turning circle dimensions in shallow water. The criteria should be fulfilled in water depths above sea level (asl) larger than 1.3 times the draught (T) of the vessel (Depth$_{asl}$/T>1.3).

Figures 8 and 9 show bathymetric surveys in the analysed area of the period of 2010-2015. If the draft of the vessel considered is 7 meters, the water depth should be higher than 9 m. As can be observed in Figure 2, there are some critical green areas where the water depth – draught ratio is lower than 1.3m: narrow to the West, North and South Quays and in the middle of the manoeuvring area. In these areas shallow waters can seriously affect navigation.

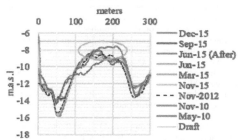

Figure 8. West quay bathymetric surveys and vessel draft. Water depth – draught ratio of the harbour basin lower than 1.3 is shown in green circle in the figure.

In the west quay, the vessel is at berth, and the speed of the vessel will be near to zero, so there is no effect related on uncontrollable motions. However, the movement of the propellers can produce significant suspended sediment in the water column during berthing and unberthing operations in these restricted areas producing important environmental problems to harbour authorities.

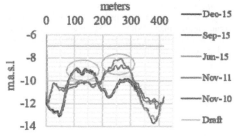

Figure 9. Central area of the harbour basin bathymetric surveys and vessel draft. Water depth – draught ratio of the harbour basin lower than 1.3 are shown in green circles in the figure.

When the vessel is sailing in the central area of the harbour, the vessel is sailing at slow ahead or slow astern speed (lower than 3 knots). At these speeds, the ratio Depth$_{asl}$/T<1.3 can disturb the vessel motions and the course keeping ability. Vessel should avoid these specific shallow waters.

Results from this section justify the need to control the sedimentation process inside a basin produced by docking and undocking manoeuvring to avoid possible unstable motions.

3.3 Adverse weather conditions for vessel manoeuvrability

Manoeuvrability at low forward speed in strong wind and, perhaps, current, is critical for vessels with large windage area, during approaching to and entering ports. Therefore low speed manoeuvrability criteria require specification of the wind speed and, sometimes, current. Quadvlieg & Coevorden (2008) recommend wind speed of 20 knots for general use and 30 knots for ferries and cruise ships, as the wind speed at which the vessel should be able to leave the quay.

The maximum wind speeds presented in Table 2 are provided by the *Puertos del Estado*. As can be seen, the maximum wind speed is always lower than 30 knots, so the vessel will be able to leave the quay at any time. In case of bad weather, vessel may need tugboats and pilot to control the vessel during berthing and unberthing manoeuvrings.

Table 2. Monthly Maximum wind speed of the studied area during 1997-2005 (http://www.puertos.es/en-us/oceanografia/)

Month	Max. Wind speed	Year
January	16.8	1997
February	13.3	2001
March	13.7	2002
April	15.8	2005
May	17.3	2002
June	15.2	2000
July	11.7	1998
August	13.1	2002
September	18.4	2005
October	20.4	1997
November	18.2	1996
December	22.1	1998

This section justifies that the RoPax vessel can maintain the scheduled sailing times considering the effect of adverse weather conditions in the analysed area.

4 CONCLUSIONS

The use of AIS for vessels can help in provide a readily available data source for traffic analysis and estimate manoeuvre plans. Is an easily available source of information about the desired manoeuvring patterns in a specific area. This paper describes the typical docking and undocking manoeuvres in a constrained area for a RoPax vessel. The analysis of the vessel track line is used to estimate the parameters of standard manoeuvres. Manoeuvres analysed define similar patterns in both berthing and unberthing manoeuvres. Due to the vessel particulars and the manoeuvring area, there are some limitations to define new alternative manoeuvres. Berthing and unberthing speed is adequate, because the vessel requires a minimum of powering. Lower speeds should affect the manoeuvrability because the rudder is not effective.

Another problem observed in this research is the sedimentation of the eroded sediment that reduces the total depth of the harbour basin. From Figures 8 and 9 we can assume that there are some critical zones caused by the interaction effects between banks (eroded sediments) produced by shallow waters and vessel motions. This situation can cause undesired and uncontrollable motions of the vessel and environmental problems as a result of the suspended sediment during manoeuvring situations in restricted areas.

From the results obtained in this paper, to decrease the effect of erosion caused by propellers during manoeuvring in shallow waters, we can conclude that current docking and undocking manoeuvres are correct (course and speed) and alternatives should be implemented. From the vessel point of view, the ship size and ship power (due to the economy of scale) should be reduced or, on the other hand, the use of tugboats in ships manoeuvring should be required to decrease the speed of the vessel.

However, AIS data alone are not enough, and a more detailed study to investigate the effect on vessel manoeuvring is useful to test these scenarios in a simulator with controlled conditions. The parameters extracted from AIS data can be used as input to a numerical navigator to mimic the behaviour of the real situation. A real-time full mission bridge simulator is a good tool for this. Further research will identify the main parameters of a manoeuvre using bridge simulators with positioning data obtained from the AIS of the same area and will identify which manoeuvring behaviour is acceptable to reduce the effect in harbour basins. The proposal of alternative manoeuvres in the same navigation area will reduce the effect of the toe scouring induced by vessel propeller. Moreover, outcomes of this analysis can be extrapolated to other harbours to improve their management.

ACKNOWLEDGMENT

This work has been supported by MINECO – FEDER, UE (the Ministry of Economy and Competition) by the Spanish Government through the project TRA2015-70473-R.

REFERENCES

Aarsæther K.G. & Moan T. 2007. Combined Maneuvering Analysis, AIS and Full-Mission Simulation. *TransNav, the International Journal on Marine Navigation and Safety of Sea Transportation*, Volume 1, Number 1.

Aarsæther K.G & Moan T. 2010. Computer Vision and Ship Traffic Analysis: Inferring Maneuver Patterns from the Automatic Identification System. *TransNav, the International Journal on Marine Navigation and Safety of Sea Transportation*, Volume 4, Number 3.

International Maritime Organization. 1974. Convention for the Safety of Life at Sea (SOLAS), as amended

International Maritime Organization. 1993. Assembly Resolution A.751: Interim Standards for Ship Manoeuvrability.

International Maritime Organization. 2003. Guidelines for the installation of a shipborne automatic identification system (AIS)

Lam, W., Hamill, G.A., Robinson, D.J., & Raghunathan, S. 2012. Semi-empirical methods for determining the efflux velocity from a ship's propeller. *Applied Ocean Research*, 35, 14-24.

Mestl T., Tallakstad K.T. & Castberg R. 2016. Identifying and Analyzing Safety Critical Maneuvers from High Resolution AIS Data. *TransNav, the International Journal on Marine Navigation and Safety of Sea Transportation*, Volume 10, Number 1.

Mujal-Colilles, A., Gironella, X., Jaquet, A., Gomez-Gesteira, R., & Sanchez-Arcilla, A. 2015. Study of the efflux velocity induced by two propellers. *Conference: International Short Course and Conference on Applied Ocean Research (SCACR)*, Florence

Mujal-Colilles, A., Gironella, X., Sanchez-Arcilla, A., Puig Polo, C. & Garcia-Leon, M. 2017. Erosion caused by propeller jets in a low energy harbour basin. *Journal of Hydraulic Research*, 55, 1, 121-128.

Quadvlieg, F.H.H.A. & van Coevorden, P. 2003. Manoeuvring criteria: more than IMO A.751 requirements alone. MARSIM International conference on marine simulation and ship manoeuvrability.

Bunkering and Fuel Consumption

Proceedings of 12th International Conference on Marine Navigation and Safety of Sea Transportation, TransNav 2017
21-23 June 2017, Gdynia, Poland

Ship Fuel Consumption Prediction under Various Weather Condition Based on DBN

X.Q. Shen, S.Z. Wang, T. Xu, C.J. Shi & B.X. Ji
Shanghai Maritime University, Shanghai, China

ABSTRACT: During a voyage, ship fuel consumption varies with different weather conditions because there are many complicated nonlinear relationships between the fuel consumption and the weather conditions so that it is difficult to directly find these correlations and predict the ship fuel consumption in the specified weather condition. In this paper, we first propose to employ Deep Belief Networks (DBN) algorithm that is able to resolve the nonlinearity problem by exploiting the latent information from the large number of the historical ship voyages and weather data, and dynamically predict the fuel consumption under various ocean meteorological condition. Then we present a system framework of predicting fuel consumption based on the DBN learning framework. Finally, the proposed approach is validated by comparing with some typical machine learning methods, e.g. SVR and BPNN. Experimental results show that the proposed method completely outperforms other methods in accuracy and efficiency, which substantially demonstrates the effectiveness of the proposed DBN-based fuel consumption prediction method.

1 INTRODUCTION

With the increase of the international trade volume year by year, marine transportation becomes more and more important for the development of the world economy. But the large amount of greenhouse gas emissions resulting from marine transportation are affecting the environment and human life. Therefore, a systematic requirement for ship energy efficiency management has been proposed by International Maritime Organization (IMO). Furthermore, both Energy Efficiency Design Index (EEDI) and Ship Energy Efficiency Management Plan (SEEMP) have been become the mandatory requirements on the ship energy efficiency management and have been added into the MARPOL convention[1-4]. The Marine Environmental Protection Committee (MEPC) [5] has also adopted mandatory measures to reduce greenhouse gas emissions in the shipping industry. Consequently, the energy saving and the emission reduction has already been an urgent task.

The control of fuel consumption not only is able to cut down the operating cost of ships, but also is an effective means to reduce gas emissions. In recent years, research institutions and scholars have carried out extensive research work on fuel consumption prediction. To begin with, many research efforts

have been dedicated to reducing fuel consumption from the point of view of transportation planning and management. For instance, Qi et al. [6] designed an optimal liner shipping schedule to minimize the total expected fuel and emissions by taking into account the uncertainty of port time and frequency requirements on the liner schedule. Secondly, Sandareka et al. [7] employed the machine learning method to predict the fuel consumption. In order to achieve the fuel consumption prediction of a long-distance fleet with multi-variable time series, they compared the random forest, gradient boosting, and neural network. The results showed that the random forest technique produced more accurate prediction results. Alonso et al.[8] used the artificial neural networks (ANNs) combined with the genetic algorithms (GAs) to optimize the feasibility of the diesel engine settings, and tried to find the parameter configurations that meets the latest stringent emission regulations so as to reduce the fuel consumption. In addition, some researchers have studied the fuel consumption prediction problem from the aspects of navigation optimization strategy[9-11], i.e., Lu et al.[12] achieved the goal of increasing the energy efficiency of the ship and reducing other greenhouse emissions through navigation optimization. The empirical fuel consumption forecasting method based on Kwon's added-resistance modeling was applied to the Suez-Max oil tanker, which established the operational performance model for each oil tanker with different load, speed and relative wave direction. The model allowed the user to query the relationship between the fuel consumption and the various sea conditions that the ship might encounter during its voyage. Bialystocki et al.[13] proposed to derive the fuel consumption and velocity curves according the ship draft and displacement, the weather resistance and direction, the hull and propeller roughness. The curve determining the operation method under different navigational conditions could help the operators save fuel and make environment-friendly decisions. Furthermore, some research work from the economic point of view studied the fuel consumption reduction [14-16]. These methods could obtain significant effect on fuel consumption prediction under the specific conditions. However, there are many complicated factors affecting the fuel consumption during a voyage, it is still difficult to reliably and accurately predict fuel consumption under various weather conditions.

At present, the deep learning of complex nonlinear relation data can achieve good results in the prediction, classification, pattern recognition and other areas. Road traffic has been gradually applied, such as Joe [18] studied stack auto-encoders deep learning network for highway vehicle speed prediction with the purpose of minimize fuel consumption. For marine transportation, in case the captains who can dynamically get the accurate fuel consumption information under a specified weather condition by using the fuel consumption prediction tool is able to make an appropriate decision to select an optimal route so as to improve energy efficiency. However, the ship fuel consumption varies with different weather conditions, furthermore, there are many complicated nonlinear relationships between the fuel consumption and the weather conditions. The traditional methods that depend on physical model simulations [17] was impossible to dynamically calculate the fuel consumption under complex ocean meteorological factors, such as wind, waves, ocean currents, etc. Therefore, this paper attempts to learn the nonlinear relationships between the weather factors and the fuel consumption using the deep learning method. We employ the Deep Belief Networks (DBN) that is able to resolve the nonlinearity problem by learning the latent information from the historical ship voyages and weather data, and then dynamically predict the fuel consumption under various ocean meteorological environments, which can assist the captain to optimize and select the optimal route, and realize the energy management and low emission navigation.

The paper is organized as follows: the DBN-based prediction model of the daily fuel consumption of the ship under different weather conditions is presented in Section 2. We evaluate the effectiveness of the proposed approach by comparing some typical algorithms in Section 3. Section 4 provides the conclusion.

2 APPROACH FOR PREDICTING SHIP FUEL CONSUMPTION UNDER VARIOUS WEATHER CONDITION

2.1 *Restricted Boltzmann Machine*

Boltzman Machine proposed by Hiton and Sejnowski is a neural network based on statistical mechanics [19]. The neurons in this network are random neurons. The outputs of neurons are only two states, namely active and inactive, which usually represent by binary 0 or 1. Boltzman Machine has strong unsupervised learning ability to learn complex rules from training dataset, but the training time is too long and it is difficult to obtain its probability distribution. To overcome this problem, Smolensky introduced a Restricted Boltzmann Machine (RBM) as shown in Figure 1.

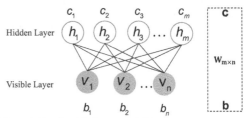

Figure 1. Restricted Boltzmann Machine

RBM is an energy-based model whose energy function is defined as follow :

$$E_\theta(\mathbf{v}, \mathbf{h}) = -\mathbf{b}^\mathsf{T}\mathbf{v} - \mathbf{c}^\mathsf{T}\mathbf{h} - \mathbf{h}^\mathsf{T}\mathbf{W}\mathbf{v} \qquad (1)$$

where, $\mathbf{v}$ and $\mathbf{h}$ respectively represent the state vector of the visible and hidden layers. $\mathbf{b}$, $\mathbf{c}$ and $\mathbf{W}$ are three real parameters which respectively represent the visible layer bias matrix, hidden layer bias matrix and the connection weight matrix of the visible layer and the hidden layer. The joint distribution of the visible layer and hidden layer is defined as Eq.2 :

$$p_\theta(\mathbf{v}, \mathbf{h}) = \frac{e^{-E_\theta(\mathbf{v}, \mathbf{h})}}{\sum_{\mathbf{v}, \mathbf{h}} e^{-E_\theta(\mathbf{v}, \mathbf{h})}} = \frac{e^{-E_\theta(\mathbf{v}, \mathbf{h})}}{Z} \qquad (2)$$

where, Z is a distribution function which is also known as the normalization factor. The most notable characteristic of the RBM is that there is no connection within the visible and hidden layers. Accordingly, if the state of the visible layer is given, the active state of each unit of the hidden layer is independent of each other, and vice versa. In the training process, the state vector of the visible layer is first mapped to the hidden layer unit; then the visible layer unit is reconstructed by the hidden layer unit; these new visible layer units are mapped to the hidden layer units again, thus new hidden units are obtained. The implementation of this iterative step is called Gibbs sampling which approximates the conditional probability distribution of the RBM, as shown in Eq.3-4.

$$P(\mathrm{h}_j = 1|v) = \frac{e^{-E_\theta(v,h)}}{\sum_h e^{-E_\theta(v,h)}} = sigmoid(c_j + w_j v) \quad (3)$$

$$P(\mathrm{v}_i = 1|h) = \frac{e^{-E_\theta(v,h)}}{\sum_v e^{-E_\theta(v,h)}} = sigmoid(b_i + w_i^T h) \quad (4)$$

We obtain the optimal RBM parameters by maximizing the likelihood function $P(V)$ of the observed data. The objective function of the network is as follows:

$$\arg\max L(W, b, c) = \frac{1}{N}\sum_{n=1}^{N} \log P\left(\mathbf{v}^{(n)}\right) \quad (5)$$

However, in this paper, we use a gradient-based Contrastive Divergence (CD) method [20] in practice instead of maximizing the likelihood function for parametric learning, because the CD algorithm is not only effective but also fast. The correlation and difference between the hidden layer units and the visible layer input are the main bases for weight update as shown in eq.6-8.

$$\Delta W = \frac{\partial \ln P(v)}{\partial w_{j,i}} \approx P(h_j = 1|v^{(0)})v_i^{(0)} - P(h_j = 1|v^{(k)})v_i^{(k)} \quad (6)$$

$$\Delta b = \frac{\partial \ln P(V)}{\partial b_i} \approx v_i^{(0)} - v_i^{(k)} \quad (7)$$

$$\Delta c = \frac{\partial \ln P(v)}{\partial c_i} \approx P(h_j = 1|v^{(0)}) - P(h_j = 1|v^{(k)}) \quad (8)$$

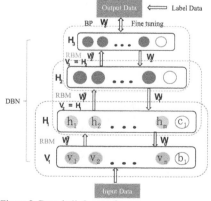

Figure 2. Deep belief networks

2.2 Deep Belief Networks

Deep Belief Networks (DBN) [21] is one of the typical deep learning methods. Greedy Layer-Wise Training methods are generally used by Deep learning [22]. There are three main steps. First of all, each layer is pre-trained using unsupervised learning [23]. Secondly, this networks training has only one layer at a time, then uses the output of the last layer as the input of the next layer. In the end, the supervised learning method is used to adjust all layers parameters. The DBN is composed of a plurality of RBMs which are stacked. The RBM is an effective generation model, which can not only excavate the deep structure information of the data, but also effectively represent the nonlinear relationship through unsupervised training. Each layer of the RBM network only ensures that the weights in the self-layer are optimal for the mapping of the feature vectors, which means that it is not able to make the feature mapping of the whole DBN achieve the optimal. Consequently, the back propagation network [24] would transmit the prediction error information from top to bottom to each layer of the RBM, then fine-tuning the whole parameters of the DBN can achieve loss function minimization like eq.9. The process of the DBN network training model can be regarded as the initialization of the weight parameters of a deep BP network, which makes DBN overcome the disadvantage that BP network randomly initializing weight parameters is easy to fall into the local optimization and the long training time. Therefore, the training process of the DBN model first does not supervise the training of each layer RBM network, which ensures that the feature vector is mapped to different feature space as much as possible to retain the feature information. Then the trained hidden layer of the RBM network is in turn stacked to constitute a DBN network. Finally, an additional BP network is added to the top layer of the DBN, and the hidden layer eigenvector of the last layer RBM is used as the input eigenvector to supervise the training of the prediction model. Assuming that the topmost layer of the DBN network is denoted by k, and $\mathbf{W}_k^T$, $\mathbf{H}_k$, $\mathbf{c}_k$ respectively represent the connected weight matrix, the top layer input matrix and the bias matrix, the predicted output is calculated as Eq.10. Figure 2 shows a DBN network architecture with three hidden layers.

$$\arg\min L(W, b, c) = \frac{1}{n}\sum_{i=1}^{n}\left(\frac{1}{2}\|\tilde{y}_i - y_i\|^2\right) \quad (9)$$

$$\tilde{\mathbf{y}} = \mathbf{W}_k^T \bullet \mathbf{H}_k + \mathbf{c}_k \quad (10)$$

where, n is the number of training samples. $\tilde{\mathbf{y}} = \begin{bmatrix} \tilde{y}_1 & \tilde{y}_2 & \dots & \tilde{y}_n \end{bmatrix}^T$ represents the predicted output matrix, and $\mathbf{y} = \begin{bmatrix} y_1 & y_2 & \dots & y_n \end{bmatrix}^T$ represents the real output matrix which is the daily fuel consumption in voyage.

2.3 *The framework of the predicting ship fuel consumption based on DBM*

There are a large number of factors affecting the fuel consumption, such as weather conditions, sea-states, main-engine status, cargo weight, and ship draft etc. Among them, weather conditions and sea-states are the major random factor for ship fuel consumption. The fuel consumption would be fluctuated under different weather conditions and sea-states. Therefore, exploring and exploiting the correlations between ocean weather and ship fuel consumption is an effective way to predict the ship fuel consumption under various weather condition and optimize the weather routing so as to reduce the ship fuel consumption and the gas emissions, and improve the ship energy-efficiency. The predicted fuel consumption combined with weather routing software can provide better navigation guidance, such as more accurate ship speed recommendations, route planning recommendations and other navigation optimization services, to achieve environmental friendly and energy saving navigation. Combined with the concept of "green shipping" and DBN deep learning model, the framework of forecasting the fuel consumption influence of ocean weather is designed. The framework consists of three parts: data preprocessing, DBN training and fuel consumption prediction, as illustrated in Figure 3.

Ship operating costs and greenhouse gas emissions are mainly determined by the fuel consumption, which is related to internal factors such as distance, load, trim, ship characteristics, and marine meteorological factors. Therefore, this data set consists of two parts. The ship sailing data from 73 ships which have a total of 486 voyage sailing basic information provided by COSCON ship navigation monitoring platform by June 2015 to May 2016, and each voyage contains 73 attributes. The weather data is derived from the National Oceanic and Atmospheric Administration(NOAA). According to the time, latitude and longitude of the navigation data, the weather information is interpolated to get the complete data set of the ship navigation.

The implication of the prediction of DBN is learning data-set probability distribution, but, in this paper the complete data set contains both meteorological data and navigation data, in addition, these data respectively represent different practical feature, the difference of each feature data are large. Therefore the data Normalization is used to eliminate the constraints imposed by different data units, and converting them into dimensionless values would facilitate comparison and weighting for indicators of different units or orders of magnitude, which can not only improve the DBN training speed, but also enhance generalization ability. According to the characteristic of a small fluctuation range within the characteristics of the navigation data, and in order to improve the difference of the normalized data, this paper proposes to using eight bits to represent each feature, as shown in Table 1.

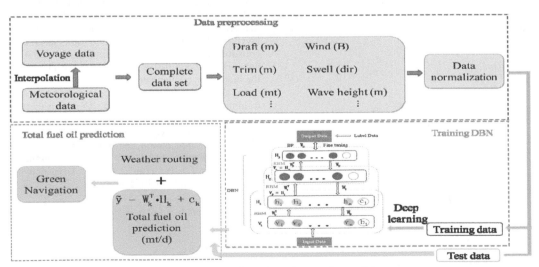

Figure 3. The framework of predicting ship fuel consumption under various weather condition.

Table 1. Data normalization

	Raw data			Normalization			
Draft(m)	Trim(m)	⋯	Swell(m)	Draft(m)	Trim (m)	⋯	Swell(m)
13.5	1.2	⋯	0.96	10110011	10010111	⋯	00100100
9.3	0.9	⋯	2.36	00101000	10001001	⋯	01100000
11.9	0.2	⋯	0.5	01111110	01101000	⋯	00010001
10.8	2	⋯	1.48	01011001	10111101	⋯	00111011

3 EXPERIMENTAL RESULTS

In order to verify the validity of the model proposed in this paper, we divide the preprocessed sailing data set into 90% training samples and 10% test samples, and use the same data to train SVR and BPNN to predict fuel consumption. In this experiment, the network structure of three hidden layers is chosen. The number of neurons in the hidden layer is 200, 200, 200, and the moderate change of the number of each layer neurons has no significant effect on the results. As shown in Figure 4-6. In general, the mean relative error can better reflect the credibility of the forecast results, so the error index using MRE is as shown in eq.11.

$$MRE = \frac{1}{n}\sum_{i=1}^{n}\frac{\left|\tilde{y}_i - y_i\right|}{y_i} \qquad (11)$$

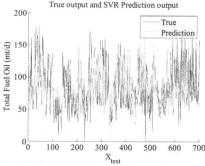

Figure 4. Total fuel oil prediction of SVR

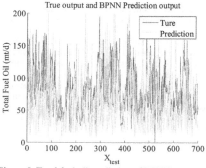

Figure 5. Total fuel oil prediction of BPNN

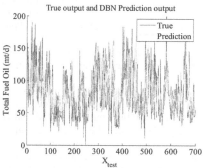

Figure 6. Total fuel oil prediction of DBN

The experimental results show that BPNN has a large fluctuation in predicting fuel consumption, and the forecast result of SVR and DBN are better than BPNN, which proves that the general artificial neural network can't achieve the desired effect with the deepening of the network layer. And DBN deep learning based on greedy layer-wise unsupervised training has significant advantages. However, the SVR and DBN predictions can't be easily distinguished from the above graphs, thus we further analyze the performance and the effectiveness using the MRE.

Table 2. Model performance

Model	Time (s)	MRE
SVR	457.2627	0.3964
BPNN	169.3768	0.6288
DBN	41.5108	0.3539

The comparison of training time and mean relative error shows that the SVR and DBN have smaller errors than BPNN. Due to the average training time of SVR is 457.2627 seconds, which means the time cost of the SVR model is the highest one. Obviously, both the training time (41.5180 s) and MRE (0.3539) of the DBN model are the minimum. Therefore, the proposed DBN-based approach shows the best performance for the fuel consumption forecast under considering the weather factors.

4 CONCLUSION

In this paper, we propose a novel approach to predict the fuel consumption under various weather condition based on DBN learning algorithm. By comparing it with the typical machine learning algorithms i.e. BPNN and SVR, the proposed DBN-based method is the most reliable and effective way for fuel consumption prediction. Combining the results of fuel consumption forecast with the weather routing software would enable the shipping operators to accurately understand the ship fuel consumption influence under the different meteorological and hydrological conditions. Thus we can calculate the weather factors and the fuel consumptions under different weather factors by using this model, and optimize the weather routing, such as the lowest fuel consumption, the shortest sailing time or the lowest total cost, which can not only improve the greenhouse gas emissions and reduce the cost of navigation, but also would realize Green Shipping. Therefore the study has great research value and wide application prospects.

In the future work, we will consider much more navigation factors according to the voyage schedule, such as required ETA, waypoints, marine traffic rules and other navigation constraints to predict the accurate fuel consumption, and calculate the ship performance speed, which is conducive to realizing the energy-saving and low-carbon voyage.

ACKNOWLEDGEMENT

The authors would like to thank the anonymous reviewers and editors for their comments and suggestions.

The research is supported by NSFC of China (No: 51379121, and 61304230) and Shanghai Shuguang Plan Project (No:15SG44).

REFERENCES

[1] IMO. 2009. 2009a Guidelines for the Development of a Ship Energy Efficiency Management Plan (SEEMP), MEPC, 1/Circ.683.

[2] IMO. 2012. 2009a Guideline for Development of a Ship Energy Efficiency Management Plan (SEEMP), MEPC, 213(63) Annex9.

[3] IMO. 2012. 2012b Guideline on Survey and Certification of the Energy Efficiency Design Index (EEDI), MEPC, 214(63) Annex 10.

[4] IMO.2012. 2012d Air Pollution and Greenhouse Gas (GHG) Emission from International Shipping, Marpol Annex VI.

[5] IMO. 2012. 2012e Guidelines for Calculation of Reference Lines for Use with the Energy Efficiency Design Index (EEDI), MEPC, 215(63) Annex 11.

[6] Qi X.T & Song D.P, 2012.Minimizing fuel emissions by optimizing vessel schedules in liner shipping with uncertain port times [J].Transportation Reseatch Part E 48: 863-880 .

[7] Sandareka W & Dilum H. M., 2016. Fuel consumption prediction of fleet vehicles using machine learning:A comparative study,2016 moratuwa engineering research conference:90-95 .

[8] Alonso J.M. & Alvarruiz F. & Desantes J.M,et.al. 2007.Combining neural networks and genetic algorithm to predict and reduce diesel engine emissions [J], IEEE Transaction on Evolutionary Computation,11(1): 46-55.

[9] Wang S & Meng Q, 2012.Sailing speed optimization for container ships in a liner shipping network, Transportation Research Part E, 48(3): 701-714.

[10] Fukasawa R. & He Q. & Santos F., 2016. A joint routing and speed optimization problem, KSCE Journal of Civil Engineering.

[11] Dulebenets M. & Mihalis M., 2016. The fleet deployment problem with variable vessel sailing speeds and port handling times, KSCE Journal of Civil Engineering.

[12] Lu R.H. & Turan Q. & Boulougouris E., 2013. Voyage optimization: Prediction of ship specific fuel consumption for energy efficient shipping [C]. Low Carbon Shipping Conference, London: 1-11.

[13] Bialystocki N. & Konovessis D., 2016. On the estimation of ship's fuel consumption and speed curve: A statistical approach[J],Journal of Ocean Engineering and Science,1(2): 157-166.

[14] Fagerholt, K. & Laporte, G. & Norstad I., 2010. Reducing fuel emissions by optimizing speed on shipping routes, J. Oper. Res. Soc., 61(3): 523-529.

[15] Norstad, I. & Fagerholt, K. & Laporte G., 2011. Tramp ship routing and scheduling with speed optimization, Transportation Research Part C, 19(5): 853-865.

[16] Psaraftis, H. N. & Kontovas C. A., 2013.Speed models for energy efficient maritime transportation: A taxonomy and survey, Transportation Research Part C, 26: 331-351.

[17] Lemieux J. & Ma Y., 2015. Vehicle speed prediction using deep learning [C]\\IEEE,Vehicle Power and Propulsion Conference, IEEE Conference Publications: 1-5.

[18] Chen Y. & WU Q. & Wang Y., 2015. A research on precise prediction method of high-speed craft resistance based on CFD[C]//IEEE.2015 8th International Conference on Intelligent Computation Technology and Automation. France : IEEE : 84-88.

[19] Ackley D. H & Hinton G.E & Sejnowski T.J., 1985. A Learning Algorithm for Boltzmann Machines [J],Congnitive Science,9: 147-169

[20] Carreira P. & Hinton G. E., 2005. On contrastive divergence learning. In Artificial Intelligence and Statistics: 33–41.

[21] Hinton G. E. & Osindero S., 2006. A fast learning algorithm for deep belief nets, Neural Comput., vol. 18, no. 7: 1527–1554.

[22] Bengio, Y. & Lamblin, P. & Popovici, D.et.al. 2007. Greedy Layer-Wise Training of Deep Networks. Advances in Neural Information Processing Systems 19: Proceedings of the 2006 Conference.

[23] Lu Y., 2016. Unsupervised learning on neural network outputs: with application in zero-shot learning, International Joint Conference on Artificial Intelligence,

[24] Fazayeli F. & Wang L.P. & Liu W., 2008. Back-propagation with chaos. 2008IEEE International Conference on Neural Networks and Signal Processing (ICNNSP2008), Zhenjiang, China, June: 7-11.

Analysis of Electric Powertrain Application to Drive an Inland Waterway Barges

A. Łebkowski

Gdynia Maritime University, Gdynia, Poland

ABSTRACT: The article discusses the issues of inland waterway transport in Poland. The historical technical solutions used in the transport at the turn of 19th and 20th century. The present state of waterways and stock used in inland shipping is reviewed. The energy consumption of various intermodal freight transport modes is analysed, including inland waterway transport, rail transport and road transport. As an alternative, the application of electric powertrains to drive a barge is presented, in the forms of an electric motor on board a barge or an electric driven land based mule, towing a barge.

1 INTRODUCTION

The necessity of decreasing the road transportation negative impact on the environment is the basis on which the concept of inland waterway transport reconstruction is considered.

In the January 2017, the AGN Convention ratification act was signed by the President of Poland, after the act was accepted by the Parliament and the Cabinet. The AGN Convention - The European Agreement on Main Inland Waterways of International Importance was done at Geneva on 19 January 1996. The AGN is indicating the inland waterway development directions in Europe [1,2]. The goal of the AGN is the introduction of legal framework which will set the coordinated plan for development and construction of a network of inland waterways of international importance and established operational and infrastructural parameters. The AGN Convention encompasses the 27 000 km of inland waterways, connecting 37 ports in European countries [3].

In Poland, there are three sections of international waterways with following designations:
- E30 – connecting the Baltic Sea from the Świnoujście port, through Oder, the future Danube-Oder-Elbe canal through Danube to Bratislava,
- E40 – connecting the Baltic sea from Gdańsk through Wisła to Warsaw, then through Narew and Bug to Brest and further through Dnepr into the Black Sea in Odessa,
- E70 – connecting the Oder from the mouth of Oder-Havel canal to the outlet of Warta in Kostrzyń, through Bydgoszcz, lower Wisła and then the Gdańsk part of Wisła or through Szkarpawa with the Vistula Lagoon, creating an European waterway between Rotterdam and Klaipeda [1-3].

Figure 1. Planned waterways in Poland [2]

By ratifying the AGN Convention, Poland has committed to adjust the main waterways to at least class IV of requirements. The table 1 presents the operational parameters required from the inland waterways. The waterways of classes Ia, Ib, II and III are classified as of regional importance, and the classes IV, Va and Vb as of international importance [4].

Table 1. Classification of inland waterways [4]

Waterway class	Trail width [m]	Transit depth [m]	Minimal clearance under bridges [m]	Clearance under power lines [m]	Max. ship/barge length [m]	Max. ship/barge beam [m]	Max. ship/barge draught [m]	Ship/barge deadweight [t]	Lock width [m]	Lock length [m]
Ia	15	1,2	3,0	8	24	3,5	1,0	<180	3,3	25
Ib	20	1,6	3,0	8	41	4,7	1,4	180	5,0	42
II	30	1,8	3,0	8	57	9,0	1,6	500	9,6	65
III	40	1,8	4,0	10	70	9,0	1,6	700	9,6	72
IV	40	2,8	5,25	12	85	9,5	2,5	1500	12,0	120
Va	50	2,8	5,25	15	110	11,4	2,8	3000	12,0	120
Vb	50	2,8	5,25	15	185	11,4	2,8	>3000	12,0	187

Presently in Poland there are about 3655km of waterways, including: 2417km of channelized, navigable rivers, 644km of canalized river sections, 336km of canals, and 259km of navigable lakes. About 92% (3365 km) of waterways are used for navigation. Unfortunately in 2015 the requirements for IV and V class were met by only 5.9% of waterways (214km). The other 94.1% (3441km) are fulfilling the I-III class requirements [5]. The Polish Government plans, as laid in ordinances, require the designers to plan new waterways to meet the highest Vb class criteria. Important elements of the inland waterway transport are the ships used for transportation.

According to the Central Statistical Office of Poland, in 2015 there were operating: 217 tugs and pushers, 89 self-propelled barges, 511 barges without propulsion and 101 passenger vessels. The majority of operated vessels: 73.0% of pushers, 48.7% of barges and 100% of self-propelled barges were built in the years 1949÷1979. The negligible use of waterways in Poland causes the inland waterway involvement in total cargo transportation in the years 2000-2015 to diminish from 0.8% down to 0.4% [5].

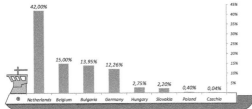

Figure 2. Volume of freight using inland waterway transport [2]

In Poland, there are plans to shift about 30% of road transport volume on distances in excess of 300km to inland waterways or rail transport before year 2030, and 50% of road transport before year 2050 [2].

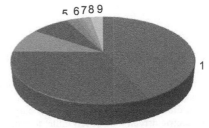

Figure 3. The structure of cargo transported via inland waterways in the year 2015, where: 1- Agricultural products; 2-Metal ores and other mining products; 3-Coal, lignite, crude oil, natural gas; 4-Recyclable materials, municipal waste, 5-non-metallic origin products; 6-Chemicals and chemical products; 7-Metals and metal products; 8-Coke and refined oil products; 9-Others.

Comparing the energy consumption of each form of transport, the most efficient is the inland waterway transport. Using the same amount of fuel powering internal combustion engines, one tonne of cargo can be transported to a distance of 370km with waterways, 300km using rail, and 100km using road transport.

Judging from the above information, the inland waterway transport has a great potential, which unfortunately needs to be stimulated by large investments in both the infrastructure and stock. Facing the threshold of rebuilding the inland waterway transport in Poland, the question arises, whether the application of electric propulsion in inland waterway transport isn't a good idea?

2 HISTORY OF ELECTRIC WATERWAY TRANSPORT

The oldest known water canals were built in Mesopotamia, about 4000 B.C., in Girnar 3000 B.C., in India 2600 B.C., in Egypt 2300 B.C., in China 500 B.C. The ancient Greeks are known to have used locks to regulate the flow of water in the third century B.C. Rivers and canals were eagerly used to transport timber and other goods, at first using human power and river current, and as the technology progressed – with steam and subsequently diesel and electric power. In the medieval period the waterway transport was faster, and several times cheaper than land transport.

In the 10th century in the Great Britain the Glastonbury Canal was built, with a length of 1750m. It was used for transportation of building stone, grain, wine and other goods, up to 14th century. In parallel, in the rest of the world canals and other hydro-technical devices were also constructed. The development of infrastructure continued up to the end 19th century, when the rapid development of rail and road transport begun. The best known canals include: the Panama Canal (1920, 80km in length), the Suez Canal (1869, 163km in length), the Kiel Canal (1895, 98km in length) and the Corinth Canal, conceived from the antiquity, but built in the late 19th century (1893, 6.3km in length).

Figure 4. Animal teams used to move barges [6].

The transportation on the waterways employed unpowered barges, towed by tugs and pushers, as well as self-propelled barges. In the beginning of inland waterway transport, the rafts, barges and boats were powered by beasts of burden in teams, which moved in in parallel to towed vessel, on the river or canal embankment.

Figure 5. Engine tractor used to move barges [6].

Along with the advances in technology, the horses and mules were replaced by engine powered tractors, and subsequently, rail tractors, powered from the overhead electric cables.

From the beginning of the 20th century, the electric barge propulsion systems can be divided into:
- Manned and unmanned locomotives – mules, powered from overhead cables, rolling on rail laid along the embankment and towing the vessel via a towline.

Figure 6. Examples of electric mules used to tow barges [6].

- Self-propelled funiculars, moving along pole-mounted cables, and powered from electric overhead line. The best known examples from that period are the Lamb and Zinzins systems.

Figure 7. An example of an unmanned electric funicular system, designed by Richard Lamb at the end of 19th century [6].

– Vessels powered directly by overhead line trolley, using on-board electric motor to tow the vessel along a submerged chain

Figure 9. Examples of electric rail locomotives used to move barges [6].

Figure 8. Examples of trolley powered submerged chain towing systems [6].

– Vessels powered directly by overhead line trolley, using on-board electric motor to power conventional propeller system

Figure 10. Examples of electric powertrain systems from the beginning of 20th century used up to this day [8,9].

Some of the presented solutions are used in France, Germany, United States or Great Britain up to this day, same as the current collector designs, developed in those years [7].

3 TRANSPORT COST COMPARISON

The research of energy consumption during operation of various road, rail and water vehicles used in intermodal freight transport confirms, that the inland waterway transport is the most energy efficient and ecological [10-12]. It is estimated that the cargo capacity of an average barge (54 TEU) with a length of 85m, beam of 9,5m and draught of 2,5m is about 15 times the capacity of a container rail car (4 TEU) and 27 times the capacity of a container carrying truck (2 TEU) [1,2,4,10-12]. According to the materials published in 2003 by the European Commission on the energy efficiency, for each litre of diesel fuel burnt, a barge can transport one ton of cargo for the distance of 127km, a train can do the same for 97km, and a truck for 50km [10,11].

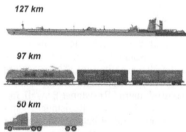

Figure 11. Efficiency of burning 1 litre of diesel fuel to transport one ton of cargo, by various means of transport [10].

11 years later, the webpage of Wasserstraßen-und Schifffahrtsverwaltung des Bundes (German Federal Waterway and Shipping Administration) presents slightly different figures. According to the data from 25.09.2014 a midland waterway ship with a length of 80÷85m and beam of 9,5m, using one litre of diesel fuel is capable of transporting one ton of cargo for 370km, the same cargo can be transported by rail for 300km, and by a road truck for 100km [12].

The Polish Ministry of Economic Development in 2016 has presented the same data, claiming the inland water barge with a length of 80÷85m and a beam of 9,5m is capable of using one litre of diesel fuel to transport a cargo with a mass of 1 ton for 370km, the rail transport can do the same task for a distance of 300km and road transport can do it for 100km [1,2,4].

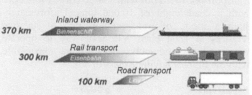

Figure 12. Transport distances for one ton of goods at the same energy expenditure [12].

Additionally, both ministries state that during the transportation of cargo by given transportation means, there is and emission of greenhouse gasses, mainly CO_2 in the following amounts: 164g/tkm for road transport, 48.1g/tkm for rail transport, and 33.4g/tkm for inland waterway transport.

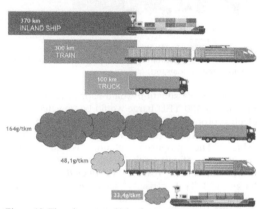

Figure 13. The advantage of inland waterway transport [1,2,4].

Basing upon independent data and performed simulation and modelling research [13-30], the analysis of energy efficiency on inland waterway, rail and road transport was performed. The results are presented in the figures 14 and 15.

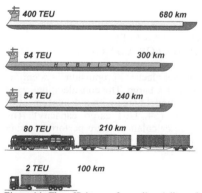

Figure 14. The efficiency of spending 1 litre of diesel fuel to transport 1 ton of cargo, by various means of transportation.

On the basis of presented results, it can be stated that the most favourable situation occurs when the transport is carried out with a large barge (400 TEU) with a deadweight of 10 000 tons. Unfortunately, such a mode of transportation due to its size, requires at least class VIb waterway to operate, whereas in Poland such waterways are unavailable, with the highest having a class of Vb.

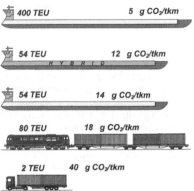

Figure 15. CO2 emission in grams per tonne-kilometre by various modes of transport.

A preferable situation would happen for a system in which a 400 TEU barge would be powered by an electric propulsion system, or would be towed by a land based tractor (mule), powered from overhead cables. The proposed solution would require construction of rail track, along with the overhead power lines. The research conducted on smaller barges (54 TEU capacity) have shown that after upgrading such a barge with hybrid powertrain the fuel efficiency increased by 20% on average, in relation to that barge before upgrade [21]. The same results were corroborated by the simulations presented in the chapter 4. Naturally, the barges move much more slowly than trucks or trains, but the cost of transport (without cargo handling) is the lowest in comparison to other modes of intermodal transport.

4 PROPOSAL OF AN ELECTRIC POWERTRAIN FOR A BARGE

A hybrid, diesel-electric propulsion system is proposed to power a barge. The considerations were performed for a model barge with a displacement of about 1850 tons (54 TEU cargo capacity). The model resistance calculations were performed using the DELFTSHIP software platform [19].

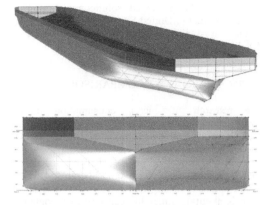

Figure 16. A project of barge hull created in DELFTSHIP [19].

For given geometric dimensions of the hull, the requirement for towing power needed to overcome the resistance at given velocity were calculated using the following relationship:

$$P_T = R \cdot V \tag{1}$$

where:
P_T – towing power [kW]
R – barge resistance [kN]
V – barge velocity [kn]

The total value of barge resistance consists of resistance values corresponding to its velocity, its dimensions and the shape of its hull. These factors influence the magnitude of resistance related to hull-water friction (Hydrodynamic Resistance), swell and river current action (Residual Resistance) and aerodynamic drag from wind effects on the surface elements of the barge (Aerodynamic Resistance).

$$R = R_H + R_R + R_A \tag{2}$$

Knowing the value of towing power and resistance, it is possible to calculate the required power output of the motor used to drive the vessel. In order to determine the power of the electric powertrain, it is required to know the drive train efficiency, which consists of: hull efficiency (0.6÷1), the shaft efficiency (0.9÷0.99), rotational efficiency (0.95÷0.99) and free running propeller efficiency (0.35÷0.75).

$$P_{EM} = \frac{P_T}{\eta \cdot \eta_H} = \frac{P_T}{\eta_S \cdot \eta_R \cdot \eta_{RP} \cdot \eta_H} \tag{3}$$

where:
P_{EM} – power output of electric powertrain [kW]
P_T – towing power [kW]
η – drive train efficiency
η_H – barge hull efficiency
η_S – shaft efficiency
η_R – rotational efficiency
η_{RP} – free running propeller efficiency

In order to properly choose the power output of electric powertrain, it is required to take into consideration the efficiency of each element used for power transmission. For the considered design, the power output of the electric powertrain will be on the order of 840 kW, with the top speed of about 22 km/h. When selecting the desired power output, it is important to keep in mind that the barge should move at a safe speed, suited to the existing navigational and atmospheric conditions. The barge speed cannot cause any danger to other vessels and users of the waterway. For this reason, there are general speed limits in place, with respect of the bank of the waterway, and when the local regulations do not say otherwise, they are 5-8 km/h upstream of channels and rivers, and 10-12 km/h downstream of channels and rivers. On the other reservoirs such as lakes or lagoons, the speed can be higher, up to 15km/h. In the event, the barge's powertrain power output is not sufficient, the travel upstream will not be possible and in addition, such vessel would present a danger to crewmembers and other vessels operating in the area.

It is worth noting, that for barge propulsion employing a land-based rail mule, in Polish environment (the maximum current speed of the fastest river in Poland – Wisła, during a surge can reach 10 km/h, and typically is 6-7 km/h), the power required to move such barge would be about 4.5 times smaller and be on the order of 180 [kW].

5 CONCLUSIONS

- The presented analysis results on energy efficiency indicate, that the lowest energy consumption among the intermodal transport modes is claimed by the barges. In addition, these figures can be further improved, by application of hybrid and electric powertrains on board, or on the land-based tractors (mules).
- The inland waterway transport is ideally suited to transportation of cargo not requiring an immediate delivery and insensitive to moisture along with oversize loads.
- The disadvantage of inland waterway transport in Poland is the low quality of waterways, lack of infrastructure, dependence on the water level as well as on weather conditions (icing during winter).
- In order to fully exploit the potential of waterway transport in Poland, it is required to invest much resources into the infrastructure and the upkeep of waterways to maintain them in good condition. The investment can stimulate the great potential present in the Polish waterways and it can return in small amount of time as an increase in economic growth, as well as growth in touristic, social and ecological fields.

REFERENCES

[1] MGMiŻŚ, Założenia do planów rozwoju śródlądowych dróg wodnych w Polsce na lata 2016-2020 z perspektywą do roku 2030. Monitor Polski, Dziennik Urzędowy Rzeczypospolitej Polskiej, Poz. 711, Uchwała Rady Ministrów z 14.06.2016r., http://isap.sejm.gov.pl/

[2] Ministerstwo Rozwoju, Założenia do planów rozwoju śródlądowych dróg wodnych w Polsce na lata 2016-2020 z perspektywą do 2030 roku, http://ungc.org.pl/wp-content/uploads/2016/04/26042016_Prezentacja_Ministerstwo_Rozwoju.pdf, (04.2016).

[3] Ratification of the Convention, in the opinion of the Polish expert 3.0 AGN Jacob Stonawskiego. www.polska-3-0.pl, (11.2016).

[4] Rozporządzenie Rady Ministrów z 07.05.2002r. w sprawie klasyfikacji śródlądowych dróg wodnych. Dz.U. Nr 77, Poz.695, załącznik 1÷3.

[5] Główny Urząd Statystyczny, Transport wodny śródlądowy w Polsce w 2015 r., http://stat.gov.pl/ (08.2016).

[6] DeDecker K., Trolley canal boats. Low-Tech Magazine, www.lowtechmagazine.com, (12.2009).

[7] Geschichte des Oberleitungsbusses. www.wikiwand.com, (02.2017).

[7] Le tunnel de Mauvages. www.bordabord.org, (12.2009).

[8] Sheller A., Le tunnel fluvial de Mauvages : un ouvrage de 4 877 mètres, accessible qu'aux plaisanciers et aux bateaux de commerce. www.allboatsavenue.com, (12.2013).

[9] Le tunnel de Riqueval et son toueur à Bellicourt. www.petit-patrimoine.com (11.2016).

[10] Inland Waterway Transport. European Commission, Energy and Transport DG, ISBN 92-894-4344-8, 2003.

[11] Piekarski L., Annex D, Statistical Approach to Inland Waterway Transport. European Conference of Ministers of Transport. ISBN 92-821-1354-x, ECMT 2006.

[12] Binnenschiff und Umwelt, Das Verkehrssystem Binnenschiff / Wasserstraße ist umweltfreundlich, kostengünstig und sicher, www.wsv.de/Schifffahrt/Binnenschiff_und_Umwelt/index.html, (25.09.2014).

[13] Nicholas L., Do CSX Trains Really Move 1 Ton of Cargo 400 Miles on 1 Gallon of Fuel?, The Center for Transportation and Livable Systems (CTLS). www.ctls.uconn.edu, (28.02.2013).

[14] Holly A., The Nation's Freight Railroads Now Average 480 Ton-miles-per-gallon. Assciation of American Railroads. www.aar.org, (22.04.2010).

[15] Fuel consumption by trucks - Specifications www.truckmania.pl/content/view/168/6/

[16] Wasilewicz W., Research electricity consumption for the purpose of traction. Przegląd kolejowy elektrotechniczny nr. 9/1974.

[17] The National Centre for Emissions Management (KOBiZE). CO2 Benchmark for Energy End-use. www.kobize.pl (11.2014)

[18] Dünnebeil F., Lambrecht U., Fuel efficiency and emissions of trucks in Germany - An overview. IFEU-Institute Heidelberg 2012

[19] Delftship Marine Software. www.delftship.net

[20] Papanikolaou A., Ship Design. Springer Netherlands 2014. DOI 10.1007/978-94-017-8751-2

[21] Bach R., Reduzierung des Dieselverbrauchs und der CO2-Emission durch Diesel-direkten Hybridantrieb. (04.2015)

[22] Flämig H., Luft- und Klimabelastung durch Güterverkehr. www.forschungsinformationssystem.de (03.2016)

[23] Mittelweseranpassung Das Großmotorgüterschiff setzt neue Maßstäbe. www.nba-hannover.wsv.de

[24] Gierusz W., Łebkowski A., The researching ship "Gdynia". Polish Maritime Research, Vol. 19, 2012, p.11-18.

[25] Gierusz W., Simulation model of the shiphandling training boat "Blue Lady". IFAC Conference on Control Applications in Marine Systems Location. IFAC Proceedings Series, 2002, p.255-260.

[26] Gierusz W., Tomera M., Logic thrust allocation applied to multivariable control of the training ship. Control Engineering Practice, Vol. 14, Issue 5, 2006, p.511-524.

[27] Lisowski J., Computational intelligence methods of a safe ship control. Procedia Computer Science, Vol. 35, 2014, p.634-643.

[28] Binnenschiff. https://de.wikipedia.org/wiki/Binnenschiff

[29] Specific CO2 emissions per tonne-km and per mode of transport in Europe, 1995-2009. www.eea.europa.eu

[30] Woolford R., McKinnon A., The Role of the Shipper in Decarconising Maritime Supply Chains. Chapter 1, Current Issues in Shipping, Ports and Logistics. Asp, Vubpress, Upa, 2011. ISBN9054878584.

The Literature Review: Bunkering and Bunkering Decisions

C. Sevgili & Y. Zorba

Dokuz Eylul University, Izmir, Turkey

ABSTRACT: As around 90% of world trade is carried by ships and fuel costs constitute over two out of three of ships' operating cost, marine bunkering is one of the most important issues for maritime sector. Thus, maritime industry is continuously trying to find a variety of solutions and alternatives in order to reduce bunker costs. Fluctuations in world trade and bunker prices make difficult and oblige to struggle with this issue. In this study, literature review of bunkering decisions and bunkering being crucial both maritime and energy sector were investigated. Basic bunkering decision criteria and research methods about bunkering in literature were examined. Results of examination, bunkering decision criteria are quite various and the most stated criteria were determined as "bunker price and price competition", "quality of bunker" and "geographical advantage of refueling area".

1 INTRODUCTION

Bunker is a common name used for ship fuels. The origin of this name comes from coal storages calling as "coal bunker" for coal-powered first steam ships. With the rise of steam ships using coal as fuel at last quarter of the 19th century, the need for supply coal was arisen beside the handling of ships in ports. This profession that supplies coal to commercial vessels in ports is called "bunkering" (Draffin, 2010).

In the first quarter of the 20th century, some attempts were made to use oil as fuel for ships. The reason for these initiatives is that oil has less labor force, high storage capacity and more efficient compared with coal. With these advantages, the use of oil in ships has increased rapidly as seen in Figure 1 (Draffin, 2008).

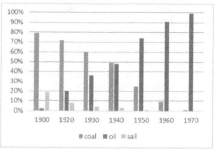

Figure 1. The change from coal to oil, (Draffin, 2008).

Maritime transportation carries approximately %80 of world's merchandise trade and accounts for %70 its value (UNCTAD, 2015). The high level of this ratio naturally also manifests itself in fuel consumption. According to OPEC reports, bunker sector consumed 4.1 million barrels of oil equivalent per day (mboe/d) in 2013. Bunker costs constitute more than %50 of the total operating cost of a ship. To reduce fuel costs in fluctuating economic and fuel market is one of the basic problems for shipping companies (Yao et al., 2012).

Recently, emissions result from ship fuels (bunker) have been one of the main environmental issues. Consumed bunker produces various emissions, primarily nitrogen oxides (NO_x), sulphur oxides (SO_x) and carbon dioxide (CO_2) (Deniz & Zincir, 2016). Maritime industry is responsible of about 2,5% of CO_2, 15% NO_x, 13% SO_x of global emissions. According to estimates of future economic and energy developments, it is stated that emissions from shipping will generally increase (3th IMO GHG Study, 2014). Various innovations and regulations are emerging to control and reduce the increasing emission values of the maritime industry. These innovations and regulations often focus on fuels that will help to reduce efficiency and emissions such as low sulphur, non-petroleum and renewable fuels.

The aim of this study is to analyze the academic studies related to bunkering through content analysis and to reveal the developments in bunkering. The studies carried out in the literature have been examined under the main headings. In addition, the bunkering decision criteria which is one of the most important issues for ship owners and ship operators are determined in the literature. The importance of the study is that such a study has not been done before in the literature.

2 METHODOLOGY

In the paper, studies on bunkering in literature were determined and examined by content analysis method. Content analysis is a systematic technique used to characterize and compare similar data or for further analysis (Altunışık et al., 2012). The articles that constitute the study were scanned from various database, mainly Dokuz Eylul University Online Library service. Literature scanning was carried out using the terms "bunkering, refueling, marine fuel" in studies. A total of 54 articles were identified, articles that are not directly related to the field are excluded from evaluation and the number of articles to be evaluated for the study was reduced to 36. The basic data of the articles is analyzed using the Statistical Package for the Social Sciences (SPSS) 22.0 packet program.

3 FINDINGS

The examined articles are divided into 5 sections according to their basic interests as bunker management, non-petroleum and renewable marine fuels, environmental, illegal bunkering and bunker services and bunkering decisions (Table 1).

Table 1. Distribution of studies to main interest.

Main interest	Number of Studies
Illegal bunkering	3
Non-petroleum and renewable fuels	9
Environmental	7
Bunker management	12
Bunker services	5

The examined articles were published in 28 different journals. Journal that most publications on the subject is European Journal of Operational Research with 3 article. When the first writers of the articles were examined, it was seen that they were from 21 different nations. The country with the highest number of publications is China with 8 articles, and South Korea following with 5 articles. 6 articles have single-author, while the others have multi-author, and when all the authors of the articles are examined, there are 8 article which are multinational studies. Half of the studies was done by authors in different institutions.

3.1 Illegal bunkering

When studies on illegal bunkering are examined, it is seen that the investigations have been made on Western African coast, especially Nigeria whose economy depends energy resources. The oil and gas sector corresponds about 35 per cent of gross domestic product, and petroleum exports revenue constitute over 90 per cent of total exports revenue (OPEC, 2016). In fact, illegal bunkering is a rare phenomenon in the world, but the events in Nigeria have a negative impact (Vreÿ, 2012). According to the information provided by local sources, hundreds of thousands of barrels of crude oil are being stolen and lost in just one day (Oyefusi, 2014).

Vreÿ (2012) reviewed illegal bunkering activities in terms of maritime aspects and security measures that would prevent illegal bunkering. Illegal bunkering activities can take a variety of forms. These may be small-scale thefts for oil production for local use, as well as thefts to tankers and barges intended to be distributed to larger scale and foreign countries. These groups who are engaged in illegal activities are organizing illegal bunkering and they are generally armed. The Nigerian government is trying to provide tighter control with private security contractors in response to increasing illegal bunkering activities. It has suggested that the regional associations such as the Economic Community of West African States, Economic Community of Central African States and The Maritime Organization for West and Central Africa be useful in preventing illegal bunkering. Also, International actors like UN, USA, France and China may aid the region to be safer (Vreÿ, 2012).

Ingwe (2015), on the other hand, mentioned the political economy and social structure of the Nigerian, phenomenal crime which effects evaluation of illegal bunkering activities in his study. Method of description for implementing was used and history of illegal bunkering and 2002-2008 data obtained from armed forces (particularly Nigerian Forces) are compared. Oyefusi (2014) handled the subject with perspective of ethno political settlement in Nigeria. According his analysis, illegal bunkering activities in Nigeria are difficult to finish in this region. It has been shown that the actors causing the conflicts in the region have recently been directed to illegal bunker activities. It has been emphasized that the political steps taken to protect the country's oil and gas resources in Nigeria should be for re-evaluating the right to property in the mining regions, establishing an efficient wealth management, and promoting broad-based growth. These steps are necessary for the sense of belonging and common fate of the ethnic groups that make up the Nigerian.

3.2 *Non-petroleum and renewable fuels*

Non-petroleum and renewable marine fuels are one of the steps taken for increased pressure and regulations which are particularly intended for certain emission control areas (ECA) towards environmental protection (Fig. 2). It is seen that these fuels are mostly focused on liquefied natural gas, liquefied biogas, methanol, bio-methanol, ethanol and hydrogen. Among these, it is seen that they are mainly working on liquid natural gas (LNG). When studies are examined, the comparison of these fuels, environmental effects and infrastructure requirements are mentioned.

Figure 2: Map of Emission Control Areas (ECA), (IMO, 2014).

Deniz & Zincir (2016) examined environmental and economic assessment of methanol, ethanol, liquefied natural gas, and hydrogen. The weights of the fuel types were established by applying Analytic Hierarchy Process (AHP) to the determined criteria. They found that LNG is the most suitable fuels among these fuels. Hydrogen can also be fuelled by alternative ships with high scores, but for this it is necessary to work on hydrogen and develop its technology. According to study, methanol and ethanol are not suitable for use as marine fuel. Brynolf et al. (2014) evaluated these marine fuels the environmental aspects. Using life cycle assessment (LCA), their purpose was the comparing the life cycle environmental performance of methane, LNG, liquefied biogas and bio-methanol. They have suggested marine fuels produced from biomass are more effective than others on climate change. Study of Schönsteiner et al (2016) focused on supply chains and calculate energy demand and GHG emissions of conventional fuels, LNG, biofuels and hydrogen in Singapore. They used well-to-tank (WTT) which is one of the variants of LCA. As in the previous study (Brynolf et al., 2014), the reduction of GHG emissions through the use of LNG is more limited than biofuels. Hydrogen is the type of fuel that must be used to completely prevent GHG. While the resource potential of LNG is adequate, production of biomass-derived fuels is limited. In Greece domestic passenger transport,

LNG and fuel oil were compared in environmental way. It has been stated that with the LNG conversion from fuel oil, there is serious decrease in air pollution, CO_2 rate, and external costs related environmental external costs. However, the use of LNG as marine fuel seems difficult to develop because the LNG bunker facilities in the Greece are insufficient (Tzannatos & Nikitakos, 2013).

Another aspect of these marine fuels is bunkering facilities. Wang & Nottebom (2015) compared the LNG bunkering project of eight ports in North European with multiple-case study to find the role of port authorities in improvement LNG bunkering. To develop LNG bunkering, ports should cooperate development policy with stakeholders, make financial incentive policy and coordinate communication policy. In the case study on the port of Busan (Yun et al., 2015) proposed the conventional design of offshore LNG bunkering terminal with analysing visiting ships, amount of needed LNG and adjusting hull structure. Criteria of conceptual design were evaluated LNG fuel supply chain and design requirements (safety, economics and principal function).

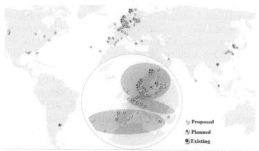

Figure 3: LNG Bunkering Infrastructure, (DNV, 2015).

Using international database, port websites and reports, Calderon et al. (2016) provided an overview of potential expansion of LNG powered ships and LNG bunkering facilities. Safety standards, logistic infrastructure and financing approaches were reviewed, and it has been found that the use of LNG as alternative marine fuel will increase, especially after 2020. Since the LNG bunkering terminal is still a very new concept, there is no common regulation regarding safety in operation, but major ports advance safety procedures themselves. Herdzik (2013) mentioned inadequacy LNG distribution and LNG bunkering terminals before the ECA. LNG barges and small carriers were presented. Blikom (2012) studied on actively working LNG powered ships reviews on bunkering operations and availability of LNG. It has been determined that the use of LNG as fuel for ships will increase, so that the main ports need to give importance to the infrastructure activities. It is also predicted that in the future the factor that determines the ship prices will be based on the fuel selection, so competition among the ship owners will increase (Fig. 4).

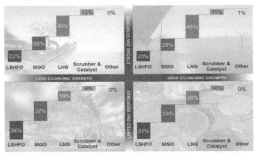

Figure 4. A scenario planning process for maritime industry, (Blikom, 2012).

3.3 *Environmental*

Studies on environmental issues generally focus on minimizing emissions, environmental costs and fuel consumption. Kim et al. (2012) used epsilon-optimal algorithm in order to solve environmental problem caused by marine fuels. The algorithm, which was developed according to the example route selected over Hong Kong, China, South Korea and western coasts of USA, evaluates ship speed. Bunker prices, taxes of carbon, ship-time costs were determined as influencing factors. Ship speed and CO_2 emission was found very sensitive to influencing factors. Corbett & Winebreak (2012) analyzed fuel cycle for container vessel using Total Energy and Environmental Analysis for Marine Systems. This method has been used for calculate emissions. They have argued that rise of CO_2 emission will be less than %1 and low sulphur fuels such as MDO and MGO may decrease CO_2 and SO_2 (about %75-85) compared to residual oil. In the meantime, as low sulphur fuels are higher prices than high sulphurs, bunker cost of ships increases (Schinas & Stefanakos, 2013).

Stikkelman et al. (2012) examined the impact of regulations on reducing emissions to the bunker market share evaluated port of Singapore and Rotterdam. Economical, technical, operational, political options and strategies to reduce emissions were discussed, and it has been stated that Rotterdam and Singapore ports will provide more benefits in case of cooperation rather than competition. Moon & Woo (2014) studied on operation costs and CO_2 emission with system dynamics. A Dynamic Liner Service Evaluation Model (DLSEM) was established as Asia–North Europe line. CO_2 emissions are seriously affecting from ship's port time. The decrease of ship's time decreased the cruising speeds. According to the calculations, a 30% reduction of ship's port time reduced the annual amount of CO_2 by about 36%.

Conversely, if port time was increased by 30%, annual CO_2 emissions increase by 30%. Therefore, it could be said that the amount of GHG emissions is largely related to the efficiency of port operations.

In a study evaluated marine fuels for emission control areas has showed that the most appropriate choice for risky ship investments will be diesel fuelled engine according to latest regulations on greenhouse gas emissions. Also, LNG fuelled engines will be viable with advancing LNG bunkering infrastructure and imposing price regulations on diesel fuels. But, dual-fuelled engines do not seem optimal because of high investment costs. As a result, LNG fuelled engine selection would be appropriate for investors who are far from riskier (Soundararajan & Han, 2014). Acomi & Acomi (2014) studied on influence of ship fuels over Energy Efficiency Operational Index (EEOI) developed by IMO for ship owners and operators to determine the level of GHG emissions from ships in operation. For case study, 4 voyages of an Aframax oil tanker were observed. HFO, LFO and MDO/MGO were compared, and results have showed that HFO has the lowest EEOI value while it is the most cost effective fuel type. Comparing with HFO, LFO and MDO/MGO had 1.14% and 2.82% higher value, respectively. Another comparison was carried out for marine heavy fuel additives. 4 samples were investigated with Fuel Combustion Analysis to determine fuel consumption and emission tendency of fuel additives. It has been found that fuel additives may have a positive and negative effect on fuel consumption. In addition, fuel additives used in study have reduced NO_x and particulate matter emission (Jang & Choi, 2016).

3.4 *Bunker management*

When studies on bunker management are examined, it is seen that the works are divided into two as on liner and tramp shipping. Tramp and liner shipping have different operating principles, so fuel management is also different. In this section, studies on bunker contracts are also examined.

It is necessary to be very careful when planning bunkers because there are serious differences in fuel prices between ports. To optimize this issue in tramp shipping, Vilhelmsen et al (2014) suggested a mixed integer programming (MIP) formulation which included time of bunkering, bunker prices, port tariffs and costs related loading rate. Likewise, Meng et al. (2015) evaluated routing and bunkering problem in tramp shipping using integer linear programming to maximize profit. As result of the tests performed, applied model has been efficient.

As variables are clearer than tramp shipping, there are more studies on liner shipping in literature. Notteboom & Vernimmen (2009) studied how to update the schedule according to increasing bunker prices in liner shipping. Cost model, whose variables were service frequency, fleet size and number of calls, showed that increasing bunker prices have significantly effecting cost per TEU on large post-panamax vessels. Yao et al. (2012), who reported that bunker fuel management consists of port selection, quantity and speed components to be taken, determined bunker management differed according to the dimensions of the container ships. When the bunkers prices increase, the speed of the vessels should be reduced in order to cut the costs. But in the case of port congestion and delays, it is necessary to increase the speed for schedule coherence (Wang & Meng, 2012). In 2013, Wang et al. proposed a number of methods (enumeration, dynamic programming, discretization, linear methods, quadratic methods, second-order cone programming) to reduce total operational costs and emissions and increase ship and port cooperation. In A study that approached the optimal bunker port selection with a hybrid Fuzzy-Delphi-TOPSIS has purposed to compare bunkering ports (Wang et al, 2014). Sheng et al. (2014) has suggested MIP model to reduce fuel consumption and determine sailing speed. They have established model of bunker prices and bunker consumption rate. According to model results, authors claim that they provide a more reliable schedule and lower cost. Wang & Meng (2015) studied on minimizing total cost in the worst case bunker consumption. a mixed-integer nonlinear optimization model has been used based on the Asia–Europe–Oceania liner shipping network, and they have found that speed variability causes higher total costs. One of the latest study focused on stochastic speed optimization with maintaining schedule reliability using dynamic program and deterministic modelling. They have suggested that liner shipping companies should determine bunkering amount and ports by taking into consideration vessel schedule, delay costs and bunker prices. Delay penalty costs also affect bunkering port decision. Results of study have showed that bunkering amount and port decision is relevant to bunker prices (Aydın et al., 2016).

As previously mentioned, bunker constitutes a major part of the operational costs of the ships. At the same time, trend of bunker price is generally fluctuating and uncertain. For this reason, service contract between vessel operator and bunker supplier may be a strategy to decrease bunker costs and supply risk especially in uncertainty. Service contract usually consists of bunker price, quantity and bunkering port. Within this scope, Ghosh et al. (2015) developed a dynamic programming model in order to determine optimal bunkering parameters and optimize total bunkering costs with service contract for liner shipping in uncertain environment. According to results of model the amount and price of contract and the damage charge multiplier should be evaluated together in determining the bunkering decisions, amount of contract should be lower than total consumption during voyage and price of contract should be less than predicted spot price. Pedrielli et al. (2015) evaluated this problem using game theory. Their findings have argued that optimum amount of contract is generally related to minimum bunker consumption and changefulness consumption and prices. They have suggested that vessel operators should raise its reservation when price variability increases. In high price volatility, vessel operators should tend to long term contracts, while bunker suppliers should prefer short term contracts for maximum profit.

3.5 Bunkering services and bunkering decisions

Bunkering services have complex structure and require experience (Chang & Chen, 2006). Delays in bunker services can extend the time spent on the port and cause problems as well as problems with cargo handling. In addition, poor quality fuel can cause maintenance costs to rise. So the service standards provided by local bunker operators are very important in terms of port image and customer satisfaction (Pinder, 1997). There are technical and management studies in the literature to optimize bunker services. These studies are generally focused on large bunker ports like the geographical distribution of petroleum used as a bunker. Despite the fact that there are approximately 400 large bunker ports in the world, several strategic ports have the biggest share (Table 2). Singapore, China, the USA, the United Arab Emirates, Netherlands and South Korea accounted for about 60% of the world's bunker demand in 2012 (WOO, 2015).

Table 2. Top 5 bunkering ports, (WOO, 2015).

Port	Annual bunker sales (million tone)	Year
Singapore	42,4	2014
Fujairah, UAE	24,0	2013
Rotterdam,Netherlands	10,6	2013
Hong Kong, China	7,4	2013
Antwerp, Belgium	6,5	2012

Chang & Chen (2006) established a knowledge based simulation model in order to effectively analyze the bunker services of Port of Kaohsiung. Expert system and discrete event simulation techniques has been used during simulation development process. Two experiments are performed as performance analysis of the current and the revised barge allocation scenarios and strategic analysis of future bunker barge fleet expansion. The current situation can be further improved according to the scenario created in the first experiments. For the second experiment, 3 proposals for future developments were tested.

Pinder (1997) assessed policy responses to the service quality of Port of Singapore. He suggests that deregulation may be quite effective to promote market share. With deregulation, current problems will be able to intervene more quickly and future plans and investigation will be more competitive. Another study on port of Singapore (Cullinane et al., 2006) stated that Singapore is vulnerable competition. But, maritime heritage of Singapore permits competitiveness up to now. They have argued that Singapore's maritime industry, generally dominated by ship broking and chartering, ship building and repair, agencies, needs to concentrate other service areas and maintain its advantage in this competitive environment.

It is seen that the studies on the bunker port competition are also made on the major bunker ports. Lam et al. (2011) compared Singapore and Shanghai. 10 variables were determined according to interviews. Most important variables specified as bunker quality, transparency (corruption-free) and bunker price competitiveness, respectively. While location of port, stability of political environment and bunker price competitiveness have top scores for appraisement of Singapore, stability of political environment, bunker quality and reliability and punctuality of bunker suppliers have the highest scores for Shanghai. Also, they stated that Singapore is better performer than Shanghai. Strategic location and liberal market structure can be cited as reasons for this. In another study on bunker port competitiveness (Acosta et al., 2011), 3 ports (Algeciras, Ceuta and Gibraltar) in Gibraltar Strait have been examined. 20 variables have been determined with interviews and questionnaires. Bunker prices and geographical advantages have been determined the most important factors. However, when the ports are compared, significant differences are seen on weight of variables.

As seen in the literature, bunkering decision is very important issue for maritime industry. Ship owners and operators struggle to maximize their profit and, one of the most important step of this issue is bunkering decisions. 28 criteria were found in literature review (Table 3). When these criteria are analyzed, the most important criteria seem "fuel price and price competition", "quality of fuel" and "geographical advantage of refueling area", respectively. Bunker price vary from port to port and, as ships take large amount of bunkers, significant saving can be provided by ship owner and operators. Bunker quality comprises of some chemical and physical properties (flash point, pour point, energy content, sulfur, vanadium, aluminum, silicon, water contents, viscosity, etc.) and it directly effects ship engine equipment and emission rates (Lam et al., 2011).

Table 3. Criteria of bunkering decisions.

	Lam et al. (2011)	Acosta et al. (2011)	Chang & Chen (2006)	Wang et al. (2014)	Yao et al. (2011)	Boutsikas (2003)	Vilhelmsen et al. (2014)
Bunker price and price competitiveness	X	X		X	X	X	X
Bunker quality	X	X		X		X	
Reliability, punctuality and safety of bunkering services	X			X			
Clear and precise information about services		X		X			
Transparency (corruption-free)	X						
Geographical advantage		X		X	X		X
Anchoring and docking availability		X					
Simplicity/accessibility to port		X					
Port tariffs		X		X			X
Location of port	X	X					X
Port congestion		X					
Port security		X					
Bunkering facilities (adequacy and efficacy)	X		X	X			
Supply waiting time		X	X				
Duration of supply			X				
Quality of bunkering services (efficiency e.g. pumping rates)	X						
Prices of complementary services for fuel supply at berth (pilotage, mooring, etc.)		X					
Prices of complementary services for fuel supply at anchorage		X					
Presence of restrictive environmental regulations		X		X			
Simplicity of crew changes		X					
Customs strictness		X					
Ship inspection thoroughness		X					
Government policies (e.g. quality control) and incentives	X						
Stability of political environment	X						
Availability of low sulphur bunkers	X						
Bunkering rules of port				X			
Weather conditions of supply region				X			
Usage and availability of barges			X				

Bunker quality get under control by some implementations and supports such as monitoring systems, international standards, testing lab and equipment, well trained bunker specialists (Lam et al., 2011). Geographical advantage of refueling area is also very important because bunkering decisions are decided on scheduling and routing of vessels. Regions where are far away from route may increase operational costs and cause schedule instability. Therefore, ship owners and operators tend to give bunkering decisions considering geographical advantage.

4 CONCLUSION

Bunker constitutes a large part of the Operational costs of ships, and bunkering is a very important issue for the maritime sector because price changes are continuous and uncertain. Despite its importance, from a literary perspective, a limited number of studies have been conducted on bunkering.

The studies have generally focused on the issues of illegal bunkering, non-petroleum and renewable fuels, environmental, bunker management and bunkering services. Illegal bunkering seems to be a situation that affects both commercial and social life, especially for the West African region. When the studies on the subject are examined, it is seen that the political and sociological situation in the region affects illegal bunkering. Increasing pressures on environmental and productivity force the maritime industry to turn to new fuel types. In literature, studies dwell on liquefied natural gas, liquefied biogas, methanol, bio-methanol, ethanol and hydrogen as alternative fuels. LNG is the most mentioned and considered as the best alternative fuel candidate for the future among non-petroleum and renewable fuels. It is confirmed that more concrete steps have been taken and applied for LNG nowadays. However, general opinion about the alternative fuels, it is thought that the transition period to alternative fuels will take a long time, especially since infrastructure and machine transformations will not take place in the short term.

It has also affected the maritime industry in response to increasing environmental problems in recent years. Measures taken by IMO, environmental organizations and governments on the issue are aimed to minimize the damage to marine fuels, especially emissions. In the studies on environmental issues, the emissions emitted by fuel types and fuel additives are examined, and the eco-friendliest and efficient fuel type is being tried to be determined.

Fuel management studies have often been based on optimizations to reduce fuel consumption and scheduling. The distinguishing feature is what type of transport the ship is carrying (liner or tramp shipping). Optimization methods are made more easily in liner shipping because schedule is especially clear. For optimal bunker management, it is important for ship owners and ship operators to pay attention to voyage plans and speeds, precautions to reduce fuel consumption. When the bunker services are examined, it is seen that the studies carried out on the biggest supply points of the world. 28 important criteria for bunkering decisions have been identified in the literature. According to the frequencies in the literature, the most important ones are "fuel price and price competition", "quality of fuel" and "geographical advantage of refueling area", respectively. It is thought that bunkering ports may get more share from the market by making improvements in line with these criteria.

It is observed that there is a limited number of studies on bunkering, and very limited research especially on illegal bunkering and bunkering services. For further studies, more studies on illegal bunkering in West Africa and other illegal bunkering areas such as Singapore are suggested. It is appropriate to undertake more comprehensive studies on developing bunkering ports and expectations of stakeholders should be determined.

REFERENCES

Acosta, M., Coronado, D., & Cerban, M. D. M. 2011. Bunkering competition and competitiveness at the ports of the Gibraltar Strait. Journal of Transport Geography, 19(4): 911916.
Acomi, N., & Acomi, O. C. 2014. The Influence of Different Types of Marine Fuel over the Energy Efficiency Operational Index. Energy Procedia, 59: 243-248.
Altunışık, R., Coşkun, R., Bayraktaroğlu, S.& Yıldırım, E. 2012. Sosyal Bilimlerde Araştırma Yöntemleri(SPSS Uygulamalı). Sakarya: Sakarya Kitabevi.
Aydin, N., Lee, H., & Mansouri, S. A. 2016. Speed optimization and bunkering in liner shipping in the presence of uncertain service times and time windows at ports. European Journal of Operational Research, 259(1): 143-154.
Blikom, L. P. 2012. Status and way forward for LNG as a maritime fuel. Australian Journal of Maritime & Ocean Affairs, 4(3): 99-102.
Boutsikas, A. 2003. The bunkering industry and its effect on shipping tanker operations (Unpuplished PhD Thesis). Massachusetts Institute of Technology, Cambridge.
Brynolf, S., Fridell, E., & Andersson, K. 2014. Environmental assessment of marine fuels: liquefied natural gas, liquefied biogas, methanol and bio-methanol. Journal of cleaner production, 74: 86-95.
Calderón, M., Illing, D., & Veiga, J. 2016. Facilities for bunkering of liquefied natural gas in ports. Transportation Research Procedia, 14: 2431-2440.
Chang, Y. C., & Chen, C. C. 2006. Knowledge-based simulation of bunkering services in the port of Kaohsiung. Civil Engineering and Environmental Systems, 23(1): 21-34.

Corbett, J. J., & Winebrake, J. J. 2008. Emissions tradeoffs among alternative marine fuels: total fuel cycle analysis of residual oil, marine gas oil, and marine diesel oil. *Journal of the Air & Waste Management Association,* 58(4): 538542.

Cullinane, K., Yap, W. Y., & Lam, J. S. 2006. The port of Singapore and its governance structure. *Research in Transportation Economics,* 17: 285-310.

Deniz, C., & Zincir, B. 2016. Environmental and economical assessment of alternative marine fuels. *Journal of Cleaner Production,* 113: 438-449.

DNV, 2015. In focus-LNG as ship fuel latest developments and projects in the LNG industry. Retrieved 19.01.2017 from https://www.dnvgl.com/maritime/lng/index.html .

Draffin, N. 2008. *An Introduction to Bunkering.* Oxfordshire: Petrospot.

Draffin, N. 2010. *An Introduction to Bunker Operations.* Oxfordshire: Petrospot.

Ghosh, S., Lee, L. H., & Ng, S. H. 2015. Bunkering decisions for a shipping liner in an uncertain environment with service contract. *European Journal of Operational Research,* 244(3): 792-802.

Herdzik, J. 2013. Consequences of using LNG as a marine fuel. *Journal of KONES,* 20(2): 159-166.

Ingwe, R. 2015. Illegal Oil Bunkering, Violence and Criminal Offences in Nigeria's Territorial Waters and the Niger Delta Environs: Proposing Extension of Informed Policymaking. *Informatica Economica,* 19(1): 77-86.

Jang, S. H., & Choi, J. H. 2016. Comparison of fuel consumption and emission characteristics of various marine heavy fuel additives. *Applied Energy,* 179: 36-44.

Kim, H. J., Chang, Y. T., Kim, K. T., & Kim, H. J. 2012. An epsilon-optimal algorithm considering greenhouse gas emissions for the management of a ship's bunker fuel. *Transportation Research Part D: Transport and Environment,* 17(2): 97-103.

Lam, J. S. L., Chen, D., Cheng, F., & Wong, K. 2011. Assessment of the competitiveness of ports as bunkering hubs: empirical studies on Singapore and Shanghai. *Transportation Journal,* 50(2): 176-203.

Maersk, 2015. *Implications of the 2015 ECA sulphur regulation.* Retrieved 10.01.2017 from http://www.maerskline.com/en-gh/countries/int/news/news articles/2014/07/sulphur-regulation .

Meng, Q., Wang, S., & Lee, C. Y. 2015. A tailored branchand-price approach for a joint tramp ship routing and bunkering problem. *Transportation Research Part B: Methodological,* 72: 1-19.

Moon, D. S. H., & Woo, J. K. 2014. The impact of port operations on efficient ship operation from both economic and environmental perspectives. *Maritime Policy & Management,* 41(5): 444-461.

Notteboom, T. E., & Vernimmen, B. 2009. The effect of high fuel costs on liner service configuration in container shipping. *Journal of Transport Geography,* 17(5): 325-337.

OPEC, 2016. *Nigeria Facts and Figures.* Retrieved 29.12.2016 from http://www.opec.org/opec_web/en/about_us/167.htm .

Oyefusi, A. 2014. Oil Bunkering in Nigeria's Post-amnesty Era: An Ethnopolitical Settlement Analysis. *Ethnopolitics,* 13(5): 522-545.

Pedrielli, G., Lee, L. H., & Ng, S. H. 2015. Optimal bunkering contract in a buyer–seller supply chain under price and consumption uncertainty. *Transportation Research Part E: Logistics and Transportation Review,* 77: 77-94.

Pinder, D. A. 1997. Deregulation policy and revitalization of Singapore's bunker supply industry: An appraisal. *Maritime Policy and Management,* 24(3): 219-231.

Schinas, O., & Stefanakos, Ch. 2013. The cost of SO$_x$ limits to marine operators; Results from exploring marine fuel prices. *TransNav, the International Journal on Marine Navigation and Safety of Sea Transportation,* 7(2): 275-281.

Schönsteiner, K., Massier, T., & Hamacher, T. 2016. Sustainable transport by use of alternative marine and aviation fuels—A well-to-tank analysis to assess interactions with Singapore's energy system. *Renewable and Sustainable Energy Reviews,* 65: 853-871.

Sheng, X., Lee, L. H., & Chew, E. P. 2014. Dynamic determination of vessel speed and selection of bunkering ports for liner shipping under stochastic environment. *OR spectrum,* 36(2): 455-480.

Soundararajan, K., & Han, E. 2014. Probabilistic Analysis of Marine Fuels in Emission Controlled Areas. *Energy Procedia,* 61: 735-738.

Stikkelman, R. M., Minnée, M. G., Prinssen, M. M. W. J., & Correljé, A. F. 2012. Drivers, Options and Approaches for Two Seaport Authorities on the Joint Reduction of Bunker Oil Related Emissions. *EJTIR,* 12(1): 132-145.

Tzannatos, E., & Nikitakos, N. 2013. Natural gas as a fuel alternative for sustainable domestic passenger shipping in Greece. *International Journal of Sustainable Energy,* 32(6): 724-734.

UNCTAD, 2015. *United Nations Conference on Trade and Development Review of Maritime Transport 2015. Retrieved 18.12.2016 from* http://unctad.org/en/PublicationsLibrary/rmt2015_en.pdf .

Vilhelmsen, C., Lusby, R. M., & Larsen, J. 2013. *Routing and Scheduling in Tramp Shipping-Integrating Bunker Optimization: Technical report.* Department of Management Engineering, Technical University of Denmark.

Vreÿ, F. 2012. Maritime aspects of illegal oil-bunkering in the Niger Delta. *Australian Journal of Maritime & Ocean Affairs,* 4(4): 109-115.

Wang, S., & Meng, Q. 2012. Liner ship route schedule design with sea contingency time and port time uncertainty. *Transportation Research Part B: Methodological,* 46(5): 615-633.

Wang, S., Meng, Q., & Liu, Z. 2013. Bunker consumption optimization methods in shipping: A critical review and extensions. *Transportation Research Part E: Logistics and Transportation Review,* 53: 49-62.

Wang, S., & Meng, Q. 2015. Robust bunker management for liner shipping networks. *European Journal of Operational Research,* 243(3): 789-797.

Wang, S., & Notteboom, T. 2015. The role of port authorities in the development of LNG bunkering facilities in North European ports. *WMU Journal of Maritime Affairs,* 14(1): 61-92.

Wang, Y., Yeo, G. T., & Ng, A. K. 2014. Choosing optimal bunkering ports for liner shipping companies: A hybrid Fuzzy-Delphi–TOPSIS approach. *Transport Policy,* 35: 358-365.

WOO, 2015. OPEC *World Oil Outlook 2015.* REtrived 21.01.2017 from http://www.opec.org/opec_web/static_files_project/media/downloads/publications/WOO%202015.pdf .

Yao, Z., Ng, S. H., & Lee, L. H. 2012. A study on bunker fuel management for the shipping liner services. *Computers & Operations Research,* 39(5): 1160-1172.

Yun, S., Ryu, J., Seo, S., Lee, S., Chung, H., Seo, Y., & Chang, D. 2015. Conceptual design of an offshore LNG bunkering terminal: a case study of Busan Port. *Journal of Marine Science and Technology,* 20(2): 226-237.

3th IMO GHG Study, 2014. *Third IMO GHG Study 2014 Executive Summary and Final Report.* Retrieved 24.12.2016 from http://www.imo.org/en/OurWork/Environment/PollutionPrevention/AirPollution/Documents/Third%20Greenhouse%20Gas%20Study/GHG3%20Executive%20Summary%20and%2 0Report.pdf

Gases Emission, Water Pollution and Environmental Protection

Proceedings of 12th International Conference on Marine Navigation and Safety of Sea Transportation, TransNav 2017
21-23 June 2017, Gdynia, Poland

Coastal Dynamics and Danger of Chemical Pollution of Southeast Sector of the Azov Sea

N.A. Bogdanov
Institute of Geography, Russian Academy of Sciences, Moscow, Russia

A.N. Paranina & R. Paranin
Herzen State Pedagogical University, St-Petersburg, Russia

ABSTRACT: Danger of pollution of water areas of the Sea of Azov and the Kerch Strait from possible technogenic catastrophes is diagnosed as a result of comparison of wind power calculations and the current state of coast of the gulf. Ideas of a uniform West Temryuk stream of deposits are disproved. Three dynamic systems in the east and in the center of a coastal arch and the unidirectional alongshore stream of pollutants and energy in the western segment of the gulf are revealed.

1 INTRODUCTION

1.1 *Relevance and purpose*

Purpose of study is to identify trends of shore dynamics to assess the risk of pollution of the Temryucsky Gulf (Fig. 1).

Figure 1. Location of Temryucsky Gulf.

The significance of this diagnosis increases due to the impending reconstruction of the port and the expansion of trans-shipment area for the food and perfume oil, petrochemical, bulk materials and lump sulfur. On the approaches to the port for large vessels wiring dredging from 5.5 to 6.4 m is planned (www.tamaninfo.ru).

1.2 *Problems of preservation of unique resources of the nature and culture*

Coasts and the waters of inland seas, mastered by ancient civilizations, are a vital source of water, biological, recreational, and other resources. Shipping provides cultural and commercial inter-regional and inter-state relations. But, on the sea accidents, especially oil-loading, transport and discharge of domestic waste water, stagnant hydrodynamic phenomena in shallow waters threaten the biodiversity of coastal marine ecosystems and reduce the quality of waters resources. Risks of this kind are typical for the Azov Sea Temryucsky Gulf, on the shores of which the village, wineries, resorts, sports, tourism and others are located. In the inner part of asymmetrically concave arc to the southeast coastline there is a famous port-fortress "Temryuk" - the main source of contaminants (pollutants) on this stretch of coastline (the mouth of the Kuban river).

2 OBJECTS AND METHODS

2.1 *Geomorphological characteristic of the coast*

The object of study - Temryucsky Gulf coast between hilar portion of spit Chushka, cape Ahilleon in the west and village Perekopskaya in the east. Abrasion-landslide and crumbling coast of the western sector of the Gulf are complicated by ravines, gullies and stacked like easily eroded loam and denser rocks on the headlands, where after the storm areas of rock and clay the bench are exposed. Shore with ledges 60 m in height are being destroyed at an average speed of 0.5-0.6 m / year – it is noted from the second half of the 20th century (Boldyrev & Nevessky 1961, Kasyan & Krylenko 2007). Shelly-sand (30-50%, and 90% detritus) beaches, narrow (5-10 m) in areas of abrasion, expand to 20-25 m on the banks of the dynamically stable sprinkling estuaries. Local abrasion in the areas of village Golubitskaya and Kurchansky liman is associated with storm surges (up to 4.2 m). Location of zones of concentration of their energy is controlled by the peculiarities of the transformation on the beach waves. Modern accumulation is seen in the following: a) shelly sand-semi-fixed dunes (up to 5 m) on the scarp near village Kuchugury (frontal section of the maximum exposure En stream); b) the extension of the sea estuarine bars and hard dismembered channels delta of river Kuban (Petrushinskoe sleeve mouth: silts, sands, shelly ground, pebbles); c) The bottom accumulation on Chaykinskoe shoal (provoked by the construction of breakwaters Glukhoy channel port) and capes Ahilleon, Kamenny, Pekly (with circulating vortex shedding). Sandy alluvium of Kuban has been involved in beach creation since 1909, when the flow of the river began to be carried directly into the sea through Petrushinskoe sleeve. By 1954, the border coastal sands with dark blue bottom silt has popped up from the depths of 4-5 m (1927) seaward – up to 8 m. Most of shelly ground is ejected from the bottom of storm erosion of silt and shelly cans (water depth of 8-12 meters) (Boldyrev & Nevessky 1961, Kasyan & Krylenko 2007).

2.2 *Assessment of environmental risk*

Study of pollution scenarios of the coastal zone as a result of accidents and ill-conceived economic activity is based on calculations of the components of the coastal stream of wave energy (En - normal, $E\tau$ - longshore, Eo - result). Mean annual and seasonal morphodynamics define their characteristics and trends redistribution along the coast waters, sediments and pollutants (Fig. 2).

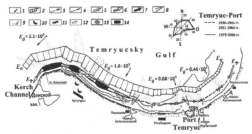

Figure 2. Trends of modern dynamics of coasts of Temryucsky Gulf: Ice-free period 1979-2006 (effective wind speed V_{10} >5 m/s). *1-7* - term average annual flow of the components of the coastal wave energy, *t/s*: *1* - resultant of coastal areas, *Eo*; *2* - cross, the diagram on the outer boundary wave-cut zone, referred to 1 running m length of the coastline, $En = (3.7\text{-}18.8)$ x 10^4; *3-7* - longshore, 1 m wide wave-cut zone $E\tau = (0.1\text{-}8.1)$ x 10^3: *3* - 0.1-0.5, *4* - 0.5-1.0, *5* - 1 -2.5, *6* – 2.5-5, *7* - 5-8.1; *8* - speculative, long-term sustainable coastal zone circulation (water, sediment and pollutants); *9* - isobath, m; *10-12* - elements of morphology and types of shores (according to [2]): *10* - semi-fixed dunes of shell detritus and quartz sand, *11* -

abrasion-landslide and landslide areas, *12* - accumulative and stable beaches; *13* - wind rose, %; *14* - village

3 DISCUSSION OF RESULTS

3.1 *Coastal dynamics of the gulf*

There is no consensus regarding the shore dynamics of the Gulf: 1) according to A.F. Flerov (1931) substance and energy flows rush to the NW and NE - from the middle of the coastal arc; 2) A.I. Simonov (1958): a single energy flow - to the west, along the shores of the entire Gulf (including the predominance of eastern and NE winds). This view was supported by V.L. Boldyrev & E.N. Nevessky (1961) (on the basis of ground surveying). In unidirectional West Temryuk stream of sediment, identified by them, (Sector East Coast arc - Spit Chushka) mechanical and mineralogical differentiation of Kuban sand alluvium was recorded (Bogdanov & Sovershaev, Zhindarev, Agapov 1989). However, it may be related to the diffusion of sediment along the shore (Bogdanov & Sovershaev, Zhindarev, Agapov 1989), but the basis of differences lays in the long-term variability of the resulting wind vector (Fig. 2).

Over the past 70 years the share of the winds from the W, SW and E increased and repeatability of other areas decreased. Among the points of the compass offshore wind exposure NE-E and SW-W-NW air flow is almost the same (35% and 37%, respectively). Configuration of the coastline, bottom topography, and the share of the north wind (up to 11%) are crucial in the direction of the generated flow of wave energy. The area of destruction of maximum waves, limited by height of the shallow reservoir, is located at a depth of 1-3.5 m (http://meteorf.ru).

3.2 *Possibility of the forecast of a rating of pollutants*

Trends of dynamics, identified by the calculations, are in good agreement with the above features of the current state of the Gulf coast. Joint analysis of relations and flows $E\tau$ En morphological and lithological features of the coastal zone, allows to predict the separation of pollutants in the water area.

The calculations used wind energy method of B.A. Popov & V.A. Sovershaev (1981), the application of which is based on: a) the absence of representative wave-measuring observations, b) the reliability and validity of the results in relation to the shores of tide-free sea, c) taking into account the relationship between wind speed and the parameters of the created waves (calculation error - no more than ± 5%) (Bogdanov & Sovershaev, Zhindarev, Agapov 1989, Popov & Sovershaev 1981, Bogdanov

& Vorontsov, Morozov 2004, Pateev 2009, About a state ... 2015, Porotov and Kaplin, Myslivets 2016).

Topographic and navigation maps of scale 1: 200,000 and 1: 500,000 (provided by V.I. Myslivec; D.I. Korzinin was involved in collecting raw data) lay the basis for preliminary constructions and angular measurements. We used the standard table of wind speeds at rhumbs repeatability for a representative period of 1976-2006. The long-term statistics of repeatability of speeds of wind on the directions is taken from the official site Roshydromet. Average annual duration of ice-free period is 295 days (HMS «Temryuk» [http://meteorf.ru]). The amendments in the calculations have been introduced for: true north, bend coastline, the roughness of the underlying surface, the height of the station above sea level. It takes into account the overlap of the offshore bank of the points of the compass and the ratio "t^0C water < t^0C air". Anemometer wind speed is given to the true - at a height of 10 m (V_{10}). The maximum length of acceleration of waves is aligned with the specific natural conditions (reservoir size curves of the coastline, the presence of shoals, islands and braid for overclocking sector, affecting the wind and wave power and other parameters.). Alluvium-moving and relief-creating effects of waves currents, coastal circulation of water, sediment and pollutants are generated by the efficient (> 5 m/s) wind [Popov & Sovershaev 1981, Bogdanov & Vorontsov, Morozov 2004, Guidance on ... 1975). Trends of dynamics that determine the risk of contamination of the Temryucsky Gulf are summarized in the following key findings (Figure 2-3)

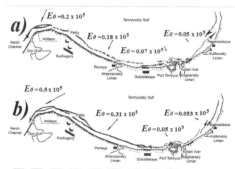

Figure 3. Trends of dynamics of coasts of Temryucsky Gulf, period 1976-2006: effective in December and storm for the year wind speed (>5 and 15 m/s, respectively). *1, 3, 4, 5, 6, 7, 8* and *9* refer to clauses *1, 3, 4, 5, 10, 11, 12* and *14* in Fig. 1; *2 – Эτ <0,1 t/s; a) E_n =(0.43-2.47) x 10^4, E_τ =(0.016-1.05) x 10^3; b) E_n =(0.45-4.05) x 10^4, E_τ =(0.004-2.2) x 10^3*

3.3 *Main regularities of dynamics of a coastal zone and rating of pollution*

The values of the components of the wave energy flux in the Gulf are comparable to those in the gulfs of inland seas (such as the Vistula gulf of the Baltic Sea) (Bogdanov & Vorontsov, Morozov 2004, Bogdanov 2009). Complex coastal circulation flows, sediment and impurities in shallow reservoirs are controlled mainly not by wind and waves acceleration distance, but by the variability of the depths to shallow coastal waters, and the configuration of the coastline.

Qualitative characteristics mean annual dynamics of the coastal zone of the Gulf are stable for the full range of effective velocities > 5 m/s, and storm winds (> 5 m/s). The differences lie in the quantitative characteristics of flows *Eo, En* and *Eτ* and whose values are from a few to tens of times lower than in the east (including Temryuk district) than in the western sector of the coastal arc. Reduction of head-loads to the NE helps to activate the removal of the water masses, suspended solids and pollutants deep into the dynamically attenuated Sea.

During the year, as in the most dynamically active December, at least three classic lithodynamic system, including the central areas of convergence and divergence Eτ flows on the flanks operate along the shores. Systems are located in areas (east - west): 1) village Perekopskaya - rerash Kulikovskii of Kurchansky liman, 2) spit Kurchansky Estuary - Seaside Kuban wellhead, 3) Glukhoy port channel - Chaykinskoe shallow water - art. Golubitskaya.

The unidirectional flow of matter and energy rushes to the west of the site of erosion of the coast in the stable localized divergence Eτ flow zone in Golubitskaya. In this area most of the year and alongshore currents carry pollutants. Some of them are coastal circulation vortices arising in capes, taken out to sea and back to the shore deflected stream En. When rounding cape Ahilleon real-energy flow splits at the offshore and longshore (inertial) branch - to the south (towards the spit Chushka) and NW (the area of the northern coast of the Kerch Peninsula), respectively.

In storm conditions the number of systems is reduced to two (supposedly Glukhoy channel port – Kulikovskoe delta arm and north of it, near village Perekopskaya). Unidirectional west flow originates at the port piers (Fig. 3).

The dynamic conditions in the port area determine holding most of the pollutants in the areas of counter alongshore migrations (in the area of village Golubitskaya – Perekopskaya). Removal of pollutants in the water area of the Gulf of discontinuous and the drain currents is possible in conditions of impaired frontal impact En flow in the coastal arc sector. In storm conditions spills of pollutants on the approaches to the port will be redistributed to the west and the delayed input longshore current in the throat of the Kerch Strait, and inertial branch - to the northern shores of the peninsula.

Possible oil spills and other pollutants in the "Temryuk" port area and in the Kuban river may cause contamination of the adjacent shores of the gulf, as well as the Kerch channel and the adjacent sector of the Azov Sea.

4 CONCLUSIONS

Results of the conducted researches can be generalized in the form of several provisions.

Along coast of Temryuk Bay three litodinamichesky systems comparable on a power engineering to the Vistula lagoon of the Baltic Sea are diagnosed. The most potent of them leads to catastrophic abrasion of coast of the western sector of the gulf and stretches to an entrance to the Kerch Strait. The weakened dynamics of a coastal zone near the port of Temryuk promotes stagnation of pollution. Technogenic catastrophes here, especially during the periods of storm, are dangerous by carrying out of pollution, both to the Sea of Azov, and to the Kerch Strait of the Black Sea. Morfodinamichesky zoning – a fundamental basis of actions for decrease in environmental risk on the sea coasts.

REFERENCES

Bogdanov, N.A., Sovershaev, V.A., Zhindarev, L.A., Agapov A.P. 1989. The evolution of ideas about the dynamics of the south-eastern shores of the Baltic Sea. *Geomorphology*, 2: 62-68.

Bogdanov, N.A., Vorontsov, A.A., Morozov, L.N. 2004. Trends of chemical pollution and the dynamics of the Vistula Lagoon. *Water resources*. 5(31): 576-590.

Bogdanov, N.A. 2009. Ecological lithodynamic analysis of the impact of development of the coastal areas: South-East Baltic. *Essays on the geomorphology urbosphere* / hole. Ed. EA Likhachev, DA Timofeev: 217-244. Moscow: Media-Press.

Boldyrev, V.L., Nevessky, E.N. 1961. West Temryuk flow of sand drifts. *Tr. OK USSR. 8*: 45-59. Moscow: Publishing House of the USSR.

Flerov, A.F. 1931. Sandy landscape of the Azov-Black Sea coast of the Caucasus, their origin and development. *Matherialy Gosudarstvennogo geograficheskogo obshchestva*, 63(1):

Guidance on research methods and calculations of sediment transport and dynamics of the coasts in engineering prospecting. 1975. Moscow: Gidrometeoizdat.

Kasyan, R.D., Krylenko, M.V. 2007. A comprehensive description of the current state of the Azov Sea coast of Krasnodar Region within Ecosystem study of Azov, Black and Caspian seas and their coasts. *TS IX.:* 315. Apatity: Publishing house of the KSC RAS.

About a state and about environmental protection of the Russian Federation in 2015. State report, Ministry of Natural Resources and Environmental Protection of the Russian Federation, 2015. Moscow: http://www.mnr.gov.ru/regulatory/list.php?part=1101

Pateev, M.R. 2009. Interphase and cross-border transfer of heavy metals in coastal and estuarial zones of the southern seas of Russia. *Abstract of the thesis ... the candidate the geogricheskikh of sciences.* Moscow: GOIN of Zubov.

Porotov, A.V., Kaplin, P.A., Myslivets, V.I. 2016. Development of accumulative coast of the northeast coast of the Black Sea in the late Holocene. Theory and methods of the modern geomorphology. *Materials of the XXXV Plenum of the Geomorphological commission of RAS, Simferopol, on October 3-8, 2016. Simferopol*, 2: 282-286.

Popov, B.A., Sovershaev, V.A. 1981. Principles for selecting input data for the calculation of wave energy flows. *The coastal sea area*: 47-52. Moscow: Nauka.

Simonov, A.I. 1958. *Hydrology of mouth area of Kuban.* Moscow: Gidrometeoizdat.

http://meteorf.ru
www.tamaninfo.ru

A New Vision to Monitoring Tank Cleaning

M. Panaitescu & F.V. Panaitescu
Maritime University, Constanta, Romania

V.A. Panaitescu
Thome Shipping, Constanta, Romania

L. Martes
Romanian Seafarers Center, Constanta, Romania

ABSTRACT: Tank cleaning involves many factors which must be considered in order to obtain good results. The new advanced technologies in tank cleaning have raised the standards in marine areas. There are many ways to realise optimal cleaning efficiency for different tanks. It is important to define the cleaning criteria, to plan the cleaning operations, to select the right methods, to select the operating parameters and the time and to choose cleaning agents which are available for most application problems. After these steps, must be verify the results and ensure that the best cleaning values can be achieved in terms of accuracy and reliability. In this paper are presented the steps of a new vision to monitoring tank cleaning according to MARPOL 73'78 (Annex I and II): monitoring the processes-cargo tank visual inspections, tank cleaning hardware, tank cleaning chemicals (MEPC 2 Annex 10), advanced devices and methods to evaluation the contaminated samples from walls of tanks. The efficiency of cleaning technologies depending on companies and their quality requirements those standards might deviate from the usual cleanliness requirement.

1 INTRODUCTION

1.1 Cargo chemical properties

The ability of a vessel to execute a successful tank cleaning operation is what makes that vessel better than its competitor. If the tank cleaning is successful, there is no waste. If the tank cleaning is not successful, everyoane feels the impact: time, manpower, cleaning chemicals, bunkers, equipment, off specification cargo and loss of earings. The basics of any tank cleaning procedure are tipically the same. There are technical publically tank cleaning guidelines, like: MIRACLE (Supplied and produced by Chemtec in Hamburg); Dr. Verway (CHEMTEC, Hamburg); MILBROS (Q88 Stamford, USA)(Sørensen O. et al.1959). All of these published guidelines basically provide the same informations in order to avoid the appearance of technological risks (Jensen B.B. et al. 2011-2012; Panaitescu F.V. et al. 2014).

One of the most important aspects of choosing the correct tank cleaning procedure is to recognise and understand the chemical properties of the previous cargo: volatility, solubility in water, viscosity, colour, drying cargo, polimerisable cargo, strong absorber. The most important aspects of understanding the chemical properties of the cargoes that are being cleaned from: volatile cargoes will tend to cvaporate, but may bc rctaincd in organic coatings; non-volatile cargoes tend to be persistent, but are not retained in organic coatings; water soluble cargoes do not need cleaning chemicals to remove them, because they are completely removed just using water; viscous cargoes will usually need warmer washing water to make them easier to remove; coloured cargoes/residues will probably need a colour remover-bleach; drying oils and polymerisable cargoes will always need ambient temperature water washing; strong absorbers are a challenge. Some additional factors strongly influence the successful outcome of any tank cleaning operation: a) outside climate conditions; b) the monitoring cleaning process; c) the pre-loading inspection specifications for the next cargo. There are three levels of pre-loading inspection: load on top (LOT); visually clean/water white standard; chemically clean/wall wash standard.

1.2 Cleaning plan

There are various parameter that contribute to effective tank cleaning (Sørensen O. et al. 1959; Sinner, H. 1959). The content of tank cleaning plan is based on the cleaning processes model: MARPOL wash (Annex I, Annex II high viscosity) or solvent;

cold/hot wash to continue; wash with chemicals; rinse (when chemical are used); repeat chemical wash/manual cleaning; steaming/fresh water; manual cleaning (CSM-L&I-IMEC.2016).

Tank cleaning operations are optimised when all cleaning steps are monitoring. After these steps, must be verify the results and ensure that the best cleaning values can be achieved in terms of accuracy and reliability.

2 MONITORING DATA

2.1 *Inspection*

Inspection of the cargo tanks is very important because it is the best indicator about whether the load port inspection will be successful or not (CSM-L&I-IMEC, 2016). If the cargo tanks and the deck look and smell clean and are well organised, the load port inspector will have a better feeling about vessel and the final results will be favorable.

Inspection of the cargo tanks includes: a) tank hatch (hatch gaskets, odour when the hatch is first opened, overall appearance of the hatch, the conditions of coating, signs of corrosion); b) deck head (sign of condensation, cargo residues, visible salt crystals, coating breakdown); c) access ladder or platforms; d) bulkheads (check for evidence of previous cargo, discoloured patches, corrosion, coating breakdown/condition of the steel, other surface debrs, tank cleaning machine "shadow" areas); e) cargo lines (first foot samples are critical, because they are the first measure of how clean the cargo lines are; always clean them in the same way that the cargo tanks are cleaned cold water/hot water/cleaning chemicals, steaming lines cleaning free from inorganic chlorides; this step includes also flexi cargo hoses which are used during cargo operations; after cleaning and prior to loading must verified that all drain valves are kept open; at the end of cleaning, the lines must be blown back to the tanks).

2.2 *Cleaning hardware*

It is important to optimise the tank cleaning process to ensure repeatable tank cleaning performance in the shortest possible amount of time (Jensen, B.B.B. et al. 2011-2012).

For this, tank cleaning process must be automated. Process control depends upon reliable real-time in-line measurements using electronic sensors to monitor and verify the performance of tank cleaning systems. It is important to choosing the right system to monitor and control tank cleaning automated system and to define the objectives for monitoring and control. These help to understand the available options and advantages (cleaning

consistency, reduced labour costs and increased production time, less downtime, higher energy savings and reduced water and cleaning fluid consumption).

2.3 *Cleaning chemicals*

The materials which can used to tank cleaning chemicals must be approved by IMO (International Maritime Organization), MEPC 2, Annex 10. Cleaning chemicals depend on remove different previous cargoes.

The most effective chemicals for removing traces of cargo residues that are insoluble in water are called detergents. Typically detergents are long chain molecules with a polar head and a non-polar tail (e.g. Linear Dodecyl Benzene Sulphonic Acid).

Generally for surface cleaning, there is a huge choce of products available, and they ar not all the same.

In the following picture are UV scans of methanol wall wash samples taken after the cleaning of zinc silicate coated test panels that were immersed in ultra-low sulphur diesel and cleaned with freshwater at 70^0 C for 2 hours with four different cleaning chemicals (product *a* to product *d*) (Figure 1). In this particular case, product *a* is the best cleaner.

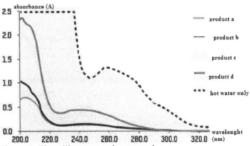

Figure 1. Zinc silicate coated test panels

For hydrocarbon removal/general tank cleaning or for zinc silicate coatings can used neutral chemicals with pH= 5.... 9 (e.g. Accell Clean, Ecosolut 24, Teepol). The chemicals with pH more 9 will be diluted to less pH (e.g. Marclean, Alkaclean safety, with sodium metasilicate, which is alkaline). Another tank cleaning chemicals are caustic based, with pH=13...14 and used for the removal of vegetable oils. The process is called saponification. Caustic based cleaners are also widely used for the removal of flue gas and inert gas generator residue. The cleaning chemicals solvent based contain strong solvents and are tipically used for removing larger volumes of oils based residues by emulsification (e.g. Seaclean, Coldwash HD, HD split, Super degreaser). Another cleaning chemicals are "environmentally friendly", or acid based (e.g.

Caretank Eco, EcoClean, GratoSol, or with citric acid, phosphoric acid).

2.4 *Methods of research*

There are three clasical methods of applying cleaning chemicals to the cargo tanks and a new vision for monitoring and test analysis of samples- VISIBLE and ULTRA-VIOLET spectroscopy (UV-VIS) (table 1). The clasical cleaning methods are: re-circulation (figure 2), injection (figure 3), direct spraying (CSM-L&I-IMEC, 2016). The important objectives for cleaning by recirculation are: the process can't stopped (once the cleaning has started, it is very straightforward to handle); work easy facilities (easy calculation for cleaning chemicals depending on work solution); easy to maintain heat during recirculation (unless the vessel is equipped with heating coils or aqn extremely efficient heater system). The important objectives for cleaning by injection are: provides a constant supply of "fresh" cleaning solution; utilises maximum washing water temperatures; depending on the efficiency of injection pump; the concentration of the cleaning chemical mixture is not really confirmed (the clening chemical concentrate is pumped into a water moving volume); can use much higher volumes of cleaning chemicals concentrate; the contact time between cleasning chemical solution and cargo tank surface is much lower (30 ... 60 minutes). Direct spraying is primarily used for spot cleaning of areas that are in the shadow of the tank cleaning machines, unless the tank cleaning machines are damaged or broken.

For monitoring and test analysis of samples in the VSIBLE region of the light spectrum, regular glass sample cells should be used. They are marked with "G'letter. The apparatus is spectrometer. For monitoring and test analysis of samples in the ULTRA-VIOLET region of the light spectrum, only sample cells made from quartz glass can be used. They are marked with "Q"letter. It is important to use the correct sample cell, otherwise the data generated might not be valid.

Tank cleaning result tests are: Permanganate Time Test-PTT (is based on the ability of potassium permanganate, $KMnO_4$, to oxidise hydrocarbon impurities that could be present in the wall wash liquid; if PTT is a reaction in a neutral solution, the value of $KMnO_4$ is small and changes its colour from pink-orange to yellow-orange)(MARPOL); Hydrocarbons Test (water miscibility)(is the qualitative detection of non-water-soluble contaminants; Chloride Test (is used the judge the presence of chlorides on bulkheads; Chloride levels vary from 0.1 ppm to 5 ppm); Colour Test (APHA); UV-Test (is used to identify hydrocarbons and chemicals); The Acid Wash Test (is used to determine the presence of Benzene, Toluene,

Xylenes, refined solvent Naphthas, similar industrial aromatic hydrocarbons, impurities in methanol); NVM (Non Volatile Matters) Test (is used to determine if there are non-volatile impurities on the tank surface by weights).

Table 1. VISIBLE and ULTRA-VIOLET spectroscopy.

Tank cleaning Test	VIS	UV
Absorbance	Units	wavelenght (nm), Absorbance (A)
1. Inorganic chlorides	ppm***	Cargo quality for potable ethanol /methanol
2. PTT*	%	SHELL CHEMICAL Analyse (alcohols, ketone)
3. APHA colour	APHA	routine wall wash inspection prior to loading potable ethanol and HMD*****
4. Hydrocarbons	FTU***	water insoluble hydrocarbons

* PTT = permanganate time test
** APHA = colour test
*** ppm = parts per miliion inorganic chlorides
*** FTU = formazine turbidity unit
*****HMD = Hexamethylene diamine

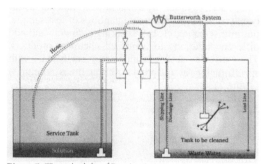

Figure 2. The principle of Re-circulation.

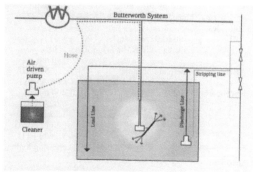

Figure 3. The principle of injection.

The test procedure for analysis may vary slightly from place to place, but the principle is always the same. In this case study are used for each test different samples where are mixed with different solution for relevant presence of residues (on chlorides, silver nitrate solution, on PTT, potassium permanganate solution, the hydrocarbons test contains traces of hydrocarbon based cargoes that are soluble in the methanol or acetone).

In the chemical tanker area, UV spectroscopy is a more widely accepted technique for analysing the quality of loaded cargo. When a UV light source is passed through a test sample, the different chemical groups that make up the sample, absorb different amounts of the UV light (absorbance) (CSM-L&I-IMEC, 2016). Each chemical substance has its own unique UV fingerprint and this permit to identify the presence of these chemical substances in the test samples.

The most commonly chemicals are: methanol, benzene, styrene monomer, phenol, fatty acid methyl ester-fame, pyrolysis gasoline, acetone, multiple hydrocarbons, unknows.

3 EXPERIMENTAL DATA AND RESULTS

3.1 *Visible spectroscopy tests*

Visible spectroscopy tests are: inorganic chlorides, PPT, APHA colour, hydrocarbons. In this paper, for each test are used different quantities of test sample, put into separate, clean measuring cylinders and add different solutions to identified chemical products. The effects and results are analysed with spectrometer, at different reaction times (minutes) (Table 2).

Table 2. Visible spectroscopy tests

Test	Add solution	Time	Results
	ml/%/g	minutes	nm, A, ppm, %, FTU
1. Inorganic	25 (5 drops silver nitrate (2%/5%)+ 5 drops 20% nitric acid)	15	420 nm 0.099 A* 1.0 ppm
chlorides			
2. PPT	500 (0.02% KMnO₄ + 0.1g solid KMnO4)	< 20 > 50	< 32% > 32% 530 nm
3. APHA colour	methanol/acetone		5...15 APHA (low levels)
4. Hydrocarbons	(solvent+water) 100	15... 20	400 nm 0.025 A 1...2 FTU (slight hydr.) 3...600 FTU (increasing hydr.)

* Absorbance

1 For visible spectroscopy
   – at inorganic chlorides test- after 15 minutes out of direct sunlight, any presence of a white colour indicates the presence of inorganic

chlorides in the test samples and using spectrometer, can be quantified these;
– at PPT , after 10 minutes visually check the appearance of each sample (typically is 50 ... 60 minutes); after 50 minutes, if was found > 32%, the result should be repored as PTT is greater than 50 minutes; if the test sample was analysed after 20 minutes and the was found < 32%, the result should be repored as PTT is less than 20 minutes. In case study, the spectrometer reading to how the colour of a methanol wall wash fades, regarding at 32% - still contains a trace of pink and regarding at 21 % - is closest to the recommended colour. Without the spectrometer, the method of determining of PTT is to prepare the colour standard described by requirements.
– at colour test, the result is in accordance with requirements in units APHA;
– at hydrocarbons test results (Table 2), can directly reading on spectrometer, in units of turbidity FTU, if it is a slight presence of hydrocarbons, or it indicates an increasing presence of hydrocarbons; in this case study, be accepted as "pass".

3.2 *UV spectroscopy tests*

The typical "aromatic"activity was tested for chemicals: methanol, benzene, styrene, phenol (figure 4, figure 5, figure 6, figure 7) with spectometer. One important aspect of UV spectroscopy is the concentration of the identified groups (i.g. to be able to quantify how much benzene is present in the sample). For this, was analysed how 10 ppm of benzene, styrene and phenol in methanol not only give different shaped peaks, but also different sized peaks as well (figure 8).

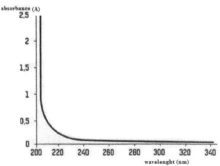

Figure 4. Pure methanol.

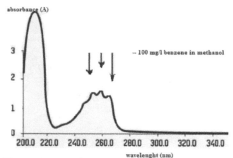

Figure 5. Benzene in methanol.

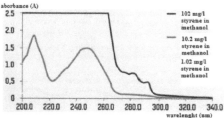

Figure 6. Styrene in methanol.

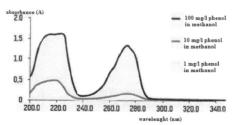

Figure 7. Phenol in methanol.

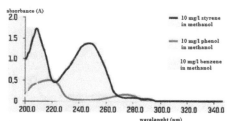

Figure 8. The concentration of the identified groups.

The grafic interpretations have shown that UV spectroscopy is very important to identify chemical groups and to determine concentartions of these groups (CSM-L&I-IMEC, 2016).

For UV spectroscopy
– for benzene in methanol there are two areas with strong activity between 200...220 nm (benzene is a pure hydrocarbon) and 240 ...260

nm (three small peaks with pure benzene and the indicative of all Aromatics);
– for styrene in methanol there is strong UV activityin two specific regions at 210 nm (hydrocarbon, ethylene group) and between 240...260 nm (Aromatics);
– for phenol in methanol there is Aromatic activity around 240...260 nm, but in this case study, it has been shifted to higher wavelenght areas, 260 .. 280 nm (because the OH group from formula of phenol is directly attached to the benzene ring (Sørensen et al. 2016).

4 CONCLUSIONS

In general, the tank cleaning monitoring procedures depend on the cargoes properties to be cleaned, on the surrounding conditions, on the available equipment and last but not least the requirements of the products to be loaded. A variety of chemical cleaning agents is available for most application problems. Cleaning agents must be IMO approved. After analise of experimental data, can confirm that:
– cargo tanks and/or washing water are free from aromatics prior loading, which provides any extra level of security that conventional pre-loading inspection specifications do not provide;
– there are more UV traces of chemical cargoes (fatty acid methyl ester-fame, pyrolysys gasoline-pygas, acetone, multiple hydrocarbons, unknows).

In literature are published more informations about additional poits for UV tests (NVM-Non-volatile- matter, AWC-acid wash colour test) and washing water analysis (http://www. alfalaval. com. 2016).

REFERENCES

CSM-L&I-IMEC, 2016. Chemical tanker training. In *Student guide-tank cleaning*, Constanta:pp. 30-58.
http://www. alfalaval. com. 2016. Tank cleaning -planning. In *Th.Ah. Tank cleaning-guide*: pp. 2-7.
Jensen, B.B. et al. 2011-2012. Tank Cleaning Technology: Innovative application to improve clean-in-place (CIP). In *EHEDG* 2012: pp. 26-30.
Panaitescu, F.V., Panaitescu, M., Voicu, I.&Panaitescu, I.I. 2014. Training for environmental risks in the Black Sea Basin. *TransNav, the International Journal on Marine Navigation and Safety of Sea Transportation*, vol.8, no.2 2014.
Sinner, H. 1959. The Sinner Circle "TACT". In *Sinner's Cleaning Philosophy,* Henkel.
Sørensen, O. & Andersen, J. Acc. 2016. Controlling hygienic tank cleaning using hygienic sensors to monitor tank cleaning. In *Alfa Laval Magazine-Thinking Ahead* :pp. 5-7.

Oil Spill Modelling with Pisces II Around Bay of Izmir

A.C. Töz, B. Koseoglu & C. Sakar
Dokuz Eylul University, Izmir, Turkey

ABSTRACT: The purpose of this study is to investigate the oil spill and predict the future accidents likely to be encountered around the Bay of Izmir. To get well informed about this fate this study makes the best possible use of trajectory model. PISCES II (Potential Incident Simulation, Control and Evaluation System) has been conducted to calculate weathering processes. Hence, in order to identify the risky areas, two scenarios have been developed. The results gained through these efforts are hoped to be useful for many organizations dealing with oil spill response operations and contribute to an effective and efficient coordination among the relevant institutions.

1 INTRODUCTION

When oil is spilled into the sea it undergoes a number of physical and chemical changes (DNV, 2011: 6), some of which lead to its removal from the sea surface, while others cause it to persist (CEDRE, 2005). The fate of spilled oil in the marine environment depends upon factors such as quantity spilled, the oil's initial physical and chemical characteristics, the prevailing climatic and sea conditions and whether the oil remains at sea or is washed ashore (ITOPF 2002).

Each process is affected by environmental changes in different period of time. As soon as oil is released into the environment, it undergoes significant property changes (Küçükyıldız, 2014: 22). For example, oil begins to spread as soon as it is spilled but it does not spread uniformly. Any shear in the surface current will cause stretching and even a slight wind will cause a thickening of the slick in the downwind direction (Lehr et.al, 2002; 192).

2 MATERIALS AND METHOD

In this section, the determination of main and intermediary processes in the light of the effects to which ship-originated oil pollution is exposed depending on the environmental variables within time and the changes on it has been aimed. The time-dependent change analysis has been aimed depending on the result outputs of these processes.

Within this scope, study subject, objective and stages, the method used in the study, the sample of the study, data collection tool, data collection, analysis and findings will be examined respectively.

2.1 *The objective of the research*

The effectiveness and success of response to marine pollution depend on the accuracy and effectiveness of operational and strategical "desk-based" decisions. The process estimation of the pollution which will be caused by the product spilt in the sea is directly associated with accurate data sources and reliable decision support members. As also stated in the first section of the study, the product spilt in the sea begins to damage nature by undergoing physical and chemical changes without losing any time. At this stage, the most significant problem which response teams encounter is the loss of time. Minimizing time loss is through simulating these processes. In this context, the basic objective of this study is to determine risky areas by revealing the pollution, possible to occur at the Bay of Izmir, which may be caused by ship-originated oil and its derivative products and of which effects will be exposed within time.

2.2 *Study Site*

Izmir, the third biggest city of Turkey, is a contemporary seaport integrated with commerce and industry. Its city center is located at the head of a long and narrow bay. The Bay of Izmir is one of the biggest natural bays of the Mediterranean with the total area of 200 km² and water capacity of 11.5 billion m³. Izmir is the biggest settlement area in the vicinity of the Bay with the area of approximately 88000 hectares (TCDD, 2013: 1). An aerial photograph of the Bay of Izmir is presented in Figure 1.

Figure 1. Aerial photograph of bay of Izmir

As understood from the photograph, the settlement of Izmir is concentrated around the bay. In this respect, the bay bears the characteristics of a waterway which runs into the inner parts of the city. When the settlement of Izmir Alsancak Port is taken into consideration, the ships which will transport to this port need to enter the inner parts of the bay. The most risky area of collision is selected for research. The study site is shown in Figure 2.

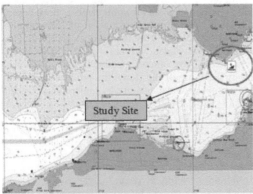

Figure 2. Study site

Some areas which can be regarded as risky especially in terms of ship transitions exist in the Bay of Izmir. Especially since the access point to the Port of Alsancak is considered to be risky especially in terms of collision. The ships which enter the bay also pose a risk in terms of other vessels in the bay.

Furthermore, the ships waiting at anchor in the bay render this narrow waterway even narrower for transition and the area riskier in terms of accidents.

2.3 Research Method

The method of simulation was used as a method in the study. Simulation is a modelling technique which ensures monitoring behaviors of the real system in the computer model under different conditions by transferring the cause-effect relations in the system into the computer (Murphy and Perera, 2001).

The simulation technique is a methodology used in the solution of the problems rather than a theory. The approach of the simulation technique to problems differs according to the structure of the system and the model to be established based on this structure (Öztürk, 2004). Similarly, oil pollution simulation models are used by estimating the areas in which the dispersion will be the most effective, in other words, estimating the winds which will direct the dispersion to the shore or nearby facilities in the most effective way. In these studies, taking only the dominant direction into consideration is not essentially sufficient in terms of the risk assessment.

2.4 Data Collection

While the type and amount of the product to be the subject of modelling are taken into consideration, shipping in the Bay of Izmir, the type and amount of cargo handled in Izmir Alsancak Port, the environmental conditions of the Bay of Izmir and meteorological data are taken into account. Within the meteorological data, the values of current, wind, sea water temperature and salinity measured as of the date of modelling in the bay are taken into consideration.

When the winds prevailing in the Bay of Izmir are taken into consideration, it is seen that the average wind velocity in the Bay of Izmir is 3.0 m/sec. In general, the surface currents are from inside the bay towards outside and they are more violent in the north of the bay. The surface currents are around 5-7 cm/sec and they can rise to the levels of 10-15 cm/sec under the effect of the wind. The currents at 10 m depth from the average water level are towards the interior of the bay and enter the cove offshore Yenikale.

Another environmental variable is air and sea water temperatures. The months in which sea water temperature, which is annually 18.5 °C on average, is the lowest are January (11 °C) and February (10.7 °C); the months in which it is the highest are July and August (26.2 °C). Sea water temperature is extremely important in terms of the viscosity of the product spilt in the sea. Depending on the changing viscosity values, the effects to which the spilt product will be exposed in the environment also alter. February water density of the Bay of Izmir is between 28.7 kg/m³ and 29.2 kg/m³. Density values in April are measured to be between 27.5 kg/m³ and 29.15 kg/m³ on the surface. Water densities in July vary between 25.4 kg/m³ and 28.6 kg/m³.

2.5 Modelling Studies

When the wind and current conditions in the region where accidents occur in the Bay of Izmir are taken into consideration, it is seen that the conditions that will affect the pollution dispersion may arise. Wind force and direction affect the dispersion depending on the amount spilt. In the simulations carried out, the seasonal prevalent wind directions, surface currents, sea water temperatures and densities have been taken into account for the most critical situation.

In accordance with this general approach, while wind directions are selected for ship maneuvering simulations, providing that it exists in the meteorological data obtained, the directions at which the highest wind velocities have been observed, if it does not, prevailing wind directions determined in accordance with the prevalence of the average wind velocities are taken into consideration. Furthermore, by taking the facility layout and, if any, the location of neighboring terminals, the condition of the coastline and bathymetry into consideration, critical courses for manoeuvring are determined so that the wind directions to be used in manoeuvring simulations are determined. Although they are not observed as frequently as the prevailing wind direction of the region, the wind directions critical for manoeuvring can be also figured out.

Similarly, oil spill models are used by estimating the areas in which the dispersion will be the most effective, in other words, estimating the winds which will direct the dispersion to the shore or nearby facilities in the most effective way. In these studies, taking only the dominant direction into consideration is not essentially sufficient in terms of the risk assessment. The related details of scenarios are shown in Table 1.

Table 1. Details of scenarios

Scenario	Pollutant	Reason to Spill	Season	Quantity (m³)
1	FUEL OIL	Collision	Summer	500
2		Collision	Winter	500
3	DIESEL OIL	Collision	Summer	1000
4		Collision	Winter	1000

PISCES II developed by TRANSAS and used in modelling studies is software which calculates weathering processes depending on time. It was first created by transferring the base of the study area

(coastline, piers and other coastal structures) into the program. According to the amount of fuel carried in tankers, how long it will continue to leak into the sea as a result of an accident was entered into the program. The simulation period for simulations was taken as 24 hours. Grids at suitable precision were created in the study area. According to these grids created and current velocities and directions, the current files were prepared. The prevailing wind velocities and directions were entered. The figures of related scenarios are shown below.

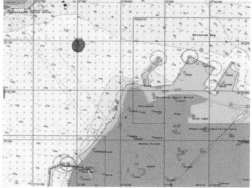
Figure 3 (a): Scenario 1 spillage

The area which 500 m³ of Fuel Oil spilt cover at the beginning is displayed in the Figure 3(a). According to this, the product spilt forms a spill film 17.9 mm in thickness. The total area that this spill covers is calculated to be 0.03 km²

As understood from the figure 3(a), the product spilt spreads on a wide area on the map. However, as much as the point at which the spill has occurred, the places on which it will have an impact and the risks that it will create also bear significance. The motion of spill after 60 minutes is shown in Figure 3(b).

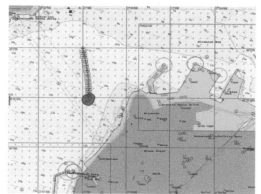
Figure 3 (b): Scenario 1/60 minutes after spillage

As understood from the Figure 3(b), the spill route which moves towards the west of the port deviates towards Alsancak pier depending on the deflection of the current. At this stage, it stands out that the variable of current is more effective when compared to the variable of wind. The reaching point of of spill to the coast is shown in Figure 3(c).

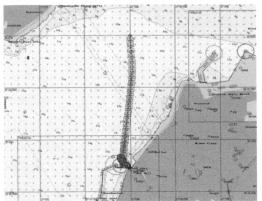

Figure 3 (c): Scenario 1/Contact with the shore

500 m³ of Fuel Oil initially spilt reach the shore after 2 hours 27 min depending on the environmental conditions in the region. As understood from the figure, the point at which the contact with the land is established is the recreational area known as the 2nd Waterfront. The thickness of the product reaches 55.8 mm and it creates 600 m³ of an oily water emulsion. At this stage, a volume increase of approximately 100 m³ is observed between the product spilt and the amount washed up onto the shore. The simulation result of spillage of Fuel Oil in December is shown in Figure 4 (a).

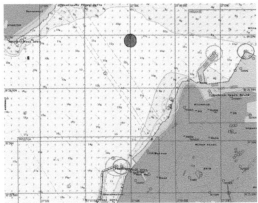

Figure 4 (a): Scenario 2/spillage

As understood from the figure, the product spilt spreads on a wide area on the map. However, as much as the point at which the spill has occurred, the places on which it will have an impact and the risks that it will create also bear significance. In addition to this region's being an important area in terms of sea traffic, it is also extremely significant in terms of the living spaces which have a coast to the Cove. The motion of spill after 60 minutes is shown in Figure 4(b).

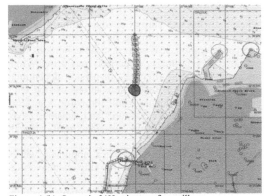

Figure 4 (b): Scenario 2/60 minutes after spillage

The change of the initial place of 500 m³ of Fuel Oil split after 60 min is shown in the figure. According to this, the thickness of the product spilt reaches 28.6 mm and the product spreads on an area of 0.033 km². In this process, 0.01% vaporization is observed in the product which creates 543 m³ of an oily water emulsion. . The reaching point of of spill to the coast is shown in Figure 4(c).

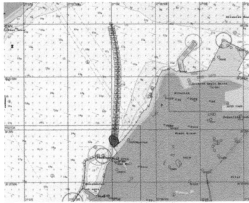

Figure 4 (c): Scenario 2/contact with the shore

500 m³ of Fuel Oil initially spilt reach the shore after 2 hours 41 min by moving under the conditions in the region. As understood from the figure, the point at which the contact with the land is established is the recreational area known as the 2nd Waterfront. At this stage, the thickness of the product reaches 63.22 mm and the amount washing

up onto the shore is 599 m³ of an oily water emulsion.

The simulation result of spillage of Fuel Oil in July is shown in Figure 5 (a). The initial area which 1000 m³ of Diesel Oil spilt cover is displayed in the figure. According to this, the product spreads on an area of 0.10 km² with a film thickness of 14.0 mm.

Figure 5 (a): Scenario 3 spillage

As understood from the figure, when 1000 m³ of Diesel Oil spills in the sea, it spreads on a much wider area compared to the Fuel Oil spill. Moreover, although the amount spilt is much more when compared to the Fuel Oil spill, the spill thickness turns out to be much lower. This situation is closely related to the density of the product, surface tension and molecular bond. The motion of spill after 60 minutes is shown in Figure 5(b).

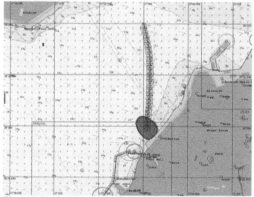

Figure 5 (b): Scenario 3/60 minutes after spillage

As understood from the figure, the spill which moves towards the west of the port deviates towards Alsancak pier depending on the deflection of the current. This movement indicates that the current force is more dominant when compared to the wind force. The reaching point of of spill to the coast is shown in Figure 5(c).

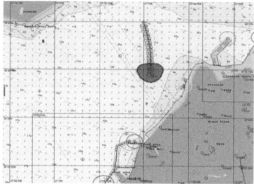

Figure 5 (c): Scenario 3/Contact with the shore

1000 m³ of Diesel Oil initially spilt reach the shore after 2 hours 9 min moving under the regional conditions. As understood from the figure, the point at which the contact with the land is established is the recreational area known as the 2nd Waterfront. The thickness of the product reaches 28.2 mm and it forms 1221 m³ of an oily water emulsion.

As understood from the figure, the first contact with the spill is the eastern coast of Alsancak ferry port. The coastal structure in this region is generally stony. Due to the coastal structure of this characteristic, this region has a low contamination coefficient.

The initial area Figure 6(a) which 1000 m³ of Diesel Oil spilt cover is displayed in the figure. According to this, the product spilt forms a spill film 14.2 mm in thickness. The total area that this spill covers is calculated to be 0.10 km²

As understood from the figure, when 1000 m³ of Diesel Oil are spilt in the sea, they spread on a much wider area compared to the Fuel Oil spill. Moreover, although the amount spilt is much more when compared to the Fuel Oil spill, the spill thickness turns out to be much lower. This situation is closely related to the density of the product, surface tension and molecular bond.

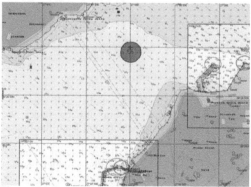

Figure 6 (a): Scenario 4 spillage

The change of the initial place of 1000 m³ of Diesel Oil split after 60 min is shown in the figure. According to this, the film thickness of the product spilt reaches 17.5 mm and the product spreads on an area of 0.1 km². In this process, 1106 m³ of oily water emulsion is formed and 0.04% vaporization and 0.01% precipitation are observed on the product.

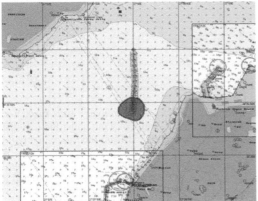

Figure 6 (b): Scenario 4/60 minutes after spillage

As understood from the figure, the spill which moves towards the west of the port deviates towards Alsancak pier depending on the deflection of the current. At this stage, it stands out that the variable of current is more effective compared to the variable of wind. The reaching point of of spill to the coast is shown in Figure 6(c).

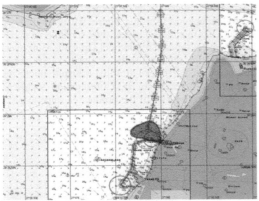

Figure 6 (c): Scenario 4/Contact with the shore

1000 m³ of Diesel Oil initially spilt reach the shore after 2 hours 28 min moving under the regional conditions. As understood from the figure, the point at which the contact with the land is established is the recreational area known as the 2nd Waterfront. The thickness of the product reaches 25.9 mm and it forms 1226 m³ of oily water emulsion.

3 CONCLUSION AND DISCUSSION

Since the specific weight of oil and its derivative products is less when compared to the sea water, in case they spill in the water, they tend to float. Since crude oil products extracted from very few reserves in the world are denser than water, in case they spill in the water, they tend to submerge.

In the study carried out, the products regarded to be risky as pollutants were chosen as Marine Fuel Oil and Marine Diesel Oil known as a bunker fuel. When both pollutants spill in the sea, they initially float and tend to disperse. The spill points were selected based on the data of the accident which occurred in the Bay of Izmir. When the seasonal conditions in which the accidents occurred were considered, modelling studies were completed by entering the environmental variables into the software.

According to this, in accordance with the results of the numerical modelling technique developed by considering the accidents possible to occur in the Bay of Izmir, the quality of the product spilt is extremely significant. The products of which density is high, viscose, surface tension force and molecular bond forces are considerable such as Fuel Oil tend to grow slower in terms of field expansion. However, the products which are lighter such as Diesel Oil tend to vaporize and precipitate quickly since their floating duration is shorter. This situation is related to the weak molecular cohesive force, low viscosity and low density.

When the act of the change of place occurring depending on the spill of both products was examined, it was concluded that the environmental variables play an extremely important role. It is seen that especially the variables such as current, wind, sea water temperature and air temperature are extremely significant in terms of the horizontal dispersion of the products. It was concluded in this study that movement directions and velocities of the products spilt on the sea are closely related to these environmental variables. It especially stands out that the direction and force of the current have a more dominant role when compared to the direction and force of the wind and that the spill mostly moves under the influence of the current and that the wind has a much lower effect on this movement.

The impact of the temperature variable on the changes of the product spilt depending on seasonal conditions has been also verified with this study. According to this, while the duration in which the product spilt reaches the shore is longer in winter depending on the temperature of air and water, this duration is observed to be shorter in summer.

If the effects to which the products spilt are exposed are to be examined, the fact that the precipitation and vaporization rate of Diesel Oil product is higher as stated in the literature compared to Fuel Oil has also been supported in modelling studies. If the vaporization percentages are to be examined, it is observed that Diesel Oil vaporizes more due to its molecular cohesive force's being less when compared to Fuel Oil and the high vapor pressure at the end of two hours. If the dispersion (precipitation) rates are to be examined, the theory that dense products are more resistant to precipitation as stated in the literature has been also supported by this modelling. The fact that within the duration of Fuel Oil's modelling no precipitation is observed but that Diesel Oil enters in the process of precipitation after 30 min stand out as the most significant outputs.

The impact of the spill related to Fuel Oil will be more severe no matter on how much narrower area the pollution spreads because this product is an extremely difficult product to remove from nature due to its high viscosity and behaviors in the process of forming an emulsion with water. The biggest danger of Diesel Oil product is the tendency of high dispersion and precipitation, as understood from the modelling. Although it is not as hard as Fuel Oil to remove, this product causes fauna and flora to be permeated more quickly depending on wider areas being contaminated and disintegrated.

When the models of the spill occurring with the risk of collision offshore Izmir Alsancak Port are considered, it stands out that the southern coasts of the bay are at risk. In the light of these results, the recreation site known as the 2nd Waterfront region is considered to be at risk. The fact that this region is a recreation site places the environmental variables to be affected in a more fragile situation. In the light of the precautions to be taken in regard to the pollution that will occur in this region, the importance of intervention equipment's being deployed in a way that will not penetrate the coast is revealed. The coastline's being rocky in this region reduces the pollution that will penetrate the interior of the coast and the severity of the contamination that will occur on the coast. The fact that the region is at an easily accessible point from land and sea for coastal operations to be carried out is pleasing in terms of the precautions to be taken.

The pollutions which two different products will cause in a small-scale area in a short time were simulated in this study. It is suggested that the modelling of the pollutions in a wider area, in a longer period of time and of different product types will be beneficial in the following studies. In the modelling studies, the observation of the free movement of the product without any intervention was ensured. The opinion that the situation analysis with intervention to the incident by deploying marine, inland and aerial intervention will be beneficial in the following studies has been formed.

REFERENCES

Beşiktepe Ş.T., Sayın, E., İlhan, T. & Tokat, E. 2011. İzmir Körfezi Akıntı Dinamiğinin Model ve Gözlem Yardımıyla İncelenmesi, *7. Coastal Engineering Conference, Conference Proceedings.* Trabzon. Turkey.

CEDRE (Centre of Documentation, Research and Experimentation on Accidental Water Pollution). 2005. Black Tide. *Learning Guide*. Paris.

DNV (Det Norske Veritas). 2011. *Report for Australian Maritime Safety Authority, Model of Oil Dispersion*, DNV Project No: PP002916.

ITOPF (International Tanker Owners Federation). 2002. *Fate of Oil Spills*. The International Tanker Owner Pollution Federation Limited. London.

Küçükyıldız. C. 2014. Petrol Tankeri Kazalarının DeniÇevresine Etkileri ve Tazmin Sistemi, *Turkish Republic Ministry of Transport, Maritime and Communication. Master Thesis. Ankara. Turkey*

Lehr, W., Jones, R., Evans, M., Beatty, D.S. & Overstreet, R..2002. Revisions of the ADIOS Oil Spill Model. *Environmental Modelling & Software,* 2002 17(2): 189-197.

Murphy C. & Perera, T. 2001. *Role of Simulation in Industries: The Definition and Potential Role of Simulation within an Aerospace Company*, In Proceedings of the 33nd Conference on Winter Simulation. . IEEE Computer Society. 829-837.

Öztürk, B. & Oral, N. 2006. The Turkish Straits, Maritime Safety, Legal and Environmental Aspects. *Turkish Marine Research Foundation Publication.*25: 120-123.

Reed M., Johansen, I., Brandvik, J., Daling, P., Lewis, A., Fiocco, R., Mackay, D. & Prentki, R. 1999. Oil Spill Modeling Towards the Close of the 20th Century: Overview of the State of the Art. *Spill Science & Technology Bulletin* 5(1): 3-16.

TCDD (Türkiye Cumhuriyeti Devlet Demiryolları- Turkish Rebuplic State Railways). 2013. *İzmir Körfezi ve Limanı Rehabilitasyon Projesi Çevresel Etki Değerlendirmesi Raporu.* İzmir. Turkey.

Evaluating Air Emission Inventories and Indicators from Ferry Vessels at Ports

G. De Melo Rodríguez & J.C. Murcia-González
Polytechnic University of Catalonia, Barcelona, Spain

E. Martín-Alcalde & S. Saurí
Center for Innovation in Transport (CENIT), Barcelona, Spain

ABSTRACT: This paper provides an estimation of exhaust emissions released by passenger ferries at the port-level. The methodology is based on the "full bottom-up" approach and starts by evaluating the fuel consumed by marine diesel engines on the basis of its individual port-activities (manoeuvring, berthing and hoteling). Specific air emissions are estimated as regards to the different propulsion sources: main engine, auxiliary engines, thrusters and boilers. The Port of Barcelona was selected as the site at which to perform the analysis, in which 25 passenger ferries operating in the Mediterranean Sea were monitored. Real-time data from the Automatic Identification System (AIS), factor emissions from engine certificates, engine loads during port time, and vessel characteristics from IHS-Sea web database were also collected for the analysis. The research findings will improve our understanding of local pollution at port-cities and help to derive appropriate measures to improve air quality.

1 INTRODUCTION

In the context of port-city areas, emissions released into the atmosphere by vessels operating in port negatively affect local communities (Dalsoren et al., 2009) since it has a significant environmental impact on the coastal communities, as 70% of the ship emissions occur within 400 km of land (Eyring et al., 2005).

Moreover, the urban character of some ports and their populated surroundings are the main focus of the negative effects of exhaust pollutants due to the associated local impacts on human health. Thus, the need to control air pollution at ports is widely acknowledged as an active policy issue by various authoritative port associations (IAPH, 2007; ESPO, 2003) as a reaction of main regulations (IMO, EC, EPA, etc.).

A fundamental requirement for emission control, assessing the impacts of growing shipping activity and planning mitigation strategies is developing accurate emission inventories for ports (ICF, 2006). These port emission inventories would aid policy makers in developing effective regulatory requirements or port environmental management systems Tzannatos (2010).

In such a context, the goal of this paper is to develop accurate emission inventories (CO_2, SO_x, NO_x and PM) and emission indicators for ports where passenger ferries are dominant by estimating the fuel consumed by each vessel on the basis of its activities in port.

The paper is organised as follows: Section 2 reviews relevant literature on the issue; Section 3 introduces the methodological approach and the formula used to estimate inventory emissions; Section 4 presents the case study results and the most relevant emission inventories; and finally, Section 5 highlights the main conclusions.

2 LITERATURE REVIEW

According to published research, which incorporates extensive reviews of ship emission estimation methodologies (Miola et al., 2010; Tichavska and Tovar, 2015), a wide variety of studies relate to emission inventories at global or regional levels but only a few do so at the port-level.

The representative approach for emission estimation in port studies was the bottom-up approach, based on port calls and estimated vessels operating at a port (Tichavska and Tovar, 2015). Furthermore, normally activity-based and/or fuel-based estimations were made because of they are more accurate than top-down methodologies that require detailed data such as routing, engine workload, ship speed, location, duration, etc. (Song, 2014).

For example, Goldsworthy and Goldsworthy (2015) have produce a model using Automatic Identification System AIS data to describe ship movements and operating modes capable of providing a comprehensive analysis of ship engine exhaust emissions in a wide region that contains numerous Australian ports. Tichavska and Tovar (2015) used Automatic Identification System AIS data and the *Ship* Traffic Assessment *Mode* STEAM emission model to calculate emissions from cruise ships and ferries in Las Palmas Port. Chang et al. (2013) calculated the emissions from ships in the port of Incheon, Korea, and compared a bottom-up approach with a top down approach and found large discrepancies.

Winnes et al. (2015) built a model that calculates GHG emissions from ships in various scenarios for individual ports and different kinds of measures for emission reductions. Maragkogianni and Papaefthimiou (2015) presented a "bottom-up" estimation based on the detailed individual activities of cruise ships in the Greek ports of Piraeus, Mykonos, Santorini, Katakolo and Corfu and Dragovic et al. (2015) estimated ship exhaust emission inventories and their externalities in the Adriatic ports of Dubrovnik (Croatia) and Kotor (Montenegro) for the period 2012–2014. The methodology for emission estimation relied on the distinction of various activity phases (manoeuvring and berth/anchorage) performed by each cruise ship call (bottom-up) as a function of energy consumption during each activity multiplied by an emission factor.

The present paper proposes a methodology based on the full bottom-up approach and begins by evaluating the fuel consumed by each vessel on the basis of its individual port-activities (manoeuvring, berthing and hoteling) and differentiating between the main vessel propulsion, auxiliary propulsion, boilers and thrusters.

Unlike previous studies, the input data model is based on empirical data and extended work field (only Cooper (2001) make emission measurements of passenger ferries, but it such case it was on-board during normal service routes) and, on the other hand, it also provides accurate emission indicators (rates per hour, per passenger, per Gross Tonnage GT or a combination of all three) for passenger ferries, which can be used by other researches to reliably and quickly estimate emission inventories in other ports at the port-level.

3 METHODOLOGY

3.1 *General approach*

The first step in the evaluation of emissions is the estimation of the fuel consumed by each vessel on the basis of its activities. Specific fuel oil consumption (measured in g/kWh) is therefore an important input to the appraisal. Once the fuel consumption is calculated, it is possible to use emission factors to estimate the emission of different pollutants.

This paper considers the full bottom-up approach but takes into account separately the fuel consumption and emissions of the following propulsion systems of ferry vessels during port operations:

– Main diesel engines as a source of power for propulsion (main propulsion);
– Auxiliary engines for providing electrical energy used during hoteling
– Transversal propulsion (thrusters) for berthing and unberthing operations
– Boilers for steam production used to heat up heavy fuel oil (HFO) fuel and modify its viscosity and for heating up water;

Then, for every vessel call the fuel consumption (based on the power consumed) and corresponding emissions will be estimated for: (a) incoming manoeuvring from the Landfall Buoy to the passenger terminal dock; (b) berthing approach; (c) stay at the terminal dock (port time); (d) unberthing operations and (e) outgoing manoeuvring from the terminal dock to the Landfall Buoy.

Figure 1 shows the methodological framework considered in this paper, in which steps 1 and 2 are related to the input data model and steps 3 to 6 are methodological aspects that are described in Section 3.2.

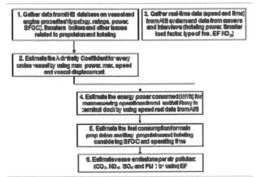

Figure 1. Methodological scheme to estimate air pollutant emissions.

3.2 Formulation

In this section, the formulation used to estimate the power and fuel consumption for each type of propulsion is introduced:

3.2.1 Main propulsion for incoming/outgoing manoeuvring

The Admiralty Coefficient method is proposed for estimating the propulsion power for manoeuvring, which is based on the basic assumption that the all resistance is frictional and that the power varies as the cube of the speed.

This method, which determines the required propulsion power according to the given ship speed and the displacement, has been used by several authors, such as Tupper (2013), Watson (1998), Taylor (1996) and Schneekluth and Bertram (1998) because of the advantages of the practicality of this methodology.

In this context, the estimation of the fuel consumption for manoeuvring is calculated as follows:

$$C_P = \sum_{ij}\left(P_{B_{ij}}t_{ij}\right)c_e \qquad (1)$$

where C_P denotes the amount of fuel consumed by the main propulsion of the vessel moving (tones); i represents those sections in which the travel distance between the dock and the Landfall Buoy is divided and velocity data is registered; j is the vessel's activity stage (incoming/outgoing manoeuvring); t_{ij} is the time (h) the vessel spends moving within the port; c_e is the specific fuel oil consumption (g/kWh) and $P_{B_{ij}}$ is the propulsion power required (kWh) during manoeuvring, which is calculated according to equation (2):

$$P_{B_{ij}} = \frac{\Delta_{ij}^{2/3}V_{ij}^3}{c_a} \qquad (2)$$

where Δ_{ij} is the real vessel displacement, V_{ij} is the vessel speed (kn) and c_a is the Admiralty Coefficient, which is related to the vessel's resistance, that is:

$$c_a = \frac{\Delta_{max}^{2/3}V_{max}^3}{P} \qquad (3)$$

in which Δ_{max} is the vessel's displacement related to the propulsion power at maximum speed, V_{max} is the maximum vessel speed and P the effective energy power (kW) which is equal to the maximum propulsion power.

3.2.2 Auxiliary engines consumption

Following the methodological approach, the fuel consumption due to the auxiliary engines (C_{AE}) during port time at the terminal and during manoeuvring is estimated as:

$$C_{AE} = c_{ec}^{AE}\left(P_{AE}t_p + P_{AE}^*t_{ij}\right) \qquad (5)$$

where t_p is the port-time at the terminal dock, P_{AE} is the auxiliary engine power (kW) and P_{AE}^* is the auxiliary engine power developed when the vessel is moving.

3.2.3 Thrusters consumption for berthing/unberthing operations

The fuel consumption required for a vessel to manoeuvre around can be estimated as:

$$C_T = \sum_{jk}\left(n_k P_k c_e\right)\left(t_{l_{kj}}r_{l_{kj}} + t_{e_{kj}}r_{e_{kj}}\right) \qquad (6)$$

where c_T is the fuel oil consumption of the thrusters (kg/h); k is the type of thruster propeller (stern and bow); n_k is the number of propellers; $t_{l_{kj}}$ is the time that each type of propeller is working on load; $t_{e_{kj}}$ is the time that each type of propeller is working empty; $r_{l_{kj}}$ is the ratio (%) corresponding to the load factor and $r_{e_{kj}}$ is the ratio (%) corresponding to the empty factor.

3.2.4 Boiler consumption

Finally, the fuel consumption provided to the boilers will be estimated as:

$$C_B = \left(\sum_{ij}t_{ij} + t_d\right)c_B \qquad (7)$$

where c_B is the fuel oil consumption of the boiler (kg/h). In this paper, this parameter is obtained through a survey completed by ship-owners. In particular, it is usually registered in the "Engine Room Log Book".

3.2.5 Total fuel consumption

Once the individual fuel consumption is estimated, the next step is to quantify vessel emissions per air pollutant by multiplying fuel consumption and emission factors (g/kWh), that is:

$$E_z = \left(C_P + C_{AE} + C_T + C_B\right)EF_z \qquad (8)$$

where z is the type of air pollutant.

Combustion emission factors (EF) vary by: engine type (main and auxiliary engines, auxiliary boilers); engine rating (Slow Speed Diesel - SSD, Medium Speed Diesel - MSD, High Speed Diesel - HSD); whether engines are pre-IMO Tier 1, or meet IMO Tier I, II or III requirements; the type of service in which they operate (propulsion or auxiliary); type of fuel (Heavy Fuel Oil - HFO, Marine Diesel Oil - MDO, Marine Gas Oil - MGO and Liquefied Natural Gas - LNG), etc.

Combustion emission factors (EF) vary by: engine type (main and auxiliary engines, auxiliary boilers); engine rating (SSD, MSD, HSD); whether engines are pre-IMO Tier 1, or meet IMO Tier I, II or III requirements; the type of service in which they operate (propulsion or auxiliary); type of fuel (HFO, MDO, MGO and LNG), etc.

Table 1 shows the EF used and data sources.

Table 1. Emission Factor (EF) and data sources used for calculations

Air pollutant	Data source and EF (g / g fuel)
CO_2	IMO regulations HFO: 3.114 g CO_2/g fuel MDO/MGO: 3.206 g CO_2/g fuel
NO_X	IMO regulations (limits set in Annex VI, MARPOL) and EIAPP (Engine International Air Pollution Prevention) certificate These EF ranges from 0.061 to 0.086 g NO_x/g fuel for main engines and from 0.051 to 0.065 g for auxiliary engines. The difference is based on engine ratings (rpm).
SO_X	IMO regulations HFO (1.5% S): 0,030 g SO_X/g fuel MDO/MGO (0.1% S): 0.002 g SO_X /g fuel
PM	IMO regulations HFO (1.5% S): 0,00426 g PM/g fuel MDO/MGO (0.1% S): 0,00097 g PM/g fuel

4 EMISSION INVENTORIES AND INDICATORS FOR FERRIES

4.1 *Data samples*

The data sample for this particular study comprises 25 passenger ferries that were monitored during 2015-16. According to the statistics of the Port of Barcelona, those 25 vessels accounted for more than 3.000 calls which represents about 86% of total passenger vessel calls in 2015 (the total number of ferries calls was 3.545).

At the same time, the data sample has been divided in two groups. The first one (G_1) includes data from 12 passenger ferries and 100 vessel calls during 2015. For every vessel call, manoeuvring and berthing time and vessel speed real-time data are obtained from the modern AIS. The second data set (G_2) is just related to the berthing activity of 13 vessels, since AIS data was not available. Therefore, we have a complete data set of 12 passenger ferries (G_1) and berthing activity data set of 13 passenger ferries (G_2).

For each vessel, engine and vessel characteristics (typology, ratings, electrical power and specific fuel consumption, GT, Length Over All LOA, draught, beam, passenger capacity) and thruster and boiler properties (power and specific fuel consumption) were obtained from IHS-Sea web database. Lastly, the load factor and working time of the thrusters, type of fuel used (HFO, MGO/MDO) and auxiliary electric power (kW) used during berthing activity are obtained through surveys and interviews of shipping companies (steps 1 and 2 from Figure 1), but only for data related to G_1.

4.2 *Annual inventory at port-level*

The total greenhouse gas GHG (CO_2) and air pollutant emissions (NO_X, SO_X and PM) for 25 passenger ferries during 2015 at the Port of Barcelona (about 3.000 vessel calls and 20.050 hoteling hours) are estimated in this section.

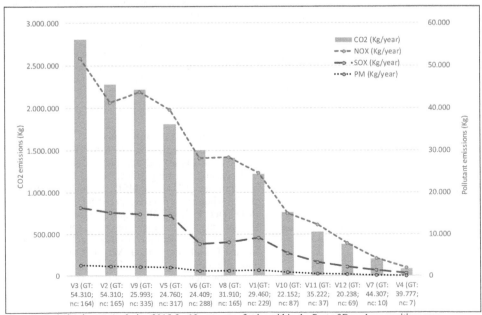

Figure 2. Emission inventory during 2015 for 12 passenger ferries within the Port of Barcelona maritime area

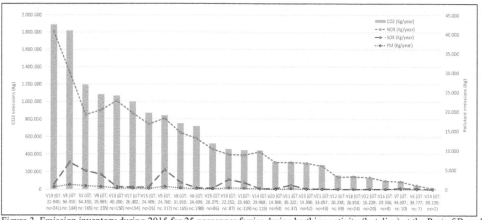

Figure 3. Emission inventory during 2015 for 25 passenger ferries during berthing activity (hoteling) at the Port of Barcelona

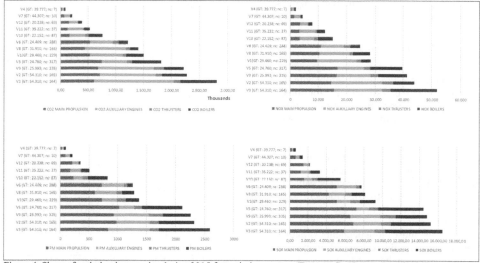

Figure 4. Share of emission inventories during 2015 for emission sources (Data set G₁-12 passenger ferries)

Figure 2 shows the yearly emission inventory within the Port of Barcelona considering the arriving-berthing-leaving activities for vessels included in the G1 data set; while Figure 3 shows the yearly emission inventory during berthing for the whole data set (25 passenger ferries). In both figures, the vertical axis shows the identification code for each chosen vessel, the number of calls in 2015 and its GT.

Then, the emission distribution per type of power used during the completely approaching-berthing-leaving cycle is depicted in Figure 4.

According to the above figures, the following first conclusions can be obtained:

1 Arriving-berthing-leaving operations (total)
  – In absolute terms, the total yearly emissions derived from the 12 passenger within the Port of Barcelona amounted to 15.210 tons of CO₂,

300 tons of NOx, 98 tons of SOx and 15 tons of PM.
  – On average, per vessel call, the estimation of emissions was: 10,25 tons of CO₂, 210 kg of NOx, 67 kg of SOx and 10 kg of PM.
  – Regarding the emissions derived from the different type of emission sources, the resulting share was:
    – Emissions CO₂: 31% main engine, 43% auxiliary engines, 20% boilers and 6% thrusters.
    – Emissions NOx: 34% main engine, 38% auxiliary engines, 22% boilers and 6% thrusters.
    – Emissions SOx: 47% main engine, 22% auxiliary engines, 21% boilers and 9% thrusters.

– Emissions PM: 43% main engine, 28% auxiliary engines, 21% boilers and 8% thrusters.

The main share's differentiation is based on the type of fuel used (HFO for manoeuvring and MGO for berthing activity) and the type of engine (main or auxiliary).

2 Berthing operations (hoteling)

– In absolute terms, the total yearly emissions derived from the 25 passenger ferries during berthing (hoteling) amounted to 15.000 tons of CO_2, 295 tons of NO_X, 18 tons of SO_X and 7,5 tons of PM.
– On average, per vessel call, the estimation of emissions was: 7,25 tons of CO_2, 145 kg of NO_X, 6,50 kg of SO_X and 3,15 kg of PM.
– Regarding the emissions derived from the different type of emission sources, the resulting averaged share was: 57% for auxiliary engines and 43% for the boilers.

4.3 Passenger ferries emission indicators

Based on the estimation of emissions represented above, the next step is to estimate indicators with the aim of extrapolating the estimations for other passenger ferry vessels based on vessel dimensions (GT and capacity) and port time (manoeuvring and berthing time).

In order to choose appropriate indicators, a regression analysis is performed between total pollutant emissions/hoteling emissions and independent variables (port time, passenger capacity and vessel GT). In case the regression model (linear regression) is deemed satisfactory, in the sense that a relationship exists among variables, then an indicator combining those independent variables will be chosen. That is, the estimated regression equation or indicator can be used to predict the emission values based on the vessel dimensions (GT) and/or port time.

From the regression analysis it can be stated that the independent variables capacity (passengers) and vessel GT cannot be individually used to predict the total emissions and hoteling emissions, as the correlation coefficient is weak, indicating that there is no relationship between the two variables. However, by combining them with the port-time variable, the regression model results indicate a better relationship.

Therefore, it can be concluded that the best independent variable to predict hoteling emissions emitted by passenger ferries at ports is the port-time – GT or port-time but only port time-GT could be used for extrapolating total port emissions since important differences arise in emission inventories. The regression indicator (Figure 5) for hoteling emissions regarding port time-GT are close to 0,85

but for total emissions at port-level are close to 0,65, which is weak.

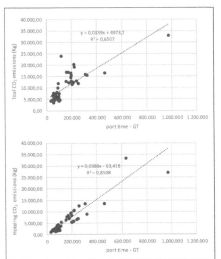

Figure 5. Example of regression analysis. CO_2 emissions as regards to port time-GT.

Finally, Table 2 lists average emission values for every selected indicator.

Table 2. Emission indicators for passenger ferries at port-level

Indicator	SO_X	NO_X	PM	CO_2
Arriving-berthing-leaving emissions within the port area				
Emissions / port-time	17,00 kg/h	53,60 kg/h	2,60 kg/h	2.600,00 kg/h
Emissions / port-time-GT	0,50 g/hGT	1,60 g/hGT	0,080 g/hGT	78,85 g/hGT
Emissions / port-time-pax	12,50 g/hpax	38,80 g/hpax	1,90 g/hpax	1.890,0 g/hpax
Berthing at terminal dock (hotelling)				
Hotelling emissions / port-time	2,90 kg/h	24,10 kg/h	0,60 kg/h	1.215,00 kg/h
Hotelling emissions / port-time-GT	0,10 g/hGT	0,75 g/hGT	0,020 g/hGT	38,00 g/hGT
Hotelling emissions / port-time-pax	2,20 g/hpax	20,65 g/hpax	0,50 g/hpax	1.020,0 g/hpax

5 CONCLUSIONS

This paper addresses the estimation of air emissions released by passenger ferries in urban ports. This is of great importance due to a significant share of emissions produced during the time vessels stay in ports and the huge amount of calls involved in regular services. In addition, this paper provides useful passenger ferries emission indicators, which could facilitate reliably estimating the in-port ship emission inventories of other ports without requiring large amounts of data.

The proposed methodology is based on the "full bottom-up" approach and begins by evaluating the fuel consumed by each vessel on the basis of its individual port-activities (manoeuvring, berthing and hoteling at the terminal dock). The methodological scheme also separately considers different types of emission sources: main propulsion (diesel engines), auxiliary propulsion for providing electrical energy for hoteling, thrusters and boilers. Once the fuel consumed is determined, the next step is estimating air emissions from vessels by employing the corresponding emission factors per air pollutant.

The methodology was implemented to a particular case in which 25 passenger ferries and 85 calls were monitored in the Port of Barcelona during 2015. The emission estimations led to the following considerations:

– Main and auxiliary engine emissions were found to be dominant (about 72%), whereas boilers and thrusters represent, on average, 28%, in which boilers were predominant (22%).
– Hoteling emissions (auxiliary engine and boilers during berthing time) represent about 70% of the total emissions for CO_2 and NO_x, whereas for SO_x and PM, it represents about 28%, since EF for MGO are smaller than LSHFO (Low-Sulphur HFO).
– According to the sample data, the average estimation of total emissions per vessel call was: 10,25 tons of CO_2, 210 kg of NO_x, 67 kg of SO_x and 10 kg of PM. And, for hoteling activity, the average amount was: 7,25 tons of CO_2, 145 kg of NO_x, 6,50 kg of SO_x and 3,15 kg of PM.

Concerning emission indicators, it was found through a regression model that the best independent variable to predict hoteling emissions was the combined variable port time – GT. Nonetheless, the variables port time – passenger is also quite robust and any variable can provide excellent results to estimate total port emissions. In relation to the indicator emission per port-time and GT, the following values could be used to estimate hoteling emissions at ports: 38.00 g CO_2/h-GT, 0.75 g NO_x/h-GT, 0.10 g SO_x/h-GT and 0.020 g PM/h-GT.

As regards to total emissions at port, the regression model analysis showed that any indicator could be used to estimate emissions with high accuracy.

With respect to the reliability of the emission indicators, it should be mentioned that information regarding vessel activities, hoteling power, engine ratings, fuel use, emission factors related to NO_x and load factors are based on empirical and real information (work field) received from shipping crew companies, which means that estimations are quite consistent.

In summary, this paper contributes to the development of emission indicators for passenger ferries, which can be extended to other ports to reliably and quickly estimate emission inventories and to calculate emission inventories, which could help to understand passenger ferry emissions when proposing environmental and policy measures.

REFERENCES

Chang, Y.T., Song, Y., Roh, Y. 2013. Assessing greenhouse gas emissions from port vessel operations at the Port of Incheon. Transportation Research Part D, 25, pp. 1–4.

Cooper, D.A. 2001. Exhaust emissions from high speed passenger ferries. Atmospheric Environment, 35, pp. 4189-4200.

Dalsoren, S.B., Eide, M.S., Endresen, O., Mjelde, A., Gravir, G., Isaksen, I.S.A. (2009). Update on emissions and environmental impacts from the international fleet of ships: the contribution from major ship types and ports. Atmospheric Chemistry and Physics, 9, pp. 2171–2194.

Dragovic, B., Tzannatos, E., Tselentis, V., Mestrovic, R., Skuric, M. (2015). Ship emissions and their externalities in cruise ports. Transportation Research Part D (in press).

ESPO (2003). Environmental code of practice. ESPO, Brussels, Belgium.

Eyring, V., Koehler, H.W., van Aardenne, J., Lauer, A. (2005). Emission from International Shipping: 1. The last 50 years. Journal of Geophysical Research, 110(D17305), pp. 1-12.

Goldsworthy, L., Goldsworthy, B. 2015. Modelling of ship engine exhaust emissions in ports and extensive coastal waters based on terrestrial AIS data – An Australian case study. Environmental Modelling & Software, 63, pp. 45–60.

IAPH (2007). Resolution on clean air programs for ports. Second plenary session. 25th World Ports Conference of IAPH. Houston, Texas, USA.

ICF (2006). Current methodologies and best practices in preparing port emission inventories. EPA, Fairfax, Virginia, USA. Final report for U.S.

Maragkogianni, A., Papaefthimiou, S. (2015). Evaluating the social cost of cruise ships air emissions in major ports of Greece. Transportation Research Part D, 36, pp. 10–17.

Miola, A., Ciuffo, B., Giovine, E., Marra, M. (2010). Regulating air emissions from ships: the state of the art on methodologies, technologies and policy options. European Commission, JRC reference reports.

Schneekluth, H., Bertram, V. (1998). Ship design for efficiency and economy. Elsevier Ltd., 2nd edition, UK.

Song, S. (2014). Ship emissions inventory, social cost and eco-efficiency in Shanghai Yangshan port. Atmospheric Environment, 82, pp. 288–297.

Taylor, D.A. (1996). Introduction to marine engineering. Elsevier Ltd. UK.

Tichavska, M., Tovar, B. (2015). Environmental cost and eco-efficiency from vessel emissions in Las Palmas Port. Transportation Research Part E, 83, pp. 126–140.

Tupper, E.C. (2013). Introduction to naval architecture. Elsevier Ltd, Fifth Edition, USA.

Tzannatos, E. (2010). Ship emissions and their externalities for the port of Piraeus – Greece. Atmospheric Environment, 44, pp. 400–407.

Watson, D. (1998). Practical ship design. Elsevier Ocean Engineering Book Series, Vol 1, Oxford, UK.

Winnes, H., Styhre, L., Fridell, E. 2015. Reducing GHG emissions from ships in port areas. Research in Transportation Business and Management, 17, pp. 73-82.

Proceedings of 12th International Conference on Marine Navigation and Safety of Sea Transportation, TransNav 2017
21-23 June 2017, Gdynia, Poland

The Structural Types of Oil Spill Response Organizations: the Comparisons of Countries on Oil Spill Response Operations

E. Kan, A.C. Töz, T. Olgaç & O. Bayazit
Dokuz Eylül University, İzmir, Turkey

Ö. Tezcan
Çanakkale Onsekiz Mart University, Çanakkale, Turkey

ABSTRACT: The lack of standardization for oil spill response industry at global level has led countries to construct different regional structures. This different approaches negatively affect the efficiency of oil spill response operations. In this study, the comparison of the marine pollution response system which is carried out by various countries will be conducted and the strengths and weaknesses of the systems will be discussed. In this context, the efficiency of the regional response system will be tried to be defined through case studies and recommendations will be made for the optimum response organization.

1 INTRODUCTION

Marine pollution encountered as a result of maritime traffic in the territorial waters or exclusive economic zones damages the economic resources of the countries and causes the extinction of marine life and degeneration of marine ecosystem in international scope. Upon occurrence of such undesired results, all the countries must be prepared for cleaning activities in order to avoid any likely harm. The response structures adopted by countries in case of any marine pollution are to be revealed and dissessed.

2 AIM OF STUDY

This study aims to determine and compare the emergency response structures adopted by certain European countries in occurrence of any marine pollution.

3 METHODOLOGY

In this study, oil spill response structures of some coastal European countries were examined. For this purpose, various regulations and institutional reports were evaluated and content analysis was carried out. In addition, relevant international agreements to which these countries are parties regarding marine pollution were also examined.

4 FINDINGS AND EVALUATION

Within the scope of this study, seventeen European countries were examined in terms of the structure of oil spill response adopted and the who has the authority and responsibility and which institution is responsible for co-ordination in case of any pollution.

4.1 *Belgium*

The structure of response against marine pollution in Belgium is based on shared authority and responsibility between two ministries and coastal states (IFREMER, 2007;10). These ministries are the Ministry of Environment and the Ministry of Interior Affairs. The Ministry of Environment is responsible for developing marine environment policies in national level (OECD, 2007). This ministry controls the North Sea in terms of application of international agreements via the department of Management Unit of the North Sea Mathematical Models (MUMM). MUMM controls marine pollution incidents by organizing aerial searches with the sub-department of Belgian Maritime Environmental Control (BELMEC) (IFREMER, 2007; 10).

4.2 *France*

The organization of struggling against marine pollution in France is completely governed by the state and the local authorities have no role in this issue. The authorization was made incompliance with POLMAR rules. The district governor is the authorized person in coastal states. Naval forces co-ordinates and distributes the duties that are going to be carried out by the public institutes. Marine pollution action plans were prepared and keeping update for three different marine zones (IFREMER, 2007; 4). There are three different types of prevention methods against marine pollution. These are; preventive measures, preparedness of quick usage of resources in case of accident and response measures to limit the consequences.

4.3 *Germany*

The German pollution response organization is one of the best examples of sharing the authority and the responsibility between the government and the states (counties). A co-established institute by government and state, Central Command for Maritime Emergencies (CCME), has 6 sub-divisions 2 of which are responsible for response organization in marine pollution (IFREMER, 2007;8).

4.4 *Ireland*

Although the Irish Coast Guard is responsible for prevention and response of marine pollution, the ministry Transportation, Tourism and Sports has also duties and responsibilities in this issue. Irish Coast Guard has authorization of initiating, controlling and managing regarding pollution response. In addition, Irish Coast Guard has responsibility in relation to marine protection issues (Bonn Agreement, 2010),(OECD 2010).

4.5 *Netherlands*

The Ministry of Transport, Public Works and Water Management is the responsible authority for the North Sea activities and the accidents occurred in coastal waters (Oral, 2013). In addition, the director of Netherlands Coast Guard is responsible for response co-ordination in case of any marine pollution (Bonn Agreement, 2010).

4.6 *Norway*

The Norwegian Coastal Administration (NCA) is responsible for co-ordination of a unit, which is consisting of government, municipal and private organizations, to respond to any marine pollution. NCA is also responsible for responding to vessel oil spills and huge pollution events from undefined sources. NCA may help to responses or takes the control in case of inadequacy in local authority (IFREMER, 2007; 12).

4.7 *Portugal*

The structure of marine pollution response in Portugal entirely belongs to the central government. An established commission according to national regulations including the Ministry of Defense and Naval Forces, port captains and port authorities determines the policies in the field of preventing and responding to marine pollution. The Ministry of Defense and Naval Forces has the authority regarding coastal pollution response issues. Port captains are responsible for coordinating the response regarding regional-scaled pollution issues (IFREMER, 2007; 11)(OECD 2011).

4.8 *Spain*

In Spain, marine pollution response organization is entirely handed over to state authority. General Directorate of Merchant Marine under the Ministry of Marine is organizing the response. The Spanish Maritime Safety Agency (SASEMAR) prepares emergency contingency plans considering pollution matters (Bonn Agreement, 2010). After the accident "Prestige", the deficiencies of response were identified. On this basis, coordination between the neighbor countries has been increased and National Coastal Response Coordination Body (CEPRECO) was established. Spain is prepared for any pollution and emergency situation by emergency contingency and response plans with CEPRECO (IFREMER, 2007; 5).

4.9 *Sweden*

Swedish Coast Guard is responsible for preventing and response matters of coastal and inland waters pollution incidents according to the national law. In local, fire brigade is responsible for responding to pollution incidents in beaches, ports and inland waters (Bonn Agreement, 2010).

4.10 *The United Kingdom*

The structure of maritime pollution response in UK is entirely organized by state. Local authorities are considered to be responsible for coastal incidents. The Maritime Coast Guard Agency is responsible for marine pollutions and any incidents related to the sea around the UK. This agency has two sub-department named as the Directorate of Maritime Safety and Pollution (MSPP) and the Directorate of Maritime Operations (DMO). These directorates are responsible for emergency response and cleaning the seas in case of any pollution. The UK is separated into four major regions by the Maritime Coast Guard Agency, and the coordination centers in each region has the structure of marine cleaning response.

4.11 *Turkey*

The responsibility to act to prevent marine pollution belongs to the Provincial Environment and Urbanization Directorates under the Ministry of Environment and Urbanization.

The responsibility for inspecting the marine pollution and applying the fine procedure, if necessary, belongs to;

1 The relevant metropolitan municipality for the coastal zones inside the limits of the metropolis.
2 The port authorities for port zones, except marinas and fishing boat shelters.
3 The Turkish Coast Guard, within the scope of the law numbered 2692, for the zones except the zones mentioned above. In addition, within the scope of the Environmental Law (No:2872), Coast Guard has authority to carry out inspections and apply administrative sanctions for the prevention of marine pollution caused by vessel in Turkish territorial waters (Turkish Coast Guard, http://www.sgk.tsk.tr, 26.12.2016).

Taking into account of these evaluations and information, the structure of marine pollution response organization of each country is different and it is coordinated by an institution. Marine pollution response responsibility may be shared between the state and the local authority according to the circumstances. Sometimes it is coordinated simultaneously by different authorities. Comparisons of adoption including responsible and the assisting organizations for pollution response with regard to these determinations are given in Table 1. In addition, distribution of the international agreements to which the relevant European countries are parties, are presented in Table 2.

Table 1. The Structural Types of Oil Spill Response Organizations: The Comparisons Of Countries

Country	Core Responsibility	Responsible Organization /Institution	Assistant Organization/Institution
Belguim	Balanced	Minister of the Interior	Management Unit of the North Sea Mathematical Models (MUMM), BELgianMaritimeEnvironmental Control (BELMEC), Coastal States
Italy	National Government	Ministry of Environment	The Central Service of SeaDefence, HarbourmasterOffices, Coast Guard.
France	National Government	Governor of a Province ve French Navy	Public Institutions in the Region
Germany	Balanced	Central Command for Maritime Emergencies	National Government and Municipal Authority' Institutions
Ireland	National Government	Irish Coast Guard	Minister for Transport, Tourism and Sport
The Netharlands	National Government	Director of the Netherlands Coastguard	The Minister for Transport, Public Works and Water Management
Norway	Balanced	The Norwegian Coastal Administration	Public Institutions in the Region, Municipal Authority' Institutions.
Portugal	National Government	Direcção-Geral da Autoridade Marítima	CapitaniasdosPortos
Spain	National Government	General Directorate of the Merchant Marine	The Spanish Maritime Safety Agency, National Coastal Response Coordination Body (CEPRECO)
Sweden	Municipal Authority	Fire Departmant of Municipal Authority	Coast Guard
United Kingdom	National Government	The Maritime and Coast Guard Agency	Directorate of Maritime Safety and Pollution Prevention (MSPP) and the Directorate of Maritime Operations (DMO)
Turkiye	National Government	Ministry of Environment And Urbanisation	Turkish Coast Guard, Port Authority, Metropolitan Municipality
Danmark	National Government	Ministry of Defence	Maritime Assistance Service
Estonia	National Government	Ministry of Internal Affairs	Estonian Border Guard
Finland	National Government	Ministry of Environment	The Finnish Environment Institute, Maritime Rescue Coordination Centre (MRCC/MRSC), Coast Guard, The Finnish Frontier Guard, the Navy, the Institute of Marine Research
Greece	National Government	Ministry of Mercantile Marine, Marine Environment Protection Division	15 Regional Marine Pollution Combating Stations (RMPCS), Coast Guard, the Navy, the Institute of Oceanographic & Fishery Research and the Environment Ministry.
Iceland	National Government	Ministry of Environment and Natural Resources, The Environmental Agency of Iceland (EAI)	Icelandic Coast Guard, Icelandic Maritime Safety Administration

Source: Adopted from Bonn Agreement (2010), IFREMER (2007), ITOPF Country Reports, , http://www.itopf.com/, 10.01.2017.

Table 2. International Agreements Relating to the Marine Pollution to which the Countries are Partied

Country	Prevention & Safety					Spill Response		Compensation						
	MARPOL ANNEXES					OPRC	OPRC-		CLC		Fund	Supp	HNS	Bunker
	73/78	III	IV	V	VI	90	HNS	69	76	92	92	Fund		
BELGIUM	√	√	√	√	√					√	√	√		√
BULGARIA	√	√	√	√	√	√				√	√			√
CYPRUS	√	√	√	√	√					√	√		√	√
DENMARK	√	√	√	√	√	√	√			√	√	√		√
ESTONIA	√	√	√	√	√	√	√			√	√	√		√
FINLAND	√	√	√	√	√	√				√	√	√		√
FRANCE	√	√	√	√	√	√	√			√	√	√		√
GERMANY	√	√	√	√	√	√	√			√	√	√		√
GREECE	√	√	√	√	√	√	√		√	√	√			√
ICELAND	√					√				√		√		√
IRELAND	√		√	√	√	√				√	√	√		√
ITALY	√	√	√	√	√	√				√	√	√		√
TURKEY	√		√			√				√	√	√		
LATVIA	√	√	√	√	√	√		√		√	√	√		√
LITHUANIA	√	√	√	√	√	√				√	√	√	√	√
MALTA	√					√	√			√	√	√		√
THENETHERLANDS	√	√	√	√	√	√	√			√	√	√		√
NORWAY	√	√	√	√			√			√	√	√		√
POLAND	√	√	√	√	√	√	√			√	√	√		√
PORTUGAL	√	√	√			√	√			√	√	√		
ROMANIA	√			√						√	√			√
SLOVENIA	√	√	√	√	√	√	√			√	√	√	√	√
SPAIN	√	√	√	√	√	√	√			√	√	√		√
SWEDEN	√	√	√	√	√	√	√			√	√	√		
UNITED KINGDOM	√	√	√	√	√	√				√	√	√		√

Source: Adopted from ITOPF Country Reports, http://www.itopf.com/, 06.02.2017.

5 CONCLUSION

With this study, the structures of response to marine pollution, that may caused by any marine accident, adopted by the European countries were examined. Most of these countries have a structure that belongs to the central government instead of local authorities. Even if the central government needs assistance from the local authorities in some cases, the responsibility and the authority are centralized. When the structural types of oil spill response organizations are examined, it is seen that local authorities, coast guard or naval forces, port authorities and ministry representatives or any institutions established for this purpose are acting against marine pollution in their region in connection with the central government. In some countries, central government and the local authority are coordinated, but in very few countries the responsibility for cleaning activities were left to the locals.

When the sensitivities of the European countries are examined, it is observed that the rate of being a party to the international agreements is found high.

REFERENCES

BonnAgreement (2010), "BonnAgreement Counter Pollution Manual" Volume 1, United Kingdom.

IFREMER Le Cedre, (2007), "ToFoster Prevention and Best Response to Accidental Marine Pollution", France

International Tanker Owners Pollution Federation, ITOPF (a),http://www.itopf.com/knowledge-resources/countries-regions/countries/estonia/

International Tanker Owners Pollution Federation, ITOPF (b)http://www.itopf.com/knowledge-resources/countries-regions/countries/finland/

International Tanker Owners Pollution Federation, ITOPF (c) http://www.itopf.com/knowledge-resources/countries-regions/countries/france/

International Tanker Owners Pollution Federation, ITOPF (d) http://www.itopf.com/knowledge-resources/countries-regions/countries/germany/

International Tanker Owners Pollution Federation, ITOPF (e)http://www.itopf.com/knowledge-resources/countries-regions/countries/iceland/

International Tanker Owners Pollution Federation, ITOPF (f)http://www.itopf.com/knowledge-resources/countries-regions/countries/latvia/

International Tanker Owners Pollution Federation, ITOPF (g)http://www.itopf.com/knowledge-resources/countries-regions/countries/poland/

International Tanker Owners Pollution Federation, ITOPF (h)http://www.itopf.com/knowledge-resources/countries-regions/countries/romania/

International Tanker Owners Pollution Federation, ITOPF (i)http://www.itopf.com/knowledge-resources/countries-regions/countries/slovenia/

International Tanker Owners Pollution Federation, ITOPF (j) http://www.itopf.com/knowledge-resources/countries-regions/countries/spain/

International Tanker Owners Pollution Federation, ITOPF (k)http://www.itopf.com/knowledge-resources/countries-regions/countries/sweden/

International Tanker Owners Pollution Federation, ITOPF (m) http://www.itopf.com/knowledge-resources/countries-regions/countries/united-kingdom/

International Tanker Owners Pollution Federation, ITOPF (n) http://www.itopf.com/knowledge-resources/countries-regions/countries/portugal/

International Tanker Owners Pollution Federation, ITOPF (o) http://www.itopf.com/knowledge-resources/countries-regions/countries/norway/

International Tanker Owners Pollution Federation, ITOPF (p) http://www.itopf.com/knowledge-resources/countries-regions/countries/italy/

International Tanker Owners Pollution Federation, ITOPF (r) http://www.itopf.com/knowledge-resources/countries-regions/countries/ireland/

International Tanker Owners Pollution Federation, ITOPF (s) http://www.itopf.com/knowledge-resources/countries-regions/countries/greece/

International Tanker Owners Pollution Federation, ITOPF (t) http://www.itopf.com/knowledge-resources/countries-regions/countries/Denmark/

International Tanker Owners Pollution Federation, ITOPF (u) http://www.itopf.com/knowledge-resources/countries-regions/countries/belguim/

OECD 2007 "EnvironmentalPerformanceReviewsBelguim" USA

OECD 2010 "Environmental Performance Reviews-Ireland" USA

OECD 2011 "Environmental Performance Reviews-Portugal" USA

Oral, Nilüfer (2013) "RegionalCo-operation and Protection of the Marine Environment Under International Law" Netherlands

Turkish Coast Guard Website http://www.sgk.tsk.tr/baskanliklar/personel/sgk_yayinlari/brosur/mavi_vatan_denizlerimiz.pdf

The Concept of "Green Ship": New Developments and Technologies

E. Çakır, C. Sevgili, R. Fışkın & A.Y. Kaya
Dokuz Eylul University, Izmir, Turkey

ABSTRACT: Ocean-borne trade has increased steadily over the past few decades and is expected to continue to play a crucial role in the globalized world economy. This growth has brought with air quality and marine pollution problems and global climate change risks unless ship emissions are controlled. Thus, International Maritime Organization (IMO) aims to guide to the sectors how to reduce carbon footprint by technical and operational measures. IMO also emphasize on the importance of a global approach in advanced improvements to energy efficiency and emission reduction. This approach led to maritime industry to seek for new technological developments to reduce ship-based CO_2, NO_x and SO_x emissions, to improve energy efficiency and to develop propulsion power such as sulphur scrubber system, LNG fuel for propulsion, advanced rudder and propulsion system and etc. The purpose of this paper is to examine aforementioned new technological developments and to discuss the advantages and disadvantages of them.

1 INTRODUCTION

The growth of the world economy and the increase in world population cause huge rise in global production and consumption. Because of this reason, there is a great need for transportation on a global scale especially in sea transport. Correspondingly, use of fossil fuels and emissions of harmful greenhouse gases have increased substantially. When compared to other modes of transport the lower fuel consumption per ton caused by sea transport. However, the emissions of greenhouse gases have been rapidly increased depending sea transport.

Looking at the figures for 2007, it is seen that the consumption of fuel for international maritime transportation reached 277 million tons. This amount is increasing every year. The distribution of greenhouse gases released to the environment after fuel consumption in 2004-2007 is given below (Tab.1):

Table 1: Distribution of exhaust emissions from marine transportation by years in million tonnes (Source: Buhaug et. al 2009).

YEAR	NOx	SOx	PM	CO	NMVOC	CO2	CH4	N2O
2005	23	13	1.6	2.3	0.7	955	0.09	0.02
2006	24	14	1.7	2.4	0.8	1008	0.10	0.03
2007	25	15	1.8	2.5	0.8	1050	0.10	0.03

Reducing greenhouse gas emissions from ships can only be achieved by using better quality fuels or by reducing fuel consumption by providing higher energy efficiency. It is possible to provide competitive advantages in terms of many economic, social, environmental benefits and businesses by reducing fuel consumption which are (Maddox, 2012):
− Economical aspects;
  − Less operational costs for ship owners and operators. More freight can be transported with less cost,
  − The end consumer will be able to reach the product with less cost,
− Social aspects;
  − Especially the reduction of respiratory diseases
  − The decrease in the number of people losing their jobs and home because of the rise of sea level
  − Decrease of the effects of natural catastrophes
− Environmental aspects;
  − Reduction of greenhouse gas emissions
  − Air quality improvement
In addition to all these, companies taking the necessary precautions in terms of energy efficiency will provide competitive advantage over those who do not.

In order to ensure a sustainable marine transportation, it is seen that many different methods have been developed especially for the prevention of air pollution. Although there are some differences in the classification of these methods, they are generally divided into operational methods, technical methods and market-based methods.

Briefly, market-based methods include fines and incentives to reduce carbon emissions; carbon taxation can be shown as an example of this. Operational and technical methods include many different approaches such as operational methods to reduce fuel consumption (such as speed optimization, trim optimization etc.), waste gas filtering techniques (sbrubbers etc.), hull and machinery design technologies, energy efficient propeller systems, and alternative fuel technologies etc. This study examine these new technologies and methods introduced to ensure sustainable maritime transport. The study will also examine the concept of green ship.

2 THE CONCEPT OF GREEN SHIP AND SUSTAINABILITY

As in most other sectors, maritime has also begun to target better energy efficiency due to environmental issues and fluctuating fuel prices (Sherbaz & Duan, 2012). Since approximately 80% of world trade is carried by maritime transportation, this transportation mode cause 2,2% of global CO_2 emissions, 14%–15% of global NO_x emissions, and approximately 7% of global SO_2 emissions (ICS,2014; 3. GHG Study, 2014). To reduce this rate and protect the environment, states, IMO (International Maritime Organization) and other environmental organizations are putting pressure on maritime stakeholders (Lee, et al., 2012). Hence maritime industry is endeavoring to curb Greenhouse Gas emissions and improve effectiveness. One of the concepts that emerged with this restructuring is the Green Ship. The Green Ship Concept is the implementation and development of ship technologies to improve energy efficiency and reduce emissions (Sherbaz & Duan, 2015). Green Ship implies developed, ecofriendly technologies which decrease emissions and air contaminant produced during voyage. These technologies are generally developed under the categories of reducing emission, energy efficiency and propulsion systems (Bai & Jin, 2016). For this reason, in this study, Green Ship Technologies are examined under these headings.

Measures to reduce emissions are generally ship design improvements and sizes, speed management and operational measures. The aim of these measures is primarily to reduce fuel consumption. As per Greenhouse Gas Study 2014 of IMO, scenarios have determined that marine CO2 emissions will increase between 50% and 250% by 2050 due to economic and energy-related developments. Nitrogen emissions are also expected to increase, but this increase is not expected to be as high as CO2. Methane emissions are also expected to rise because of increased use of LNG. Depending on requirements on fuel Sulphur content, SOx emission will decrease by 2050. Given the emission estimates, the development of the concept of green ship is of vital importance (UNCTAD, 2015).

On energy efficiency, IMO has established a series of baselines for the amount of fuel each type of ship burns for a certain cargo capacity. In accordance with energy efficiency regulations, current ships now have to have management and technical measures such as advanced voyage planning, waste heat recovery systems. With this type of measures, by 2025, ships will be about %30 more energy efficent than ships of today (IMO, 2016).

There are various propulsion systems such as steam propulsion, diesel propulsion, nuclear propulsion, gas turbine propulsion, wind propulsion, fuel-cell propulsion, bio-diesel propulsion, solar propulsion, gas fueled propulsion, integrated power systems, etc (Molland et al. 2011; Marine in Sight, 2016). The propulsion systems for green ship applications in among the mentioned propulsion systems will be examined next section of this paper. More environmentally friendly approaches have been adopted in propulsion systems with the development of technology.

The concept of sustainability has been discussed definitions related to the concept in the literature (Solak Fışkın & Deveci, 2015). It is widely agreed that definition of sustainability is 'Sustainable development meets the needs of the present without compromising the ability of future generations to meet their own needs.' and it is a three-pillar concept; economic, environmental and social sustainability (WCED, 1987). Sustainable shipping is a new concept and some of the issues sustainable shipping can be shown as Table 2.

Table 2. Sustainable Shipping (Cabezas-Basurko et al., 2008)

Environmental Sustainability	Economic Sustainability	Social Sustainability
Protect water, air, land environments and ecosystems from any kind of pollution	Present cost-effective	Create new jobs
Maintain marine fisheries quality and quantity.	Present economically feasible technology	Present a safe workplace and prevent work fatalities
	Provide economic growth to local communities	Prevent accidents.

In the last decade, some proactive attempts to promote environmental management improvements in maritime industry which are called sustainable or green shipping initiatives have emerged. These initiatives can be classified as research and innovation, corporate social responsibility and marketing, Awareness raising and environmental education, shipping and international regulation/compliance, voluntary class notations and certifications, economic benefits and environmental protection (WWF, 2012).

It is clear that if Green Ship Technologies could be integrated existing ships now, they could be cleaner and more environmental friendly. With the emergence of new technologies and the development of existing technologies, more than %33 eco friendliness would be accomplished that heading to the zero emissions ship in the near future.

3 NEW TECHNOLOGIES AND DEVELOPMENTS FOR GREEN SHIP

Shipping plays an important role in global trade, but the negative environmental impact of shipping cannot be ignored. For instance, oil spill from ships have often had an impact on the marine ecosystem. Air pollution from ships is one of the major apprehension of the global environmental movement. Ballast water may pose ecological, economic and health issues because of various marine species. These potential effects led to introduce a new concept "green shipping". The concept is accepted as the solution for environmental problem (Ahmad, 2014).

The shipping companies and classification societies such as DNV, NYK, Volvo, Evergreen have been investing technological and research solution for green shipping for so long. Some of the key projects focusing on high responses to these issues and operated by these companies and societies as follows: Eco Ship Project: NYK (Japan), Volvo Penta-Led Swedish (Sweden), Post-Panamax ships – S-class: Evergreen (Taiwan), Rotor Sails: Greenwave Wind Engines (UK), Fellow SHIP Programme: DNV (Norway), Air Cavity System (ACS): DK Group (The Netherlands), Sea Water Scrubbing System: Hamworthy Krystallon (UK) (Pike et al., 2011: 12).

A new concept green ship is a crucial for green shipping and sustainable transport. This concept requires exploring and implementing technology on ships to enhance and improve energy efficiency and reduce emission (Sherbaz & Duan, 2012). Thus, in this section of the study, new technologies for green ships is examined to reveal recent developments. The section is divided into three subtitle as reducing emissions, energy efficiency and propulsion systems.

3.1 Reducing emissions

Selective Catalytic Reduction (SCR) technology is an operation to reduce the NOx concentration owing to combustion exhaust that includes the injection of aqueous solution of urea in the tail pipe of a four stroke, constant speed diesel engine (Praveen and Natarajan, 2014). The SCR technology enables the NOx reduction reaction to occur in an oxidizing atmosphere. The system is called selective since the catalytic reduction of NOx with ammonia (NH_3), urea, trimethylamine, cyanuric acid, monomethylamine, dimethylamine, carbamates, ammonium carbonate, ammonium bicarbonate, etc. as a redundant take place favorably to oxidation of NH_3 with oxygen with oxides of nitrogen (Jiang et al., 2010 as cited in Praveen and Natarajan, 2014). Applying a SCR technology to reduce emission a number of elements such as operating temperature window, flue gas pressure drop, flue gas contaminants, physical constraints, catalyst disposal, ammonia storage and etc. have been identified that should be considered. The elements are based on information get from various manufacturers and operating data from numerous installations both in Japan and the USA (Cobb et al., 1991). Yang et al. (2012) conducted a survey by applying TOPSIS and AHP to reveal preferred techniques for NOx and SOx emission reduction. The study concluded that SCR technique is the least preferred method to reduce emission because of its investment and operational cost although the most effective technique with a reduction rate of up to 95%.

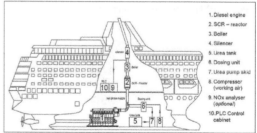

Figure 1. The Main Principle of the SCR System (www.yara.ie, 2016).

Humid Air Motor (HAM) technology is a water based technique to reduce emission. The use of water to prevent NOx during the combustion progress is prevalent. The HAM method is a promising solution comparing to other technology. The system includes evaporating large quantities of water at the compressor so as to bring the charge air close to saturation. The system reduces the formation NOx by roughly 70% by saturating the inlet air with water vapor (Riom et al., 2001). The system was tested by Riom et al. (2001). The test results showed that the system is efficient to reduce

NOx emission, have simple operation and low operating cost, contribute the engine operation optimization and also stable and responsive.

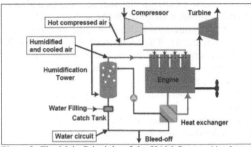

Figure 2. The Main Principle of the HAM System (Axelsson, 2009).

Direct Water Injection (DWI) technology used for reduction of NOx emission by the injection of water directly into the combustion chamber via separate nozzle can reduce 50%-60% NOx emission. The combined injection valve through which both fuel and water are injected is the key element in the design of the system. One needle in the nozzle is tasked for fuel injection, and the other one for water injection. The water injection before fuel injection to cool down the combustion space is important to ensure low NOx formation before fuel ignition (Wartsila, 2007). Sarvi et al. (2009) conducted a study to show reduction in NOx emission by DWI system. The study shows that a significant reduction in NOx emission for HFO and LFO up to 35%, slight for HC emission reduction but for Particulate Matter (PM) is not very effective.

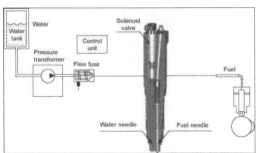

Figure 3. The Main Principle of the DWI System (www.civilengineeringhandbook.tk, 2016).

Scrubber is a device installed in the exhaust system in order to remove most of the SOx from the exhaust and reduce PM. All scrubber technologies create a waste stream comprising the substance used for the cleaning process plus the SOx and PM removed from the exhaust. There are two type application of scrubber system commonly proposed for ship: dry scrubbers and wet scrubbers. Dry scrubbers do not use any liquid to do the scrubbing

process whereas wet scrubbers pass the exhaust gas through a liquid media so as to remove the SOx compounds from the gas. The most common liquid for the task are untreated sea water or chemically treated fresh water (ABS, 2013).

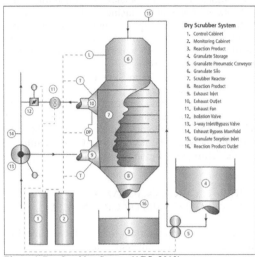

Figure 4. Dry Scrubber System (ABS, 2013).

3.2 *Energy efficiency*

The International Maritime Organization (IMO) has extensively seeking ways to reduce greenhouse gases (GHG) emissions. For this reason, the Energy Efficiency Design Index (EEDI), indicated by the quantity of CO_2 allowed during transportation, in units of freight per distance, was brought into force by IMO in 2013 (Bai & Jin, 2016). Under the energy-efficiency regulations, existing ships now have to have a Ship Energy Efficiency Management Plan (SEEMP) in place, looking at new practices like improved voyage planning, cleaning the underwater hull of the ship and the propeller more frequently, developing technical innovations such as waste heat recovery systems, or even fitting a new propeller (IMO, 2016). In this context, one of the new technologies and mostly used applications related to ship energy efficiency are given below:

The waste heat recovery system is accepted one of the best energy saving methods to make a more efficient utilization of fuels to achieve economic and environmental improvement (Senary et al., 2016). While sailing in water, diesel engines onboard have efficiency about 48–51% and the rest of the total heat energy of the Internal Combustion Engine is thrown back to the atmosphere without any usage (Dzida and Mucharski, 2009). By installing the ship's main propulsion plant with a waste heat recovery system (WHRS), effective utilization of wasted thermal energy can enhance the plant efficiency (Sing and Pedersen, 2016).

Figure 5 shows the main principles WHRS are that by using a boiler, the wasted amount of energy is used to generate steam. This steam can drive a steam turbine and so, with a generator, produce electricity on board which allows to pre heat the fuel before the combustion and reduce the greenhouse emissions. According to study of Baldi and Gabrielli (2015), depending on the sources of waste heat employed (exhaust gas, charge air cooling, various types of cooling systems), on the type of complementary auxiliary generation system (shaft generator or auxiliary engines), and on the exergy efficiency of the recovery system, fuel savings from 5% to 15% can be achieved for medium range tanker.

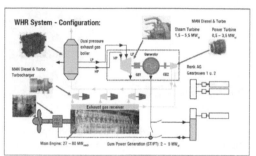

Figure 5. Waste Heat Recovery System Principles (MAN, 2013).

The ship during voyage at actual sea will be confronted with external weather loads such as wind, wave etc. and thus causes the resistance increase, which result in speed reduction if the power never changes, or alternatively, requires a power increasing in order to maintain a certain speed (Luo et al., 2016). In other words, increased resistance on the ship decreases the ship energy efficiency and lead to use more machine power to compensate the lost energy. Air lubrication system, hull form optimization and hull coating are the primary techniques/technologies to reduce the resistance on the hull of a ship and to increase the ship energy efficiency.

Air Lubrication Systems (ALS) technology is an original system which saves energy and reduces emission. By covering the ship's bottom using a blower, the system reduces frictional resistance between the ship hull and seawater as the ship cruises. With its energy-saving benefit for heavy cargo ships already verified, Air Lubrication System is designed for installation on cargo ships and passenger ships (MHI, 2016). Also known as the "bubble technology", it works on the principle of supplying air of the ship's underside so as to create layer of tiny bubbles that would help in reducing the friction between hull and the seawater. The system is expected to reduce CO_2 emissions up to 15%

together with saving of fuel (Marine in Sight, 2016b).

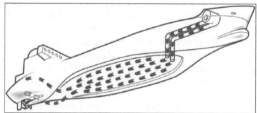

Figure 6. Air Lubrication System (Intertanko, 2009).

Hull form optimization continues to be recognized as a growing field within the marine community as a means to increase energy efficiency of ships (ABS, 2013). Depending on the capital costs and expected gains in ship energy efficiency, three options are available for the owner in terms of hull form optimization. The first option is to operate the ship in optimal conditions without any change in the hull form (c.g. service speed, ballast condition and ship capacity). The second option enables optimization of the design for service conditions (e.g. a number of expected operating draft, trim and speed combinations with their associated service durations). This alteration mostly involves modifications to the fore-body design and aft-body design. The third option is the development of a new design, where hull particulars, propulsion system, and power plant will be adjusted to reach the maximum efficiency (ABS, 2013). According to Svensen (2012), reductions in resistance of up to 20% can be expected if hull optimization is done at the beginning of the design process with optimal speed/draft combinations.

Traditionally, hull coatings have been used to reduce hydrodynamic drag and to control the growth of organisms on the hull (Fathom, 2013). The growth of these organisms has a significant effect on frictional resistance for a ship's hull that reduces the fuel efficiency of the vessel. The other adverse effect of such accumulations is that those organisms can be transferred as invasive aquatic species (Madsen et al., 2014). Tributyltin (TBT) based coatings was widely used to hinder the growth of organisms on the ship's hull but despite of the effectiveness of TBT, has now been banned by many countries and IMO because of the toxic damages to the marine environment. (MARAD, 2016). After technological developments and the prohibition of the usage of TBT based coatings that silicone based, self-polishing and foul-release types of coatings have been started to be used. According to several test of ABS (2013), when applied in combination with appropriate hull cleaning/maintenance, a high quality coating can provide an average reduction in propulsion fuel consumption by up to 4%.

3.3 Propulsion systems

The energy source for the propulsion of ships has experienced remarkable transformations over the last 150 years, starting with sails (renewable energy) through the use of coal to heavy fuel oil (HFO) and marine diesel oil (MDO), now the dominant fuel for this sector (IRENA, 2015). However, the extreme consumption of fossil fuels enforced IMO to restrict these kinds of energy sources to protect environment. As a consequence, these restrictions have led to companies and institutions to find alternative energy sources and propulsion systems for ships which are eco-friendlier and energy efficient. Some of the various types of new propulsion systems used in ships are given below:

The number of *electric propulsion ships* being built is expected to increase because of their ability to meet societal demands such as simplifying the running of a ship, decreasing fuel consumption during travel, and reducing CO2 emissions. On the other hand, this propulsion system has some disadvantages such as the efficiency of electrical plant is less than conventional propulsion systems and the installation cost of electric propulsion system is much higher (Ajioka and Ohno, 2013). Figure 7 shows the basic idea of a marine electric propulsion system is quite simple: an electric motor is driven by a battery pack. That battery pack can be charged by plugging it into the on-shore power grid while docked, by a renewable energy source located on-shore, or by a renewable energy source on-board such as solar panels or wind turbines (NSBA, 2015).

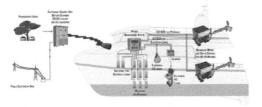

Figure 7. Schematic of Eco-Friendly Electric Propulsion Boat (Spagnolo et al., 2011)

Wind-assist propulsion technologies has been one of most invested industry last decade because of the potentially offering double-digit fuel savings (MARAD, 2016). Wind-assisted propulsion is the use of a device, such as a wingsail, soft sail, and kite or Flettner rotor, to capture the energy of the wind and generate forward thrust (LR, 2016). Each sail type has advantages and disadvantages based on many factors such as impact on operation, performance, flexibility and installation cost etc. and they provide different fuel-savings capacity (MARAD, 2016). This difference is due not only to the varied technology types, but also to the different options for implementation and the impact of

operational factors such as weather conditions and the ship's route (LR, 2015). Among the wind-assist propulsion devices, Kite is the most efficient and advantageous sail technology (MARAD, 2016). Kite attached to the bow of the ship operate at altitude to maximize wind speeds as shown in Figure 8. According to study of Naajen & Koster (2008), by using Kite of 500 m^2 attached to a 300 m towing line at Beaufort 7 with stern quartering wind can yield up to 50% fuel savings for 50.000 dwt tanker theoretically. On the other hand, applicability restrictions due to ship structures, high installing and maintenance cost and unpredictable effect on ship stability are the main disadvantages of wind-assisted propulsion systems (ABS, 2013).

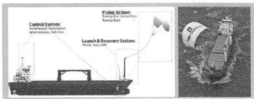

Figure 8. Kite-Assisted Sail (Skysails GmbH, 2016)

Fuel cells are regarded to be a good option for propulsion due to their higher efficiency and less emissions than conventional propulsion systems (Luckose et al., 2009). Fuel cells, like a battery, produce energy from an electro-chemical process by using hydrogen or methane (RAENG, 2013). The conversion of hydrogen or methane to DC power is very clean and efficient (50-70% efficiency in converting fuel to DC power), with the only emits water if hydrogen used and CO2 if hydrocarbon used like methane (MARAD, 2016). Furthermore, fuel cell propulsion system works with low vibration, low noise and do not degrade in course of time (Luckose, et al., 2009).

Hybrid system is alternative propulsion for ships where one or more modes can be utilized to optimize performance for economic, environmental or operational reasons (RAENG, 2015). New advanced hybrid propulsion systems use two types of propulsion combined a conventional main diesel engine with an active-front-end variable speed electric drive with an induction motor, that can operate as both a motor and a generator to propel ship (Simmonds, 2016). Hybrid propulsion systems have some advantages such as longer engines service life and consequently reduction of maintenance cost, significant reduction of CO2 and NO$_x$, low noise and vibrations and excellent fuel savings (Auxilia, 2016).

4 CONCLUSION

In recent years, a number of new methods and approaches have emerged that improve energy efficiency and reduction of greenhouse gases emission in ships. Furthermore, there are noteworthy efforts to enhance existing energy efficiency and greenhouse gases emission reduction methods. All these facts show that, "sustainable maritime transport" and "green ship" have become very important concepts for maritime industry. The methods that increase energy efficiency in ships are already used by many ship owning companies all over the world because they are cost effective. But there are some obstacles to the implementation of these methods. Green ship practices could be implemented more effectively, in case of elimination of market-based barriers and the effective coordination of the maritime industry's stakeholders. Some methods are used by ship owning companies very limited that are not energy efficient and cost effective but have potential to reduce GHG emissions such as scrubber practices. There are a lot of studies to increase the implementation of these methods by cutting costs, but these efforts are currently insufficient. A number of incentives and penalties should be used to increase the implementation of these methods.

REFERENCES

ABS – American Bureau of Shipping. 2013. Exhaust gas scrubber systems: status and guidance. http://ww2.eagle.org/content/dam/eagle/publications/2013/Scrubber_Advisory.pdf Accessed: 05.01.2017.

ABS – American Bureau of Shipping. Ship Energy Efficiency Measures: Status and Guidance. 2013. https://www.eagle.org/eagleExternalPortalWEB/ShowProperty/BEA%20Repository/References/Capability%20Brochures/ShipEnergyEfficiency Accessed: 08.01.2017.

Ahmad, M. 2014. Green Ships Fuelled by LNG: Stimulus for Indian Coastal Shipping. *India Quarterly: A Journal of International Affairs,* 70(2): 105-122.

Ajioka, Y. & Ohno, K. 2013. Electric Propulsion Systems for Ships. *Hitachi Review,* 62(3): 231.

Auxilia. 2016. Hybrid drive for commercial vessels. http://www.auxilia propulsion.com/images/pdf/Auxilia_%20commercial.pdf Accessed: 12.01.2017

Axelsson, M. 2009. Introduction of water to reduce NOx emissions. http://www.econologie.info/share/partager3/144467016792Duad.pdf Accessed: 06.01.2017.

Bai, Yong. & Jin, W. 2016. *Marine Structural Design (Second Edition).* USA: Elsevier.

Baldi, F. & Gabrielli, C. 2015. A feasibility analysis of waste heat recovery systems for marine applications. *Energy,* 80: 654-665.

Buhaug, Ø., Corbett, J.J., Endresen, Ø., Eyring, V., Faber, J., Hanayama, S., Lee, D.S., Lee, D., Lindstad, H., Mjelde, A., Pålsson, C., Wanquing, W., Winebrake, J.J. & Yoshida, K. 2009. Second IMO Greenhouse Gas Study. International Maritime Organization, London.

Cabezas-Basurko, O., Mesbahi, E. & Moloney, S.R. 2008. Methodology for sustainability analysis of ships. *Ships and Offshore Structures,* 3(1): 1-11.

Cobb, D., Glatch, L., Ruud, J. & Snyder, S. 1991. Application of selective catalytic reduction (SCR) technology for NOx reduction from refinery combustion sources. *Environmental progress,* 10(1): 49-59.

Dzida, M. & Mucharski, J. (2009). On the possible increasing of efficiency of ship power plant with the system combined of marine diesel engine, gas turbine and steam turbine in case of main engine cooperation with the gas turbine fed in parallel and the steam turbine. *Polish Maritime Research,* 59(16): 47–52.

Fathom Focus. 2013. Hull Coatings for Vessel Performance / The Important Role of the Hull in Ship Efficiency. http://www.civilengineeringhandbook.tk/fuel-injection/waterbased-techniques.html Accessed: 06.01.2017.

Harding, R. 2006. Ecologically sustainable development: origins, implementation and challenges. *Desalination,* 187(1): 229-239. http://www.yara.ie/nox-reduction/nox-reduction-in-ships/scr-systems-for-marine-vessels/ Accessed: 06.01.2017. http://www.skysails.info/english/ Accessed: 1 0.01.2017.

IMO, International Maritime Organization, http://www.imo.org/en/MediaCentre/HotTopics/GHG/Pages/default.aspx Accessed: 25.12.2016.

Intertanko. 2009. Green ship of the future. https://www.intertanko.com/upload/2.4%20GSF%20presentation%20TEF%202009-09-02.pdf Accessed: 05.01.2017.

IRENA- International Renewable Energy Agency. 2015. Renewable Energy Options for Shipping: Technology Brief.

Jiang, L., Ge, Y., Shah, A. N., He, C. & Liu, Z. (2010). Unregulated emissions from a diesel engine equipped with vanadium-based urea-SCR catalyst. *Journal of Environmental Sciences,* 22(4): 575-581.

Lloyd's Register (2015). Wind-powered shipping: A review of the commercial, regulatory and technical factors affecting uptake of wind assisted propulsion. http://www.ussrail.eu/wpcontent/uploads/2015/12/Wind_powered_shipping-Lloyds-Register.pdf Accessed: 12.01.2017.

Luckose, L., Hess, H. L. & Johnson, B. K. 2009. Fuel cell propulsion system for marine applications. *Proceedings of Electric Ship Technologies Symposium,* p. 574-580.

Luo, S., Ma, N. & Hirakawa, Y. 2016. Evaluation of resistance increase and speed loss of a ship in wind and waves. *Journal of Ocean Engineering and Science,* 1(3): 212-218.

Madsen, R. T., White, E. E, & Larsen, D. W. 2014. Regional Class Search Vessel Design: Green Ship Initiatives. https://ceoas.oregonstate.edu/ships/rcrv/pdf/12100-054-01_Green_Ship_Initiatives_Rev_P3.pdf Accessed: 18.01.2017

Maddox Consulting. 2012. Analysis of market barriers to cost effective GHG emission reductions in the maritime transport sector. Final report.

MAN Diesel & Turbo. 2013. Waste Heat Recovery System (WHRS) for Reduction of Fuel Consumption, Emissions and EEDI. http://marine.man.eu/docs/librariesprovider6/technical-papers/waste-heat-recovery-system.pdf?sfvrsn=10 Accessed: 08.01.2017.

MARAD- Maritime Administration. 2016. Ship Operations Cooperative Program: Energy Efficiency White Paper.

Marine in Sight. 2016. Different types of marine propulsion systems used in the shipping world. http://www.marineinsight.com/main-engine/different-types-of-marine-propulsion-systems-used-in-the-shipping-world/ Accessed: 29.12.2016.

Marine in Sight. 2016b. Seven technologies to reduce fuel consumption of ships. http://www.marineinsight.com/

tech/7-technologies-to-reduce-fuel-consumption-of-ships/ Accessed: 06.01.2017.

Mitsubishi Heavy Industries – MHI. 2016. Mitsubishi Air Lubrication System (MALS). https://www.mhigloal.com/products/detail/engineering_mals.html Accessed: 06.01.2017.

Molland, A.F., Turnock, S.R. & Hudson, D.A. 2011. *Ship Resistance and propulsion practical estimation of ship propulsive power*. New York: Cambridge University Press.

Naaijen, P. & Koster, V. 2007. Performance of Auxiliary Wind propulsion for Merchant Ship Using a Kite. 2nd International Conference on Marine Research and Transportation, p. 45-53.

NSBA- Nova Scotia Boat builders Association. 2015. Review of All-Electric and Hybrid-Electric Propulsion Technology for Small Vessels.

Pike, K., Butt, N., Johnson, D. & Walmsley, S. 2011. *Global sustainable shipping initiatives*. Audit and overview 2011. A Report for WWF.

Praveen, R. & Natarajan, S. (2014). Experimental study of selective catalytic reduction system on CI engine fueled with diesel-ethanol blend for NOx Reduction with Injection of Urea Solutions. *International Journal of Engineering and Technology*, 6(2): 895-904.

RAENG-Royal Academy of Engineering. 2013. Future ship powering options: Exploring alternative methods of ship propulsion. http://www.raeng.org.uk/publications/reports/future-ship-powering-options Accessed: 11.01.2017.

Riom, E., Olsson, L. O. & Hagestron, U. 2001. *NOx emission reduction with the humid air motor concept*. CIMAC, Hamburg.

Sarvi, A., Kilpinen, P. & Zevenhoven, R. (2009). Emissions from large-scale medium-speed diesel engines: 3. Influence of direct water injection and common rail. *Fuel Processing Technology*. 90(2): 222-231.

Senary, K., Tawfik, A., Hegazy, E. & Ali, A. 2016. Development of a waste heat recovery system onboard LNG carrier to meet IMO regulations. *Alexandria Engineering Journal*. 55(3): 1951-1960.

Sherbaz, S. & Duan, W. 2012. Operational options for green ships. *Journal of Marine Science and Application*, 11(3): 335-340.

Sherbaz, S., Maqsood, A. & Khan, J. 2015. Machinery Options for Green Ship. Journal of Engineering Science and Technology Review 8(3): 169-173.

Simmonds, O. J. 2016. Advanced hybrid systems and new integration challenges. 13th International Naval Engineering Conference (INEC).

Singh, D. V. & Pedersen, E. 2016. A review of waste heat recovery technologies for maritime applications. *Energy Conversion and Management*. 111: 315-328.

Solak Fışkın, C. & Deveci, D.A. 2015. Perceptions of Female and Male University Students on Sustainable Maritime Development Concept: A Case Study from Turkey, *Journal of ETA Maritime Science*, 3(1): 23-36.

Spagnolo, G.S., Papalillo, D. & Martocchia, A. 2011. Eco Friendly Electric Propulsion Boat. Proceedings of the 10th International Conference on Environment and Electrical Engineering (EEEIC'11), p. 1–4.

Svensen, T. 2012. *Hull design boosts fuel economy of ultra large container ships*. SMM, Hamburg.

UNCTAD Review of Maritime Transport. 2015. United Nations Conference on Trade and Development, http://unctad.org/en/PublicationsLibrary/rmt2015_en.pdf , Accessed: 24.12.2016.

Wartsila. 2007. Direct water injection. Wartsila Encyclopedia of Marine Technology. http://www.wartsila.com/encyclopedia/term/direct-water-injection-(dwi), Accessed: 05.01.2017.

WCED. 1987. World Commission on Environment and Development Our Common Future, Oxford University Press, Oxford.

WWF. 2012. World Wildlife Fund - Shipping and Sustainability Based on the Global Sustainable Shipping Initiatives Report for WWF. http://www.wwf.at/de/view/files/download/showDownload/?tool=feld=download &sprach_connect Accessed: 26.12.2016.

Noise Reduction in Railway Traffic as an Element of Greening of Transport

J. Kozyra, Z. Łukasik & A. Kuśmińska-Fijałkowska
University of Technology and Humanities in Radom, Radom, Poland

ABSTRACT: Discussion on railway noise has become very important issue in a few European countries, especially in those, where extent of railway transport is increasing and is playing more and more important role in greening of transport. Railway noise is a problem mainly in central and western Europe. In this article, the authors presented percentage share of emission of railway noise expressed by rates LAeqD and LAeqN for 86 points in railway traffic. Classification of vehicles, tracks and supporting structures for measurements of noise level and locations of equal sources of noise was presented. The division of emitted noise depending on mechanism of emission was made and noise of turning, traction, aerodynamical, squeals and noise caused by acoustic influence of the objects of infrastructure, for example, bridges and viaducts was distinguished. The authors presented new solutions of process of implementation of quiet rolling stock, which is an element of strategy of railway noise reduction

1 INTRODUCTION

Noise is defined as sound unwanted or harmful for human health, caused by means of transport: road traffic, railway traffic, air traffic and from the areas of industrial activity. Ensuring the best possible acoustic climate of environment helps to protect against noise, that is, acoustic phenomena in a given area, in particular, by maintaining noise level below permissible value or on this level and reducing noise level to at least permissible. On the one hand, there is an increase in traffic noise risk in Poland (Brdulak J., Zakrzewski B. 2016; Grad B., Krajewska R. 2011). On the other hand, increase is reduced and there are downward trends in the field of industrial noise (GUS, Departament Badań Regionalnych i Środowiska. 2015; Łukasik Z., Kuśmińska-Fijałkowska A., Kozyra J. 2016).

Discussion on railway noise is becoming more and more important, because the extent of railway transport is increasing and it is playing more and more important role in greening of transport (Brdulak J., Zakrzewski B. 2016; Krysiuk, C. 2011). To achieve goals of sustainable development, formulated in 2011 in the European Commission's White Paper on transport and regulations on greening of transport, the impact of railway on environment must be minimized (Kozyra J. 2009; Łukasik Z., Kuśmińska-Fijałkowska A., Kozyra J.

2016). It is necessary to maintain its position as an ecological means of transport and promote railway transport, in order to reduce general impact of transport on environment.

In order to analyse the noise situation in Europe, following current EC legislation, the Member States have to provide noise maps and noise action plans. Noise action plans describe the measures taken to lower environmental noise for identified affected inhabitants. However, legal conditions differ widely across Europe as Member States have different limits or threshold limits for environmental noise emissions, and usually these limits are tested only when building new infrastructure or during major redevelopment (EU Parliament, Transport end Tourism. Reducing railway noise pollution. 2012).

2 THE INFLUENCE OF RAILWAY NOISE IN MEASURING AND STATISTICAL TERMS

Noise is an annoying phenomenon, contaminating the environment and adversely affecting the health of people exposed to high ambient noise levels above 70 dB(A) – or even less.

Noise calculations shall be defined in the frequency range from 63 Hz to 8 kHz Frequency band results shall be provided at the corresponding frequency interval. Permissible noise level for the roads and railway lines expressed by daytime rates L_{AeqD} (dB) and nightly rates L_{AeqN} (dB) is presented in the publication. For residential and single-family and farmstead buildings, connected with temporary or permanent stay and in central zone of cities above 100 thousand inhabitants, permissible daytime noise L_{AeqD} (dB) is between $50 \div 68$ dB, in the night time, the rate L_{AeqN} (dB) is $45 \div 60$ dB.

Calculations are performed in octave bands for road traffic, railway traffic and industrial noise, except for the railway noise source sound power, that uses third octave bands. For road traffic, railway traffic and industrial noise, based on these octave band results, the A-weighted long term average sound pressure level for the day, evening and night period, as defined in Annex I and referred to in Art. 5 of Directive 2002/49/EC, is computed by summation over all frequencies (Dyrektywa 2002/49/WE). Long-term sound level L_{Aeq} from all sources is expressed by a dependency (1):

$$L_{Aeq,T} = 10 \times \lg \sum_{i=1} 10^{\left(L_{eq,T,i} + A_i\right)/10} \quad (1)$$

where :

A_i – weighting curve A in accordance with definition of a norm IEC 61672-1;

I – frequency band,

T – time corresponding to daytime, evening or night.

In 2014, measurements of noise in 86 points were taken by railway lines. Emission was mainly measured in regional, interregional and local lines. Table 1 presents the number of measuring points in particular provinces.

Table 1. The number of points measuring railway noise in 2014 for the provinces (IOŚ – PIB. 2014)

Province	The number of measuring points
łódzkie	12
małopolskie	8
mazowieckie	22
opolskie	2
podlaskie	15
pomorskie	3
śląskie	10
świętokrzyskie	2
wielkopolskie	11
zachodniopomorskie	1

Based on measuring data from Ehalas in 2014, percentage share of emission of railway noise was measured at a distance of 10 and 20 meters from railway tracks. Long-term sound level was expressed by daytime rate L_{AeqD} (dB) and for night time L_{AeqN} (dB). Ehalas base is a register run in accordance with art. 120 a of environmental protection act and decree of the Minister of the Environment of April 25, 2008 on requirements concerning register containing information about acoustic state of environment (Dz. U. [Journal of Laws] No. 82, item 500). Daytime emission in 2014

did not exceed 70 dB (IOŚ – PIB. 2014). Fig. 1 presents percentage share of emission of railway noise expressed by daytime rate.

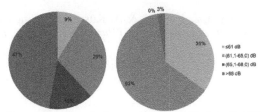

Figure 1. Emission of railway noise expressed by L_{AeqD} (dB) for particular ranges of emission at a distance of a) 10 meters for 34 measuring points, b) 20 meters for 40 measuring points (Dyrektywa (UE) 2015/996)

Fig. 2 presents the percentage share of emission of railway noise expressed by night time rate.

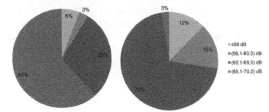

Figure 2. Emission of railway noise expressed by L_{AeqN} (dB) for particular ranges of emission at a distance of a) 10 meters for 34 measuring points, b) 20 meters for 40 measuring points (Dyrektywa (UE) 2015/996)

Measuring range at a distance of 20 meters will be more reliable for measurements taken on protected areas around railway lines. It results from the line of residential buildings, which runs at a distance not shorter than 20 meters from railway tracks. Accepting such condition for the analysis, presented graphs show that exceeding permissible noise levels during the daytime is not expected. Whereas, in the night time, they were exceeded in 90% of measurements.

3 CLASSIFICATION OF RAIL VEHICLES, TRACKS AND SUPPORTING STRUCTURES FOR MEASUREMENTS OF NOISE LEVEL

In the measurements of noise level, rail vehicles are defined as a separate subunit of a rail set (usually locomotive, actuating member, pulling or freight carriage), which can be independently moved and detached from the set. Some specific circumstances may occur for sub-units of a train that are a part of a non-detachable set, e.g. share one bogie between them. For the purpose of this calculation method, all these sub-units are grouped into a single vehicle.

Table 2. Classification and descriptors for railway vehicles (Baza EHALAS)

Digit	1	2	3	4
Descriptor	Vehicle type	Number of axles per vehicle	Brake type	Wheel measure
Explanation of the descriptor	descriptor A letter that describes the type	The actual number of axles	A letter that describes the brake type	A letter that describes the noise reduction measure type
Possible descriptors	**h** high speed vehicle (> 200 km/h) **m** self-propelled passenger coaches **p** hauled passenger coaches **c** city tram or light metro self-propelled and non-self-propelled coach **d** diesel loco **e** electric loco **a** any generic freight vehicle **o** other (i.e. maintenance vehicles etc.)	**1** **2** **3** **4** **etc.**	**c** cast-iron block **k** composite or sinter metal block **n** non-tread braked, like disc, drum, magnetic	**n** no measure **d** dampers **s** screens **o** other

Table 3. Terminology to the track types (Baza EHALAS)

Digit	1	2	3	4	5	6
Descriptor	Track base	Railhead Roughness	Rail pad type	Additional measures	Rail joints	Curvature
Explanation of the descriptor	Type of track base	Indicator for roughness	Represents an indication of the 'acoustic' stiffness	A letter describing acoustic device	Presence of joints and spacing	Indicate the radius of curvature in [m]
Codes allowed	**B** Ballast **S** Slab track **L** Ballasted bridge **N** Non-ballasted bridge propelled coach **T** Embedded track **O** Other	**E** Well maintained and very smooth **M** Normally maintained **N** Not well maintained **B** Not maintained and bad co	**S** Soft (150-250 MN/m) **M** Medium (250 to 800 MN/m) **H** Stiff (800-1 000 MN/m)	**N** None **D** Rail damper **B** Low barrier **A** Absorber plate on slab track **E** Embedded rail **O** Other	**N** None **S** Single joint or switch **D** Two joints or switches per 100 m **M** More than two joints or switches per 100 m	**N** Straight track **L** Low (1 000- 500 m) **M** Medium (less than 500 m and more than 300 m) **H** High (less than 300 m)

Table 2. defines a common language to describe the vehicle types. It presents the relevant descriptors to be used to classify the vehicles in full. These descriptors correspond to properties of the vehicle, which affect the acoustic directional sound power per metre length of the equivalent source line modelled.

The number of vehicles for each type shall be determined on each of the track sections for each of the time periods to be used in the noise calculation. It shall be expressed as an average number of vehicles per hour, which is obtained by dividing the total number of vehicles travelling in a given time period by the duration in hours of this time period (e.g. 24 vehicles in 4 hours means 6 vehicles per hour). All vehicle types travelling on each track section shall be used.

The existing tracks may differ because there are several elements contributing to and characterising their acoustic properties. The track types are listed in table 3.

Some of the elements have a large influence on acoustic properties, while others have only secondary effects. In general, the most relevant elements influencing the railway noise emission are: railhead roughness, rail pad stiffness, track base, rail joints and radius of curvature of the track. Alternatively, the overall track properties can be defined and, in this case, the railhead roughness and the track decay rate according to ISO 3095 are the two acoustically essential parameters, plus the radius of curvature of the track

A track section is defined as a part of a single track, on a railway line or station or depot, on which the track's physical properties and basic components do not change.

4 THE SOURCES OF NOISE IN RAIL VEHICLES

The different equivalent noise line sources are placed at different heights and at the centre of the track. All heights are referred to the plane tangent to the two upper surfaces of the two rails. The equivalent sources include different physical sources. These physical sources are divided into different categories depending on the generation mechanism, and are:

- rolling noise (including not only rail and track base vibration and wheel vibration but also,

where present, superstructure noise of the freight vehicles);
- traction noise;
- aerodynamic noise;
- impact noise (from crossings, switches and junctions);
- squeal noise
- noise due to additional effects such as bridges and viaducts.

Equivalent noise sources position is shown in Fig 3.

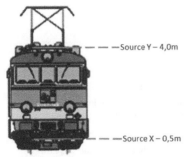

Figure 3. Position of the equivalent sound sources

The roughness of wheels and railheads, through three transmission paths to the radiating surfaces (rails, wheels and superstructure), constitutes the rolling noise. This is allocated to h = 0,5 m (radiating surfaces X) to represent the track contribution, including the effects of the surface of the tracks, especially slab tracks (in accordance with the propagation part), to represent the wheel contribution and to represent the contribution of the superstructure of the vehicle to noise (in freight trains).

The equivalent source heights for traction noise vary between 0,5 m (source X) and 4,0 m (source Y), depending on the physical position of the component concerned. Sources such as gear transmissions and electric motors will often be at an axle height of 0,5 m (source X). Louvres and cooling outlets can be at various heights; engine exhausts for diesel-powered vehicles are often at a roof height of 4,0 m (source Y). Other traction sources such as fans or diesel engine blocks may be at a height of 0,5 m (source X) or 4,0 m (source Y). If the exact source height is in between the model heights, the sound energy is distributed proportionately over the nearest adjacent source heights. For this reason, two source heights are foreseen by the method at 0,5 m (source X), 4,0 m (source Y), and the equivalent sound power associated with each is distributed between the two depending on the specific configuration of the sources on the unit type.

Aerodynamic noise effects are associated with the source at 0,5 m (representing the shrouds and the screens, source Y), and the source at 4,0 m

(modelling all over roof apparatus and pantograph, source X). The choice of 4,0 m for pantograph effects is known to be a simple model, and has to be considered carefully if the objective is to choose an appropriate noise barrier height. Impact, squeal and bridgenoise is associated with the source at 0,5 m (source X).

4.1 *Rolling noise*

The vehicle contribution and the track contribution to rolling noise are separated into four essential elements:
- wheel roughness,
- rail roughness,
- vehicle transfer function to the wheels.

The vehicle contribution and the track contribution to rolling noise are separated into four essential elements: wheel roughness, rail roughness, vehicle transfer function to the wheels and to the superstructure (vessels) and track transfer function. Wheel and rail roughness represent the cause of the excitation of the vibration at the contact point between the rail and the wheel, and the transfer functions are two empirical or modelled functions that represent the entire complex phenomena of the mechanical vibration and sound generation on the surfaces of the wheel, the rail, the sleeper and the track substructure. This separation reflects the physical evidence that roughness present on a rail may excite the vibration of the rail, but it will also excite the vibration of the wheel and vice versa. Not including one of these four parameters would prevent the decoupling of the classification of tracks and trains.

Rolling noise is mainly excited by rail and wheel roughness in the wavelength range from 5-500 mm. The roughness level L_r is defined as 10 times the logarithm to the base 10 of the square of the mean square value r^2 of the roughness of the running surface of a rail or a wheel in the direction of motion (longitudinal level) measured in μm over a certain rail length or the entire wheel diameter, divided by the square of the reference value r_0. Determined according to dependency (2).

$$L_r = 10 \times \lg \left(\frac{r}{r_0} \right)^2 \qquad (2)$$

where :
$r_o = 1 \mu m$
r – r.m.s. of the vertical displacement difference of the contact surface to the mean level

The roughness level Lr is typically obtained as a spectrum of wavelength λ and it shall be converted to a frequency spectrum $f = v/\lambda$, where f is the centre band frequency of a given 1/3 octave band in Hz, λ is the wavelength in m, and v is the train speed in km/h. The roughness spectrum as a function of frequency shifts along the frequency axis for different speeds. In general cases, after conversion to the frequency spectrum by means of the speed, it is necessary to obtain new 1/3 octave band spectra values averaging between two corresponding 1/3 octave bands in the wavelength domain. To estimate the total effective roughness frequency spectrum corresponding to the appropriate train speed, the two corresponding 1/3 octave bands defined in the wavelength domain shall be averaged energetically and proportionally.

Vehicle, track and superstructure transfer function speed-independent $L_{H,TR,i}$ $L_{H,VEH,i}$ oraz $L_{H,VEH,SUP,i}$, are defined: the first for each j-th track section and the second two for each t-th vehicle type. They relate the total effective roughness level with the sound power of the track, the wheels and the superstructure respectively.

The superstructure contribution is considered only for freight wagons, therefore only for vehicle type „a" (table.2). For rolling noise, therefore, the contributions from the track and from the vehicle are fully described by these transfer functions and by the total effective roughness level. When a train is idling, rolling noise shall be excluded.

For sound power per vehicle the rolling noise is calculated at axle height, and has as an input the total effective roughness level $L_{R,TOT,i}$ as a function of the vehicle speed v, the track, vehicle and superstructure transfer functions $L_{H,TR,i}$, $L_{H,VEH,i}$ and $L_{H,VEH,SUP,i}$, and the total number of axles N_a for source X. Sound power with three speed-independent transfer functions, $L_{H,TR,i}$ $L_{H,VEH,i}$ and $L_{H,VEH,SUP,i}$, is defined equations (3), (4), (5)

$$L_{W,0,Tr,i} = L_{R,TOT,i} + L_{H,TR,i} + 10 \text{ x } \lg(N_a)\qquad(3)$$

$$L_{W,0,VEH,i} = L_{R,TOT,i} + L_{H,VEH,i} + 10 \text{ x} \lg(N_a)\qquad(4)$$

$$L_{W,0,VEHSUP,i} = L_{R,TOT,i} + L_{H,VEHSUP,i} + 10 \text{ x} \lg(N_a)\qquad(5)$$

where:
$L_{R,TOT,i}$ – The total and effective roughness level for wave-number band i
N_a – is the number of axles per vehicle for the t-th vehicle type.

The total and effective roughness level for wave-number band i ($L_{R,TOT,i}$) is defined as the energy sum of the roughness levels of the rail and that of the wheel plus the A3(λ) contact filter to take into account the filtering effect of the contact patch between the rail and the wheel, and is in dB:

$$L_{R,TOT,i} = 10 \text{ x } \lg\left(10^{L_{r,TR,i}/10} + 10^{L_{r,VEH,i}/10}\right) + A_{3,i}\qquad(6)$$

where:
$L_{r,TR,i}$ – The rail roughness level (track side roughness) for the i-th wave-number band

$L_{r,VEH,i}$ – The wheel roughness level vehicle side roughness)for the i-th wave-number band

A minimum speed of 50 km/h (30 km/h only for trams and light metro) shall be used to determine the total effective roughness and therefore the sound power of the vehicles (this speed does not affect the vehicle flow calculation) to compensate for the potential error introduced by the simplification of rolling noise definition, braking noise definition and impact noise from crossings and switches definition.

4.2 Impact noise

Impact noise can be caused by crossings, switches and rail joints or points. It can vary in magnitude and can dominate rolling noise. Impact noise shall be considered for jointed tracks. For impact noise due to switches, crossings and joints in track sections with a speed of less than 50 km/h (30 km/h only for trams and light metro), since the minimum speed of 50 km/h (30 km/h only for trams and light metro) is used to include more effects according to the description of the rolling noise chapter, modelling shall be avoided. Impact noise modelling shall also be avoided under running condition c = 2 (idling).

Impact noise is included in the rolling noise term by (energy) adding a supplementary fictitious impact roughness level to the total effective roughness level on each specific j-th track section where it is present. In this case shall be used $L_{R,TOT + IMPACT,i}$, and it will equation (7):

$$L_{R,TOT,i+IMPACT,i} = 10 \text{ x} \lg\left(10^{L_{R,TOT,i}/10} + 10^{L_{R,IMPACT,i}/10}\right)\qquad(7)$$

$L_{R,IMPACT,i}$ is a 1/3 octave band spectrum (as a function of frequency). To obtain this frequency spectrum, a spectrum is given as a function of wavelength λ and shall be converted to the required spectrum as a function of frequency using the relation $\lambda = v/f$, where f is the 1/3 octave band centre frequency in Hz and v is the s-th vehicle speed of the t-th vehicle type in km/h.

Impact noise will depend on the severity and number of impacts per unit length or joint density, so in the case where multiple impacts are given, the impact roughness level to be used in the equation above shall be calculated as follows:

$$L_{R,IMPACT,i} = L_{R,IMPACT,i - SINGLE,i} + 10 \text{ x} \lg\left(\frac{n_i}{0,01}\right)\qquad(8)$$

where:
$L_{R,IMPACT - SINGLE,i}$ – is the impact roughness level as given for a single impact and n_l is the joint density

The default impact roughness level is given for a joint density $n_l = 0,01$ m^{-1}, which is one joint per each 100 m of track. Situations with different numbers of joints shall be approximated by adjusting the joint density n_l. It should be noted that when modelling the track layout and segmentation, the rail joint density shall be taken into account, i.e. it may be necessary to take a separate source segment for a stretch of track with more joints. The $L_{W,0}$ of track, wheel/bogie and superstructure contribution are incremented by means of the $L_{R,IMPACT,i}$ for +/– 50 m before and after the rail joint. In the case of a series of joints, the increase is extended to between – 50 m before the first joint and + 50 m after the last joint. The applicability of these sound power spectra shall normally be verified on-site. For jointed tracks, a default n_l of 0,01 shall be used.

4.3 Squeal

Curve squeal is a special source that is only relevant for curves and is therefore localised. As it can be significant, an appropriate description is required. Curve squeal is generally dependent on curvature, friction conditions, train speed and track-wheel geometry and dynamics. The emission level to be used is determined for curves with radius below or equal to 500 m and for sharper curves and branch-outs of points with radii below 300 m. The noise emission should be specific to each type of rolling stock, as certain wheel and bogie types may be significantly less prone to squeal than others.

The applicability of these sound power spectra shall normally be verified on-site, especially for trams. Taking a simple approach, squeal noise shall be considered by adding 8 dB for R < 300 m and 5 dB for 300 m < R < 500 m to the rolling noise sound power spectra for all frequencies. Squeal contribution shall be applied on railway track sections where the radius is within the ranges mentioned above for at least a 50 m length of track.

4.4 Traction noise

Although traction noise is generally specific to each characteristic operating condition amongst constant speed, deceleration, acceleration and idling, the only two conditions modelled are constant speed (that is valid as well when the train is decelerating or when it is accelerating) and idling. The source strength modelled only corresponds to maximum load conditions and this results in the quantities $L_{W,0,const,i} = L_{W,0,idling,i,fghhj}$. Also, the $L_{W,0,idling,i}$, corresponds to the contribution of all physical sources of a given vehicle attributable to a specific height.

The $L_{W,0,idling,i}$ is expressed as a static noise source in the idling position, for the duration of the idling condition, and to be used modelled as a fixed point. It shall be considered only if trains are idling for more than 0,5 hours.

These quantities can either be obtained from measurements of all sources at each operating condition, or the partial sources can be characterised individually, determining their parameter dependency and relative strength. This may be done by means of measurements on a stationary vehicle, by varying shaft speeds of the traction equipment, following ISO 3095:2005. As far as relevant, several traction noise sources have to be characterised which might not be all directly depending on the train speed:

- noise from the power train, such as diesel engines (including inlet, exhaust and engine block), gear transmission, electrical generators, mainly dependent on engine round per minute speed (rpm), and electrical sources such as converters, which may be mostly load-dependent,
- noise from fans and cooling systems, depending on fan rpm; in some cases fans can be directly coupled to the driveline,
- intermittent sources such as compressors, valves and others with a characteristic duration of operation and corresponding duty cycle correction for the noise emission.

As each of these sources can behave differently at each operating condition, the traction noise shall be specified accordingly. The source strength is obtained from measurements under controlled conditions. In general, locomotives will tend to show more variation in loading as the number of vehicles hauled and thereby the power output can vary significantly, whereas fixed train formations such as electric motored units (EMUs), diesel motored units (DMUs) and high-speed trains have a better defined load. There is no a priori attribution of the source sound power to the source heights, and this choice shall depend on the specific noise and vehicle assessed. It shall be modelled to be at source X and at source Y.

4.5 Aerodynamic noise

Aerodynamic noise is only relevant at high speeds above 200 km/h and therefore it should first be verified whether it is actually necessary for application purposes. If the rolling noise roughness and transfer functions are known, it can be extrapolated to higher speeds and a comparison can be made with existing high-speed data to check whether higher levels are produced by aerodynamic noise. If train speeds on a network are above 200 km/h but limited to 250 km/h, in some cases it may not be necessary to include aerodynamic noise, depending on the vehicle design. The aerodynamic noise contribution is given as a function of speed defined by the equations (9), (10):

$$L_{W,0,i} = L_{W,0,1,i}\,(v_0) + \alpha_{1,i} \times \lg\left(\frac{v}{v_0}\right) \quad \text{for source X} \quad (9)$$

$$L_{W,0,i} = L_{W,0,2,i}\,(v_0) + \alpha_{2,i} \times \lg\left(\frac{v}{v_0}\right) \quad \text{for source Y} \quad (10)$$

where:

v_0 – is a speed at which aerodynamic noise is dominant and is fixed at 300 km/h

$L_{W,0,1,i}$ – is a reference sound power determined from two or more measurement points, for sources at known source heights, for example the first boogie

$L_{W,0,2,i}$ – is a reference sound power determined from two or more measurement points, for sources at known source heights, for example the pantograph recess heights

$\alpha_{1,i}$ – is a coefficient determined from two or more measurement points, for sources at known source heights, for example the first boogie

$\alpha_{2,i}$ – is a coefficient determined from two or more measurement points, for sources at known source heights, for example the pantograph recess heights.

4.6 Additional effects

In the case where the track section is on a bridge, it is necessary to consider the additional noise generated by the vibration of the bridge as a result of the excitation caused by the presence of the train. Because it is not simple to model the bridge emission as an additional source, given the complex shapes of bridges, an increase in the rolling noise is used to account for the bridge noise. The increase shall be modelled exclusively by adding a fixed increase in the noise sound power per each third octave band. The sound power of only the rolling noise is modified when considering the correction and the new $L_{W,0,rolling-and-bridge,i}$.

$$L_{W,0,rolling-and-bridge,i} = L_{W,0,rolling-only,i} + C_{bridge} \qquad (11)$$

where:

C_{bridge} – is a constant that depends on the bridge type,

$L_{W,0,rolling-only,i}$ – is the rolling noise sound power on the given bridge that depends only on the vehicle and track properties.

Various sources like depots, loading/unloading areas, stations, bells, station loudspeakers, etc. can be present and are associated with the railway noise. These sources are to be treated as industrial noise sources (fixed noise sources) and shall be modelled, if relevant, according to the following chapter for industrial noise. These sources are to be treated as industrial noise sources (fixed noise sources) and shall be modelled, if relevant for industrial noise.

5 THE SOLUTIONS ELIMINATING RAILWAY NOISE

Railway noise is largely a problem of freight trains and trains containing older wagons or engines, and is a particularly severe problem during the night.

To reduce railway noise pollution, passive measures at the place of disturbance can be distinguished from active measures at the noise source. The most important passive methods used to reduce the impact of railway noise on the environment are noise protection walls and insulating windows. However, they are only locally effective, requiring huge investments to protect wider parts of railway networks.

In contrast, source-driven measures lower noise across the whole railway system if they are widely introduced. According to the current Technical Standard for Interoperability (TSI Noise), rolling stock which was introduced since the year 2000 (including engines and passenger coaches or passenger power cars) are required to lower noise emissions by about 10 dB(A) compared to the equipment of the 1960s and 1970s

Noise emitted by freight carriages can be reduced by using new carriages and their modernization in accordance with art.14 section 3 of the directive 2001/16/WE. Permissible values of noise are between 82dB (A) and 87 dB (A).

An effective solution in the running systems of freight carriages, reducing noise emission is to add disc brakes in a carriage. Using Y37-type freight carriages with disc brakes can reduce the level of noise emission by 8dB(A) (Gąsowski W., Sobaś M. 2013). The problem of loud freight carriages can also be solved by replacement of phosphoric cast-iron brake pads with composite ones. In this case, two developmental trends can be distinguished by using:
– plastic K-type inserts
– plastic LL-type inserts.

The use of K-type and LL-type inserts is regulated by regulations of UIC (International Union of Railways). Replacing phosphoric cast-iron inserts can significantly reduce noise emission, which is estimated about 10 dB(A) at 100 km/h. K-type brake inserts have higher friction coefficient and their value is more stable than for phosphoric cast-iron inserts. These inserts can be used in new or modernized freight carriages. LL-type inserts have friction coefficient similar to phosphoric cast-iron inserts, therefore, the can be replaced instead of cast-iron inserts without additional modernization expenditure.

Conducted exploitation tests have shown that significant reduction of noise level during a ride of a freight train can be achieved, when there is 30% of carriages with phosphoric cast-iron inserts in the set (Gąsowski W., Sobaś M. 2013).

Another method of reduction of noise level is the use of wheels with reduced noise emission. Using spring wheels with suppressing elements enables to reduce noise by $5 \div 6$ dB(A). Additional reduction of noise emission having its source in freight carriages can be achieved by using a wheel with a noise suppressor. Such suppressor is made of composite materials of „sandwich-type" structure and it is built in a structure of a wheel using metal ring. The better solution would be to apply a noise suppressor on both sides. The results of measurements of noise have shown that expected reduction of noise emission is: 2÷3 dB(A) at 80 km/h.

The reduction of noise level can also be achieved by using cover in an aerodynamic pantograph and reduction of their number.

6 CONCLUSIONS

Noise reduction is expensive technically difficult. Implementation of constructional means reducing noise emission is necessary to maintain competitiveness of railway transport. The share of environmentally-friendly rolling stock on the railway market has been decreasing in last years. Within the space of particular years in Poland, freight transports significantly decreased from 54 million tonne-kilometres of freight in 2000 to 47 million tonne-kilometres in 2014. The limitation for development of modern constructional means, reducing noise emission is money. In order to make this transport competitive, rolling stock must be successively modernized and replaced. Public aid should also be actively involved in such modernization, because it costs a lot of money that domestic carriers may not have. In case of building new freight carriages, it will be necessary to use running systems with plastic inserts, whereas, in case of exploited freight carriages, replacing phosphoric cast-iron P10 inserts with LL-type inserts will be necessary. In the second case, the costs of modernization are not high. Modernization of exploited carriages, that is, adding K-type inserts instead of phosphoric cast-iron inserts, is expensive and requires modernization of brake system. Therefore, this developmental direction is inhibited by economic considerations. It is estimated that the number of carriages necessary to run transport activity in 2014 was 60 thousand carriages. In case of rebuilding of brake systems for K-type inserts – the cost for one carriage will be about EUR 10 thousand, overall cost for 60 thousand carriages will be about EUR 0,6 billion, replacing hoop wheels with monobclok ones in about 50 thousand carriages, overall cost will be about EUR 0,55 billion. Further exploitation of carriages equipped with composite brake inserts is by 16% more expensive than exploitation of louder carriages. All these costs will be reflected in transport rates and rise in operational costs in other sectors of economy that make use of railway transport, among others, mining and power industry. In some cases, the use of railway carriages may turn out to be completely unprofitable. Moreover, high costs of modernization of existing rolling stock will be limited by investments in other areas such as purchase or modernization of locomotives. In Poland, replacement of rolling stock into quieter one has been made in a natural way since 2006 as a result of standard replacement of rolling stock, and all new carriages meet requirements of noise emission. Therefore, investing in new rolling stock than modernization is more rational.

REFERENCES

GUS Departament Badań Regionalnych i Środowiska: „Ochrona środowiska 2015. Informacje i opracowania statystyczne", Warszawa 2015

Łukasik Z., Kuśmińska-Fijałkowska A., Kozyra J.: Emission of acoustic sources of noise in the industrial plant. International Journal of Engineering Research and General Science 4, pp. 522-529, 2016

Kozyra J.: Wykorzystanie biopaliw w transporcie. Logistyka 3/2009

Łukasik Z., Kuśmińska-Fijałkowska A. , Kozyra J.: Eco-friendly technology to reduce CO2 emissions of passenger cars based on innovative solutions. Przegląd Elektrotechniczny 8/2016

Directorate general for internal policies. Policy departament B: Structural and cohesion policies. Transport end Tourism: „Reducing railway noise pollution". Brussels 2012

Dyrektywa 2002/49/WE Parlamnetu Europejskiego i Rady z dnia 25 czerwca 2002 r. odnosząca się do oceny i zarządzania poziomem hałasu w środowisku

Wyniki badań hałasu szynowego w roku 2014. Instytut Ochrony Środowiska – Państwowy Instytut Badawczy

Dyrektywa (UE) 2015/996 z dnia 19 maja 2015 r. ustanawiająca wspólne metody oceny hałasu zgodnie z dyrektywą 2002/49/WE

Baza EHALAS

Rozporządzenie Ministra Środowiska w sprawie dopuszczalnych poziomów hałasy w środowisku z dnia 15 października 2013 poz.112

Gąsowski W., Sobaś M.: Kostrukcyjne sposoby zmniejszania hałasu w układach biegowych wagonów towarowych. Pojazdy szynowe NR 1/2013, Kwartalnik naukowo-techniczny poświęcony zagadnieniom konstrukcji, budowy i badań taboru szynowego ISSN-0138-0370, Instytut Pojazdów Szynowych „TABOR"

Brdulak J., Zakrzewski B. Zarządzanie infrastrukturą techniczną Paneuropejskiego Korytarza Transportowego Nr II – problemy teoretyczne Problemy programowania inwestycji infrastrukturalnych w transporcie pp. 14-27 Instytut Transportu Samochodowego, 2016

Kuśmińska-Fijałkowska A., Lukasik Z. The Land Trans-Shipping Terminal In Processes Flow Stream Individuals Intermodal Transportion. TransNav, the International Journal on Marine Navigation and Safety of Sea Transportation 5.3, (2011)

Krysiuk, C. Nowy instrument finansowania infrastruktury transportowej w UE. Biuletyn Informacyjny ITS (2011): 5-6

Grad B., Krajewska R. Wpływ kryzysu na przewozy pasażerów transportem kolejowym w krajach Unii Europejskiej. Logistyka 3 (2011).

Noise in Road Transport as a Problem in European Dimension

A. Kuśmińska-Fijałkowska, Z. Łukasik & J. Kozyra
University of Technology and Humanities in Radom, Radom, Poland

ABSTRACT: In the last decade, activities aiming at reduction of negative impact of transport on natural environment in Poland have been intensified, especially after accession to the European Union. Solving the problem of transport noise, which is called pollution of the 21st century is important. Traffic noise has a significant impact on quality of natural environment. The authors of this article presented traffic noise and growing number of passenger cars and trucks as a problem in European dimension. The following factors were distinguished in traffic flow: ambient noise caused by mutual influence of tyres and road surface, noise emitted by drive units (engine, exhaust system) and others. Therefore, various protective solutions must be applied in the areas of emission, sound propagation. The most popular method of anti-noise protection is noise barriers. The authors presented solutions eliminating traffic noise, using photovoltaics integrated with acoustic barriers, and present intermodal transport as an alternative traditional road transport.

1 INTRODUCTION

The problem of transport noise in road traffic is one of the biggest threats of our civilization in the period of permanent economic growth (Gallasch E., Raggam, Reinhard B.,Cik M. 2016; Krysiuk C., Nowacki, G., Zakrzewski B. 2015) . Noise has impact on lowering the quality of life and health (Baliatsas Ch., van Kamp I., Swart W. 2016). It is the second largest ecological problem of European Union, apart from air pollution. It is estimated that:

- nearly 67 million people in the European Union is exposed to excessive traffic noise (exceeding 55dB L_{DWN}),
- nearly 21 million people in the European Union live in places, in which excessive noise at nights has negative impact on their health.

Noise, by disturbing sleep, makes people more aggressive, tired and distracted and, as a result, contributes to increasing number of accidents (www.obserwatoriumbrd.pl; Łukasik Z., Kuśmińska -Fijałkowska A., Kozyra J. 2016). The problem of exposing population to noise was perceived by the European Union a dozen or so years ago, which contributed to establishment of legal regulations within this scope (Directive 2007/46/WE). Nowadays, noise is treated as environmental pollution, therefore, similar rules, obligations and attitudes to noise and other fields of environmental protection were accepted (Ruiz-Padillo A., Ruiz Diego P., Torija, Antonio J. 2016; Zbyszyński M., Kamiński T., Krysiuk C. 2015; Banak M., Brdulak, J., Krysiuk, C. 2014). Therefore, European Union promotes intermodal transport as an alternative to traditional road transport, it may play a significant role in achieving goals of all-European policy of sustainable development of transport, because it may help to solve current and future transport problems in Europe. In Poland, intermodal transport network is a part of trans-European transport network and includes linear and nodal infrastructure (trans-shipment terminals) allowing transport and trans-shipment of intermodal transport units. Poland has good conditions for development of intermodal transport due to its geographical situation at the intersection of main European transport corridors (Krajewska R., Łukasik Z. 2010).

2 SOURCES OF NOISE, CATEGORIES OF VEHICLES ON THE ROAD

Noise calculations shall be defined in the frequency range from 63 Hz to 8 kHz. Frequency band results shall be provided at the corresponding frequency interval (1).

$$L_{Aeq,T} = 10 \times lg \sum_{i=0} 10^{\frac{(L_{Aeq,T,i} + A_i)}{10}} \qquad (1)$$

where:

A_i denotes the A-weighting correction according to IEC 61672-1

i frequency band index

T is the time period corresponding to day, evening or night.

The road traffic noise source shall be determined by combining the noise emission of each individual vehicle forming the traffic flow. Fig.1.

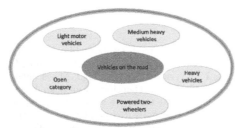

Figure 1. Road noise

These vehicles are grouped into five separate categories with regard to their characteristics of noise emission Table 1, Fig. 2:

Figure 2. Classification of road vehicles

Each vehicle (category 1, 2, 3, 4 and 5) is represented by one single point source radiating uniformly into the 2-π half space above the ground. The first reflection on the road surface is treated implicitly.-This point source is placed 0,05 m above the road surface. (Fig. 3)

The traffic flow is represented by a source line. In the modelling of a road with multiple lanes, each lane should ideally be represented by a source line placed in the centre of each lane. However, it is also acceptable to model one source line in the middle of a two way road or one source line per carriageway in the outer lane of multi- lane roads.

Figure 3. Location of equivalent point source on light vehicles (category 1), heavy vehicles (categories 2 and 3) and two-wheelers (category 4)

The sound power of the source is defined in the 'semi-free field', thus the sound power includes the effect of the reflection of the ground immediately under the modelled source where there are no disturbing objects in its immediate surroundings except for the reflection on the road surface not immediately under the modelled source (De Coensel B., Brown A. L. Tomerini, D. 2016).

The noise emission of a traffic flow is represented by a source line characterised by its directional sound power per metre per frequency. This corresponds to the sum of the sound emission of the individual vehicles in the traffic flow, taking into account the time spent by the vehicles in the road section considered. The implementation of the individual vehicle in the flow requires the application of a traffic flow model.

Table 1. Categories of vehicles on the road (Directive 2007/46/WE).

Category	Name	Vehicle category in EC Whole Vehicle Type Approval	Description
1	Light motor vehicles	M1 and N1	Passenger cars, delivery vans ≤ 3,5 tons, SUVs, MPVs including trailers and caravans
2	Medium heavy vehicles	M2,M3 or N2,N3	Medium heavy vehicles, delivery vans > 3,5 tons, buses, motorhomes, etc. with two axles and twin tyre mounting on rear axle
3	Heavy vehicles	M2 i N2 with trailer, M3,N3	Heavy duty vehicles, touring cars, buses, with three or more axles
4	Powered two-wheelers	L1, L2, L6, L3, L4, L5, L7	Two-, Three- and Four-wheel Mopeds Motorcycles with and without sidecars, Tricycles and Quadricycles
5	Open category	Not applicable	To be defined according to future needs

If a steady traffic flow of Q_m vehicles of category m per hour is assumed, with an average speed v_m (in km/h), the directional sound power per metre in frequency band i of the source line $L_{W',eq,line,i,m}$ is defined by (2):

$$L_{W',eq,line,i,m} = L_{W,i,m} + 10 \times \lg\left(\frac{Q_m}{1000 \times v_m}\right) \qquad (2)$$

where $L_{W',i,m}$ is the directional sound power of a single vehicle. $L_{W',m}$ is expressed in dB (ref. 10– 12 W/m). These sound power levels are calculated for each octave band i from 125 Hz to 4 kHz.

Traffic flow data Q_m shall be expressed as yearly average per hour, per time period (day-evening-night), per vehicle class and per source line. For all categories, input traffic flow data derived from traffic counting or from traffic models shall be used.

The speed v_m is a representative speed per vehicle category: in most cases the lower of the maximum legal speed for the section of road and the maximum legal speed for the vehicle category. If local measurement data is unavailable the maximum legal speed for the vehicle category shall be used.

In the traffic flow, all vehicles of category m are assumed to drive at the same speed, i.e. v_m, the average speed of the flow of vehicles of the category. A road vehicle is modelled by a set of mathematical equations representing the two main noise sources:

- Rolling noise due to the tyre/road interaction; (point2.)
- Propulsion noise produced by the driveline (engine, exhaust, etc.) of the vehicle. (point 3.)

Aerodynamic noise is incorporated in the rolling noise source. For light, medium and heavy motor vehicles (categories 1, 2 and 3), the total sound power corresponds to the energetic sum of the rolling and the propulsion noise. Thus, the total sound power level of the source lines $m = 1, 2$ lub 3 is defined by (3):

$$L_{W,i,m}(v_m) = 10 \times lg\left(10^{L_{WR,i,m}(v_m)/10} + 10^{L_{WP,i,m}(v_m)/10}\right) \quad (3)$$

where:
$L_{W,i,m}$ is the sound power level for rolling noise and $L_{W,P,i,m}$ is the sound power level for propulsion noise.

This is valid on all speed ranges. For speeds less than 20 km/h it shall have the same sound power level as defined by the formula for $v_m = 20$ km/h. For two-wheelers (category 4), only propulsion noise is considered for the source (4):

$$L_{W,i,m=4}\left(V_{m=4}\right) = L_{W,P,i,m=4}\left(V_{m=4}\right) \qquad (4)$$

This is valid on all speed ranges. For speeds less than 20 km/h it shall have the same sound power level as defined by the formula for $v_m = 20$ km/h.

The source equations and coefficients are valid for the following reference conditions:
- a constant vehicle speed,
- a flat road,
- an air temperature $\tau_{ref}. = 20$ °C,
- a virtual reference road surface, consisting of an average of dense asphalt concrete 0/11 and stone mastic asphalt 0/11, between 2 and 7 years old and in a representative maintenance condition,
- a dry road Surface,
- no studded tyres.

3 ROLLING NOISE

The rolling noise sound power level in the frequency band i for a vehicle of class m = 1, 2 or 3 is defined as: (5):

$$L_{WR,i,m} = A_{R,i,m} + B_{R,i,m} \times lg\left(\frac{v_m}{v_{ref}}\right) + \Delta L_{WR,i,m} \qquad (5)$$

The coefficients $A_{R,i,m}$. i $B_{R,i,m}$ are given in octave bands for each vehicle category and for a reference speed $v_{ref} = 70$ km/h. $\Delta L_{WR,i,m}$ corresponds to the sum of the correction coefficients to be applied to the rolling noise emission for specific road or vehicle conditions deviating from the reference conditions (6):

$$\Delta L_{WR,i,m} = \Delta L_{WR,road,i,m} + \Delta L_{studdedtyres,i,m} + \Delta L_{WR,acc,i,m} + \Delta L_{W,temp} \quad (6)$$

where:
$\Delta L_{WR,road,i,m}$ accounts for the effect on rolling noise of a road surface with acoustic properties different from those of the virtual reference Surface. It includes both the effect on propagation and on generation;

$\Delta L_{studdedtyres,i,m}$ is a correction coefficient accounting for the higher rolling noise of light vehicles equipped with studded tyres;

$\Delta L_{WR,acc,i,m}$ accounts for the effect on rolling noise of a crossing with traffic lights or a roundabout. It integrates the effect on noise of the speed variation;

$\Delta L_{W,temp}$ is a correction term for an average temperature τ different from the reference temperature $\tau_{ref} = 20$ °C.

3.1 *Correction for studded tyres*

In situations where a significant number of light vehicles in the traffic flow use studded tyres during several months every year, the induced effect on rolling noise shall be taken into account. For each vehicle of category $m = 1$ equipped with studded tyres, a speed-dependent increase in rolling noise emission is evaluated by (7)

$$\Delta_{stud,i}(v) = \begin{cases} a_i + b_i \times \log(\dfrac{50}{70}) & when\, v < 50\, km/h \\ a_i + b_i \times \log(\dfrac{v}{70}) & when\, 50 \le v \le 90\, km/h \\ a_i + b_i \times \log(\dfrac{90}{70}) & when\, v > 90\, km/h \end{cases}$$

where coefficients a_i and b_i are given for each octave band.

The increase in rolling noiseemission shall only be attributed according to t proportion of light vehicles with studded tyres and during a limited period T_s (in months) over the year. If $Q_{stud,ratio}$ is the average ratio of the total volume of light vehicles per hour equipped with studded tyres during the period T_s, (in months), then the yearly average proportion of vehicles equipped with studded tyres p_s is expressed by: (8):

$$p_s = Q_{stud,ratio} \times \frac{T_s}{12} \qquad (8)$$

The resulting correction to be applied to the rolling sound power emission due to the use of studded tyres for vehicles of category $m = 1$ in frequency band i shall be: (9):

$$\Delta L_{studdedtyres,i,m} = 10 \times lg\left[(1 - p_s) + p_s 10^{\frac{\Delta_{stud,i,m=1}}{10}} \right] \qquad (9)$$

For vehicles of all other categories no correction shall be applied: (10):

$$\Delta L_{studdedtyres,i,m \ne 1} = 0 \qquad (10)$$

3.2 Effect of air temperature on rolling noise correction

The air temperature affects rolling noise emission; the rolling sound power level decreases when the air temperature increases. This effect is introduced in the road surface correction. Road surface corrections are usually evaluated at an air temperature of $\tau_{ref} = 20$ °C. In the case of a different yearly average air temperature °C, the road surface noise shall be corrected by (11):

$$\Delta L_{W,temp,m}(\tau) = K_m \times (\tau_{ref} - \tau) \qquad (11)$$

The correction term is positive (i.e. noise increases) for temperatures lower than 20 °C and negative (i.e. noise decreases) for higher temperatures. The coefficient K depends on the road surface and the tyre characteristics and in general exhibits some frequency dependence. A generic coefficient $K_{m=1} = 0,08$ dB/°C for light vehicles (category 1) and $K_{m=2} = K_{m=3} = 0,04$ dB/°C for heavy vehicles (categories 2 and 3) shall be applied for all road surfaces. The correction coefficient shall be applied equally on all octave bands from 63 to 8 000 Hz.

4 NOISE COMING FROM THE PROPULSION UNIT

The propulsion noise emission includes all contributions from engine, exhaust, gears, air intake, etc. The propulsion noise sound power level in the frequency band i for a vehicle of class m is defined as (12):

$$L_{WP,i,m} = A_{P,i,m} + B_{P,i,m} \times \left(\frac{v_m - v_{ref}}{v_{ref})} \right) + \Delta L_{WP,i,m} \qquad (12)$$

where:
The coefficients $A_{P,i,m}$ i $B_{P,i,m}$ are given in octave bands for each vehicle category and for a reference speed $v_{ref} = 70$ km/h.

$\Delta L_{WP,i,m}$ corresponds to the sum of the correction coefficients to be applied to the propulsion noise emission for specific driving conditions or regional conditions deviating from the reference conditions (13):

$$\Delta L_{WP,i,m} = \Delta L_{WP,road,i,m} + \Delta L_{WP,grad,i,m} + \Delta L_{WP,acc,i,m} \qquad (13)$$

where:
$\Delta L_{WP,road,i,m}$ accounts for the effect of the road surface on the propulsion noise via absorption.

$\Delta L_{WP,grad,i,m}$ and $\Delta L_{WP,acc,i,m}$ account for the effect of road gradients and of vehicle acceleration and deceleration at intersections.

4.1 Effect of road gradients

The road gradient has two effects on the noise emission of the vehicle: first, it affects the vehicle speed and thus the rolling and propulsion noise emission of the vehicle; second, it affects both the engine load and the engine speed via the choice of gear and thus the propulsion noise emission of the vehicle. In this section, it was assumed constant speed, examined only the impact on engine noise.

The effect of the road gradient on the propulsion noise is taken into account by a correction term $\Delta L_{WP,grad,m}$ which is a function of the slope s (in %), the vehicle speed v_m (in km/h) and the vehicle class m. In the case of a bi-directional traffic flow, it is necessary to split the flow into two components and correct half for uphill and half for downhill. The correction term is attributed to all octave bands equally (14-17):

for $m = 1$

$$\Delta L_{WP,grad,i,m}\left(v_m\right) = \begin{cases} \dfrac{Min\left(12\%;-s\right)-6\%}{1\%} & when \quad s < -6\% \\ 0 & when -6\% \le s \le 2\% \\ \dfrac{Min\left(12\%;s\right)-2\%}{1,5\%} \times \dfrac{v_m}{100} & when \quad s > 2\% \end{cases} \quad (14)$$

for $m = 2$

$$\Delta L_{WP,grad,i,m=2}\left(v_m\right) = \begin{cases} \dfrac{Min\left(12\%;-s\right)-4\%}{0,7\%} \times \dfrac{v_m-20}{100} & when \quad s < -6\% \\ 0 & when -6\% \le s \le 2\% \\ \dfrac{Min\left(12\%;s\right)-2\%}{1,5\%} \times \dfrac{v_m}{100} & when \quad s > 2\% \end{cases} \quad (15)$$

for m=3

$$\Delta L_{WP,grad,i,m}\left(v_m\right) = \begin{cases} \dfrac{Min\left(12\%;-s\right)-6\%}{1\%} & when \quad s < -6\% \\ 0 & when -6\% \le s \le 2\% \\ \dfrac{Min\left(12\%;s\right)-2\%}{1,5\%} \times \dfrac{v_m}{100} & when \quad s > 2\% \end{cases} \quad (16)$$

dla m=4

$$\Delta L_{WP,grad,i,m=4}\left(v_m\right) = 0 \qquad (17)$$

The correction $\Delta L_{WP,grad,m}$ implicitly includes the effect of slope on speed.

5 EFFECT OF THE ACCELERATION AND DECELERATION OF VEHICLES

Before and after crossings with traffic lights and roundabouts a correction shall be applied for the effect of acceleration and deceleration as described below. The correction terms for rolling noise, $\Delta L_{WR,acc,m,k}$ and for propulsion noise $\Delta L_{WP,acc,m,k}$ are linear functions of the distance x (in m) of the point source to the nearest intersection of the respective source line with another source line. They are attributed to all octave bands equally (18), (19):

$$\Delta L_{WR,acc,m,k} = C_{R,m,k} \times Max\left(1-\dfrac{|x|}{100};0\right) \qquad (18)$$

$$\Delta L_{WP,acc,m,k} = C_{P,m,k} \times Max\left(1-\dfrac{|x|}{100};0\right) \qquad (19)$$

The coefficients $C_{R,m,k}$ and $C_{P,m,k}$ depend on the kind of junction k ($k = 1$ for a crossing with traffic lights; $k = 2$ for a roundabout) and are given for each vehicle category. The correction includes the effect of change in speed when approaching or moving away from a crossing or a roundabout. Note that at a distance $|x| \ge 100\ m$, $\Delta L_{WR,acc,m,k} = \Delta L_{WP,acc,m,k} = 0$

5.1 *Effect of the type of road surface*

For road surfaces with acoustic properties different from those of the reference surface, a spectral correction term for both rolling noise and propulsion

noise shall be applied. The road surface correction term for the rolling noise emission is given by (20):

$$\Delta L_{WR,road,i,m} = \alpha_{i,m} + \beta_m \times \log\left(\dfrac{v_m}{v_{ref}}\right) \qquad (20)$$

where:

$\alpha_{i,m}$ is the spectral correction in dB at reference speed v_{ref} for category m (1, 2 or 3) and spectral band i.

β_m is the speed effect on the rolling noise reduction for category m (1, 2 or 3) and is identical for all frequency bands.

The road surface correction term for the propulsion noise emission is given by (21):

$$\Delta L_{WP,road,i,m} = min\left\{\alpha_{i,m};0\right\} \qquad (21)$$

Absorbing surfaces decrease the propulsion noise, while non-absorbing surfaces do not increase it.

The noise characteristics of road surfaces vary with age and the level of maintenance, with a tendency to become louder over time. In this method the road surface parameters are derived to be representative for the acoustic performance of the road surface type averaged over its representative lifetime and assuming proper maintenance.

6 THE MEASUREMENTS OF NOISE EMISSION AND SOLUTIONS ELIMINATING TRAFFIC NOISE

Growing number of passenger cars and trucks driving on the roads causes the increase in volume of street noise (Kuśmińska-Fijałkowska, A., Łukasik Z. 2011). Due to these threats, limitations to traffic of trucks have been implemented in many EU countries, not only on weekends, but also at nights. The norms of acceptable noise and exhaust fumes emitted by vehicles, as well as fees for using the roads and ecological taxes have also been implemented. Some countries also implemented administrative restrictions limiting transit traffic of goods vehicles. Switzerland and Austria are leading in such restrictions. Therefore, European Union promotes intermodal transport, moving of goods transports from road transport into alternative ones and more natural environmentally-friendly technologies of transport: railway, short sea shipping, inland waterway shipping and combined transport.

In 2014, the measurements of noise emission on Polish roads were taken in 189 sections of roads (Fig. 4-7) (www.gios.gov.pl):
– the measurements were taken in 54 sections of national roads,
– 83 sections of provincial roads,

– 52 sections of remaining roads (including district, commune and local roads)

Figure 4. The measurements of traffic noise emission expressed by $L_{Aeq,D}$

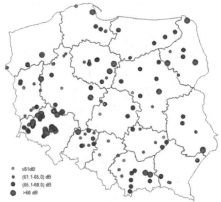

Figure 5. The measurements of traffic noise emission expressed by $L_{Aeq,N}$

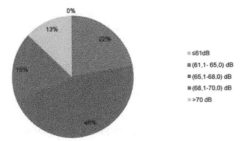

Figure 6. The measurements of sound level taken on national roads during the day

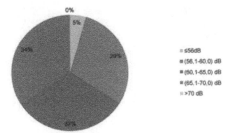

Figure 7. The measurements of sound level taken on national roads at night

In 2014, long-term measurements of emission were taken in 61 measuring points located in uniform sections of roads (Fig. 8-10) (www.gios.gov.pl):
– the measurements were taken in 20 sections of national roads,
– 25 sections of provincial roads,
– 16 sections of remaining roads (including district, commune and local roads).

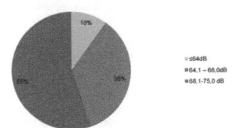

Figure 8. The measurements of sound level taken on national roads for L_{DWN}

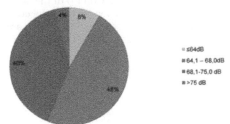

Figure 9. The measurements of sound level taken on provincial roads for L_{DWN}

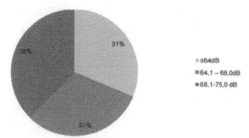

Figure 10. The measurements of sound level taken on remaining roads for L_{DWN}

70 dB is accepted in many European countries as an alarm limit and above this limit, noise is hazardous for health. This value was also accepted operationally, more than 10 years ago, in national noise monitoring as a value of noise threat (which functionally corresponds with alarm level).

The regulations require from road construction workers to reduce street noise to the level required by norms and orders (Directive 2007/46/WE). It concerns both newly designed and built roads, as well as existing roads if they are to be extended or modernized. When road is too close to the buildings, it is required to build acoustic barriers reducing street noise (Oltean-Dumbrava C., Miah A. 2016). Building noise barriers near the roads is a method of reduction of street noise roads (Ruiz-Padillo A., de Oliveira, Thiago B. F.; Alves M. 2016; Venkatram A. Isakov V., Des'hmukh P. 2016).

Noise barriers are important part of environmental protection and they can soon be even more important because new type of noise barriers are being designed in Holland. They will protect against noise and generate energy from the sun. Fig. 11 Recently, it has been observed that devices used for conversion of solar energy are more and more popular. High interest in renewable sources of energy causes that technologies used to acquire energy from renewable sources are more and more popular and they are constantly improved. Photovoltaics is one of the most promising technologies of production of electric energy (Łukasik Z., Kuśmińska-Fijałkowska A., Kozyra J. 2016).

Figure 11. Acoustic screens integrated with photovoltaics (http://pvportal.pl)

A prototype of a noise barrier, which will be not shorter than 450 m and 6 meters high, an innovation in this structure is the use of photovoltaic panels for production of energy. The structure will not be made of concrete, but of transparent panels to make road users not isolated. Photovoltaic panels will be place on both sides of the screen. Photovoltaic elements will be used in sections in south-east direction, which will enable to acquire energy in the morning, photovoltaic panel get rays of sunshine in the evening. The biggest challenge for such structure is to prevent pollination and soiling of panels by exhausts of vehicles, dust caused by road traffic and potential overshadowing by trucks. Therefore, the screen will be at least 6 meters high, the panels with a special coating will be mounted on the fourth meter from the top, whereas, the remaining 2 meters of panels will be covered with glass (http://pvportal.pl). Solar energy acquired along highways and expressways is a future. It is not hard to imagine charging stations of electric vehicles on the roads powered by photovoltaic panels.

7 CONCLUSIONS

Short-term measurements of noise emission on Polish roads were taken in 189 measuring points. Emission was higher than 65 dB during the day in 58 % of measuring points. Emission higher than 56 dB at night was in 41% of roads . Long-term measurements of traffic noise emission were taken in 61 measuring points, in 2/3 of them, emission higher than 64 dB for L_{DWN}. Long-term measurements of traffic noise has shown that in 55% of national roads, 44 % of provincial roads and 38 % of remaining roads, emission was higher than 68 dB for L_{DWN}. Short-term and long-term measurements have shown that noise emission on national roads is between 68,1 and 75 dB, and emission on other roads is mostly lower than 68 dB. The reduction of noise level can be achieved through reduction of emission from the source, for example, applying new technical solutions or new technologies. In case of transport noise, efforts are put to lower noise of an engine and the whole drive unit and to create appropriate surfaces of tyre treads. The research works on appropriate properties of surface layers of the roads, which may reduce street noise. Efficient organization and optimisation of supply chains may enable to reduce specific consumption of energy and may have positive impact on natural environment.

REFERENCES

Gallasch E., Raggam, Reinhard B.,Cik M.; et al. Road and rail traffic noise induce comparable extra-aural effects as revealed during a short-term memory test. Noise & Health, Volume 18, Issue 83, Pages: 206-213 Published: Jul-Aug 2016

Baliatsas Ch., van Kamp I., Swart W.; et al. Noise sensitivity: Symptoms, health status, illness behavior and co-occurring environmental sensitivities. Environmental Research, Volume 150, Pages 8-13, Published Oct 2016

Zagrożenie hałasem w ruchu drogowym https://www. obserwatoriumbrd.pl/pl/analizy_brd/problemy_brd/halas/

Łukasik Z., Kuśmińska-Fijałkowska A., Kozyra J. Emission of acoustic sources of noise in the industrial plants. International Journal of Engineering Research and General Science 4, pp. 522-529, 2016

Ruiz-Padillo A., Ruiz Diego P., Torija, Antonio J.; et al. Selection of suitable alternatives to reduce the environmental impact of road traffic noise using a fuzzy multi-criteria decision model. Environmental Impact Assessment Review, Volume 61, Pages 8-18, Published Nov 2016

De Coensel B., Brown A. L. Tomerini, D.; A road traffic noise pattern simulation model that includes distributions of vehicle sound power levels. Applied Acoustics, Volume 111, Pages 170-178, Published Oct 2016

Kuśmińska-Fijałkowska, A., Łukasik Z. The Land Trans-Shipping Terminal In Processes Flow Stream Individuals Intermodal Transportion. TransNav, the International Journal on Marine Navigation and Safety of Sea Transportation 5.3 (2011): pp. 395-400

Directive 2007/46/WE Parlamentu Europejskiego i Rady z dnia 5 września 2007 r. ustanawiająca ramy dla homologacji pojazdów silnikowych i ich przyczep oraz układów, części i oddzielnych zespołów technicznych przeznaczonych do tych pojazdów (Dz.U. L 263 z 9.10.2007, s. 1).

Oltean-Dumbrava C., Miah A.; Assessment and relative sustainability of common types of roadside noise barriers.

Journal of Cleaner Production, Volume 135, Pages 919-931, Published Nov 1 2016

Ruiz-Padillo A., de Oliveira, Thiago B. F.; Alves M. ; et al. Social choice functions: A tool for ranking variables involved in action plans against road noise. Journal of Environmental Management, Volume 178, Pages 1-10, Published Aug 1 2016

Venkatram A. Isakov V., Des'hmukh P.; et al. Modeling the impact of solid noise barriers on near road air quality. Atmospheric Environment, Volume 141, Pages 462-469, Published Sep 2016

Łukasik Z., Kuśmińska-Fijałkowska A., Kozyra J. Eco-friendly technology to reduce CO2 emissions of passenger cars based on innovative solutions. Przegląd Elektrotechniczny 8/2016

http://pvportal.pl/nowosci/6050/ekrany-dzwiekochlonne-zintegrowane-z-fotowoltaika

Bujak A. Wybrane aspekty transportu intermodalnego s. 115–122, https://tirynatory.pl/wp-content/uploads/2012/01/TECHNOLOGIE_TRANSPORTU_INTERMODAL.pdf

http://www.gios.gov.pl

Zbyszyński M., Kamiński T., Krysiuk C., Mitraszewska I., Odachowska E., Zakrzewski B. Ekonomiczna jazda samochodem i jej wpływ na środowisko. Instytut Transportu Samochodowego 2015

Banak M., Brdulak, J., Krysiuk, C., Pawlak, P. Kierunki rozwoju infrastruktury transportu samochodowego w Polsce. Wydawnictwo Instytutu Transportu Samochodowego, Warszawa. (2014).

Krysiuk C., Nowacki, G., Zakrzewski B. Rozwój miast w Polsce, czynnik transportu. Logistyka 4 (2015), pp.7813-7822.

Krajewska R., Łukasik Z. Ocena atrakcyjności sektora usług TSL w Polsce. Logistyka 2 (2010).

Occupational Accidents

Investigation of Occupational Accidents on Board with Fuzzy AHP Method

Ü. Özdemir
Mersin University, Mersin, Turkey

I. Altinpinar & F.B. Demirel
Karadeniz Technical University, Trabzon, Turkey

ABSTRACT: There are many legislations, agreements and practices to obtain series of security measures in order to ensure safety and security of seafarers. Causes of on-board occupational accidents need to be evaluated in a correct manner to regulate more functional practices and also to lower the on-board accident rates. However, causes of on-board accidents can be extremely complex. Therefore, scientific methods should be used to evaluate the causes and to determine the measures to be taken. In this study, factors have been identified that lead to seafarers' occupational accidents on board and we tried to present alternative solutions which can be applied on this issue. Severity of the reasons that led to the accidents and their relationships with each other are identified to be able to sort through the alternative solutions with a model using the fuzzy AHP (Analytic Hierarchy Process) method approach.

1 INTRODUCTION

Marine traffic, which is a natural consequence of growing world trade, not only increased the number of ship accidents but also caused more frequent in-ship occupational accidents (Özdemir and Güneroğlu 2015, Uğurlu 2016). There is no doubt that maintaining the seafarers' safety is utterly important. According to the International Labour Organization (ILO), seafarer is any person who works in any position on board a seagoing ship or vessel engaged in commercial maritime navigation, whether publicly or privately owned, other than a ship of war. Being a seafarer includes professionally difficulties in addition to the harsh environmental factors. Unlike other professions, they spend 24 hours a day at work. Therefore, when dangerous occupations are listed, the risk of accident is rather high for seafarers (Roberts, 2002). There are many different reasons for the occupational accidents on board. Hanson's study carried out a survey that showed that mortal injuries among Danish seafarers were 11.5 times higher than average rates among the Danish male workers ashore between 1986 and 1993 (Hansen, 1996). Accidents on ships may pose great risks for personnel and the environment, according to the nature of the accident (Portela 2005, Uğurlu et al. 2016). Accident can be described as an unexpected event that results in a personal injury or loss of property, or both. We see that ship related

occupational accidents can be inspected under three main headings. These can be listed as auxiliary causes, sudden causes and other causes (Hansen 1996, Hansen et al. 2002). Certain operations on board may cause extremely dangerous working conditions. Inadequate safety of ship operations are not only inclined by the material precautions but also human factors such as seafarers' behaviors, habits, lack of attentions and occupational educations (Roberts 2000, Martins and Maturana 2010, Uğurlu et al. 2016). Most of marine accidents are caused by some form of human error as well as incidents (Havold 2000, Rothblum 2000, Toffoli et. al. 2005, Hetherington et al. 2006, Grech et. al. 2008, Talley 2009, Özdemir and Güneroğlu 2015). Previous researches demonstrate that for each serious accident in the maritime industry, or in any other area, there are a larger number of incidents, a big number of near-misses and huge number of safety-critical events and unsafe acts (Grech et. al. 2008). Example of workers' unsafe acts include the determination to proceed with work in unsafe conditions, disregarding standard safety procedures such as not wearing safety equipment, working while intoxicated, occupational illiteracy, working with insufficient sleep and fatigue (Abdelhamid and Everett 2000). Personal injuries are much worse when they are on board, due to the fact that seafarer's health care opportunities are poorer than those ashore. A critical conflict in the treatment of seafarer at sea is that medical care on-board is applied by a medical health officer who is not medical professional (Oldenburg et al. 2010).

Main reasons of ship related accidents can be listed as; human factors (psychological, physical, human relations, team work, communication); machinery/equipment factors (incorrect machinery and equipment layout, absent or defective protectors, inadequate standardization, inadequate control and maintenance, inadequate engineering services); environmental factors (inadequate knowledge, improper working methods, improper working environment) and management factors (inadequate management organization, incomplete rules and regulations, inadequate security management plan, educational inadequacies, inadequate health controls, employment of incompetent personnel, etc.) (O'Neil 2003, Portela 2005, Hetherington and Flin 2006, Özdemir and Güneroğlu 2015, Uğurlu 2016)

In this study, we used a quantitative application to examine the reasons of work related accidents in ships and tried to discover alternative solutions for the matter. Work related accidents on ships have several interconnected reasons. For the solutions of such problems in which various factors and criteria must be analysed and evaluated, using a "fuzzy multi-criteria decision-making model" (FMCDM) can be a positive approach. In this study, we determined the criteria that cause work related accidents and offered an appropriate methodology for the solution of the problem by using a fuzzy multi-criteria decision-making model. Fuzzy AHP (Analytic Hierarchy Process) is used to determine and rank the accident-related criteria according to their importance.

2 METHODOLOGY

The problem of the causes of occupational accident selection criteria on ship is considered as a multi-dimensional complex issue that can be resolved by a MCDM approach. In such a complex problem, the availability of many choices and their relative impact on the final solution are always risky as they may be misleading the decision maker if there is not a reliable tool in hand (Özdemir and Güneroğlu 2015, Uğurlu 2015, Güneroğlu et. al 2016). In this study, causes of occupational accident selection criteria' weighting was implemented using Fuzzy AHP following Buckley (1985) by pairwise comparisons of the experts' scores were applied for ranking and evaluating the criteria.

The steps of the applied technique and related case study are presented in following sub-sections.

2.1 Fuzzy AHP

The first step of implemented methodology is endured on Buckley's AHP technique. Actually, this technique is a magnified version of the original AHP technique by Saaty (2006) and instead of classic rational numbers, it uses fuzzy comparison ratios. The notable virtue of the Buckley's Fuzzy AHP is the expansion of the statements to Fuzzy environment with relatively easy efforts as well as the guarantee of one absolute value. Complex, laborious and error-prone computational requirements are the main disadvantages of the same technique. The Fuzzy AHP technique by Buckley (1985) can be summarized as followed (Buckley, 1985; Kafalı et al., 2014; Özdemir and Güneroğlu,2017);

The first step of the technique contains defining the main criteria that potentially affects the problem under investigation by experts and decision makers. Afterwards, in order to convert the expert evaluations to the fuzzy numbers, a predefined linguistic scale is used. Then, expert evaluations were received as pairwise comparisons that converted to the fuzzy numbers with in matrix form as shown in Eq.1,

$$
\tilde{A}^k = \begin{bmatrix} 1 & \tilde{A}_{12} & \tilde{A}_{1m} \\ \tilde{A}_{21} & 1 & m \\ \cdot & \cdot & \cdots & \cdot \\ \cdot & \cdot & & \cdot \\ \cdot & \cdot & & \cdot \\ \tilde{A}_{m1} & \tilde{A}_{m2} & 1 \end{bmatrix} \quad (1)
$$

where "$\tilde{A}^k$" is the response matrix by each expert. Linguistic scale and corresponding fuzzy numbers used in the Buckley's technique were given in Table 1 (Xu and Yager 2008, Kafalı et al. 2014).

Then all data received as result of experts' evaluations is compiled by using weighted mean formula as given in Eq.2,

$$
\tilde{A}_{xy} = \frac{Z_1 A_{xy}^1 + Z_2 A_{xy}^2 + \ldots + Z_k A_{xy}^k}{Z_1 + Z_2 + \ldots + Z_k} \quad (2)
$$

In Eq.2 "$\tilde{A}_{mn}$", is the joined comparison value of the critieria "x" and "y"; "Z_k" is the weighted value of "k."expert; "A_{xy}^k" is the comparison value of "k." expert evaluations corresponding to "x" and "y" criteria. The decision matrix formed by weighted means of all experts' scores can be shown in following matrix form (Eq.3)

$$\tilde{A} = \begin{bmatrix} 1 & \tilde{A}_{12} & \dots & \tilde{A}_{1m} \\ \tilde{A}_{21} & 1 & \dots & \tilde{A}_{2m} \\ \vdots & & \dots & \vdots \\ \vdots & \vdots & \dots & \vdots \\ \tilde{A}_{m1} & \tilde{A}_{m2} & \dots & 1 \end{bmatrix} \qquad (3)$$

After obtaining the decision matrix, the weight of each criterion can be calculated in two steps. The first step is computing the geometric mean of each row in decision matrix, as shown in Eq.4

$$\tilde{b}_i = \left(\widetilde{a_{i1}} \otimes \widetilde{a_{i2}} \otimes \dots \otimes \widetilde{a_{in}} \right)^{1/n} \qquad (4)$$

where "n" stands for total number of criteria, "a_{in}" is the fuzzy comparison value between two criteria "i." and "n." and "b_i" is the fuzzy geotmetric mean of the all compared criteria. As a second step, a fuzzy weight of each criterion is calculated by applying Eq.5.

$$\widetilde{w_i} = \tilde{b}_i \otimes \left(\tilde{b}_1 + \tilde{b}_2 + \dots + \tilde{r}_n \right)^{-1} \qquad (5)$$

where, "$\widetilde{w_i}$" is the fuzzy weight of criterion "i.". The remaining part of the Buckley's Fuzzy AHP technique requires convertion of the fuzzy numbers to corresponding absolute values and calculation of relative weights among all criteria.

Where,"B" is referred to triangular fuzzy number, defuzzification of "B" can be applied using Eq.6.

$$\tilde{B} = \frac{\tilde{b}_1 + \tilde{b}_2 + \tilde{b}_3}{3} \qquad (6)$$

For better evaluation of the obtained values, normalization is applied on the main criteria as it is in Eq.7.

$$\left(w_i^R \right)^N = \frac{w_i^N}{\sum_{i=1}^{n} w_i^N} \qquad (7)$$

where "$\left(w_i^R \right)N$"is referred to normalized weight of each criteria and "n" is the total number of the criteria. Similarly, normalization of sub-criteria is also performed for all elements of the matrix.

Finally, relative fuzzy and absolute weights of main and sub-criteria are computed by multiplying main and related each sub-criteria in the matrix as shown in Eq.8 and Eq.9.

$$\left(\tilde{w}_i^R \right)^{SN} = \left(\tilde{w} \right)^N \otimes \left(\tilde{w}_i \right)^{SN} \qquad (8)$$

where, "$\left(\tilde{w}_i^R \right)SN$" is the fuzzy relative weight of "i." sub-criterion, "$\left(\tilde{w} \right)N$" is the fuzzy weight of the related main criterion and "$(\tilde{w}_i)SN$" is the fuzzy weight of the same sub-criterion.

$$\left(w_i^R \right)^{SN} = \left(w^R \right)^N \times \left(w_i^R \right)^{SN} \qquad (9)$$

In Eq.9, "$\left(w_i^R \right)SN$" is the normalized relative absolute weight of the "i." sub-criterion, "$\left(w^R \right)N$" is the normalized absolute weight of the related main criterion and "$\left(w_i^R \right)SN$" is the normalized absolute weight of the same sub-criterion.

3 CASE STUDY

With an aim of obtaining data sets by establishing criteria within the framework of research model; ship crews, company officials, academicians, analysts and casualties/casualties' relatives were interviewed. Also, safety reports were evaluated and criteria were established by considering the agreements and conventions published for maritime safety. As a result of the evaluations made, it was decided that 5 main criteria shown in Table 1, would be studied.

Table 1. Criteria determined for the study.

#	Criteria
1	Environmental factors (Sea condition, weather condition etc.) - C1
2	Lack of management (Lack of ship rules, failure to take measures, lack of management, lack of communication etc.) – C2
3	Shipborne Troubles (Ship age, ship condition, condition of equipment, equipment inadequacy, poor lighting etc.) - C3
4	Human Factor (Lack of training, unawareness, carelessness, occupational willies, fatigue, dangerous movements etc.) – C4
5	Cargo Troubles (inappropriate loading, dangerous cargoes etc.) - C5

Maritime management is comprised of 9 people (Oceangoing master/3, ship casualties/3, academician/3) and these people have experiences about ship's crew. Questionnaire forms of compiled criteria shown in Table 1 were applied to the participants with an aim of obtaining opinions of decision makers. Paired comparison matrixes of each expert related to all the criteria were obtained in the form of verbal statements as a result of the evaluation of all the criteria. Due to the fact that all the questionnaire data, which were collected from the experts, were in verbal forms, they need to be converted to triangular fuzzy numbers in accordance with fuzzy number equivalents of linguistic scale, which was specified before (Xu and Yager, 2008; Özdemir and Güneroğlu, 2017; Özdemir and Çetin, 2017). The values in Table 2 were used in these conversions.

Incorporated fuzzy decision matrixes of the data, which were obtained as a result of paired comparison of main criteria by using formula 10 and 11, were calculated as in Table 3.

$$\tilde{C}_{ij} = \left(1 / N\right) \otimes \left(\tilde{c}_{ij}^{1} \oplus \tilde{c}_{ij}^{2} \oplus \dots \oplus \tilde{c}_{ij}^{N}\right) \qquad (10)$$

$$\tilde{D}_{ij} = \left(1 / N\right) \otimes \left(\tilde{d}_{ij}^{1} \oplus \tilde{d}_{ij}^{2} \oplus \dots \oplus \tilde{d}_{ij}^{N}\right) \qquad (11)$$

Table 2. Linguistic terms used for Buckley's Fuzzy AHP (Chen et al., 2005; Kafalı et al., 2014; Özdemir, 2015; Özdemir and Güneroğlu, 2016; Özdemir and Güneroğlu, 2017).

Linguistic Terms	Fuzzy Number
Slightly more important (Row)	(1, 3, 5)
Strongly more important (Row)	(3, 5, 7)
Highly more important (Row)	(5, 7, 9)
Absolutely more important(Row)	(7, 7, 9)
Equally important	(1, 1, 3)
Absolutely more important (Column)	(0.111, 0.111, 0.143)
Highly more important (Column)	(0.111, 0.143, 0.200)
Strongly more important (Column)	(0.143, 0.200, 0.333)
Slightly more important (Column)	(0.200, 0.333, 1.000)

Table 3. Aggregated fuzzy decision matrix for the criteria

	C1	C2	C3	C4	C5
C1	1.000	0.125	0.324	0.621	1.458
	1.000	2.354	1.000	0.388	2.642
	1.000	3.652	2.024	0.547	2.033
C2	0.256	1.000	0.387	0.254	2.010
	0.745	1.000	2.457	0.541	2.354
	2.354	1.000	1.874	2.456	2.247
C3	0.385	0.845	1.000	0.451	1.687
	1.845	1.354	1.000	0.365	1.347
	2.214	2.410	1.000	1.354	4.024
C4	1.651	2.874	0.343	1.000	2.410
	1.033	3.120	1.687	1.000	3.025
	0.333	1.018	2.374	1.000	4.024
C5	0.897	0.985	0.852	1.241	1.000
	0.624	0.458	0.349	0.314	1.000
	1.025	1.303	1.541	0.652	1.000

The step after calculating incorporated fuzzy decision matrixes calculations was the calculation of criterion weights according to Buckley approach and it is carried out in the second step. As a first step, geometric average of each line of incorporated fuzzy decision matrixes is calculated. This process is expressed in formula 4. In the second step, geometric average of matrix is calculated and then its fuzzy weight value is calculated with the help of formula 5. This process was applied for all main and sub-criteria. Weighted fuzzy decision matrix, which was calculated for main criteria, was shown in Table 4.

According to Buckley approach, the next step is conversion of fuzzy values into absolute values. According to this, defuzzification and normalization processes are carried out. Formula 6 was used for this calculation. Formula 7 was benefited in order to evaluate absolute weights in a better way. The results, in which calculated defuzzification process was included for the criteria, were shown in Table 4 and normalization results were shown in Table 5.

Table 4. Weighted fuzzy decision matrix for criteria (a), defuzzified criteria weights (b).

(a)	Weighted Fuzzy Decision Matrix		
C1	0.034	0.398	0.403
C2	0.204	0.452	0.182
C3	0.024	0.065	0.035
C4	0.287	0.078	0.802
C5	0.136	0.304	0.246
(b)	Defuzzified criteria weights		
C1	0.026	0.228	0.542
C2	0.125	0.309	0.021
C3	0.203	0.065	0.520
C4	0.028	0.122	0.033
C5	0.077	0.201	0.217

Table 5. Normalized criteria weights

Criteria	Normalization (Crisp)	Percentage Value (%)
C1(5)	0.1248	12,48
C2(2)	0.2438	24,38
C3(3)	0.1878	18,78
C4(1)	0.2598	25,98
C5(4)	0.1838	18,38
Total	1.000	100

4 RESULTS

According to the results of the study, the reasons that occupational accident on board are specified as follows in order of priorities: human factors (C4), lack of management (C2), shipborne Troubles (C3), Cargo Troubles (C5), and Environmental factors (C1). According to the results of the study, criteria of human factors such as lack of training, unawareness, carelessness, occupational willies, fatigue, dangerous movements etc., which is ranked in the first place, that main reasons of occupational accident on board were experienced. When the criteria are examined generally, we see that C2 and C4 are directly human-induced while C3 and C5 are indirectly effected by human error. C1 stands out as the distinctive criteria that differentiates the maritime occupation from other occupational fields. Safety culture affects behavioral pattern of people against dangerous situations (Cox and Flin 1998, Cooper 2000, Dursun 2013). Reiman ve Oedewald (2002) collected the criteria of 'good safety culture' in the literature under the main headings such as: security policies; apparent sagacity of management for security; democratic applications and competencies; positive values with security tendencies; open definitions of responsibilities and necessities; security priority operations; balance of security and production; competent workers and education; high motivation and work-satisfaction; mutual trust and fair approach between management and workers; update of quality, rules and regulations; regular machinery maintenance; proper and regular reporting of every incident and interpretation; healthy information flow from different managerial levels and positions; adequate funds and constant development; proper design and business relations with authority. Perception of safety culture of individuals is an important factor in the accidents that originate from such human errors. Enhancing the perception of safety culture will contribute to prevent accidents that results from human-induced errors.

When the previous studies about privateering are considered, it is thought that this study will significantly contribute to the literature and future studies due to the lacking number of quantitative studies about work accidents on ship, which are accepted as a major problem for maritime industry.

REFERENCES

Abdelhamid, T. S., & Everett, J. G. (2000). Identifying root causes of construction accidents. Journal of construction engineering and management, 126(1), 52-60.

Accident prevention on board ship at sea and in port. An ILO code of practice. Geneva, International Labour Office, 2nd edition, 1996 ISBN 92-2-109450-2.

Buckley, J.J. (1985), "Fuzzy Hierarchical Analysis", Fuzzy Sets and Systems, Vol. 17, pp. 233- 247.

Cooper Ph. D, M. D. (2000). Towards a model of safety culture. Safety science, 36(2), 111-136.

Cox, S., & Flin, R. (1998). Safety culture: philosopher's stone or man of straw?. Work & stress, 12(3), 189-201.

Dursun, S. (2013). İş güvenliği kültürünün çalışanların güvenli davranışları üzerine etkisi. Sosyal Güvenlik Dergisi (SGD), 3(2).

Grech, M., Horberry, T., & Koester, T. (2008). Human factors in the maritime domain. CRC Press.

Hansen, H. L. 1996. "Surveillance of Deaths on Board Danish Merchant Ships 1986-93."Occupational&Environmental Medicine 53: 269–275.

Hansen, H. L., D. Nielsen, and M. Frydenberg. 2002. "Occupational Accidents aboard Merchant Ships." Occupational&Environmental Medicine 59: 85–91.

Havold, J. I. (2000). Culture in maritime safety. Maritime Policy & Management, 27(1), 79-88.

Hetherington, C., R. Flin, and K. Mearns. 2006. ""Safety İn Shipping: The Human Element." Journal of Safety Research 37 (4): 401–411. doi:10.1016/j.jsr.2006.04.007.

Kafalı, M., Özkök, M. and Çebi, S. (2014), "Evaluation of Pipe Cutting Technologies In Shipbuilding", Brodogradnja, Vol. 65, No. 2, pp. 33-48.

Martins, M. R., and M. C. Maturana. 2010. "Human Error Contribution İn Collision and Grounding of Oil Tankers." Risk Analysis 30 (4): 674–698. doi: 10.1111/risk.2010.30.issue-4.

O'Neil, W. A. 2003. "The Human Element İn Shipping." Journal of Maritime Affairs 2 (2): 95–97. doi:10.1007/BF03195037.

Oldenburg, M., Baur, X., & Schlaich, C. (2010). Occupational risks and challenges of seafaring. Journal of occupational health, 52(5), 249-256.

Özdemir, Ü. and Çetin, M.S. (2017). Investigation of An Quantitative Research on the Causes of Sea Piracy. 12th International Conference on Marine Navigation and Safety of Sea Transportation, Gdynia, Poland, in progress.

Özdemir, Ü. and Güneroğlu, A. (2016), "Cargo Type Selection Procedure Using Fuzzy AHP and Fuzzy TOPSIS Techniques: "The Case of Dry Bulk Cargo Ships", International Journal of Shipping and Transport Logistics (in progress).

Özdemir, Ü. and Güneroğlu, A. (2017). Quantitative Analysis of the World Sea Piracy by Fuzzy AHP and Fuzzy TOPSIS Methodologies. International Journal of Transport Economics, in progress.

Portela, R. 2005. "Maritime Casualties Analysis as a Tool to Improve Research about Human Factors on Maritime Environment." Journal of Maritime Research 2: 3–18.

Reiman, T., & Oedewald, P. (2002). The assessment of organizational culture: A methodological study. VTT Research Notes 2140. Espoo, Finland.

Roberts, S. 2000. "Occupational Mortality among British Merchant Seafarers (1986-1995)." Maritime Policy & Management 27 (3): 253–265. doi:10.1080/030888300411095.

Roberts, S. E. (2002). Hazardous occupations in Great Britain. The Lancet, 360(9332), 543-544.

Rothblum, A. M. (2000, October). Human error and marine safety. In National Safety Council Congress and Expo, Orlando, FL.

Talley, W. K. 2009. "Determinants of the Probability of Ship Injuries." The Asian Journal of Shipping and Logistics 25 (2): 171–188. doi:10.1016/S2092-5212(09)80001-1.

Toffoli, A., Lefevre, J.M., Bitner-Gregersen, E., Monbaliu, J., 2005. Towards the identi- fication of warning criteria: analysis of a ship accident database. Applied Ocean Research 27 (6), 281–291.

Uğurlu Ö. 2016. "A case study related to the improvement of working and rest hours of oil tanker deck officers", MARITIME POLICY & MANAGEMENT, vol.43, pp.524-539.

Uğurlu Ö., Köse E., Yildirim U., Yüksekyildiz E., 2015. "Marine Accident Analysis for Collision and Grounding In Oil Tanker Using Fta Method", MARITIME POLICY & MANAGEMENT, no.2, pp.163-185.

Uğurlu Ö., Kum S., Aydoğdu Y.V., 2016. "Analysis of Occupational Accidents Encountered by Deck Cadets in Maritime Transportation", Maritime Policy & Management, vol.7, pp.1-19.

Uğurlu Ö.,2015. "Application of Fuzzy Extended AHP Methodology for Selection of Ideal Ship for Oceangoing Watchkeeping Officers", International Journal of Industrial Ergonomics, vol.47, pp.132-140.

Xu, Z. and Yager, R. R. (2008), "Dynamic Intuitionistic Fuzzy Multi-Attribute Decision Making", International Journal of Approximate Reasoning, Vol. 48, No. 1, pp. 246–262.

A Research on Occupational Accidents Aboard Merchant Ships

E. Çakır & S. Paker
Dokuz Eylul Universty, Izmir, Turkey

ABSTRACT: The aim of this study to investigate the frequency, to explore the causes of the occupational accidents and to identify the risks which occur during the daily routine works and dangerous works to be fulfilled. After analyzing the main causes and risk factors of the occupational accidents clearly, it would be easier to take correct preventive measures. The data for this study were obtained from occupational accidents reports which published by countries' marine accident investigation branches/units such as Marine Accident Investigation Branch (MAIB), Marine Safety Investigation Unit and etc. The findings showed that fall on board/falls overboard is the most common accident type and able seamen are more prone to be injured when compare with the other members of crew. As a conclusion, occupational accidents aboard merchant ships is massive problem to be taken care of and the analysis showed that still more consideration should be given to human and organizational factors.

1 INTRODUCTION

Merchant seafarers work in a dangerous environment which comprises the physical, ergonomic, chemical, biological, psychological and social elements which could lead to occupational accidents, injuries and diseases (ILO, 2014). As a result of this dangerous environment, merchant seafarers exposure to extreme weather conditions (Curry, 1996; Buxton & Cuckson, 1997), hazardous enclosed spaces, noisy mechanical equipment and toxic cargoes (Nielsen, 1999a:121). Furthermore, because of the nature of the maritime profession, seafarers have been faced stringent working conditions such as isolation from everyday life, long hours of work, rigid organizational structures and high levels of stress and fatigue (ILO, 2014).

Merchant shipping has, in some countries, fatality rates which are more than 20 times that of the average of the respective country's shore based industries (Roberts, 1996; Roberts & Hansen, 2002). According to study of Saarni (1989) which comprises Finnish seafarers, the rate of occupational accidents was found close to the rate of for the whole working population in terms of non-fatal injuries. However, there is a lack of statistics in the area of maritime occupational safety and health (MOSH) because of the poor recording and significant differences in data collection

methodologies of occupational accidents, incidents and diseases in flag States (ILO, 2014). Especially, near-accidents, or accidents not causing loss of work-time or sick-leave, are seldom reported from ships (ILO, 1993).

The aim of this study is to investigate the frequency, circumstances, and causes of occupational accidents aboard merchant ships and to identify the risks which occur during the daily routine works and dangerous works to be fulfilled. After analyzing the main causes and risk factors of the occupational accidents clearly that would be easier to take correct preventive measures.

2 MATERIAL AND METHODS

2.1 Inclusion criteria

Included in this study were all occupational accidents occurred on board merchant ships which trade internationally, has 500 gross registered tons and over and happened between 2005 and 2015. This study excluded non-crew members such as passengers, pilots, cargo inspectors and dock workers. Also, occupational accidents occurred on board fishing vessels and passenger ships were excluded due to these vessels require quite different work experiences in terms of technical demands (Mayhew, 1999:10).

2.2 Data collection

Data used in this study obtained various occupational accident reports issued by countries' accident investigation units or maritime authorities such as Marine Accident Investigation Unit (MAIB), Marine Safety Investigation Unit (MSIU) and Australian Transport Safety Bureau (ATSB). Table 1 shows the number of occupational accident reports obtained from each investigation unit or maritime authority. A total of 253 reports satisfied the inclusion criteria for the eleven-year period from 2005 to 2015. Hong Kong, Malta, Denmark, Germany, Australia and UK provided 80% of the all accident reports used in this study.

Table 1. Number of occupational accident reports according to preparatory Accident Investigation Unit

Investigation Unit	Country	Number of Reports
MAIB	UK	24
AIBN	Norway	6
JTSB	Japan	8
BSU	Germany	27
ATSB	Australia	25
DMAIB	Denmark	22
MSIU	Malta	40
TAIC	New Zealand	3
BEA-MER	France	1
MARDEP	Hong Kong	66
HBMCI	Greece	4
DSB	Holland	9
TSBC	Canada	3
IMMA	Isle of Men	3
TBMA	Bahama	2
SIA	Finland	6
MSA	China	2
PKBWM	Poland	2
Total		253

2.3 Method

Accidents reports were examined through content analysis in the light of the variables obtained from the literature. Variables were divided into three categories as ship-related variables, accident-related variables and victim-related variables. Ship-related variables included ship flag, ship age and ship type. Accident-related variables included occupational accident type, worksite of accident occurred on board ship, accident severity, working situation at time of the accident and cause of accident. And, the variables related to the victim included seafarer age, seafarer experience, seafarer nationality, rank of seafarer and time on board spent by seafarer when accident took place.

An occupational accident type means the mode in which a seafarer on board was injured or killed. In this context, accident, accident not related to ship operations, illness, suicide and homicide can be regarded as occupational accident type (EMSA, 2016). Some studies which examined occupational mortality at sea (Nielsen, 1999a, b; Nielsen & Roberts, 1999; Roberts,2006; Oldenburg et al., 2010; Hansen, 1996; Borch et al., 2012) divided cause of deaths into two categories as deaths from disease and deaths from external causes. Deaths from external causes comprised of accidents, suicide, homicide and unknown circumstances. And accidents which caused deaths at sea examined under three headings as maritime disasters, occupational accidents and off-duty accidents. However, occupational accident types were not used in the analysis or accident type included to study as the cause of accident in the aforementioned studies.

In this study, only included occupational accidents occurred on board ship while homicide, suicide and maritime disaster caused to death or injury excluded. And, the occupational accidents were classified according to ILO (1996) in terms of accident type (e.g., falls of persons, struck by falling objects and exposure to or contact with harmful substances).

The study employed descriptive statistics to gain an overall understanding of occupational accidents occurred on board ship. The statistical calculations and analysis were performed using SPSS version 23 (SPSS Inc., Chicago, IL, USA).

3 RESULTS

Table 2 shows the numbers and percentages of the occupational accidents according to accident severity. A total of 253 accident reports were examined which yielded 311 victims. Out of 311 victims, 63% suffered very serious marine casualties (loss of life), 32.5% suffered serious injuries and 4.5% suffered non-serious injuries. According to IMO (2008), a very serious marine casualty means a marine casualty involving the total loss of the ship or death or severe damage to the environment. A serious injury means an injury which sustained by a seafarer, resulting in incapacitation where the seafarer is unable to function normally for more than 72 hours, commencing within seven days from the data when the injury was suffered (IMO, 2008). Serious injuries includes any fracture, any loss of a limb or part of a limb, loss of sight, whether temporary or permanent and any other injury leading to hypothermia or unconsciousness etc. When the incapacitation is less than 72 hours, it is classified as a non-serious injury (EMSA, 2015).

Asian seafarers accounted for the almost half of the occupational accidents (51%) and European seafarers were second place with 20%. Besides, nearly 30% of seafarers were different countries all over the world. Of the 311 victims, seventy-two (23.2%) were Filipino, sixty-eight (21.9%) were Chinese, eighteen (5.8%) were Ukrainian and eleven (4.2) were Turkish (Table 3).

Table 2. Number and percentage of seafarers who suffered occupational accidents according to accident severity

Accident Severity	N (%)
Very Serious Marine Casualty (Loss of life)	196 (63)
Serious Injury	101 (32.5)
Non-serious Injury	14 (4.5)
Total	311 (100)

The mean age of 286 seafarers was 38 years (SD= 10.6; range= 18-66) and 25 seafarers' age was not specified in the accident reports. 9% were aged 25 or under, 33.8% were aged 26-35, 24.4 were aged 36-45, 19% were aged 46-55 and 5.8% were 56 or over. (Table 3).

Table 3. Number and percentage of seafarers who suffered occupational accidents according to seafarer nationality, age, sea experience and time aboard when accident took place

Nationality	N (%)
Filipino	72 (23.2)
Chinese	68 (21.9)
Ukrainian	18 (5.8)
Turkish	13 (4.2)
Indian	12 (3.9)
Polish	12 (3.9)
Russian	11 (3.5)
Japanese	5 (1.6)
Korean	5 (1.6)
Other	93 (29.9)
Missing Value	2 (0.6)
Seafarer age	
≤ 25	28 (9.0)
26-35	105 (33.8)
36-45	76 (24.4)
46-55	59 (19.0)
56 ≥	18 (5.8)
Missing Value	25 (8.0
Sea Experience (years)	
≤ 5	87 (28.0)
6-10	71 (22.8)
11-15	42 (13.5)
16-20	35 (11.3)
21-25	21 (6.8)
25 ≥	25 (8.0)
Missing Value	30 (9.6)
Time aboard when accident took place (days)	
≤ 15	19 (6.1)
16-30	26 (8.4)
31-60	33 (10.6)
61-90	48 (15.4)
91 ≥	146 (46.9)
Missing Value	39 (12.5)
Total	311 (100)

The mean sea experience of seafarers who had occupational accident was 11.59 years (SD= 8.75; range: 1 month-44 years) and were not given any information related to sea experience of 30 seafarers in the accident reports. While sea experience of seafarers was increasing, numbers of occupational accidents was declining as seen from Table 3. The seafarers had sea experience five years or less and 6-10 years comprised of almost half of the 281 cases.

However, the seafarers with 25 years or more and 21-25 years sea experience accounted for nearly 15% of the cases.

Time on board means the day of the accident from the day the seafarer signed on. Seafarers are signed on the day they arrive on board and the first they is usually used for travel and because of that they have not much time for work on board. In the light of this explanation, almost half of the seafarers (46.9%) who exposed to an occupational accident was serving on board 91 days and more. This percentage declined to 15.4 for 61-90 days, 10.6 for 31-60 days, 8.4 for 16-30 days and 6.1 for 15 days or less (Table 3). It was observed that as the time spent on board increased, the numbers of seafarers exposed to an occupational accident increased.

Table 4 shows the detailed numbers of seafarers suffered occupational accident according to rank. Able seamen (n= 70) and ordinary seamen (n= 70) accounted for 45% of the all seafarers who suffered occupational accident. Boatswains (n= 32), chief officers (n= 26) and second officers (n= 19) were the other ranks were prone to suffer from occupational accident. A majority of the occupational accidents involved deck ratings (56%) and more than three-quarters (76%) of the seafarers were employed at deck department.

Table 4. Number and percentage of seafarers who suffered occupational accident according to rank

Rank of seafarer	N (%)
Able seaman	70 (22.5)
Ordinary Seaman	70 (22.5)
Boatswain	32 (10.3)
Chief Officer	26 (8.4)
Second Officer	19 (6.1)
Chief Engineer	15 (4.8)
Welder	14 (4.5)
Third Officer	13 (4.2)
Master	10 (3.2)
Second Engineer	10 (3.2)
Forth Engineer	7 (2.3)
Third Engineer	5 (1.6)
Deck Cadet	5 (1.6)
Electrician	4 (1.3)
Oiler	3 (1.0)
Electrical Officer	2 (0.6)
Engine Cadet	2 (0.6)
Wiper	1 (0.3)
Chief cook	1 (0.3)
Steward	1 (0.3)
Mechanic	1 (0.3)
Total	311 (100)

Falls on board/falls overboard was found the most common type of occupational accident that 128 seafarers suffered and 96 of them lost their lives, 24 of them exposed to serious injuries and 8 of them suffered non-serious injuries. Exposure to harmful substances and asphyxiation in cargo holds and tanks was also a more common form of occupational accident among seafarers with 44 cases. 27 and 16 of 44 cases resulted with deaths and serious injurious, respectively. "Struck by rope or chain" (9.3%), "caught between objects" (9.3%) and "striking against or struck by objects" (8.4%) were the other more common types of accident that caused 17, 18 and 12 fatalities respectively (Table 5).

Table 5. Number and percentage of seafarers who suffered occupational accidents according to type of accident

Type of accident	N (%)
Falls on board/ Falls overboard	128 (41.2)
Exposure to or contact with harmful substances/ Asphyxiation	44 (14.1)
Struck by rope or chain	29 (9.3)
Caught or in between objects	29 (9.3)
Striking against or struck by objects	26 (8.4)
Slips, stumbles and falls	11 (3.5)
Exposure to or contact with extreme temperatures	14 (4.5)
Exposure to or contact with electric current	5 (1.6)
Struck by falling objects	7 (2.3)
Exposure to fire/explosion	17 (5.5)
Contact with working machine	1 (0.3)
Total	311 (100)

Table 6. Number and percentage of seafarers who suffered occupational accidents according to working situation at time of accidents

Working situation at time of accident	N (%)
Maintenance on deck	55 (17.7)
Loading/Unloading cargo	45 (14.5)
Mooring operations	34 (10.9)
Boat drills	31 (10.0)
Cleaning in tank/hold	30 (9.6)
Entrance to enclosed spaces	22 (7.1)
Walking from one place to another	19 (6.1)
Rigging and taking in gangways and pilot ladders	16 (5.1)
Engine maintenance at sea	15 (4.8)
Duty on bridge/Watchkeeping	8 (2.6)
Lashing and unlashing of cargo	7 (2.3)
Routine tasks	7 (2.3)
Control of tank or hold	4 (1.3)
Anchoring operations	4 (1.3)
Opening and closing of hatches	3 (1.0)
Working in enclosed space	3 (1.0)
Gas welding and cutting	3 (1.0)
Reading draughts	2 (0.6)
Working with electrically powered equipment	2 (0.6)
Working at funnel	1 (0.3)
Total	311 (100)

In Table 6, all occupational accidents have been classified based on the activity of seafarers at the time of the accident. Most of the 311 seafarers (272, 87.5%) had been working on deck areas at the time of the accident and a further 28 (9%) were in the engine room. The most frequent work situation at the time of the accident were maintenance and repair work (36 fatalities and 18 serious injuries), loading and unloading cargoes (32 fatalities and 12 serious

injuries), mooring operations (19 fatalities and 13 serious injuries), boat drills (11 fatalities and 12 serious injuries), cleaning in tanks or holds (20 fatalities and 9 serious injuries) and entrance to enclosed spaces (16 fatalities and 5 serious injuries).

The highest frequency of occupational accidents was found on bulk carriers, container ships and general cargo ships with 68, 65 and 62 cases, respectively. Out of 253 ships involved occupational accidents, 51 ships (16.4%) were Hong Kong registered, 44 ships (14.1%) were Malta registered, 20 ships (6.4%) were Germany registered and 20 ships (6.4%) were flying the Denmark flag (Table 7). Besides, 39% of ships which occurred occupational accident on, were sailing under the flag of convenience (e.g. Panama, Liberia, Malta), excluded ships in other category. Also, 69 ships involved occupational accident were aged 5 or less, 56 ships were aged 6-10 and 52 ships were aged 20 or more (Table 7).

Table 7. Number and percentage of ships involved occupational accident according to type, flag and age

Ship Type	N (%)
Bulk Carrier	68 (21.9)
Container Ship	65 (20.9)
General Cargo	62 (19.9)
Chemical Tanker	27 (8.7)
Ro-Ro	8 (2.6)
Oil/Petrol Tanker	7 (2.3)
LNG/LPG	5 (1.6)
Reefer	3 (1.0)
Other	8 (2.6)
Ship Flag	
Hong Kong	51 (16.4)
Malta	44 (14.1)
Germany	20 (6.4)
Denmark	20 (6.4)
Panama	17 (5.5)
China	12 (3.9)
UK	11 (3.5)
Holland	9 (2.9)
Liberia	7 (2.3)
Norway	7 (2.3)
Bahama	5 (1.6)
Marshall Islands	5 (1.6)
Isle of Men	5 (1.6)
Other	40 (12.9)
Ship Age	
≤ 5	69 (22.2)
6-10	56 (18.0)
11-15	41 (13.2)
16-20	34 (10.9)
20 ≥	52 (16.7)
Total	253 (100)

The most common causal factors led to occupational accidents were examined in ten categories which obtained from occupational accident reports. Dangerous work practices and ignorance of rules and instructions (48.6%) was found the most common cause of accident (e.g., working in an inappropriate place on board, not using personal protective equipment, dangerous work habits, negligence of snap back zones). Other common factors contributing to the accidents were insufficient risk assessment and hazard identification (e.g., default on risk assessment or underrating the significance of identified hazards), deficiencies in instruction and guidance (e.g., shortcomings in safety management manuals, unforeseeable situations), machine/equipment malfunction (e.g., design failure, poor maintenance), inappropriate education, experience and training and unsafe working environment (e.g., rough weather, poor illumination, insufficient ventilation) and lack of communication and team work (Table 8).

Table 8. Main causes of occupational accidents occurred on board ships

Causal and contributing factors	N (%)
Dangerous work practices and ignorance of rules and instructions	123 (48.6)
Insufficient risk assessment or hazard Identification	29 (9.1)
Deficiencies in instruction and guidance	20 (6.3)
Machine/equipment malfunction	19 (6.0)
Inappropriate education, experience and training	19 (6.0)
Unsafe working environment	16 (5.0)
Lack of communication and team work	14 (4.4)
Inadequate supervision	7 (2.2)
Inappropriate working plane or duty	3 (0.9)
Inappropriate physical or mental condition	3 (0.9)
Total	253 (100)

4 DISCUSSION

This study only covers occupational accidents occurred on board merchant ships which investigated and after that prepared an investigation report related to accident, by maritime authorities or marine accident investigation units. The accident investigations were conducted in accordance with international treaties and instruments. These include the Safety of Life at Sea (SOLAS) Convention (specifically the Casualty Investigation Code), relevant International Maritime Organization (IMO) resolutions and Article 94(7) of the United Nations Convention on the Law of the Sea (UNCLOS) (ATSB, 2017) and these reports are crucial for the dissemination of safety lessons and the prevention of similar future accidents and incidents (MTI, 2017).

According to IMO (2008), a marine safety investigation shall be conducted into every "very serious marine casualty and also, other marine casualties and incidents strongly advised to be investigated. However, non-fatal occupational accidents, non-serious injuries and near misses are rarely investigated and reported by marine accident investigation units. In this context, 253 occupational accident reports included to the study covering the

period 2005-2015. As predicted, the majority of the occupational accident reports (177, 70%) were related to very serious marine casualties while serious marine injuries (69, 27.3%) and non-serious marine injuries (7, 2.8%) comprised 30% of the accident reports. Because of the insufficient accident reports related to serious injuries, non-serious injuries and near misses, the results of this study cannot be generalized about occupational risks to seafarers. On the other hand, this study enables us to gain overall understanding of occupational accidents occurred on board merchant ships and to compare the findings of this study with other studies in terms of type of accident, seafarer age, rank of seafarer and etc.

In this study, only occupational accidents included which led to deaths, serious injuries or less serious injuries while suicides, homicides and maritime disaster were excluded. In this sense, when comparing this study with other studies, taking into account inclusion and exclusion criteria would be beneficial. Also, the accident rate was not calculated in this study due to insufficient accident reports and would not be reliable to compare with other studies' accident rate.

More than half of the fatal occupational accidents (116, 59%) affected deck ratings, which is consistent with other studies internationally that have identified the deck department as carrying the highest risk of on-duty accidents (Nielsen & Panayides, 2005; Nielsen, 1999; Hansen, 1996; Roberts, 2006; Roberts & Marlow, 2005; Roberts & et al., 2004). For non-fatal injuries, deck department (74, 73%) has still the largest exposure to highest risk due to heavy works on deck and in holds. However, non-fatal injuries are more evenly distributed among the ratings and officers when compared with fatal accidents.

Falls on board or falls overboard merchant ships caused almost 40% of the occupational accidents which is the same as reported by Roberts & Marlow (2005) for seafarers employed in British merchant shipping. They largely involved falls into or inside holds during cargo operations, off ladders or down stairways, falls from heights while working at the ship's side for painting or cleaning, and falls overboard when rigging or de-rigging gangways and pilot ladders. Fatalities and serious injuries through falls overboard or falls on board can substantially be prevented by the more widespread use of self-inflating life vests or safety harnesses when engaged in potentially hazardous operations at great heights or along the sides of ships.

Asphyxiation and exposure to harmful substances in enclosed spaces during entry in holds and tanks for cleaning cargo residue, maintenance and routine inspection have also been reported as one of the most common type of occupational accident among seafarers (Roberts et al., 2014; Nielsen & Panayides,

2005; Nielsen, 1999; Hansen, 1996; Roberts, 2006; Roberts & Marlow; 2006). Hansen (1996) pointed that oxygen contents enclosed spaces were not properly checked and safety lines were not rigged which caused to delay the rescue. Prevention of enclosed space accidents can be achieved through adherence to recommended procedures and checklists when entering enclosed spaces, which have been extended progressively over time in shipping (IMO, 2010).

While mooring and anchoring operations, being struck by mooring or towing ropes and chain of anchor (17 fatalities, 12 serious injuries) is one type of occupational accident caused fatal and serious injuries among seafarers. Hazards often arise when the ropes become tense due to sudden unpredictable ship movements, use of worn ropes/wires and negligence of snap back zones. Also, weak communication among ship's crew and between ship and shore personnel was one of the main cause of mooring accidents. This type of accidents can be avoided by maintenance and replacement of worn ropes regularly and caution towards standing in proximity to ropes under stress. Besides, before planning the mooring and anchoring operations, considering the weather condition factors such as wind and current is crucial for prevention of unforeseeable ship movements.

The most critical cause of occupational accidents was found the dangerous work practices and ignorance of rules and instructions (e.g., not using personal protective equipment, dangerous work habits, and negligence of snap back zones) in this study. In the studies of Oldenburg et al. (2010) and Uğurlu et al. (2016), it was also found that not using or improper use of personal protective equipment was the main causal factor leading to deaths or serious injuries. Also, insufficient risk assessment or hazard identification (9.1%), deficiencies in instruction or guidance (6.3%), machine/equipment malfunction (6%) and inappropriate education, experience and training (6%) were the other main causes led to occupational accidents on board merchant ships.

REFERENCES

ATSB- Australian Transport Safety Bureau. https://www.atsb.gov.au/marine/marine-safety/Accessed: 08.02.2017

Borch, D. F., Hansen, H. L., Burr, H. & Jepsen, J. R. 2012. Surveillance of maritime deaths o board Danish merchant ships, 1986-2009.International Maritime Health,63(1):7-16.

Buxton, I.L. & Cuckson, B.R., 1997. Ship susceptibility, loss risk and marine insurance. *Transactions of the Royal Institution of Naval Architects*, 139A, 98-116.

Curry, R., 1996. Merchant ship losses 1934-1993: an overview. *Transactions of the Royal Institution of Naval Architects*, 138A, 1-20.

EMSA- European Maritime Safety Agency. 2015. *Annual overview of marine casualties and incidents 2015.*.

EMSA- European Maritime Safety Agency. 2016. *Annual overview of marine casualties and incidents 2016.*

Hansen, H.L. 1996. *Occupation-related morbidity and mortality among merchant seafarers with particular reference to infectious diseases*, (PhD Thesis), South Jutland University Centre, Esbjerg/ Denmark.

Hansen, H.L. 1996. Surveillance of deaths on board Danish merchant ships 1986–93: Implications for prevention. *Occupational and Environmental Medicine*,53(4):269–276.

IMO- International Maritime Organization. 2008. *Adoption of the code of the international standards and recommended practices for a safety investigation into a marine casualty or marine incident (casualty investigation code).* Resolution MSC.255(84). (adopted on 16 May 2008).

IMO- International Maritime Organisation. 2010. *ISM Code and Guidelines on Implementation of the ISM Code 2010.* London, UK.

ILO- International Labour Organization. 1993. *Occupational accidents among seafarers resulting in personal injuries, damage to their general health and fatalities.*

ILO- International Labour Organization. 1996. *Reporting and notification of occupational accidents and disesases.*

ILO- International Labour OrganizIMOation. 2014. *Guidelines for implementing the occupational safety and health provisions of the Maritime Labour Convention, 2006.*

Mayhew, C. *Work-related traumatic deaths of British and Australian Seafarers: What are the causes and how can they be prevented?* Seafarers International Research Centre, Cardiff University.

MTI- The Ministry for Transport and Infrastructure of Malta. https://mti.gov.mt/en/Pages/MSIU/Marine-Safety-Investigation-Unit.aspx Accessed: 08.02.2017.

Nielsen, D. 1999a. Death at sea-a study of fatalities on board Hong-Kong-registered merchant ships (1986-95). *Safety Science*, 32(2-3): 121-141.

Nielsen, D. 1999b. Occupational accidents at sea. *WIT Transactions on The Built Environment*, 45.

Nielsen, D. & Roberts, S. 1999. Fatalities among the world's merchant seafarers(1990-1994).*Marine Policy*,23(1):71-80.

Nielsen D. & Panayides P.M. 2005. Causes of casualties and the regulation of the occupational health and safety in the shipping industry. *WMU J Marit Affairs*, 4(2): 147–167.

Oldenburg, M., Baur, X. & Schlaich, C. 2010. Occupational risks and challenges of seafaring. *Journal of Occupational Health*, 52: 249-256.

Roberts, S. E. 2006. Surveillance of work related mortality among seafarers employed on board Isle of Man registered merchant ships from 1986 to 2005. *International Maritime Health*, 57: 1-4.

Roberts, S. E. & Marlow, P.B. 2005. Traumatic work related mortality among seafarers employed in British merchant shipping, 1976–2002. *Occupational & Environmental Medicine*; 62 (3):172–180.

Roberts, S. E., Nielsen, D., Kotlowski, A. & Jaremin, B. Fatal accidents and injuries among merchant seafarers worldwide. *Occupational Medicine*, 64: 259-266.

Roberts, S.E. & Hansen, H.L. 2002. An analysis of the causes of mortality among seafarers in the British merchant fleet (1986–1995) and recommendations for their reduction. *Occupational Medicine*, 52(4): 195–202.

Saarni, H. 1989. Industrial accidents among Finnish seafarers. *Travel Medicine International*, 7: 64-68.

Uğurlu, Ö., Kum, S. % Aydoğdu, Y. V. 2016. Analysis of occupational accidents encountered by deck cadets in maritime transportation. Maritime Policy & Management , 1-19.

Supply Chain of Blocks and Spare Parts

Simulation-based Modeling of Block Assembly Area at Shipyards

M. Özkök
Karadeniz Technical University, Trabzon, Turkey

İ.H. Helvacıoğlu
Istanbul Technical University, Istanbul, Turkey

ABSTRACT: Shipbuilding process contains many activities and these activities are performed in various work stations. Among these activities, block assembly operations play an important role in ship production. The blocks, which are subject to assembly processes in accordance with mounting sequence, are moved to slipway by using cranes and these bocks are erected at slipway. In this way, whole vessel steel hull structure is built. The lateness which may occur in block assembly area causes the blocks to be moved tardy to slipway. This is an unwanted thing in terms of shipyard. For this reason, the fabrication of blocks is needed to be completed as soon as possible. In this study, the steel structure production of blocks is considered. The production process of block steel structure is modeled on basis of simulation environment. By determining the operation processing times for the activities, simulation model will be created for block assembly area. Then, these times will be altered. In this way, it will be determined how long it takes to finish the production of the steel portion of the block and what actions cause bottlenecks.

1 INTRODUCTION

More recently, it is a common thing that the shipyard activities have been done quickly and efficiently and the block production has been completed as soon as possible. The reason for this is that there is a tough competition in the global scale and this competition is pushing the shipyards to faster delivery. Blocks are the structures forming the vessel and the faster they are fabricated, the faster they are delivered to the ship owners. So, the operations in block assembly area have to be completed as soon as possible and the blocks fabricated in this area have to be transported to the slipway in a short time. In other words, makespan of the block has to be shortened. In this way, the block mounting area can be emptied and the blocks coming from behind can be unloaded.

Block assembly area is the place where the block structures are completed and ready for transporting to slipway. Here, various steel works of the block are also performed. Blocks are fabricated on flat jigs in this area. First, major sub assembly structure is laid on flat jig. Then, the flat panel structure is placed on major sub assembly structure and, in this position, the two block structures are linked together by tack welding. Then, the curvilinear plates forming the bilge turn are welded. The block

standing in a reverse position is brought to a flat position and the tack- welded places are grounded first, then gas welding process is performed and the welding process is completed. In addition, curvilinear stiffeners are welded on curvilinear plates. In this way, steel operations of the blocks in the block assembly area are completed.

(Seo et al. 2007) developed a process planning system based on drawings and CAD model. (Cho et al. 1998) identified an integrated process planning and scheduling system for block mounting in ship production. (Koh et al. 1999) introduced a manufacturing scheduling system for block assembly work unit at a shipyard. (Zheng et al. 2011) developed a spatial scheduling algorithm reducing makespan at block assembly area in order to achieve the better performance and higher efficiency. (Lee et al. 2009) performed a material flow analysis to increase process productivity at panel block assembly shop by using simulation technology. (Cha & Roh 2010) proposed a simulation framework and this framework is applied to block erection process ship production. (Song et al. 2009) calculated assembly lead time based on detailed block assembly procedure by using simulation environment. (Zhou et al. 2011) rationalized block placement and and tried to improve long-term assembly area utilization by using discrete-event simulation. (Ryu et al. 2008) presented a spatial planning and scheduling system in order to support planning system in block assembly shop and balance the workload. (Liu et al. 2011) proposed a look-ahead scheduling system in order to solve the dynamic spatial scheduling problem by using simulation-based framework. (Kim et al. 2015) presented a vision-based system for monitoring block assembly.

2 METHOD

In this study, firstly, the general activities performed on the block assembly area are defined and the average end times of these activities are determined by asking the field engineers. The data were taken according to the triangular distribution. After the data were collected, the work activities performed on the block assembly area were modeled by using simulation software. After the simulation model of the work activities is built up, the relevant data is entered on the simulation model one by one. After the data entry was completed, the model was run for 1 year and some results were obtained. By evaluating the results obtained, the work activities that could cause the bottleneck were determined.

3 SIMULATION MODEL

After the simulation model of the block assembly area was created, various scenarios were defined on the model. Within each scenario, 10 different situations are dealt with. In Scenario 1, the welding times of curved plates are reduced regularly while the duration of the block gas welding, the duration of the grinding after welding, and the welding time of the curved stiffeners are kept constant. After this improvement, it was determined how many blocks were fabricated on the block assembly area. In Scenario 2, the gas welding time of curved plates, the grinding duration after gas-welding, the duration of welding of curved stiffeners are kept constant, and the time of the block gas welding is reduced systematically. Scenarios 3 and 4, on the other hand, regularly reduce the grinding time after gas welding and the welding time of curved stiffeners, while other variables are kept constant.

While simulation model of the block assembly area is building up, some processing data have to be entered to the model. Table 1 shows the processing times of various activities performed in block assembly area.

Besides, in block assembly area, there are many operations in addition to the activities indicated in Table 1 and these activities are also entered to the simulation model. Table 2 presents these activities.

Table 1. Activity definitions and processing times performed in block assembly area

Activity Definitions	Processing time (min.)
Major sub assembly is adjusted on flat jig	Random.Triangular (25,30,35)
Flat panel structure is placed on major sub assembly structure	Random.Triangular (40,50,60)
Flat panel and major sub assembly structures are tack-welded	Random.Triangular (125,160,200)
Curved plates are gas-welded on block	Random.Triangular (250,350,400)
Block is lifted	Random.Triangular (5,10,15)
Block is upturned	Random.Triangular (40,45,50)
Tack-welded locations are grinded	Random.Triangular (100,120,150)
Locations to be welded are gas-welded	Random.Triangular (350,400,450)
Gas-welded locations are grinded	Random.Triangular (150,200,250)
Curved stiffeners are tack-welded	Random.Triangular (120,150,180)
Tack-welded locations of stiffeners are grinded	Random.Triangular (60,80,110)
Curved stiffeners are mounted by gas metal arc welding	Random.Triangular (200,240,300)
Gas-welded stiffeners are grinded	Random.Triangular (80,120,160)

Table 2. Operations carried out in block assembly area.

Operations	Time(min.)
The trailer carries the plate batch to the block assembly area	Random.Triangular (0.5,1,1.5)
The trailer unloads the plate batch to the block assembly area	Random.Triangular (10,15,20)
The trailer goes to profile cutting station	Random.Triangular (0.6,1,1.5)
The trailer loads stiffener batch	Random.Triangular (10,15,20)
The trailer carries the profile batch to the block assembly area	Random.Triangular (0.8,1,1.3)
The trailer unloads the profile batch to the block assembly area	Random.Triangular (10,15,20)
Low bed transporter unit loads major sub assembly structure	Random.Triangular (15,20,25)
Low bed transporter unit carries major sub assembly structure to block assembly area	Random.Triangular (2,5,7)
Crane loads major sub assembly over low bed transporter unit	Random.Triangular (15,20,25)
Major sub assembly structure is adjusted on flat jig	Random.Triangular (25,30,35)
Low bed transporter unit goes to steel work station to pick up flat panel structure	Random.Triangular (1,2,3)
Low bed transporter unit loads flat panel structure	Random.Triangular (15,20,25)
Low bed transporter unit carries flat panel structure to block assembly area	Random.Triangular (2,5,7)
Crane picks up flat panel structure from low bed transporter unit	Random.Triangular (7,10,16)
Crane lowers flat panel on major sub assembly	Random. Triangular (25,30,35)
Placing weight on flat panel.	Random. Triangular (10,15.18)

After the durations indicated in Table 1 and 2 are entered to the model, simulation model of the block assembly area depicted in Figure 1 is created.

3.1 *Experiment 1*

The results obtained from Experiment 1 are shown in Figure 2 and 3. Accordingly, improvements that can be made to the curved plate source will not change the number of blocks that are manufactured within 1 year of block assembly.

3.2 *Experiment 2*

The results obtained from Experiment 2 are shown in Figure 4 and 5. Accordingly, improvements that can be made to the gas welding in block will increase the number of blocks that are manufactured within 1 year of block assembly.

3.3 *Experiment 3*

The results obtained from Experiment 3 are shown in Figure 6 and 7. Accordingly, improvements that can be made to the grinding after gas welding in block will not change the number of blocks that are manufactured within 1 year of block assembly.

3.4 *Experiment 4*

The results obtained from Experiment 4 are shown in Figure 8 and 9. Accordingly, improvements that can be made to the grinding after gas welding of curved stiffener will not change the number of blocks that are manufactured within 1 year of block assembly.

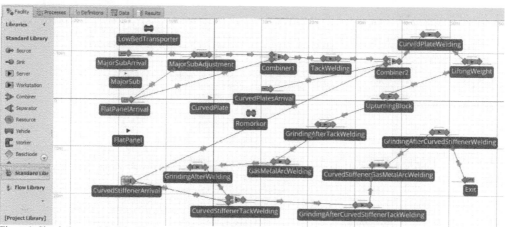

Figure 1. Simulation model of the block assembly area at shipyard.

Name	Status	Required	Completed	ProcessTime_GasMetalArcWeld (Minutes)	ProcessTime_GrindingAfterWeld (Minutes)	ProcessTime_CurvedStiffenerWeld (Minutes)	ProcessTime_CurvedPlateWeld (Minutes)	ExitNumberEntered
Case1	Completed	10	10 of 10	Random.Triangular(350,400,450)	Random.Triangular(150,200,250)	Random.Triangular(200,240,300)	Random.Triangular(250,350,400)	307
Case2	Completed	10	10 of 10	Random.Triangular(350,400,450)	Random.Triangular(150,200,250)	Random.Triangular(200,240,300)	Random.Triangular(240,340,390)	307,1
Case3	Completed	10	10 of 10	Random.Triangular(350,400,450)	Random.Triangular(150,200,250)	Random.Triangular(200,240,300)	Random.Triangular(230,330,380)	306,6
Case4	Completed	10	10 of 10	Random.Triangular(350,400,450)	Random.Triangular(150,200,250)	Random.Triangular(200,240,300)	Random.Triangular(220,320,370)	307,5
Case5	Completed	10	10 of 10	Random.Triangular(350,400,450)	Random.Triangular(150,200,250)	Random.Triangular(200,240,300)	Random.Triangular(210,310,360)	306,5
Case6	Completed	10	10 of 10	Random.Triangular(350,400,450)	Random.Triangular(150,200,250)	Random.Triangular(200,240,300)	Random.Triangular(200,300,350)	306,6
Case7	Completed	10	10 of 10	Random.Triangular(350,400,450)	Random.Triangular(150,200,250)	Random.Triangular(200,240,300)	Random.Triangular(190,280,340)	307,9
Case8	Completed	10	10 of 10	Random.Triangular(350,400,450)	Random.Triangular(150,200,250)	Random.Triangular(200,240,300)	Random.Triangular(180,270,330)	307,3
Case9	Completed	10	10 of 10	Random.Triangular(350,400,450)	Random.Triangular(150,200,250)	Random.Triangular(200,240,300)	Random.Triangular(170,260,320)	307,1
Case10	Completed	10	10 of 10	Random.Triangular(350,400,450)	Random.Triangular(150,200,250)	Random.Triangular(200,240,300)	Random.Triangular(160,250,310)	307,5

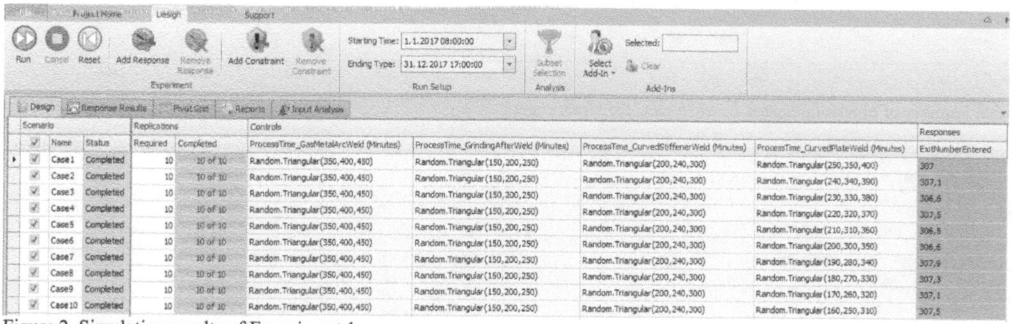

Figure 2. Simulation results of Experiment 1.

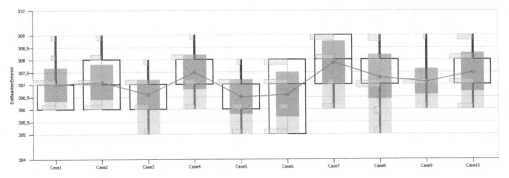

Figure 3. Number of blocks fabricated for each case for Experiment 1.

Scenario		Replications		Controls				Responses
Name	Status	Required	Completed	ProcessTime_GasMetalArcWeld (Minutes)	ProcessTime_GrindingAfterWeld (Minutes)	ProcessTime_CurvedStiffenerWeld (Minutes)	ProcessTime_CurvedPlateWeld (Minutes)	ExitNumberEntered
Case1	Compl...	10	10 of 10	Random.Triangular(350,400,450)	Random.Triangular(150,200,250)	Random.Triangular(200,240,300)	Random.Triangular(250,350,400)	307
Case2	Compl...	10	10 of 10	Random.Triangular(340,390,440)	Random.Triangular(150,200,250)	Random.Triangular(200,240,300)	Random.Triangular(250,350,400)	314,9
Case3	Compl...	10	10 of 10	Random.Triangular(330,380,430)	Random.Triangular(150,200,250)	Random.Triangular(200,240,300)	Random.Triangular(250,350,400)	323,8
Case4	Compl...	10	10 of 10	Random.Triangular(320,370,420)	Random.Triangular(150,200,250)	Random.Triangular(200,240,300)	Random.Triangular(250,350,400)	331,9
Case5	Compl...	10	10 of 10	Random.Triangular(310,360,410)	Random.Triangular(150,200,250)	Random.Triangular(200,240,300)	Random.Triangular(250,350,400)	341,6
Case6	Compl...	10	10 of 10	Random.Triangular(300,350,400)	Random.Triangular(150,200,250)	Random.Triangular(200,240,300)	Random.Triangular(250,350,400)	351,5
Case7	Compl...	10	10 of 10	Random.Triangular(290,340,390)	Random.Triangular(150,200,250)	Random.Triangular(200,240,300)	Random.Triangular(250,350,400)	360,2
Case8	Compl...	10	10 of 10	Random.Triangular(280,330,380)	Random.Triangular(150,200,250)	Random.Triangular(200,240,300)	Random.Triangular(250,350,400)	368,7
Case9	Compl...	10	10 of 10	Random.Triangular(270,320,370)	Random.Triangular(150,200,250)	Random.Triangular(200,240,300)	Random.Triangular(250,350,400)	368,8
Case10	Compl...	10	10 of 10	Random.Triangular(260,310,360)	Random.Triangular(150,200,250)	Random.Triangular(200,240,300)	Random.Triangular(250,350,400)	367,7

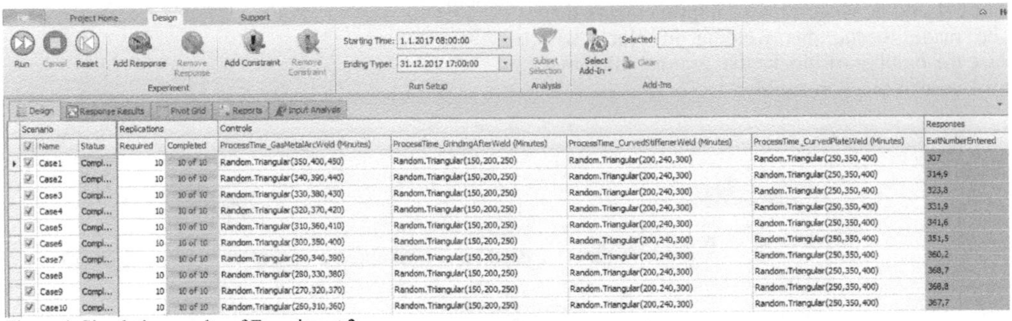

Figure 4. Simulation results of Experiment 2.

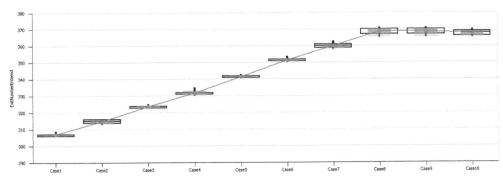

Figure 5. Number of blocks fabricated for each case for Experiment 2.

Scenario								Responses
Name				…eld (Minutes)	ProcessTime_GrindingAfterWeld (Minutes)	ProcessTime_CurvedStiffenerWeld (Minutes)	ProcessTime_CurvedPlateWeld (Minutes)	ExitNumberEntered
Case1	Compl...	10	10 of 10	Random.Triangular(350,400,450)	Random.Triangular(150,200,250)	Random.Triangular(200,240,300)	Random.Triangular(250,350,400)	307
Case2	Compl...	10	10 of 10	Random.Triangular(350,400,450)	Random.Triangular(140,190,240)	Random.Triangular(200,240,300)	Random.Triangular(250,350,400)	307
Case3	Compl...	10	10 of 10	Random.Triangular(350,400,450)	Random.Triangular(130,180,230)	Random.Triangular(200,240,300)	Random.Triangular(250,350,400)	307,4
Case4	Compl...	10	10 of 10	Random.Triangular(350,400,450)	Random.Triangular(120,170,220)	Random.Triangular(200,240,300)	Random.Triangular(250,350,400)	306,9
Case5	Compl...	10	10 of 10	Random.Triangular(350,400,450)	Random.Triangular(110,160,210)	Random.Triangular(200,240,300)	Random.Triangular(250,350,400)	306,9
Case6	Compl...	10	10 of 10	Random.Triangular(350,400,450)	Random.Triangular(100,150,200)	Random.Triangular(200,240,300)	Random.Triangular(250,350,400)	307,2
Case7	Compl...	10	10 of 10	Random.Triangular(350,400,450)	Random.Triangular(90,140,190)	Random.Triangular(200,240,300)	Random.Triangular(250,350,400)	307,4
Case8	Compl...	10	10 of 10	Random.Triangular(350,400,450)	Random.Triangular(80,130,180)	Random.Triangular(200,240,300)	Random.Triangular(250,350,400)	307,4
Case9	Compl...	10	10 of 10	Random.Triangular(350,400,450)	Random.Triangular(70,120,170)	Random.Triangular(200,240,300)	Random.Triangular(250,350,400)	307,3
Case10	Compl...	10	10 of 10	Random.Triangular(350,400,450)	Random.Triangular(60,110,160)	Random.Triangular(200,240,300)	Random.Triangular(250,350,400)	307,2

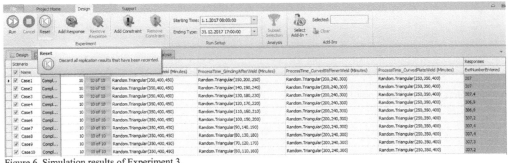

Figure 6. Simulation results of Experiment 3.

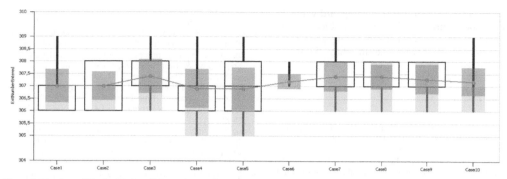

Figure 7. Number of blocks fabricated for each case for Experiment 3.

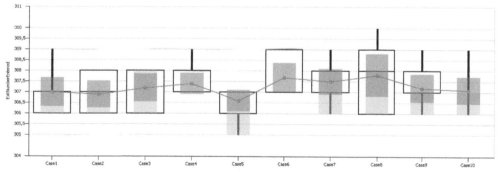

Figure 8. Simulation results of Experiment 4.

Figure 9. Number of blocks fabricated for each case for Experiment 4.

4 CONCLUSIONS

In this paper, it was purposed to determine the effects of the main activities on the number of fabricated blocks in block assembly area. For this aim, a simulation model of the block assembly area was built up and some experiments were performed on the model and some results were achieved. It was found that gas welding process has an effect on the number of blocks fabricated as can be seen from Table 3.

Table 3. Comparison of experiments based on cases.

Case No	Experiment1	Experiment2	Experiment3	Experiment4
1	307	307	307	307
2	307,1	314,9	307	306,9
3	306,6	323,8	307,4	307,2
4	307,5	331,9	306,9	307,4
5	306,5	341,6	306,9	306,5
6	306,6	351,5	307,2	307,7
7	307,9	360,2	307,4	307,5
8	307,3	368,7	307,4	307,8
9	307,1	368,8	307,3	307,2
10	307,5	367,7	307,2	307,1

So, this study shows that if the shipyard wants to enhance the number of fabricated blocks, the improvement has to be on gas welding process in block structure. As this is done, the throughput can be maximized.

REFERENCES

Sea, Y., Sheen, D., Kim, T. 2007. Block assembly planning in shipbuilding using case-based reasoning, *Expert Systems with Applications*, Vol. 32, No. 1, 245-253.

Cho, K.K., Oh, J.S., Ryu, K.R., Choi, H.R. 1998. An Integrated Process Planning and Scheduling System for Block Assembly in Shipbuilding, *CIRP Annals-Manufacturing Technology*, Vol. 47, No. 1, 419-422.

Koh, S-G., Park, J-C., Choi, Y-S., Joo, C-M. 1999. Development of a Block Assembly Scheduling System for Shipbuilding Company, *IE Interfaces*, Vol.12, No. 4, 586-594.

Zheng, J., Jiang, Z., Chen, Q., Qunting, L. 2011. Spatial scheduling algorithm minimising makespan at block assembly shop in shipbuilding, *International Journal of Production Research*, Vol. 49, No. 8, 2351-2371.

Lee, K., Shin, J-G., Ryu, C.H. 2009. Development of simulation-based production execution system in a shipyard: a case study for a panel block assembly shop, *Production Planning & Control*, Vol. 20, No. 8, 750-768.

Cha, J-H. & Roh, M-II. 2010. Combined discrete event and discrete time simulation framework and its application to the block erection process in shipbuilding, *Advances in Engineering Software*, Vol. 41, No. 4, 656-665.

Song, Y-J., Lee, D-K., Choe, S-W., Woo, J-H., Shin, J-G. 2009. A Simulation-Based Capacity Analysis of a Block-Assembly Process in Ship Production Planning, *Journal of the Society of Naval Architects of Korea*, Vol. 46, No. 1, 78-86.

Zhuo, L., Chua, D., Huat, K., Wee, K-H. 2011. Scheduling dynamic block assembly in shipbuilding through hybrid simulation and spatial optimization, *International Journal of Production Research*, Vol. 50, No. 20, 5986-6004.

Ryu, C., Shin, J.G., Kwon, O.H., Lee, J.M. 2008. Development of integrated and interactive spatial planning system of assembly blocks in shipbuilding, *International Journal of Computer Integrated Manufacturing*, Vol. 21, No. 8, 911-922.

Zhuo, L., Chua, D., Huat, K., Wee, K-H. 2011. A Simulation Model for Spatial Scheduling of Dynamic Block Assembly in Shipbuilding, *Journal of Engineering, Project, and Production Management*, 1(1), 3-12.

Kim, M., Choi, W., Kim, B-C., Kim, H., Seol, J. H., Woo, J., Ko K-H. 2015. A vision-based system for monitoring block assembly in shipbuilding, *Computer-Aided Design*, Vol. 59, 98-108.

Exploring the Potential of 3D Printing of the Spare Parts Supply Chain in the Maritime Industry

E. Kostidi & N. Nikitakos
University of the Aegean, Chios, Greece

ABSTRACT: Maritime assets have notable characteristics similar to other industries with moving assets such as aircraft/aerospace, defense units, and automotive. The aim of this work is to explore the introduction of additive manufacturing in the supply chain of the spare parts of the ships.   A literature review of the available technology (methods, materials), and lessons learned from the application in industries with similar to shipping characteristics (industries with moving assets), reveals the potential of applying it in the shipping industry. The existing situation process is modeled including the involved stakeholders (ship, the shore, suppliers, manufacturer etc.) in order to get an understanding of it. An alternative new technology future scenario is proposed for supplying the ships with spare parts.

1 INTRODUCTION

Additive manufacturing (AM) or 3d printing as it is commonly known has been already implemented in various sectors (industrial products, consumer products, automotive, aerospace, etc.). It is based on the principle of the construction in layers by directly converting the 3D data into physical objects, can produce functionally integrated components (including spare parts) in a single production step, in small batches. The main benefit of this technology is that it allows the flexible production of customized products at no extra cost in terms of manufacturing. This is achieved by directly converting the 3D data into physical objects, without the need of additional tools or molds. Furthermore, the principle of the construction in layers can produce functionally integrated components in a single production step, thereby obviating the need for the assembly stage. In comparison to conventional manufacturing, it has a number of advantages in terms of better energy efficiency, cutback in emissions, better design handling and lower manufacturing lead time.

Maritime assets are capital-intensive and their out of service time has economic consequences. They usually operate away from the home base at remote locations and are on continuous move. Other sectors with similar characteristics are aircraft/aerospace, defense units, and automotive. Based on literature, we will try to explore how lessons learned from the other sectors, could be applied in shipping.

Furthermore, we conducted interviews with people working in the shipping industry, in order to get an understanding of the supply chain of the spare parts of the ships, and get an idea of how this can be changed with the introduction of additive manufacturing.

2 THE AVAILABLE TECHNOLOGY

Additive manufacturing is the official industry standard term (ASTM F2792) for all applications of the technology. It is defined as the process of joining materials to make objects from 3D model data, usually layer upon layer, as opposed to subtractive manufacturing methodologies. Figure 1. shows the principle of building the part from the basic construction unit (voxel).

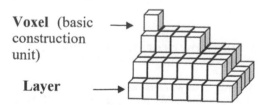

Voxel (basic construction unit)

Layer

Figure 1. Building the part layer upon layer

Synonyms are additive fabrication, additive
Synonyms are additive fabrication, additive

processes, additive techniques, additive layer manufacturing, layer manufacturing, and freeform fabrication.

Under the umbrella of AM there are many processes. ASTM groups them in seven types (Piazza & Alexander, 2015): 1) Binder jetting (3D printing)– AM process where a liquid bonding agent is deposited to join powdered materials together. 2) Direct energy deposition (direct manufacturing)– AM process where thermal energy fuses or melts materials together as they are added. 3) Material extrusion (fused deposition modeling)– AM process that allows for depositing material via a nozzle. 4) Material jetting-AM process where droplets of material are deposited. 5) Powder bed fusion (laser sintering)–AM process where thermal energy fuses or melts material from a powder bed. 6) Sheet welding (e-beam welding, laminated object manufacturing)- AM process where sheets of materials are Bonded together and 7) Vat photo-polymerization (digital light processing)-AM process where liquid photopolymer in vat is cured by light.

In some processes the material is squirted, squeezed or sprayed and in others fused, bind or glued. The power source is thermal, high-powered laser beam, electron beam, ultraviolet laser, or photo curing.

All of the aforementioned techniques rely on the application of gravity to assist in the construction process. While additive manufacturing can build a wide variety of products in a controlled and static environment the use of such techniques afloat creates questions about the viability of the process. This should not eliminate AM from consideration for the marine industry. However it does seem that at this time it will be contained to construction and repair facilities or platforms with little to no relative motion (Strickland, 2016).

The raw materials for the process are: polymers, metals, ceramics, composites, and biological materials. The starting materials could be liquid, filament/paste, powder, or solid sheet. Currently, the most common metallic materials are steels (tool steel and stainless), pure titanium and titanium alloys, aluminum casting alloys, nickel-based super alloys, cobalt-chromium alloys, gold, and silver (Frazier, 2014).

3 THE GROWTH OF THE ADDITIVE MANUFACTURING INDUSTRY

According to Wohlers Report 2016, the additive manufacturing (AM) industry grew 25.9% (CAGR – Corporate Annual Growth Rate) to $5.165 billion in 2015. Frequently called 3D printing by those outside of manufacturing circles, the industry growth consists of all AM products and services worldwide.

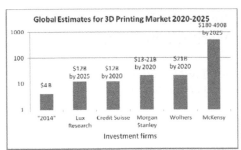

Figure 2 Global Estimates for 3D Printing Market 2020-2025 (Modified from ARK Invest, 2016)

The CAGR for the previous three years was 33.8%. Over the past 27 years, the CAGR for the industry is an impressive 26.2% (McCue, 2016).

The ARK Invest (2016) summarizes in a chart the growth projections from reputable firms.

As it can be seen in figure 2 McKensy estimates that 3D printing market could reach $180-490 by 2025. The 3D printing industry has one of the highest projections for economic growth.

4 IMPACT OF ADDITIVE MANUFACTURING ON PRODUCTION, SUPPLY CHAIN AND TRANSPORT

Market surveys predict that the introduction AM will have a major impact on industries and manufacturing (Weller, Kleer, & Piller 2015). The new technology eliminated stages of production (e.g. assembly) and thus simplify production line. The place of production moves closer to demand. The sale is made before the production of the product, upsetting the known production process.

The production facilities can now be located closer to the customer in Europe or North America, where it is more direct response to market needs (Manners-Bell & Lyon 2012).

The concept of constructing products in large complex facilities could become obsolete as companies adopt the more flexible model of additive manufacturing (Cottrill 2011).

3D printing is expected to have a significant impact on domestic and international freight operators, in particular regarding the reduction of the importance of some transport paths, and possibly lead to the opening of new ones. A recent analysis (Tipping, Schmahl, & Duiven 2015) for Strategy& about two dozen industry sectors, found that up to 41 percent of the air cargo business and 37 per cent of businesses container ocean carriers is at risk because of 3D printing.

Ye et al., (2015) based on a model, conclude that, in the next two decades, 3D printing is not likely to pose a threat, on the concept of significant production capacity, or reduce the transport flow, in terms of global container traffic. As the GDP of the world's population is not likely to decline over the next 50 years, world trade will probably continue to cause high demand for transport.

5 CHARACTERISTICS OF THE MARITIME SECTOR

Maritime assets are capital intensive and downtime has financial consequences. Usually operate away from the base in remote areas and are in constant movement. Other sectors with similar characteristics are aircraft / aerospace, defense units and road transport. Ships are operating under random environments in isolation from repair facilities and spare parts storage.

Maintenance networks are involving many actors, such as the owners of the assets, systems integrators, original equipment manufacturers (OEM), the service providers and their logistics service providers. The International Maritime Organization and classification societies impose rules that the ships have to follow (such as periodic inspections, mandatory equipment).

The spare parts inventory planning and supply chain include decisions such as determining the appropriate spare parts procurement policies, quantification and distribution of stocks of spare parts and design of service networks, taking into account, for example, emergency transport, side transshipments and joint spare pool (Eruguz, Tan, & van Houtum 2015).

The repair can be executed and stocks of spare parts can be stored on the ship itself, ashore by asset owners, system integrators, service providers, or makers. Assets can be classified as long-lived, since they have a useful life of about 25 years. The vessel is in an environment that is highly corrosive, with turbulence and vibrations.

6 LESSONS LEARNED FROM THE INTRODUCTION OF ADDITIVE MANUFACTURING IN OTHER INDUSTRIES

The published work regarding the observations from the actual implementation of additive manufacturing as it is finding its way into mainstream manufacturing industry reveals its benefits and challenges.

There are many active investments by various industries for utilization of AM parts to capitalize on the value-added properties provided by AM as shown by Seifi, Salem, Beuth, Harrysson, & Lewandowski, (2016), which highlights some industrial examples for AM parts. In particular, General Electric (GE) has received Federal Aviation Administration (FAA) certification for fuel nozzle

implementation in the GE LEAP engine. In this case, AM reduced the total part count and replaced more complex brazing of multiple components to create a lighter, simpler, and more durable product.

Compared to conventional manufacturing, it has a number of advantages in terms of better energy efficiency, cutback in emissions, better design handling and lower manufacturing lead time.

Mokasdar's work (2012) evaluated additive manufacturing impact on the aircraft spare parts supply chain. Conversely, the manufacturing lead time is small compared to these conventional processes, and hence the author with his work tries to advocate this feature of additive manufacturing to demonstrate how the total inventory of spare parts held in an aircraft spare parts supply chain, can be significantly reduced using additive manufacturing.

Other authors (Liu, Huang, Mokasdar, Zhou, & Hou, 2014) studied the impact of AM in the aircraft spare parts industry, with an emphasis on the use of distributed manufacturing strategy to reduce inventory cost. They concluded that on-demand and centralized production of spare parts is most likely to succeed.

The feasibility of localized manufacturing is also explored by Khajavi, Partanen, & Holmström, (2014), who studied the fabrication of spare parts through a quantitative cost-based assessment. It was demonstrated that currently AM is both capital and labor intensive, making centralized production preferable on financial measures.

Eyers & Potter, (2015) suggest that, it is necessary to bridge research in e-commerce, AM, and supply chain management. In order to understand better the way in which eCAM may be applied in the supply chain, Their research based on the interviews with localized manufacturers of aerospace spare parts is suggesting that costs of machines and operators, together with issues regarding quality assurance and material supply chain co-ordination would further consideration before wide-scale adoption of this eCAM model, for cost reductions and increased efficiency that may not automatically follow.

In a project (Sterkman, 2015) focused on the impact that Additive Manufacturing can have on the after sales services supply chains in the aerospace industry the conclusion was that at this moment, AM can better be outsourced. This is more favorable because: AM machines are still expensive, low utilization cannot justify these investment costs, rapid technology developments are expected and there will be need for specialized personnel.

The most important goal in the defense is to secure the supply of spare parts, followed by respectively improving service and reducing costs as Balistreri, (2015) states.

Augustsson & Becevic, (2015) investigated the inventory costs for low turnover spare parts for a truck manufacturer. They concluded that costs can be lowered, but still offer the same availability by using additive manufacturing in the automotive industry. This could implicate that the main benefit of using additive manufacturing is a big increase in customer service. Daimler, which owns the Mercedes-Benz brand, and has more than 100,000 printed prototype parts, and according to Reuters(Taylor & Cremer, 2016) it will expand production using 3D printing methods.

Abbink, Karsten, & Basten, (2015) based on modeling choices, concludes that AM is typically not beneficial for low demand, single-item situations. This is the case for both in-house printing, as well as outsourced printing. For a multi-item situation however, where the printer is sufficiently utilized, AM can be cheaper than traditional means of manufacturing.

Table 1. Lessons learned from other industries

Automotive (truck manufacturer)

The inventory costs for low turnover spare parts can be lowered.
Increase in customer service (Augustsson & Becevic2015).

Spare parts for capital goods

AM is typically not beneficial for low demand, single-item situations (Abbink, 2015).

Aircraft spare parts industry

Centralized production of spare parts is most likely to succeed (Liu, et al., 2014)

Aerospace industry

Better to outsource AM (Sterkman, 2015)

Aeronautics industry

Centralized production preferable on financial measures (Khajavi et al., 2014)

Aircraft companies and operators

The total inventory of spare part can be significantly be reduced using additive manufacturing (Mokasdar, 2012)

Aerospace

The use of e-commerce with AM has often been oversimplified (Eyers & Potter, 2015)

Defense

The most important goal is to secure the supply of spare parts, followed by respectively improving service and reducing costs (Balistreri, 2015).

Summing up the literature review of actual case studies one can conclude that additive manufacturing is a promising technology. The inventory costs for low turnover spare parts can be lowered and at the same time increase in customer service. AM could be beneficial for low demand, single-item situations, if it is difficult to make it otherwise. The centralized production of spare parts is most likely to succeed. That is also preferable on financial measures. The total inventory of spare part can be significantly reduced using additive manufacturing. The most important goal is to secure the supply of spare parts, followed by respectively improving service and reducing costs.

7 ADDITIVE MANUFACTURING IN/AND THE MARITIME INDUSTRY

Maritime will not be left out of technological developments on the IT. Although there are no published case studies of application in real situations, there are initiatives in place. Apart from general prototyping applications, there are about parts maker tests and application in ships, both in the defense sector, and the commercial.

A pilot project '3D Printing Marine Spares', (2016) was initiated by Innovation Quarter, the Port of Rotterdam Authority and RDM Makerspace with the participation of 28 businesses and agencies. The consortium partners selected and redesigned parts, had them printed and tested the results. Making use of three different production processes, the advantages of the various methods for additive manufacturing and the maturity of the technology was experienced. Thus the project brought a wealth of information on the current and near future state of 3D printing as an alternative method for producing maritime parts. The conclusion was that 3D printing indeed holds promises for a number of parts, and that product requirements can be met in a number of cases. Also the business case can be positive, especially when time to market is essential. On the other hand the findings also indicate that extra work needs to be done to get regulations adjusted to be able to qualify 3D printed parts (Zanardini, Bacchetti, Zanoni, & Ashourpour, 2016).

Green Ship of the Future (2016) and 20+ industry partners have explored the opportunity space of 3D printing and additive manufacturing, in order to assess and comprehend the potential of the technology and derived opportunities for the maritime industry. They end up with the need to explore how shipping and the maritime industry can be on the forefront of development and be part of the disruption.

US Navy has already tested the technology for maintenance activities (Scheck et al., 2016). The maintenance has given the Navy the time needed to permanently install, and test out a 3D printer on board. In the meantime, the crewmembers on board the ship have been busy printing out anything from plastic syringes, to oil tank caps, to model planes used for the mock-up of the flight deck. The US Navy argues that they are still several years away from being able to print out actual spare parts for aircraft or the ship itself, but it is certainly a good starting point. The reason why AM technologies are under evaluation is the possibility to reduce the time to supply spare parts and components to remote zone, eliminating unnecessary actors and lead time.

One of the world's largest container shipping companies, Maersk, explored 3D printing as a way to fabricate spare parts on container ships. In June 2015, the company revealed that will install 3D printers on board. The printers were capable of printing a small amount of components, according to the materials used, ABS thermoplastics. However, the company considers the possible utilization of powder based metal laser sintering printers in order to enhance the range of printed components. The main advantage is related to the possibility to immediately repair broken components, instead to be supplied with a spare part when the vessel is moored in a port (Zanardini et al., 2016). However, official information of results has not been published yet.

Published maritime cases are rare, since most examples come from the air industry. The pilot project '3D Printing Marine Spares', (2016) culminated in the printing practical appliance of seven maritime parts: a propeller, cooled valve seat, spacer ring, hinge, T-connector, seal jig and manifold. Tru-Marine (Loke, D. W. S., 2014), has developed a proprietary AM process. AM is used in actual industry level for the repair of turbocharger nozzle rings.

8 THE SPARE PARTS SUPPLY CHAIN IN THE MARITIME INDUSTRY

Maritime industry is characterized by heavy utilization of equipment and machinery and by really specific operating conditions. Ships work in a very unique operational context, and that makes the requirements of reliability and safety particularly critical (Nenni & Schiraldi, 2013).

The type and quantity of the spare parts that must be on board a ship is imposed by the authorities for its safety, or suggested by the original equipment manufacturer (OEM) in order to avoid unexpected breakdowns and ship downtime, or even by experience.

Spare parts inventory is necessary, but it costs (mainly in capital, and in some cases in available space). Various optimization techniques are used. Nenni & Schiraldi, (2013) propose an approach to calculate the optimum level of inventory for spare parts of ship equipment. Eruguz, Tan, & van Houtum, (2015) consider an integrated maintenance and spare part optimization problem for a single critical component of a moving asset for which the degradation level is observable.

We conducted interviews (semi structured) with people working in the maritime industry, in order to get an understanding of the supply chain of the spare parts of the ships, and get an idea of how this can be changed with the introduction of additive manufacturing.

The need for a replacement may occur either because the predetermined stock has fallen below the threshold, or before a predetermined maintenance or because of an extraordinary damage. If the replacement is not in stock at the ship, then a request is send to the land office (usually by the chief engineer). In the land office, after approval from the technical department, the request passes it to the procurement department.

The purchasing process is pretty much typical (Purchase Order, Request Quotations, Receive Quotations, Select the supplier, Order, Receive order, Invoice).

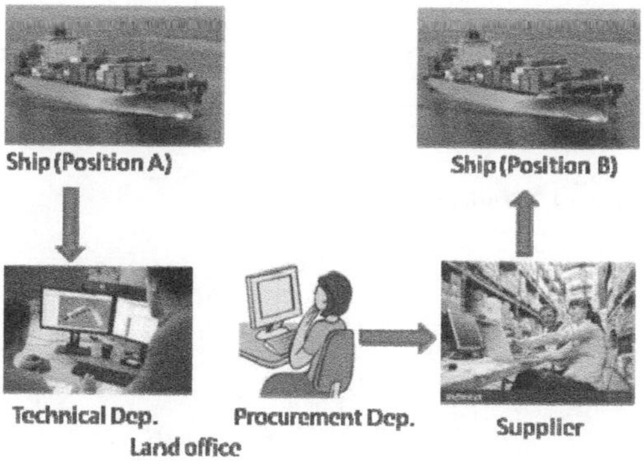

Figure 3. The purchasing process.

In this simplified diagram (fig. 3), one must note that the ship is away from the base and changes location. The spare must be timely delivered at the next port that the ship will reach. There is also an option to purchase an imitation of the spare part, or order it at a local workshop. If the requested spare part is out of stock in the chosen supplier's inventory, that must be requested from the regional warehouse, the peripheral warehouse, or finally at the OEM. If it is out of stock at the OEM, then it will be manufactured (as soon as there is economic batch).

Most of the people we talked had an idea of what 3D printing is (we did not ask about AM). Almost all had a positive attitude for the new technology and the rest were skeptical, but not negative. Their main concern was if the spare part made by the AM is comparable with the part made by the traditional method. Another concern was the cost of the AM machine, and the cost to build the part.

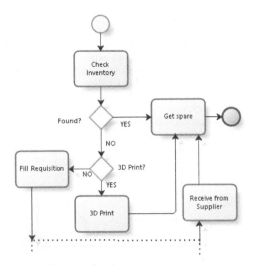

Figure 4. 3D printing decision

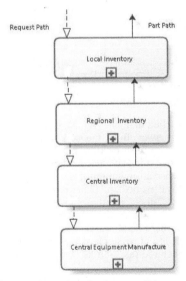

Figure 5. Places where that inventory is kept.

9 A FUTURE SCENARIO

In a future scenario the needed spare part could be made by the end user at the place it is needed, the time it is needed, avoiding the part inventory. What one would need is the proper machine, a file with the information to instruct the machine, and the raw material. With the push of a button the machine will start to make the part. The decisions that must be made are (fig. 4): 1) are the needed (machine, file, and raw material) available? 2) Is it more economic to make than buying the part?

In every place that is kept inventory (fig. 5) of spare parts to meet the part demand (on board the ship, at the supplier, the local, the regional, the central warehouse or the OEM), the AM process could take its place, in order to get the new technology benefits.

Paradigms show that it can be done today (sections 6 &7). But there are challenges pointed out in the literature, and by the people we interviewed.

10 DISCUSSION

It is enraging that most of the people we interviewed had an idea of what 3D printing is, and almost all had a positive attitude although skeptical.

Their concern if the spare part made by the AM is comparable with the part made by the traditional method could be overcome by the development of standard methods to test processes and parts (Monzón, Ortega, Martínez, & Ortega, 2014). For the time been there are standards to test the row materials, and the parts could be tested with the ways the conventional parts are tested.

As far as the cost is concerned, as the market advances, patents expire, and demand grows, the machine cost as well as the production cost will fall.

Obviously there are more issues to be phased. As it was mentioned in section 2, there are too many AM processes available. Which of the processes best suits the installation on board a ship? How will the intellectual rights be protected? How the required files will be distributed? Where in the supply chain is optimum to have the AM of the parts? How will the personnel be trained in the new technology, taking into consideration that the salesman at the supplier will be made manufacturer.

11 CONCLUSIONS

Among the benefits of AM is the flexible production of customized products, in small batches. The direct transformation of the three-dimensional data stored in a file, simply by supplying the raw materials to the machine and the production of natural objects, obviating the need for the assembly step can be applied to the manufacture of spare parts. The consequence will be change in the supply chain. This will ensure the supply of spare parts, with consequent improvement in the provided services and cost reduction.

The case studies of AM implementation, in industries with similar to maritime characteristics, have to offer several lessons. The cost of inventories for low use parts can be reduced while improving customer service time. When the printer is utilized sufficiently, the AM may be cheaper than the production applying traditional means. The AM could be beneficial for low demand situations of individual parts, if it is difficult to manufacture them otherwise. The central production of parts is more likely to succeed. This is also preferable in economic basis. The total stock of spare parts can be greatly reduced by the use of AM. The most important goal is to ensure the supply of spare parts, followed by an improvement in services and cost reductions.

As far as people we interviewed are concerned, AM is a promising technology and it should be seriously taken into account by the maritime industry.

Forthcoming standards will assure that there will be methods to ensure processes and test parts. The more the market advances, the patents expire, and the demand grows, the more the machine cost as well as the production cost will fall. The maritime industry can learn from other industries that already adapted AM in one way or the other, but further study that will take into consideration the special characteristics, is needed.

REFERENCES

Abbink, R., Karsten, F. J. P., & Basten, R. J. I. (2015). The Impact of Additive Manufacturing on Service Supply Chains. Retrieved from http://alexandria.tue.nl/extra2/afstversl/tm/Abbink_2015.pdf

ASTM(2013)AM_Standards_Development_Plan_v2.docx(n.d.).Retrieved from http://www.astm.org/COMMIT/AM_Standars_Development_Plan_v2.docx

Augustsson Robert, & Becevic Denijel. (2015). Implementing Additive Manufacturing for Spare Parts in the Automotive Industry- A case study of the use of additive manufacturing for spare parts [Master's Thesis].,Retrieved from http://studentarbeten.chalmers.se

Balistreri, G. (2015). Potential of Additive Manufacturing in the after-sales service supply chains of ground based military systems. Retrieved from http://essay.utwente.nl/67745/

Cottrill, K. (2011). Transforming the future of supply chains through disruptive innovation. *MIT Center for Transportation and Logistics, Working Paper, Spring.* Retrieved from http://www.misi.edu.my/student/spv1/assets/Disruptive_Innovations4_1.pdf

Eruguz, A. S., Tan, T., & van Houtum, G.-J. (2015). A survey of maintenance and service logistics management: Classification and research agenda from a maritime sector perspective. Retrieved from http://purl.tue.nl/24194911834760.pdf

Eyers, D. R., & Potter, A. T. (2015). E-commerce channels for additive manufacturing: an exploratory study. *Journal of Manufacturing Technology Management, 26*(3), 390–411.

Frazier, W. E. (2014). Metal Additive Manufacturing: A Review. *Journal of Materials Engineering and Performance, 23*(6), 1917–1928. https://doi.org/10.1007/s11665-014-0958-z

Green Ship of the Future (2016), The opportunity space of 3D printing and additive manufacturing in maritime, Retrieved 24 February 2017, from http://greenship.org/wp-content/uploads/2017/01/The-maritime-opportunity-space-of-3D-print.pdf

Khajavi, S. H., Partanen, J., & Holmström, J. (2014). Additive manufacturing in the spare parts supply chain. *Computers in Industry, 65*(1), 50–63. https://doi.org/10.1016/j.compind.2013.07.008

Liu, P., Huang, S. H., Mokasdar, A., Zhou, H., & Hou, L. (2014). The impact of additive manufacturing in the aircraft spare parts supply chain: supply chain operation reference (scor) model based analysis. *Production Planning & Control,25*(13–14),1169–1181. https://doi.org/10.1080/09537287.2013.808835

Loke, D. W. S. (2014). U.S.Patent Application No. 14/464,709.

Manners-Bell, J., & Lyon, K. (2012). The implications of 3D printing for the global logistics industry. *Transport Intelligence*, 1–5.

McCue, T. J. (2016). Wohlers Report 2016: 3D Printing Industry Surpassed $5.1 Billion. Retrieved 1 February 2017, from http://www.forbes.com/sites/tjmccue/2016/04/25/wohlers-report-2016-3d-printer-industry-surpassed-5-1-billion/

Mokasdar, A. S. (2012). *A Quantitative Study of the Impact of Additive Manufacturing in the Aircraft Spare Parts Supply Chain.* University of Cincinnati. Retrieved from https://etd.ohiolink.edu/ap/10?0::NO:10:P10_ACCESSION_NUM:ucin1352484289

Monzón, M. D., Ortega, Z., Martínez, A., & Ortega, F. (2014). Standardization in additive manufacturing: activities carried out by international organizations and projects. *The International Journal of Advanced Manufacturing Technology,76*(5–8),1111–1121. https://doi.org/10.1007/s00170-014-6334-1

Nenni, M. E., & Schiraldi, M. M. (2013). Validating virtual safety stock effectiveness through simulation. *International Journal of Engineering Business Management, 5.* Retrieved from http://search.proquest.com/openview/d6d614d4a0ee29df9e60b6c107213317/1?pq-origsite=gscholar

Piazza, M., & Alexander, S. (2015). Additive Manufacturing: A Summary of the Literature. Retrieved from http://engagedscholarship.csuohio.edu/urban_facpub/1319/

Port of Rotterdam, Innovation Quarter, & RDM Makerspace, (2016). 3D Printing Marine Spares (Final Report). Retrieved from http://www.innovationquarter.nl/sites/default/files/InnovationQuarterFinal%20Report%203D%20Printing%20-Marine%20Spares%20.pdf

Scheck, C. E., Wolk, J. N., Frazier, W. E., Mahoney, B. T., Morris, K., Kestler, R., & Bagchi, A. (2016). Naval Additive Manufacturing: Improving Rapid Response to the Warfighter. *Naval Engineers Journal, 128*(1), 71–75.

Seifi, M., Salem, A., Beuth, J., Harrysson, O., & Lewandowski, J. J. (2016). Overview of Materials Qualification Needs for Metal Additive Manufacturing. *JOM, 68*(3), 747–764. https://doi.org/10.1007/s11837-015-1810-0

Sterkman, C. (2015). Logistical impact of additive manufacturing on the after-sales service supply chain of a spare part provider. Retrieved from http://essay.utwente.nl/67741/

Strickland, J. D. (2016). Applications of Additive Manufacturing in the Marine Industry. *Proceedings of PRADS2016, 4*, 8th.

Taylor, E., & Cremer, A. (2016, July 13). Daimler Trucks to use 3D printing in spare parts production. Retrieved 13 August 2016, from http://uk.reuters.com/article/us-daimler-3dprinting-idUKKCN0ZT201

Tipping, A., Schmahl, A., & Duiven, F. (2015). 2015 Commercial Transportation Trends. Retrieved 12 June 2016, from http://www.strategyand.pwc.com/perspectives/2015-commercial-transportation-trends

Weller, C., Kleer, R., & Piller, F. T. (2015). Economic implications of 3D printing: Market structure models in light of additive manufacturing revisited. *International Journal of ProductionEconomics,164*,43–56. https://doi.org/10.1016/j.ijpe.2015.02.020

Ye, M., Tavasszy, L. A., Van Duin, J. H. R., Wiegmans, B., Halim, R. A., TU Delft: Technology, Policy and Management: Transport and Logistics, & TU Delft, Delft University of Technology. (2015, March 12). The Impact of 3D Printing on the World Container Transport. Retrieved from http://resolver.tudelft.nl/uuid:f16ee590-5804-4beb-b72c-a32346d0f175

Zanardini, M., Bacchetti, A., Zanoni, S., & Ashourpour, M. (2016). Additive Manufacturing Applications in the Domain of Product Service System: An Empirical Overview. *Procedia CIRP,47*,543–548. https://doi.org/10.1016/j.procir.2016.03.048

Modelling of Short Sea Shipping Tanker in Black Sea

E. Kose, M. Özkök, E. Demirci & E. Peşman
Karadeniz Technical University, Trabzon, Turkey

ABSTRACT: Short sea shipping poses significant problems for all seafarers, particularly for officers employed in oil tanker as the chief officer bears the highest workload. This study examines working times on short sea shipping. This study employs a simulation using the Simio software package to evaluate the effects of voyage time, speed, etc. on ships economy. The results shows time spend in maneuvering, cleaning, maintenance and resting in detail.

1 INTRODUCTION

The transfer of oil from one port to another is a very important task. Billions barrels of oil a day are transported to different destinations by Oil Tankers.

1.1 Short Sea Shipping

Short Sea, or Coastal Shipping as it is also called, is the transportation of commodities in short distances, along a coast, without crossing an ocean and it is usually serviced by small vessels. On the other hand, Deep Sea Shipping is the transportation of commodities in longer distances mainly crossing an ocean and it is usually utilized by bigger vessels (i.e. supramax, panamax, postpanamax, capesizes) in order to achieve economies of scale.

1.2 Blacksea Oil Transport

Historically, oil transported mainly from its two main terminals at Novorossiysk and Tuapse, across the Black Sea, supplemented by the Ukrainian port of Odessa.[8] In addition, there are two additional Black Sea ports at Supsa and Batumi in Georgia.[9] The total capacity of these ports is around 50-60 million tons per annum. Most of the oil tanker cargo transits the Turkish Straits,[10] and the rest is off-loaded in Black Sea ports for further pipeline transportation or direct consumption.

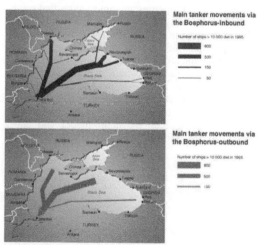

Figure 1. Main oil transport routes at Blacksea

2 METHODOLGY

In this investigation 3 ships with different particulars were used in simulation. These ships are working between the following ports;

Table 1. Loading and Discharching ports

Loading port	Discharging port	Distance (Mile)
Yarımca (Turkey)	Ambarlı (Turkey)	60
Tuapse	Samsun (Turkey)	235
Batum	Ambarlı	690
Novorsosky	Samsun (Turkey)	250
Novorsosky	Trabzon (Turkey)	280

2.1 Simulation Method

Simulation is used in nearly every field such as engineering, science, and technology, and simulation systems have been increasingly adapted for a wide variety of applications. The techniques are currently employed to design new systems, analyze existing systems, train for all types of activities, and as a form of interactive entertainment (Smith, 1999). In the maritime industry, simulations are generally used to model marine accidents, maritime traffic, the ship construction processes, and to model and evaluate port operational efficiency, among others (Kim et al., 2004; Goerlandt and Kujala, 2011; Uğurlu, et al. 2015; Demirci, et al., 2000; Lee et al. 2003; Hirsch, et al., 1998; Köse et al., 2003; Demirci, 2003; Başar, 2010).

The Simio is used to simulate 3D animated models– e.g. factories, supply chains, emergency departments, airports, and service systems (SIMIO, 2014).

Simio uses combined objects to represent physical components. An object (or model) is defined by its properties, states, events, external view, and logic, the key concepts for building and using objects. Properties are input values that can be specified by the object's user. For example, an object representing a server might have a property that specifies the service time, which should be specified in the facility model.

Models are defined within a project. A project may contain any number of models and associated experiments. A project will typically contain a main model and an entity model. A new project will automatically include the main model and entity model to the project, and it is possible to add additional models to create sub-models that are then used to build the main model (Kelton et al., 2013).

Three ships with the following particular are use in simulations.

Table 2. Ship Particulars

	Ship A	Ship B	Ship C
LOA (m)	107	100	122
Draft(m)	5.80	7.90	6.35
Speed (knots)	12.5	12	10
Cargo capacity (cub-m)	6544	6783	7213

2.2 Simulations

In order to simulate tanker movement in Blacksea, three ships started from three ports: Novorossiysk, Samsun, Ambarlı. Figure 2. Shows the main page of Simio simulation of the system.

In order to simulate the system in Simio, three sources were defined at Novorossiysk, Samsun, Ambarlı and ship arrival time at these station arranged to get only one ship in the system from each source during the simulation. Then, server were defined at each port. Servers and their loading/unloading (processing) times are given in table 3.

Table 3. Servers and processing times

Servers	Processing time (hours)
Novorossisyk	Normal(20,2)
Samsun	Triangular (10,12,13)
Ambarlı	Triangular (3,4,5)
Trabzon	Triangular (10,15,20)
Batumi	Normal(15,2)
Tuapse	Normal(15,2)
Yarımca	Normal(15,2)

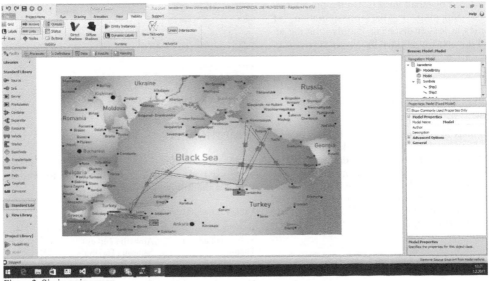

Figure 2. Simio main screen

Table 4.Sequence table for Ships

Ship A	Ship B	Ship C
Input@Ambarlı	Input@Novorossiysyk	Input@Yarimca
Input@Batumi	Input@Trabzon	Input@Ambarlı
Input@Samsun	Input@Tuapse	Input@Batumi
Input@Novorossiysyk	Input@Samsun	Input@Samsun
Input@Trabzon	Input@Novorossiysyk	Input@Tuapse
Input@Novorossiysyk	Input@Yarimca	Input@Trabzon
Input@Ambarlı		Input@Novorossiysyk
		Input@Samsun
		Input@Novorossiysyk
		Input@Ambarlı

Then, sequence of the ships going to ports are defined by sequence tables for each ship. Thjs tables are given in Table 4.

Then, to obtain statistics from the ports, process are defined in Simio. Defined tally statistics are given in Figure 3. As seen from this figure, these statistics give time in system for each ship.

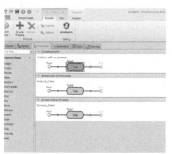

Figure 3. Tally statistics

As seen from Figure 3, Simio allow user to collect statistics for any variable. User module can define where and which variable is to be followed.

3 CONCLUSIONS

Simulations were run for 60 days and the following results were obtained.

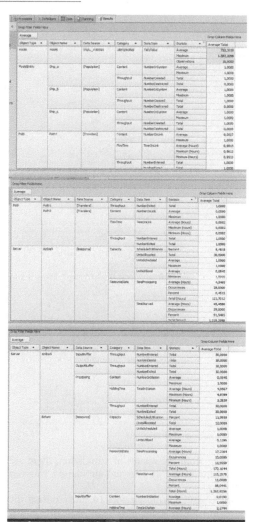

Figure 4. Simio simulation results

As seen from the results of simulation, there are only 3 ships in the system and simulation was run for 60 days.

Ship A visited Ambarlı port 14 times during the simulation and spent average 691.90 hours, maximum 1414, and minimum 49 hours at the port. Ship B was at Ambarlı 18 times and spent 751, 1439 and 39 hours average, maximum and minimum respectively. Similarly, Ship C called Ambarlı port 8 times and stayed at the port average 298, maximum 683 and minimum 26 hours.

If ports are taken in to consideration, it can be summarized that Ambarlı port utilizations is 8.45%

and total of 30 ships were handled. Usage of Batumi port is 11.96% and total number of ship handled is 12. Minimum utilizations were Trabzon and Novorossiysk port as 0.019 and 0.22 respectively. Maximum usage of the ports were Samsun 23% and Batumi 11.5%

ACKNOWLEDGE

Authors would like to thank Simio LLC for given us Academic version of Simio Simulation software.

REFERENCES

Başar, E. (2010). Investigation into marine traffic and a risky area in the Turkish Straits system: Canakkale Strait. *Transport, 25*, 5-10.

Demirci, E., Araz, T. & Köse, E. (2000). Trabzon Port Simulation Model *Journal of Industrial Engineering, 11*, 2-14.

Demirci, E. (2003). Simulation modelling and analysis of a port investment. *Simulation, 79*, 94-105.

Goerlandt, F., & P. Kujala (2011). Traffic simulation based ship collision probability modeling. *Reliability Engineering & System Safety, 96*, 91-107.

Hirsch, B.E., Kuhlmann, T., & Schumacher, J. (1998). Logistics simulation of recycling networks. *Computers in Industry, 36*, 31-38.

SIMIO (2014). SIMIO. Accessed September 5, 2014. http://www.simio.com/index.php

Kelton, W. D. and Simith, J. S. and Sturrock, D. T. (2013). Simio and Simulation: Modeling, Analysis, Applications Third Edition, Simio LLC.

Kim, H., Park, J. H., Lee, D., & Yang, Y. S. (2004). Establishing the methodologies for human evacuation simulation in marine accidents.*Computers & Industrial Engineering, 46*, 725-740.

Köse, E., Başar, E., Demirci, E., Güneroğlu, A., & Erkebay, Ş. (2003). Simulation of marine traffic in Istanbul Strait. *Simulation Modelling Practice and Theory, 11*, 597-608.

Smith, R. D. (1999). Simulation: The engine behind the virtual world. *GEN, 1*, 72.

Uğurlu, Ö., Köse, E., Başar, E., Yüksekyıldız, E., & Yıldırım, U. (2012). Investigation of Working Hours of Watchkeeping Officers on Short Sea Shipping: A Case Study in an Oil Tanker. *IAME 2012*, Taipei.

Uğurlu, Ö., Yüksekyıldız, E., & Köse, E. (2014). Simulation Model on Determining of Port Capacity and Queue Size: A Case Study for BOTAS Ceyhan Marine Terminal. *TransNav, the International Journal on Marine Navigation and Safety of Sea Transportation, 8*, 143-150.

Uğurlu, Ö. (2015a). Application Of Fuzzy Extended AHP Methodology For Selection Of Ideal Ship For Oceangoing Watchkeeping Officers. *International Journal of Industrial Ergonomics, 47*, 132-140.

Uğurlu, Ö. (2015b). A case study related to the improvement of working and rest hours of oil tanker deck officers. *Maritime Policy & Management*.

Uğurlu, Ö., Köse, E., Yildirim, E., & Yüksekyıldız, E. (2015c). Marine accident analysis for collision and grounding in oil tanker using FTA method. *Maritime Policy & Management, 42*, 163-185.

Electrotechnical Problems

The Impact of Electromagnetic Interferences on Transport Security System of Certain Reliability Structure

P. Dziula
Gdynia Maritime University, Gdynia, Poland

J. Pas
Military University of Technology, Warsaw, Poland

ABSTRACT: The article presents the impact of electromagnetic interferences of certain frequency range, on transport security systems of certain reliability structures. Intended and unintended (stationary and mobile), electromagnetic interferences, impacting on items and components constituting transport system at wide area (seaport, airport, etc.), cause changes of its vulnerability, resistance and durability. Diagnostics of interferences sources (amplitude, frequency range, radiation characteristics, etc.), appearing within transport environment, and usage of appropriate technical solutions of systems (i.e. shielding, reliability structures), allows for safe implementation of safety surveillance of human beings, properties and communication means.

1 INTRODUCTION

Transport security systems are highly equipped with different kinds of appliances, that are subject of continuous miniaturization, caused by rapid evolution taking place in the field of electronics. The appliances are used in stationary and mobile objects, i.e. fire detection systems (microprocessor detectors, control panels, power modules). Electronic components and systems operate under lower voltage supply and reduced power consumption. Therefore, electromagnetic fields of lower intensity, can interfere with electronic devices, causing disruptions in transport security systems functionality, or even resulting with their failure.

Following safety states of transport security systems can be distinguished, taking into account the impact of interferences [Dyduch & Paś & Rosiński 2011, Paś 2015b, Rosiński 2015a]:

1 transport security system not affected by inside and outside interferences – the intensity of interferences too low, their permitted level not exceeded, the system stay in its actual operation state;

2 the appliances, operating within transport security systems, automatically compensate interferences by means of passive or active filters, or other devices securing their functionality;

3 the appearance of interferences causes transition of transport security system, from the state of full operational ability, into the state of restricted operational ability – return to the full ability state demands intervention of service [Rosiński 2015b, Siergiejczyk & Paś & Rosiński 2016, Siergiejczyk & Paś & Rosiński 2015];

4 the appearance of electromagnetic interferences within transport security system results with its partial or complete damage – state of unreliability of safety.

To analyse electromagnetic interferences influence on transport security systems, following criteria should be taken into account:

– system's resistance to interferences, defined as ability to continue proper functionality of its appliances, under interferences appearance;

– system's susceptibility to interferences – meaning response of functioning system to inside and outside interferences;

– system's durability to interferences – understood as its ability to maintain initial conditions until interferences discontinue.

According to European Directive 89/336/EEC, concerning electromagnetic compatibility, all components of transport security systems should ensure appropriate level of resistance to interferences, allowing them to maintain proper functionality in particular electromagnetic environment [Ott 2009]. This applies not only to electronic security systems, but also to others, that are built of electronic components [Krzykowska & Siergiejczyk & Rosiński 2015, Siergiejczyk & Krzykowska & Rosiński 2015a, Siergiejczyk & Krzykowska & Rosiński 2015b, Siergiejczyk &

Rosiński & Dziula & Krzykowska 2015, Weintrit & Dziula & Siergiejczyk & Rosiński 2015].

Transport security systems, used on the vast areas, usually operate in hard conditions, caused by fluctuations of air temperature and humidity, precipitation, climate changes, water intrusion, vibrations [Burdzik & Konieczny & Figlus 2013], service abilities [Laskowski & Łubkowski & Pawlak & Stańczyk 2015]. Sources of intentional electromagnetic fields appearing in ports and aboard ships (radar and radio-navigational stations, stationary and mobile radio-communication stations, etc. [Kaniewski & Lesnik & Susek & Serafin 2015, Paszek & Kaniewski 2016]) and of non-intentional ones (industry power supplies, generators and electric engines, power cables and lines, etc.), are contributing to formation of an "electromagnetic smog", able to decrease working conditions of components of security systems, and also other ICT systems [Kasprzyk & Rychlicki 2014, Lewiński & Perzyński & Toruń 2012, Sumiła & Miszkiewicz 2015]. The quality of information does also mean, in case of mentioned exploitation conditions [Stawowy 2015, Stawowy & Dziula 2015]. Transport security systems, operating in hard environmental conditions, should meet the requirements, specified in respective regulations concerning their exploitation.

When projecting transport security systems, it is necessary to consider environmental conditions of transport system's area, and select appropriate appliances, that are going to be included in the systems (i.e. power supplies [Rosiński 2015c]). To meet the requirements concerning electromagnetic compatibility, for proper project works on transport security systems, it is necessary to investigate electromagnetic environment within particular area [Paś & Siergiejczyk 2016, Siergiejczyk & Paś & Rosiński 2015, Siergiejczyk & Rosiński & Paś 2016].

To determine indicators of interferences influence on transport security systems, within the area of electromagnetic interferences impact, it is required to know characteristics of radiation sources. For this purpose, electromagnetic interferences are divided into two sub-ranges: in regard to frequency range, and following properties of electromagnetic field:

1 the way of propagation of interferences within the transport area;
2 attenuation of propagation of interferences within the transport area;
3 shielding of interferences within the transport area by walls, building and metal partitions of various thickness, vegetation, etc.;
4 impact of interferences on transport security system;
5 the way of protecting against interferences impacting on particular components of transport security systems – Fig. 1;

6 strength of intentional and non-intentional interferences, occurring within restricted transport area.

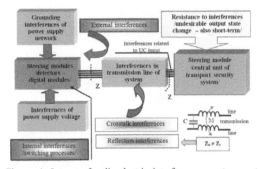

Figure. 1. Impact of radio-electric interferences on transport security system: Z_o, Z_c – impedance of power-load and centre, M – factor of inductive coupling, UC – digital device (control panel)

2 TRANSPORT SECURITY SYSTEM CONSISTING OF ONE TRANSMISSION BUS AND MODULES BASED ON RS-485 INTERFACE

The system of above structure is shown in Fig. 2, in distracted version. The modules are connected to the main unit by bus with RS-485 interface. The interface used, enables to connect components of the system at distance up to 1200 metres. The system structure is modular [Rosiński 2008, Rosiński 2011], that can be built of system enhancing modules, controllers, and monitoring and control computer [Duer & Zajkowski & Duer & Paś 2012, Paś 2015a].

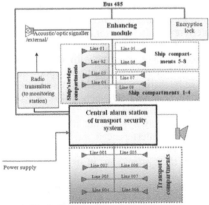

Figure. 2. Block diagram of the distracted system, consisting of one transmission bus and modules based of RS-485 interface

Transport security systems are ones, whose purpose is detection of threats appearing within the transport process (for both stationary and mobile objects) [Paś 2015c, Paś 2016]. The usage of the systems, in transport process, increases, and they are supporting safety of [Dziula & Kołowrocki & Soszyńska-Budny 2013]:

1 human beings (i.e. monitoring systems installed in stationary objects of airports, railway stations, ports, etc.);
2 cargo stored in stationary objects (i.e. logistic centres, land and sea cargo handling terminals, etc.);
3 cargo transported by mobile objects (railway, maritime and air transport, where, if supported by GPS system, allow to monitor cargo conditions and track movement of transport means).

Transport security systems function in various climate conditions and various surrounding electromagnetic environment, that can cause occurrence of interferences. Proper functionality of transport security system depends on:
– reliability of particular components, the system is built of;
– internal reliability structure of transport security system;
– undertaken strategy of exploitation of transport security system;
– electromagnetic interferences impacting on system operation process.

3 TRANSPORT SECURITY SYSTEM CONSISTING OF TWO TRANSMISSION BUSES AND MODULES – EXPLOITATION INDICATORS

Fig. 3 shows safety states defined for exploitation process of transport security system, consisting of two transmission buses and modules, in case of impact of electromagnetic interferences. The impact of interferences has been indicated by transition intensity lines γ_Z and γ_{1Z}. In case of occurrence of electromagnetic interferences of very high level, i.e. atmospheric discharge (catastrophic event), the system changes state from $R_0(t)$ into $Q_B(t)$ - unreliability of safety.

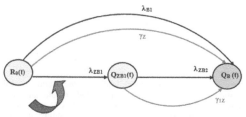

Figure 3. Transport security system consisting of two transmission buses and modules.

The problem of electromagnetic interferences has appeared at the early stage of radio-communication

evolution. There were already, before the Second World War, public services involved in interferences, in many countries (i.e. in England – respective service was established in 1920). Rapid evolution of radio-communication and television after 1945, and usage of higher and higher transmission frequencies, caused the number of complaints regarding interferences in England, over the years 1947-1956, increased up to 160 thousand per year (to compare – number of the complaints in 1934 was around 34 thousand). The researches on influence of interferences on radio reception, started in Poland in 1935. Two years later, special services involved in eliminating of interferences, were established. Polish Electric Standard PN/E-58: „Recommendations for eliminating of interferences within radio reception, caused by different electric devices", was issued in 1935. Actually, within transport area, analog and digital electronic equipment is used, that themselves, during operation, emit non-intentional electromagnetic field, and is also exposed to external fields emitted by external devices.

The system shown in Fig. 3 can be described by following Chapman – Kolmogorov equations:

$$R_0'(t) = -\lambda_{B1} \cdot R_0(t) - \gamma_Z \cdot R_0(t) - \lambda_{ZB1} \cdot R_0(t)$$

$$Q'_{ZB1}(t) = \lambda_{ZB1} \cdot R_0(t) - \lambda_{ZB2} \cdot Q_{ZB1}(t) - \gamma_{1Z} \cdot Q_{ZB1}(t)$$

$$Q'_B(t) = \lambda_{B1} \cdot R_0(t) + \gamma_Z \cdot R_0(t) + \gamma_{1Z} \cdot Q_{ZB1}(t) + \lambda_{ZB2} \cdot Q_{ZB1}(t)$$

Assuming initial conditions:

$$R_0(t) = 1$$

$$Q_{ZB1}(0) = Q_B(0) = 0$$

And by means of Laplace transform, following linear set of equations is obtained:

$$s \cdot R_0^*(s) - 1 = -\lambda_{B1} \cdot R_0^*(s) - \gamma_Z \cdot R_0^*(s) - \lambda_{ZB1} \cdot R_0^*(s)$$

$$s \cdot Q^*_{ZB1}(s) = \lambda_{ZB1} \cdot R_0^*(s) - \lambda_{ZB2} \cdot Q^*_{ZB1}(s) - \gamma_{1Z} \cdot Q^*_{ZB1}(s)$$

$$s \cdot Q^*_B(s) = \lambda_{B1} \cdot R_0^*(s) + \gamma_Z \cdot R_0^*(s) + \gamma_{1Z} \cdot Q^*_{ZB1}(s) + \lambda_{ZB2} \cdot Q^*_{ZB1}(s)$$

Then, by means of schematic approach:

$$(R_0, Q_{ZB1}, Q_B) \rightarrow \begin{bmatrix} \dfrac{1}{a} \\ \dfrac{\lambda_{ZB1}}{b \cdot a} \\ \dfrac{\lambda_{B1} \cdot b + \gamma_{1Z} \cdot \lambda_{ZB1} + \lambda_{ZB2} \cdot \lambda_{ZB1}}{s \cdot a \cdot b} \end{bmatrix}$$

where:

$$a = s + \lambda_{ZB1} + \gamma_Z + \lambda_{B1}$$

$$b = s + \lambda_{ZB2} + \gamma_{1Z}$$

Marks „*" and „s", associated with probabilities of system stay at particular safety states R_0, Q_{ZB1}, and Q_B, have been skipped when forming final above result.

By means of the inverse transform, following results are obtained:

$$R_0(t) = e^{-(\lambda_{ZB1} + \gamma_Z + \lambda_{B1}) \cdot t}$$

$$Q_{ZB1}(t) = \lambda_{ZB1} \cdot \left[\frac{\exp[-(\lambda_{B1} + \gamma_Z + \lambda_{ZB1}) \cdot t] - \exp[-(\lambda_{ZB2} + \gamma_{1Z}) \cdot t]}{(\lambda_{B1} + \gamma_Z + \lambda_{ZB1} - \lambda_{ZB2} - \gamma_{1Z})} \right]$$

$$Q_B(t) = \begin{bmatrix} \left[\dfrac{\lambda_{ZB1} \cdot \lambda_{ZB2} + \lambda_{B1}\lambda_{ZB2} + \lambda_{ZB1} \cdot \gamma_{1Z}}{(\lambda_{B1} + \lambda_{ZB1}) \cdot (\lambda_{ZB2} + \gamma_{1Z})} + \lambda_{ZB1} \cdot \lambda_{ZB2} \cdot \right] \\ \cdot \left[\dfrac{\exp[-(\lambda_{B1} + \lambda_{ZB1}) \cdot t] - \exp[-(\lambda_{ZB2} + \gamma_{1Z}) \cdot t]}{(\lambda_{B1} + \lambda_{ZB1}) \cdot (\lambda_{B1} + \lambda_{ZB2} - \lambda_{ZB1} - \gamma_{1Z})} \right] + \\ + \lambda_{B1} \cdot \left[\dfrac{\exp[-(\lambda_{B1} + \lambda_{ZB1}) \cdot t] \cdot \lambda_{ZB2} - \exp[-(\lambda_{B1} + \lambda_{ZB1}) \cdot t] \cdot}{\dfrac{\lambda_{B1} - \exp[-(\lambda_{B1} + \lambda_{ZB1}) \cdot t] \cdot \lambda_{ZB1}}{(\lambda_{B1} + \lambda_{ZB1}) \cdot (\lambda_{B1} + \lambda_{ZB1} - \lambda_{ZB2} - \gamma_{1Z})}} \right] + \\ + \lambda_{ZB1} \cdot \gamma_{1Z} \cdot \left[\dfrac{\exp[-(\lambda_{B1} + \lambda_{ZB1}) \cdot t]}{(\lambda_{B1} + \lambda_{ZB1}) \cdot (\lambda_{B1} + \lambda_{ZB1} - \lambda_{ZB2} - \gamma_{1Z})} \right] + \\ + \gamma_{1Z} \cdot \left[\dfrac{\lambda_{B1} \cdot \exp[-(\lambda_{ZB2} + \gamma_{1Z}) \cdot t] - \lambda_{ZB1} \cdot \exp[-(\lambda_{ZB2} + \gamma_{1Z}) \cdot t]}{(\lambda_{B1} + \lambda_{ZB1}) \cdot (-\lambda_{B1} + \lambda_{ZB2} + \gamma_{1Z} - \lambda_{ZB1})} \right] \end{bmatrix}$$

4 INFLUENCE OF INTERFERENCES ON TRANSPORT SECURITY SYSTEMS – CERTAIN OPERATION STATES

The influence of electromagnetic interferences on transport security system results with change of probability value of the state of full operational ability $R_0(t_b)$ – Figure 4. Increase of interferences level (amplitude), results with change of $R_0(t_b)$ parameter value. I.e. for parallel system reliability structure, $R_0(t_b)$ decreases linearly, obtaining zero value for indicator $\gamma = 1$ (occurrence of electromagnetic interferences of high amplitude, resulting with catastrophic failure – induction of overvoltage in power supply line, caused by an atmospheric discharge). For series-parallel system reliability structure, increase of interferences level (amplitude) to certain value, does not result with change of $R_0(t_b)$ – electromagnetic interferences of low amplitude, acceptable by system items and components. Increase of interferences level above the certain value, results with rapid decrease of $R_0(t_b)$ value for this structure.

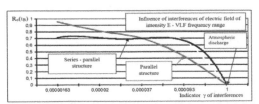

Figure 4. Change of probability value of the state of full operational ability $R_0(t_b)$, under the influence of interferences

Figure 5 indicates changes of probabilities of the state of full operational ability $R_0(t_b)$, state of restricted operational ability $Q_{ZB1}(t_b)$, and the state of unreliability of safety $Q_B(t_b)$, of the series-parallel transport security system. Transport means, having transport security system installed, are under influence of interferences of different frequency ranges – i.e. interferences of induction B of magnetic field within VLF frequency range (2 – 100 kHz). For determined Γ level of safety of transport security system functionality, meaning system's resistance to mentioned frequency range, equals to 0,05 for the function $R_0(t_b)$ leaning. The Γ level of safety of transport security system functionality, is permissible value of the amplitude of interferences, which is function of: frequency range of interferences, spectrum of interferences, amplitude of interferences, values of particular components of electromagnetic field – magnetic or electric field. Increase of level (amplitude) of interferences γ, results with the decrease of function $R_0(t_b)$ value. While the increase of values of functions $Q_{ZB1}(t)$ and $Q_B(t)$ takes place. Similar character of changes of parameters of probability functions of safety states, takes place for the system having parallel reliability structure – Figure 6.

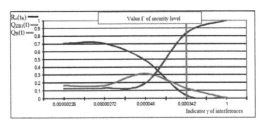

Figure 5. Changes of probabilities of the state of full operational ability $R_0(t_b)$, state of restricted operational ability $Q_{ZB1}(t_b)$, and the state of unreliability of safety $Q_B(t_b)$, for the transport security system having series-parallel reliability structure (impact of interferences - induction B of magnetic field within VLF frequency range)

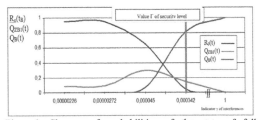

Figure 6. Changes of probabilities of the state of full operational ability $R_0(t_b)$, state of restricted operational ability $Q_{ZB1}(t_b)$, and the state of unreliability of safety $Q_B(t_b)$, for the transport security system having parallel reliability structure (impact of interferences - induction B of magnetic field within VLF frequency range)

The character of changes of functions of unreliability of safety $Q_B(t)$, and of the state of restricted operational ability $Q_{ZBi}(t)$, of the transport security system having series-parallel reliability structure, has been shown in Figure 7.

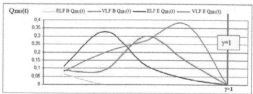

Figure 7. Changes of probability values of the state of restricted operational ability $Q_{ZBi}(t)$, of the transport security system having series-parallel reliability structure (impact of interferences of ELF (5-2000 Hz) frequency range, and VLF (2-100 kHz) frequency range)

Transport security system is most resistant to interferences of VLF frequency range – both magnetic and electric field components. For the indicator of interferences $\gamma = 1$ value (interferences caused by signal of very high amplitude – catastrophic failure), all functions determining unreliability of safety of transport security system reach $Q_B(t) = 1$ value. Values of probability of the state of restricted operational ability $Q_{ZBi}(t)$, for the system of series-parallel reliability structure, for the indicator $\gamma = 1$ value, also reach the zero level.

5 CONCLUSIONS

Basing on analysis performed in the article, following conclusions and observations are specified:
– the value of indicator of the safety level (meaning resistance to interferences), of the transport security system, in case of impacting electromagnetic interferences, from certain frequency range, depends on chosen reliability structure of items and components the system is built of – series, parallel, series-parallel, etc., the structures are also influencing on safety and reliability of the system;
– the highest value of reliability indicator R_S, and resistance to interferences, is ensured by transport security system having parallel structure, with control centres concentrated, or spread;
– electromagnetic interferences impacting on transport security system, that is used within wide transport area, result with increase of the system electronic components failure intensities, independently of reliability structure, induced, radiated or conducted interferences cause unintended increase of i.e. power supply voltage, change of the operation point of an active item, increase of non-linear distortions, etc.;
– transport security system having series structure is the most vulnerable one to interferences;

– transport security systems installed within wide areas, in case of interferences, are having the lowest value of intended operating time, because the electromagnetic interferences are not damped by building structures;
– transport security systems installed inside buildings, located within transport areas, are less vulnerable to interferences, due to shielding influence of i.e. lightning grids, reinforced concretes, walls, metal roof covers;
– the value of resistance to electromagnetic interferences, of transport security system located inside buildings within wide transport areas, depends on the size of a lightning grid spacing (especially in case of interferences of low frequencies);
– values of particular probabilities of the transport security system stay at the operation states, depend on properties of impacting electromagnetic field, and which component – field vector (electric or magnetic), is dominant;
– the highest influence on probabilities of the system stay at certain operation states, is caused by the B induction of the magnetic field of the frequency range ELF, the damping of the B induction by lightning grid is the lowest one for this frequency range;
– the lowest influence on probabilities of the system stay at certain operation states, is caused by the E intensity of the electric field of the VLF frequency range (2 – 100 kHz). For this electromagnetic field component, values of probabilities $R_0(t)$, $Q_{ZBi}(t)$ and $Q_B(t)$, reach maximum values for different sources of interferences;
– for low values of amplitude of electromagnetic interferences, impacting on transport security system, the probability function of the system stay at the state of full operational ability, remains at the constant level;
– permissible (limit) levels of interferences influence on transport security system, can be determined by means of resistance indicator, of particular items and components of the system, to certain frequency ranges;
– the permissible level of interferences can be defined as transport security system resistance to impact of electromagnetic interferences of certain frequency range;
– the maximum resistance, of transport security system, to impact of interferences of certain frequency range, depends on minimum resistance of all system items and components, it is built of;
– transport security system resistance depends also on properties of impacting electromagnetic field – if its character is mainly magnetic (E/H < 337 [Ω]) or electric (E/H > 337 [Ω]);
– the level Γ of resistance of transport security system, is the function depending on:

- properties of electromagnetic field – values of its particular field components E and H for certain frequency ranges;
- particular items and components of transport security system, having abilities for shielding of electromagnetic interferences;
- locations of items and components of transport security system – within open area (environment), or in building structures, damping interferences;
- the lowest value, of the Γ resistance level, of the transport security system appears for the B induction of magnetic field of the ELF frequency range;
- transport security systems, installed within wide transport areas, where an atmospheric discharge takes place, move to the state of unreliability of safety at the moment t₀ [$R_o(t)=0$, $Q_B(t)=1$], if there is no counteraction to direct or induced electric impulses (no overvoltage means securing the system);
- in case of installation of overvoltage means and components, potential equalization sources, other available solutions protecting power supply lines, transmission buses, detection zones, etc., of transport security system, there is certain time period, allowing to counteract to consequences of atmospheric discharge (there is then an ability to start so-called emergency process, preventing against failure of whole system or its most important components, i.e. control panel).

REFERENCES

Burdzik, R. & Konieczny, Ł. & Figlus, T. 2013. Concept of on-board comfort vibration monitoring system for vehicles. In J. Mikulski (ed.), *Activities of Transport Telematics*, TST 2013, CCIS 395: 418-425. Heidelberg: Springer.

Duer, S. & Zajkowski, K. & Duer, R. & Paś, J. 2012. Designing of an effective structure of system for the maintenance of a technical object with the using information from an artificial neural network. *Neural Computing & Applications*. DOI: 10.1007/s00521-012-1016-0.

Dyduch, J. & Paś, J. & Rosiński, A. 2011. *Basics of maintaining electronic transport systems*. Radom: Publishing House of Radom University of Technology.

Dziula, P. & Kołowrocki, K. & Soszyńska-Budny J. 2013. Maritime Transportation System Safety - Modeling and Identification. *TransNav, the International Journal on Marine Navigation and Safety of Sea Transportation, Vol. 7, No. 2*, pp. 169-175.

Kaniewski, P. & Lesnik, C. & Susek, W. & Serafin, P. 2015. Airborne radar terrain imaging system. In *16th International Radar Symposium (IRS)*, Dresden, Germany, pp. 248-253.

Kasprzyk, Z. & Rychlicki, M. 2014. Analysis of phiysical layer model of WLAN 802.11g data transmission protocol in wireles networks used by telematic systems. In: *Proceedings of the Ninth International Conference Dependability and Complex Systems DepCoS-RELCOMEX*, given as the monographic publishing series

– „Advances in intelligent systems and computing", Vol. 286. Springer, pp. 265-274.

Krzykowska, K. & Siergiejczyk, M. & Rosiński, A. 2015. The concept of the SWIM system in air traffic management. In A. Weintrit (ed.), *Activities in Navigation - Marine Navigation And Safety Of Sea Transportation*: 255-259. London: CRC Press/Balkema, Taylor & Francis Group.

Laskowski, D. & Łubkowski, P. & Pawlak, E. & Stańczyk, P. 2015. Anthropotechnical systems reliability. In: the monograph „*Safety and Reliability: Methodology and Applications - Proceedings of the European Safety and Reliability Conference ESREL 2014*", editors: Nowakowski T., Młyńczak M., Jodejko-Pietruczuk A. &Werbińska–Wojciechowska S. CRC Press/Balkema, London, Paper: 399-407.

Lewiński, A. & Perzyński, T. & Toruń A. 2012. The analysis of open transmission standards in railway control and management. In *Communications in Computer and Information Science*, vol. 329, Berlin Heidelberg, Germany: Springer-Verlag, pp. 10-17.

Ott, H.W. 2009. *Electromagnetic Compatibility Engineering*. Wiley.

Paszek, J. & Kaniewski, P. 2016. Simulation of random errors of inertial sensors. In *13th International Conference on Modern Problems of Radio Engineering, Telecommunications and Computer Science (TCSET)*, Lviv-Slavske, Ukraine, 153-155.

Paś, J. & Siergiejczyk, M. 2016. Interference impact on the electronic safety system with a parallel structure, Diagnostyka, Vol. 17, No. 1, ISSN 1641-6414, Paper: 49-55.

Paś, J. 2015a. Analysis of exploitation access control system selected object. *Przegląd elektrotechniczny*, vol no 10, ISSN 0033-2097, R. 91, Paper: 219 – 224.

Paś, J. 2015b. *Operation of electronic transportation systems*. Radom: Publishing House University of Technology and Humanities.

Paś, J. 2015c. Selected methods for increases reliability the of electronic systems security, *Journal of KONBiN*, 3(35), ISSN 1895-8281, Paper: 147–156, DOI 10.1515/jok-2015-047.

Paś, J. 2016. Shock a disposable time in electronic security systems. Warszawa: *Journal of KONBiN* 2(38). Paper: 5–31, DOI 10.1515/jok-2016-0016, ISSN 1895-8281.

Rosiński, A 2008. Design of the electronic protection systems with utilization of the method of analysis of reliability structures. In *Proceedings Nineteenth International Conference On Systems Engineering ICSEng 2008*, Las Vegas, USA. Paper: 421-426.

Rosiński, A. 2011. Reliability analysis of the electronic protection systems with mixed m–branches reliability structure. In *Proceedings International Conference European Safety and Reliability ESREL 2011*, Troyes, France. Paper: 2064-2071.

Rosiński, A. 2015a. *Modelling the maintenance process of transport telematics systems*. Warsaw: Publishing House Warsaw University of Technology

Rosiński, A. 2015b. Rationalization of the maintenance process of transport telematics system comprising two types of periodic inspections. In Henry Selvaraj, Dawid Zydek, Grzegorz Chmaj (ed.), *Proceedings of the Twenty-Third International Conference on Systems Engineering*, given as the monographic publishing series – „Advances in intelligent systems and computing", Vol. 1089. Paper: 663-668. Springer.

Rosiński, A. 2015c. Reliability-exploitation analysis of power supply in transport telematics system. In Nowakowski T., Młyńczak M., Jodejko-Pietruczuk A. & Werbińska–Wojciechowska S. (ed.), *Safety and Reliability: Methodology and Applications - Proceedings of the*

European Safety and Reliability Conference ESREL 2014. Paper: 343-347. London: CRC Press/Balkema, London.

Siergiejczyk M. & Paś J. & Rosiński A. 2015. Train call recorder and electromagnetic interference. *Diagnostyka*, vol. 16, no. 1, Paper: 19-22.

Siergiejczyk M. & Paś J. & Rosiński A. 2016. Issue of reliability–exploitation evaluation of electronic transport systems used in the railway environment with consideration of electromagnetic interference, *IET Intelligent Transport Systems*, Vol. 10, Issue 9, November 2016. pp. 587 – 593.

Siergiejczyk, M. & Krzykowska, K. & Rosiński, A. 2015a. Parameters analysis of satellite support system in air navigation. In Henry Selvaraj, Dawid Zydek, Grzegorz Chmaj (eds) Proceedings of the Twenty-Third International Conference on Systems Engineering: 673-678, given as the monographic publishing series – „Advances in intelligent systems and computing", Vol. 1089. Springer.

Siergiejczyk, M. & Krzykowska, K. & Rosiński, A. 2015b. Reliability assessment of integrated airport surface surveillance system. In W. Zamojski, J. Mazurkiewicz, J. Sugier, T. Walkowiak, J. Kacprzyk (ed.), *Proceedings of the Tenth International Conference on Dependability and Complex Systems DepCoS-RELCOMEX*, given as the monographic publishing series – „Advances in intelligent systems and computing", Vol. 365. 435-443. Springer.

Siergiejczyk, M. & Rosiński, A. & Dziula, P. & Krzykowska K. 2015. Analiza niezawodnościowo – eksploatacyjna autostradowych systemów telematyki transportu. *Journal of Konbin, 33 (1)*, pp. 177-186.

Siergiejczyk, M. & Paś, J. & Rosiński, A. 2015. Modeling of process of exploitation of transport telematics systems with regard to electromagnetic interferences. In *Tools of transport telematics*, given as the monographic publishing series – „Communications in Computer and Information Science", Vol. 531: 99-107. Springer.

Siergiejczyk, M. & Rosiński, A. & Paś, J. 2016. Analysis of unintended electromagnetic fields generated by safety system control panels. *Diagnostyka* Vol. 17, No. 3: 35-46.

Stawowy, M. 2015. Model for information quality determination of teleinformation systems of transport. In: Nowakowski T. & Młyńczak M. & Jodejko-Pietruczuk A. & Werbińska–Wojciechowska S. (eds) Safety and Reliability: Methodology and Applications - Proceedings of the European Safety and Reliability Conference ESREL 2014: Paper: 1909–1914. London: CRC Press/Balkema.

Stawowy, M. & Dziula, P. 2015. Comparison of uncertainty multilayer models of impact of teleinformation devices reliability on information quality. In *Proceedings of the European Safety and Reliability Conference ESREL 2015*: 2685-2691. Zurich.

Sumiła, M. & Miszkiewicz, A. 2015. Analysis of the problem of interference of the public network operators to GSM-R. In J. Mikulski (ed.), *Tools of Transport Telematics*, given as the monographic publishing series – „Communications in Computer and Information Science", Vol. 531: 76-82.

Weintrit, A. & Dziula, P. & Siergiejczyk, M. & Rosiński A. 2015. Reliability and exploitation analysis of navigational system consisting of ECDIS and ECDIS back-up systems. In A. Weintrit (ed.): *Activities in Navigation - Marine Navigation And Safety Of Sea Transportation*, , CRC Press/Balkema, Taylor & Francis Group, London. pp. 109-115.

Increasing Energy Efficiency of Commercial Vessels: by Using LED Lighting Technology

E. Eyüboğlu, S. Yıldız, Ö. Uğurlu, U. Yıldırım & F.B. Demirel
Karadeniz Technical University, Trabzon, Turkey

ABSTRACT: From past to present improvement of energy efficiency in maritime trade has been a common goal for all parties of the maritime industry, including shipping companies first. In order to reach this goal; vessels, ports and shipping companies have undergone adaptations in line with the scientific and technological developments. Adaptations have accelerated by published international conventions and national regulations. Despite all the developments occurred in the maritime sector; energy efficiency has still been the focal point of maritime industry. Therefore, ever-growing world merchant fleet requires improving energy efficiency, and taking new measures to reduce emissions.

This study has been aimed at the reduction of emissions and energy consumed by vessels. For this purpose, energy savings has been demonstrated, in case of the replacement of conventional lighting systems used on ships with Light Emitting Diode (LED) technology. As a result, the amount of energy savings and emission savings has been calculated for 24 hour period for a sample merchant vessel.

1 INTRODUCTION

Illumination has always been an important issue in the transportation sector, as construction, decoration, healthcare, information technology, defense industry and many other industries. Previously, lighting devices used to improving living/working standards and increasing efficiency in human life; by the technology developed along with the industrial revolution the usage area of lighting devices has expanded. The energy consumed to ensure the desired enlightenment has always been an important research topic. Ever since the development of daylight incandescent bulbs in the 1800s, methods of producing more efficient white light have been sought to increase lighting efficiency and reduce cost (Steele, 2001). The modern lighting period, which started with incandescent lamp and fluorescent lamp, continued to evolve by a radio technician named Oleg Vladimirovich Losev noticed that the diodes used in radio receivers were emitting light and in 1927 published a manuscript about a light-emitting diode (LED) in a Russian newspaper (Zheludev, 2007, Novikov, 2004). In the modern sense, the first LED was produced in 1962 using gallium arsenide phosphate, a compound semiconductor alloy (Steigerwald et al., 2002, Holonyak Jr & Bevacqua, 1962). After 1962, with the widespread use of semiconductors in LED technology, production costs have begun to decrease (Steigerwald et al., 2002). In the early 2000s, LED lighting devices reached a sales value of up to 600 million dollars in the world lighting market (Mills, 2003), and in 2007 the size of global LED market reached 4,6 billion dollars (Zheludev, 2007). As LED lighting technologies have evolved, they have become an important component of today's technology (Zheludev, 2007), replacing old lighting technologies (Steigerwald et al., 2002, Steele, 2001).

Due to its low energy consumption and high efficiency, LED lighting systems have been adapted to many industries (telecommunication, construction, automotive, aviation, railway etc.) where lighting systems are used (Krames et al., 2007, Ji et al., 2009, Yeh & Chung, 2009). However, there is very little practice in maritime transport (Cizek, 2009, Freymiller, 2009, Ji et al., 2009), which has a large share in world trade (Uctad, 2016). As long as the ships are cruising, the electrical energy they need is met by generators. If this generator is a shaft generator, electricity is generated from the shaft that is driven by the main engine of the vessel. If it is a diesel generator, electricity is generated from diesel oil that the generator consumes. In both cases, fossil fuels are used to provide electricity for the vessel's lighting.

Therefore, the savings in internal and external lighting of the vessel will reduce the fuel costs as well as indirectly reduce exhaust gas emissions.

This case study is a theoretical work on a sample container ship's lighting equipment. The amount of savings that will be incurred as a result of replacing all the lights in the ship complying with SOLAS standards with the equivalent LED lights were calculated. As a result, if such a replacement is made, it is aimed to provide a perspective on how the cost-benefit ratio might be.

consumes 85% less energy than incandescent lamps and lasts 30 times longer. Compared to fluorescent lighting, LED consumes half as much energy and lasts five times longer (Cizek, 2009). Despite these advantages, the number of studies related to LED applications in the maritime area is limited in the literature. For this purpose, in this study, the effect of LED lighting systems on shipboard energy efficiency was examined with a sample vessel. As a result of the study, the energy efficiency to be achieved with led lighting systems was calculated.

2 LITERATURE REVIEW

In their study, Ji et al., (2009), analyzed the characteristics of the light sources in order to determine if LED lighting is feasible for shipboard illumination. They made an experimental study and produced a prototype of LED luminaire to replace the existing fluorescent and incandescent lamps onboard. As a result, LED luminaires has higher luminance with less power consumption than the existing lamps, and by using these LED luminaires 14% and 95% of energy may be saved.

Banks et al., (2013) identified the main trends of the ship operating profiles for different ship types, and investigated the effect of energy efficiency on ship operations. As a result of the study, they concluded that different ship types have different operating profiles and there may be differences in operating costs and energy consumption. They determined that the main reason of the difference in these profiles were duration of the voyage and the period of stay at port. They also stated that logistics and economic costs have significant effects on ship operating profiles.

Ship Energy Efficiency Management Plan (SEEMP), is one of the major international regulation that is expected to affect short-term increased CO2 emissions in maritime transport. In their study, Johnson et al., (2013) tried to determine the shortcomings in SEEMP. They compared the existing gaps in the SEEMP rules with Safety Management Systems (SMS) in maritime transport companies, International Standard for Energy Management Systems (EMS), ISO 50001 and International Safety Management (ISM) code. They pointed out that SEEMP is lack of some important regulations such as oversight of policy and management that should be included in a typical management system standard.

Regulations on energy efficiency in maritime trade are critical to improving the profile of maritime environment (Acciaro et al., 2013). Renewable energy and energy efficiency (REEE) policies have a wide range of impacts such as safe energy, climate change, economic competitiveness, environmental pollution (Lo, 2014). LED lighting

3 MATERIAL AND METHOD

The number and location of lighting equipment in ships vary according to ship type, ship size and construction characteristics. The required minimum standards for this lighting equipment were determined in the SOLAS Convention and COLREG, which are the main sources of maritime sector. After the construction of the ship, the lighting equipment are fitting according to the international standards and all fitted equipment are documenting on the ship's illumination plan. In the study, a lighting plan of a feeder container vessel built in 2007 was taken as an example, which sailing on international waters. Ship particulars are summarized in Table 1.

Table 1. Particulars of the vessel

Type of vessel	Container
Length overall	184 m
Breadth overall	25 m
Gross tonnage	17687
Year of built	2007
Voyage type	International

The situation of installing LED lights (replaced) in place of all incandescent, compact fluorescent (Energy saving bulb-CFL), halogen and fluorescent lamps (existing) and in the ship's plan has been examined. While replacing the existing equipment, only lamps were changed and other fittings and sockets were remained untouched. Thus, it is aimed to avoid extra conversion costs. Due to the lack of proper light intensity and socket, it is necessary to change floodlights' and searchlight's armature.

4 REPLACEMENT OF LEDS

Criteria below are considered while doing market research and for determining the appropriate LED replacement of the existing lighting equipment on the illumination plan (Figure 1).
- Socket type: P28s, E27, G4, G13, G23 etc.
- Electrical working ranges: 230V~AC, 24V-DC
- Light intensity: required luminous value checked in order to provide sufficient lighting

Type of lamp	Existing (E)	Type of LED	Replaced (R)
Fluorescent lamp	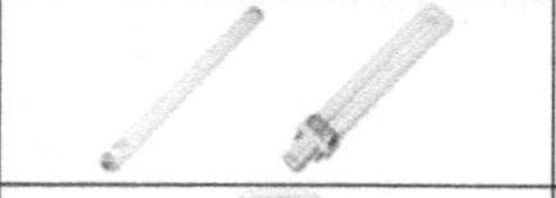	Led tube	
CFL		Led bulb	
Incandescent lamp	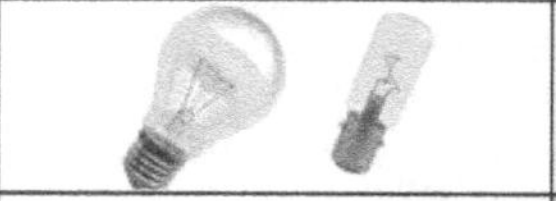	Led bulb	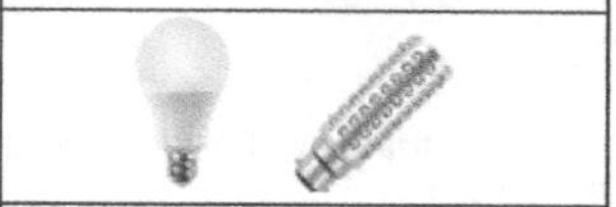
Halogen lamp	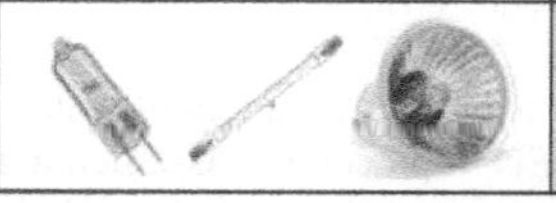	Led G4, R7s, PAR38	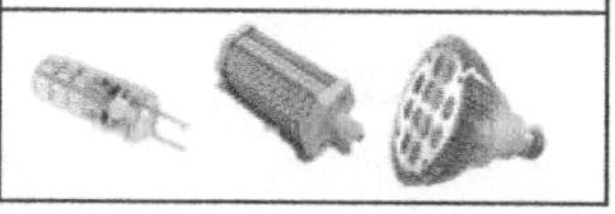

Figure 1. LEDs to be replaced by existing lighting equipment

Ship's illumination equipment table is formed based upon the mentioned criteria and existing plan (Table 2). Layout, standards and numbers of existing lighting equipment are shown in Table 2. Besides, replaced LEDs' electrical power and light intensity are also shown in the Table 2 (see p.6-7).

The main reason of creating a 'range' for the same intensity value while determining the light intensity that corresponds to intensity value of the existing lighting equipment is the different production standards of various producers. Light intensity of two different products, which have same power value, may vary according to the color temperatures. When two same brand 6W E27 socket LEDs are compared, we see that one with 6500°K color temperature is 510 lumens whereas the one with 3000°K is 490 lumens.

On ships, each of the lighting equipment has a different working duration. Therefore, for the theoretical efficiency study, formulas below are used in order to calculate the amount of energy that is consumed hourly.

See general equation of consumed electrical energy below:

$$E = P.t \qquad (1)$$

where E = electrical energy (kWh); P = power of the electrical appliance (W); and t = time of electricity consumption (h).

See equation of consumed electrical energy for existing/replaced illumination equipment below:

$$ET_X = \sum_{i=1}^{n} P_{X_i}.t.Q_i \qquad (2)$$

where ET_X = total consumed electrical energy of lighting equipment (kWh); X = type of illumination equipment (Existing (E) or Replaced/LED (R)); i = ID for each different type of lighting equipment (from 1 to 25); P_{X_i} = power of each different lighting equipment (W); t = time of electricity

consumption (h); Q_j = total quantity of each different lighting equipment (pcs); and n = total ID number of different lighting equipment (25). (t value was taken as '1' to calculate the consumed energy for unit value (1 hour).)

Energy saving ratio is calculated after determining the energy consumption of existing and LED-replaced situations. We used Formula 3 in order to calculate the energy saving ratio. Beside of the total energy savings, Formula 2 is used to calculate the energy consumption according to 'place of use' and 'lighting equipment types' (incandescent, fluorescent, CFL and halogen).

See equation to calculate energy saving from LED replacement below:

$$EnergySaving = \frac{ET_E - ET_R}{ET_E} x100 \qquad (3)$$

where ET_E = total consumed electrical energy of existing lighting equipment (kWh); and ET_R = total consumed electrical energy of replaced (LED) lighting equipment (kWh).

5 RESULTS

In this study, a container vessel was examined that contains 1390 lighting lamps. When we examined the placing and intended purpose, it is seen that the 59% of the lamps is Surface Mounting Lights that have to be on for long periods of time (Figure 2). Floodlights, HNAs, Recessed Ceiling Lights and Surface Mounting Lights comprise the 92% of total lighting equipment and have broad usage areas.

When we examined the lighting equipment according to the lamp type, it is observed that 81% of the total lights are fluorescent (Figure 3). Main reason behind this is the preference of energy saving and robust fluorescent lamps and CFLs over incandescent lamps in accommodation.

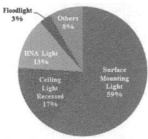

Figure 2. Quantitative distribution of lighting equipment by place of usage (n=1390)

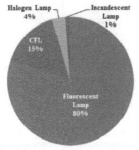

Figure 3. Quantitative distribution of lighting equipment by type of lamp (n=1390)

When the energy consumption of the existing and replaced lamps according to the place of usage is observed, it is seen that the surface mounting lights and floodlights are the highest energy consuming groups in both situations (Figure 4). Even though the percentage of floodlight numbers to the total lamp numbers is low (3%), they consume more energy because of being electrically powerful. Therefore, LED replacement decreases the floodlight's energy consumption by 23%.

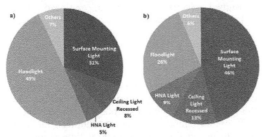

Figure 4. Quantitative distribution of lighting equipment by type of lamp (n=1390)

Energy consumption according to the lamp type (incandescent, fluorescent, CFL and halogen) is given in Table 3. Despite the fewer numbers (4%), halogen lamps consume highest energy 52%. Fluorescent lamps with highest percentage of use (80%) consume 40% of the total energy.

Table 3. Energy consumption values of existing lighting equipment by type of lamp

Type of lamp	Energy consumption of existing lighting equipment [Wh]	Energy consumption of replaced lighting equipment [Wh]	Energy saving ratio between existing and replaced lamp
Fluorescent Lamp	20066 (40%)	12419	38%
CFL	2968 (6%)	2002	32.5%
Halogen Lamp	25680 (52%)	5635	78%
Incandescent Lamp	1080 (2%)	136	87%
Total energy consumption [Wh]	49794 (%100)	20192	

As a result of the study, total hourly savings by the LED replacement is calculated as 59,5% (29,6 kWh). This value is calculated according to the hourly consumptions of existing (ET_E) and replaced (ET_R) illuminations. When energy saving ratio between existing and replaced lamps is taken into consideration, it is understood that the biggest saving has been achieved for incandescent and halogen lamps after the replacement. The reason of this can be explained by the parallelism of the illumination technology and energy efficiencies. Therefore, it is possible to state that the literature confirms the study's calculated data. Also, LED replacement causes the values of electrical currents to decrease parallel to power ratings, resulting in a positive impact for the energy loss due to the resistivity of the power cables. One-year illumination of the lights at their maximum would result in an electrical energy saving of 259,3 MWh.

If we consider the price of LED and armature replacement; for this vessel the total replacement price will be around $30420 according to initial calculations. Prices for new LED bulbs obtained from the current market (Dose, 2017). As known, LED prices are changing according to the brand and quality standards. In order to standardize this fluctuation, we have considered average market price for each LED. On board the vessel, 2 diesel generators are operating for 24 hours. These generators consume diesel oil and feeding average 1000 kW of load. When 2 generators are working together they consume about 3 metric tons (MT) of diesel oil in 24 hours. The hourly fuel consumption of the generators is calculated as 0,125 MT. When the bunker index for Marine Diesel Oil (MDO) is examined, 1 MT MDO is around $550 (Bi, 2017). In this case, the monetary value of the consumed fuel is $0,06875 in order to produce 1 kWh of electricity. The monetary value of the 29,6 kWh energy gain obtained after 1 hour burning of all lights is $2,035. As a result, the equipped LED lights can cover the cost of purchasing when lit up 14940 hours (623 days).

Table 2. Lighting equipment table of the vessel

ID (i)	Place of usage	Voltage and frequency	Standards of the armature	Type of lighting equipment		Qty of armature	Qty of bulb per armature	Total qty (pcs) (Q_i)	Power per unit (P_X)	Light intensity per unit [lumens]
1	Ceiling Light Recessed	230V, 60Hz	IP44, G13 Socket	Existing	Fluorescent Lamp, Warm White T8	9	2	18	18W	2100-2200
				Replaced	Led Tube, T8				11W	
2	Ceiling Light Recessed	230V, 60Hz	IP20, G13 Socket	Existing	Fluorescent Lamp, Warm White T8	84	2	168	18W	2100 - 2200
				Replaced	Led Tube, T8				11W	
3	Ceiling Light Recessed	230V, 60Hz	IP44, G23 Socket	Existing	Fluorescent Lamp, Warm White	28	2	56	11W	850 – 900
				Replaced	Led Tube				9W	
4	Surface Mounting Light	230V, 60Hz	IP55, G13 Socket	Existing	Fluorescent Lamp, T8	297	2	594	18W	2100 - 2200
				Replaced	Led Tube, T8				11W	
5	Surface Mounting Light	230V, 60Hz	IP66, G13 Socket	Existing	Fluorescent Lamp, T8	97	2	194	18W	2100 - 2200
				Replaced	Led Tube, T8				11W	
6	Surface Mounting Light	230V, 60Hz	IP55, G13 Socket	Existing	Fluorescent Lamp, T8	13	2	26	36W	2500 – 3000
				Replaced	Led Tube, T8				24W	
7	Cold Store Light	230V, 60Hz	IP67, G13 Socket	Existing	Fluorescent Lamp, T12	3	2	6	40W	2000 – 3000
				Replaced	Led Tube, T8				24W	
8	Corridor Edge Light	230V, 60Hz	IP20, G13 Socket	Existing	Fluorescent Lamp, Warm White, T8	15	1	15	18W	1050 – 1100
				Replaced	Led Tube, T8				11W	
9	Treatment Light	230V, 60Hz	E27 Socket	Existing	Halogen Lamp, PAR38	1	1	1	120W	1750 - 1800
				Replaced	Led, PAR38				20W	
10	Emergency Exit-Lamp	230V, 60Hz	IP44	Existing	Fluorescent Lamp, T5	23	1	23	8W	340 – 430
				Replaced	Led Tube, T5				4W	
11	Wall Light	230V, 60Hz	IP20, E27 Socket	Existing	CFL (Energy Saving)	3	1	3	9W	360 – 495
				Replaced	Led Bulb				5W	
12	Wall Light	230V, 60Hz	IP20, G23 Socket	Existing	CFL (Energy Saving)	3	1	3	9W	510 – 600
				Replaced	Led Bulb				6W	
13	Wall Light	230V, 60Hz	IP20	Existing	CFL (Energy Saving)	3	2	6	13W	870 – 900
				Replaced	Led Bulb				9W	
14	Berth Light	230V, 60Hz	IP20, G23 Socket	Existing	CFL (Energy Saving)	19	1	19	7W	400 – 425
				Replaced	Led Bulb				6W	
15	Mirror Light	230V, 60Hz	IP44, G13 Socket	Existing	Fluorescent Lamp, Warm White, T8	12	1	12	18W	1050 – 1100
				Replaced	Led Tube, T8				11W	
16	Table Light	230V, 60Hz	IP20, E27 Socket	Existing	CFL (Energy Saving)	3	1	3	11W	530 – 660
				Replaced	Led Bulb				7W	
17	Table Light	24V	IP20, G4 Socket	Existing	Halogen Lamp	2	1	2	20W	280 – 370
				Replaced	Led G4				5W	
18	Table Light	230V	IP20, E27 Socket	Existing	Incandescent Lamp	4	1	4	75W	850 – 960
				Replaced	Led Bulb				10W	
19	HNA Light (Merchant Ship Standard Attachment)	230V, 60Hz	IP56, E27 Socket	Existing	CFL (Energy Saving)	178	1	178	15W	800 - 970
				Replaced	Led Bulb				10W	
20	Explosion Proof Light	230V, 60Hz	IP56, G13 Socket, EEx ed IIC T4 Ignition Class	Existing	Fluorescent Lamp	2	2	4	18W	2100 – 2200
				Replaced	Led Tube				11W	
21	Floodlight	230V, 60Hz	IP66, R7s Socket	Existing	Halogen Lamp	33	1	33	500W	7200–10500
				Replaced	Led R7s				100W	

22	Floodlight	230V, 60Hz	IP66, R7s Socket	Existing Replaced	Halogen Lamp Led R7s	8	1	8	1000W 250W	20000–26000
23	Daylight Signaling Lamp	24V DC	IP55, G4 Socket	Existing Replaced	Halogen Lamp Led G4	1	1	1	20W 5W	280 - 370
24	Searchlight	230V, 60Hz	IP56, GX9.5Socket	Existing Replaced	Halogen Lamp Led GX9.5	1	1	1	1000W 300W	20500–26000
25	Navigation Lantern	230V~ 60Hz, 24V DC	IP56, P28s/ B22 Socket	Existing Replaced	Incandescent Lamp Led Bulb	12	1	12	65W 8W	750 - 850

6 CONCLUSION

As a result, we observed that adapting LED illumination for the model ship decreases the energy consumption by approximately 60%. Thus, adapting LED illumination can increase the energy efficiencies and eventually decrease the emissions in the shipping industry. Beside of its positive impacts on energy efficiencies, long-life, low working temperatures and robustness towards environmental effects such as vibration and weather conditions make it suitable for the maritime usage. Consequently, a more detailed future study is necessary involving actual working durations and fitting costs. Concordantly, annual energy savings' impact on generator's fuel consumption must be investigated in order to determine the actual effect on running costs. Beside of the ships, adaptation of LEDs to ports, which are considered an important part of the shipping industry, may result in substantial savings. Also, for the situations in which light filters are used to achieve certain colored lights such as navigational lanterns, LEDs come across various problems that need to be investigated. Another disadvantage is the cost; as equivalent lamps are still cheaper despite the recent technological developments that reduce the LED prices.

REFERENCES

Acciaro, M., Hoffmann, P. N. & Eide, M. S. 2013. The energy efficiency gap in maritime transport. *Journal of Shipping and Ocean Engineering*, 3, 1.

Banks, C., Turan, O., Incecik, A., Theotokatos, G., Izkan, S., Shewell, C. & Tian, X. Understanding ship operating profiles with an aim to improve energy efficient ship operations. *Proceedings of the Low Carbon Shipping Conference 2013*, 2013 London/UK. Low Carbon Shipping & Shipping in Changing Climates (LCS&SCC), 1-11.

Bi 2017. Bunker Index MDO. Access date: 25.01.2017. http://www.bunkerindex.com/prices/bixfree.php?priceindex _id=4

Cizek, C. J. 2009. *Shipboard LED lighting a business case analysis.* Master, California. Naval Postgraduate School, 1-104.

Dose 2017. Net Prices 2017 – LED Lighting. Access date: 24.01.2017. http://www.karl-dose.de/pdf/dose_preisliste_ 2017_LED.pdf

Freymiller, A. T. 2009. *LED shipboard lighting: A comparative analysis.* Master, California. Naval Postgraduate School, 1-50.

Holonyak Jr, N. & Bevacqua, S. F. 1962. Coherent (visible) light emission from Ga (As1– xPx) junctions. *Applied Physics Letters*, 1, 82-83.

Ji, H. K., Jung, K. S., Jang, U. Y., Kil, G. S. & Kim, S. Y. *Application of High Brightness LED for Shipboard Light Source.* Advances in Maritime and Naval Science and Engineering, 2009, Romania. WSEAS Press, 82-87.

Johnson, H., Johansson, M., Andersson, K. & Södahl, B. 2013. Will the ship energy efficiency management plan reduce CO2 emissions? A comparison with ISO 50001 and the ISM code. *Maritime Policy & Management*, 40, 177-190.

Krames, M. R., Shchekin, O. B., Mueller-Mach, R., Mueller, G. O., Zhou, L., Harbers, G. & Craford, M. G. 2007. Status and Future of High-Power Light-Emitting Diodes for Solid-State Lighting. *Journal of Display Technology*, 3, 160-175.

Lo, K. 2014. A critical review of China's rapidly developing renewable energy and energy efficiency policies. *Renewable and Sustainable Energy Reviews*, 29, 508-516.

Mills, A. 2003. Solid state lighting - a world of expanding opportunities at LED 2002. *The Advanced Semiconductor Magazine*, 16, 30-33.

Novikov, M. A. 2004. Oleg Vladimirovich Losev: Pioneer of Semiconductor Electronics (celebrating one hundred years since his birth). *Physics of the Solid State*, 46, 5-9.

Steele, R. V. High-brightness LED market overview. *Solid State Lighting and Displays*, 2001, San Diego, CA, USA. SPIE, 1-4.

Steigerwald, D. A., Bhat, J. C., Collins, D., Fletcher, R. M., Holcomb, M. O., Ludowise, M. J., Martin, P. S. & Rudaz, S. L. 2002. Illumination with solid state lighting technology. *IEEE journal of selected topics in quantum electronics*, 8, 310-320.

Unctad 2016. Review of Maritime Transport 2016. United Nations Conference on Trade and Development (UNCTAD). New York and Geneva: United Nations Publication.

Yeh, N. & Chung, J.-P. 2009. High-brightness LEDs - Energy efficient lighting sources and their potential in indoor plant cultivation. *Renewable and Sustainable Energy Reviews*, 13, 2175-2180.

Zheludev, N. 2007. The life and times of the LED - a 100-year history. Nature Photonics, 1, 189-192

Electromagnetic Compatibility of the Radio Devices in Maritime Shipping

Z. Łukasik, T. Ciszewski, Z. Olczykowski & J. Wojciechowski
Kazimierz Pulaski University of Technology and Humanities in Radom, Radom, Poland

ABSTRACT: ElectroMagnetic Compatibility – EMC is a feature of devices, describing their resistance and electromagnetic emissivity. This issue is particularly important in maritime shipping, where the radio devices must be resistant to any interference. In the article the authors have presented the analysis of selected protection solutions which provide electromagnetic compatibility of the radio devices. In the article the authors have also presented the measurements results of the electromagnetic compatibility for selected radio devices on sailing boat. These results were analysed and discussed.

1 INTRODUCTION

Watercraft are separate and isolated objects for a wide variety of reasons. Such condition frequently arises from the lack of bailout systems and appliances, reducing the crew safety level and forcing all existing systems and appliances to boost their reliability. Such circumstances require an individual approach to numerous issues, one involving the electromagnetic compatibility of all electronics, navigational and marine communication instruments in particular. Their proper operation is vital to safe and effective navigation. A commonly applied practice, electromagnetic compatibility testing forms part of the responsibilities of appliance manufacturers. Yet often as not it may be deemed insufficient – laboratory testing does not account for actual operational conditions. The situation is particularly crucial for difficult navigational conditions. The impact of environmental circumstances of navigation on appliances, electronic and power instruments included, that laboratory attestation cannot be recognized as sufficient (Venskauskas, K.K. & Krestianinov, V.V., 1984, Marin, G., 2014). In an attempt to highlight the need for test the electromagnetic compatibility of navigational and communication instruments in natural operating conditions, the authors have made their study results part of the paper. Said results shall form a basis for future activities – in research and legislation alike.

2 THE ELECTROMAGNETIC COMPATIBILITY ISSUE

2.1 *Electromagnetic Compatibility – EMC*

Electromagnetic compatibility is a set of features of the given electronic or power appliance vital to its interaction with other devices and its internal operating functionalities (Clayton, P. R., 1992, González, V.H.J., 2013). In particular, these features include the following:

– zero sensitivity to the electromagnetic impact of other appliances,
– zero electromagnetic impact on other appliances,
– zero electromagnetic impact on the appliance itself.

A failure to meet specified electromagnetic compatibility levels results in a threat of disruption in the operation of the given appliances. In terms of the disruptions trans-mission route, emission of or resistance to conducted disturbance and radiated disturbance.

Directive 2004/108/EC, Directive 2006/42/EC and Directive 2006/95/EC are fundamental laws regulating the issue of electromagnetic compatibility in European Union. Individual European Union member states have adapted their own requirements to aforementioned Directives, putting appropriate standards and legal regulations in place. In Poland, relevant standards chiefly norms PN-EN 61000-6 parts1-4 and PN-EN 61000-4-6.

2.2 Threat of Electromagnetic Interference on Watercrafts

The threat of electromagnetic interference on watercraft has two aspects. Firstly, the electromagnetic field threat, tying in with the operation of individual appliances and lightning. Secondly, the power supply quality in electric installations. While all aspects are crucial and interesting, the impact of lightning-related electromagnetic field is of fundamental importance. Although incidents of vessels being struck by lightning are relatively rare, consequences may be dire, to sailboats in particular. In many countries, the risk factor has been recognized sufficiently high to justify the introduction of identical lightning protection requirements for boats and buildings on land (NFPA 780: Standard for the Installation of Lightning Protection Systems), (Johns, D., 2016).

In an analogy to general discharge, lightning strikes can be classified as spark or partial discharges. The latter are specifically related to navigation – sailors were the first to notice them on their vessels' masts and give them the name of St. Elmo's fire. Both discharge types may be downward leaders or upward streamers. Downward leaders travel from thunder-clouds towards the earth. Upward streamers travel from tall objects on land toward the clouds. Spark discharges – thunder strikes – carry immense doses of energy (according to NASA's satellite BS 6651 - 50% over 28kA); not only is such energy capable of destroying or damaging electronic appliances aboard a watercraft – it can also become a major threat to the entire vessel. For sample values I=28kA, U=500kV, R=10Ω and t=50μs, we can calculate P=7.84 GW and E=392kWs – for a single discharge.

This is why, in case of sailboats in particular, appropriate lightning protection systems should be applied. Yet while a lightning protection system will protect the vessel on the one hand, it may become an upward streamer threat on the other. Thus, the use of sophisticated lightning protection system is most definitely justified.

The problem of alleviating potential lightning strike effects is usually resolved by developing a protection system referencing the Faraday cage aboard, accompanied by earthing and surge protection systems.

Sailboat masts are usually the only tall objects in the vicinity – as such, they are natural locations of lightning protection systems mounting. Other Faraday cage development points involve connecting all shrouds, stays, and backstays to water (Ewen, M. T., 1991). Making a lightning protector part of the system will eliminate the upward streamer threat. Yet such including lightning arrester into electrical circuit does not ensure full protection against the strike of an electromagnetic wave capable of destroying navigational and communication instruments. This is why authors of this paper have focused their research on the analysis of electromagnetic compatibility connected with the energy of an indirect lightning strike.

Figure 1. Modern lightning protection on sailing yacht and motorboat - Air Terminals, grounding plates with down, side flash, equalization conductors, loop conductors, catenary conductors (Coté, J., 2016).

2.3 Electromagnetic Compatibility Assessment

In order to ensure safety for the crew and for all electronic and power appliances onboard, electromagnetic compatibility levels have to be assessed. Such assessment can be carried out in laboratory or under real conditions. In Poland, norm PN-EN 55011 regulates the issue of measurement methodology (conducted emissions), with PN-EN 55016-2-3 regulating radiated emissions. Measurements of individual values are taken as function of time or as frequency function. Measurements as a function of time refer to the stationarity of the environment. Measurements as a function of frequency provide information on the spectrum of the electromagnetic field. Two frequency ranges are important to radiated emissions analyses: a) 9kHz-30MHz, b) over 30MHz. Compatibility parameter analyses involve sophisticated electromagnetic field meters (used by authors to take measurements presented in part 3 hereto) and meters with dedicated frequency bands. The latter involve dedicated ELF-10^{2}Hz – frequencies resembling those used when operating power appliances, VLF-10^{4}Hz – frequencies resembling those used when operating selected navigational and communication appliances, and

HF-10^7Hz – frequencies resembling those used when operating marine radio communication systems.

In case of the former of the aforementioned frequency ranges (9kHz - 30MHz), the magnetic field intensity (H) is metered under conditions of its highest value. Measurements are taken for three orthogonal directions of field intensity vectors forming a resultant field calculated with the following formula applied (Kaiser, K.L., 2005):

$$H = \sqrt{H_x^2 + H_y^2 + H_z^2} \qquad (1)$$

where H = the resultant magnetic field intensity; H_x, H_y, H_z = component magnetic field intensity for orthogonal directions of x, y, and z.

3 EMC – LABORATORY TESTING

3.1 *The Testing Procedure*

In the introduction to this paper, the authors emphasised that their actions had been driven by the insufficiency of testing the electromagnetic compatibility of appliances in laboratory conditions. While this part of the article presents laboratory test results indeed, they have been based on a re-enactment of natural environmental conditions of electrical discharges. The reconstruction is partial only, as actual lightning is rather difficult to re-enact for technical reasons (Nicolopoulou, E.P. et al., 2016). All tests involved a high voltage laboratory workstation, comprising:
- high-voltage impact generator (test transformer 100kV, spark gap, automation system),
- electromagnetic compatibility measurement system (meter, probes, oscilloscope).

Figure 2. View of the metering workstation - electromagnetic discharge generator and a metering system for EMC parameters.

A spectrum analyser of the following parameters was used:
- frequency range: 100kHz to 3GHz,
- resolution bandwidths: (-3dB): 10kHz to 1MHz in 1–3 steps, 200kHz,
- amplitude measurement range: -104 to +20dBm
- sweep time: f_{span}=0Hz - 2 to 100ms; f_{span}> 0Hz - 20ms to 1000s.

As show by the analyser's parameters, its lower operational frequency range is 100kHz - LF. This is the starting operational frequency range of numerous appliances, such as GPS, short- and long-range radars (marine, weather, satellite communication, coastguard communication, marine radio systems, AM- and FM-frequency radios, ham radios). Devices used in testing procedures analyse compatibility for all frequency ranges required (LF, MF, HF, VHF, UHF).

The studies were conducted for partial discharges and spark discharges. Partial discharges were obtained in the needle-needle electrode arrangement and spark discharges in the sphere-sphere electrode arrangement. Application of sphere-sphere electrode arrangement gave us the ability to extinction of the arc and its re-ignition. Measurements were carried out for several voltage values, from the minimum value to the ignition of the electric arc. The measurements were taken for the entire spectrum of the signal and the selected ranges. For partial discharge electrodes, working in needle-needle arrangement, were located at a distance of 3.5cm. The probe for measuring the electromagnetic field was set at a distance of 40 cm from the centre of the electrodes. The measurements were taken for the following voltages: 5, 10, 15, 20, 25 and 30kV. Partial discharges start at voltage of 20kV. Measurements taken for the entire spectrum of the signal (100kHz to 3GHz) allowed us to determine the range of the most significant changes. It was a range of 100kHz to 1MHz. The measured spectrum in the range of 100kHz to 1MHz for two cases: absence of discharges and partial discharges were shown in Figures 3-4.

The increase in the number of harmonics in the whole range of presented results is visible. You may also notice an increase in the amplitudes of the harmonics.

To obtain the spark discharges sphere-sphere electrode arrangement operating at a distance of 1cm from each other were used. The probe for measuring the electromagnetic field was set at a distance of 40cm from the electrodes. Measurements taken for the entire spectrum of the signal allowed us to determine the range of the most significant changes. As in the case of partial discharge it was a range of 100kHz to 1MHz. Within this scope we observed more than two-fold increase in amplitude if we compare it to the rest of the measuring range.

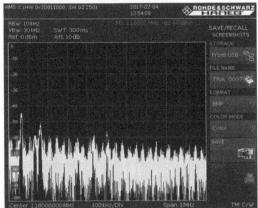

Figure 3. The spectrum of the electromagnetic field in the absence of partial discharges for a range of 100kHz to 1MHz (needle-needle electrode arrangement).

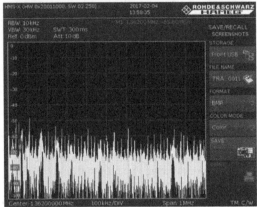

Figure 4. The spectrum of the electromagnetic field for partial discharges for a range of 100kHz to 1MHz (needle-needle electrode arrangement).

Figure 5. The spectrum of the electromagnetic field in the absence of spark discharges for a range of 100kHz to 1MHz.

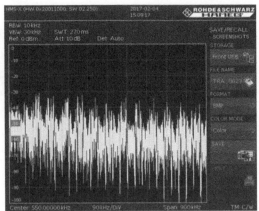

Figure 6. The spectrum of the electromagnetic field for spark discharges for a range of 100kHz to 1MHz.

The measured spectrum in the range of 100kHz to 1MHz for two cases: absence of discharge and spark discharges was shown in Figures 5-6.

The increase in the number of harmonics in the whole range of presented results is visible. You may also notice an significant increase in the amplitudes of the harmonics. In case of the absence of discharge they had values below -60dBm for virtually the entire range. In the case of discharge the values were raised up to -45dBm for the whole range.

4 EMC – CASE STUDY

The authors of this paper have, in sea voyage conditions, researched selected electromagnetic compatibility parameters. Measurements were taken during the period of January 14th through 21st, along the entire route covered (as shown in Figure 7).

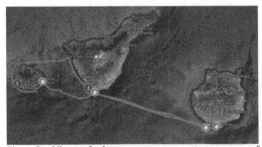

Figure 7. View of the sea voyage route – area of electromagnetic compatibility parameters metering.

All testing was carried out aboard s/y Dufour 412 Grand Large, carrying the following navigational and communication instruments: multifunction navigational unit, radar, autopilot, fishfinder, instruments and VHF communication unit. The yacht was not fitted with a lightning protection system. All electrical appliances were powered by a direct current electric installation with a feed from a battery set and an electric power generator interfaced with a combustion engine. When berthing in port, the yacht was using the harbour power grid as its source of electricity.

The magnetic field intensity (H) was the selected electromagnetic compatibility parameter. Measurements were taken with the use of an electromagnetic field intensity meter (tested frequency range: up to 20kHz, average testing time: 20ms). The decision to use such meter arose from measurement conditions, i.e. unfavourable conditions as sea. Measurements were continuous, the maximum field intensity values were recorded.

The purpose of metering the electromagnetic field intensity was to estimate its variations during period of weather change in the researched basin. In January, the storm risk in the basin in question is low – no storm was recorded. Weather change of potential impact on parameters measured primarily involve changes in atmospheric pressure. Figure 8 shows field intensity metering results along with atmospheric pressure readings for a single day selected.

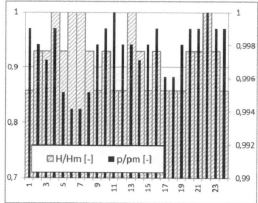

Figure 8. Changes to magnetic field intensity and atmospheric pressure on January 15th (worst weather conditions recorded).

Furthermore, electromagnetic compatibility comprises power quality tests from the electrical installation system. As aforementioned, sailboats are supplied with DC – U_n=12V direct current systems – thus, popular AC grid standards will not apply. In case of DC installations, the two basic energy parameters are voltage and disturbances (Ruciński, M. et al., 2015, Musznicki, P., 2013). Voltage decreases as batteries discharge during power appliances operation. Once voltage drops below a prespecified minimum acceptable value (U=10V), the yacht's generator kicks in. When berthing in port, batteries are charged from the harbour power supply. Possible disturbances are caused by transient states that occur when switching on and off electrical appliances. Given their relatively low power (the marine radar excepted) and the strong damping effect of the electric installation, transient states are of negligible importance.

5 CONCLUSION

The authors undertook in their research, the issue of electromagnetic compatibility of electronic and electrical equipment on watercrafts. The results of qualitative test of electromagnetic field emissions for arcing partial discharges and for spark discharges are presented in this paper. The studies used the phenomenon of electric arc AC 50Hz. The authors focused in their studies on this type of electric arc, treating it as a specific variable component of discharges. With regard to the electromagnetic compatibility, this variable component plays a crucial role in the process of interference of the functioning radio and navigation devices. The studies have indicated a significant increase in the intensity of the electromagnetic field during experimental arc discharges. The largest increases were recorded in the frequency range of 100kHz to 1MHz. This range corresponds to the range of work VLF, LF and MF devices (sonar, transoceanic communication system, LW radio, radio navigation beacon system, marine radio).

Conducted studies show the great need to continue work in the field of electromagnetic compatibility on watercrafts.

REFERENCES

Directive 2004/108/EC of the European Parliament and of the Council of 15 December 2004 on the approximation of the laws of the Member States relating to electromagnetic compatibility and repealing Directive 89/336/EEC L 390/24 Official Journal of the European Union EN.

Directive 2006/42/EC of the European Parliament and of the Council of 17 may 2006, on machinery and amending Directive 95/16/EC.

Directive 2006/95/EC of the European Parliament and of the Council of 12 December 2006 on the harmonisation of the laws of Member States relating to electrical equipment designed for use within certain voltage limits.

PN-EN 61000-6-1:2008; Electromagnetic compatibility (EMC) - Part 6-1: Generic standards - Immunity for residential, commercial and light-industrial environments.

PN-EN 61000-6-2:2008; Electromagnetic compatibility (EMC) - Part 6-2: Generic standards - Immunity for industrial environments.

PN-EN 61000-6-3:2008; Electromagnetic compatibility (EMC) - Part 6-3: Generic standards - Emission standard for residential, commercial and light-industrial environments.

PN-EN 61000-6-4:2008; Electromagnetic compatibility (EMC) - Part 6-4: Generic standards - Emission standard for industrial environments.

PN-EN 61000-4-6:2014; Electromagnetic compatibility (EMC) - Part 4-6: Testing and measurement techniques - Immunity to conducted disturbances, induced by radio-frequency fields.

PN-EN 55011:2011; Industrial, scientific and medical equipment - Radio-frequency disturbance characteristics - Limits and methods of measurement.

EN 55016-2-3:2014; Specification for radio disturbance and immunity measuring apparatus and methods - Part 2-3: Methods of measurement of disturbances and immunity - Radiated disturbance measurements.

Clayton, P. R. (1992). Introduction to Electromagnetic Compatibility. *John Wiley & Sons Inc.*, ISBN 0-471-54927-4.

Ruciński, M. et al. (2015). Analysis of electromagnetic disturbances in DC network of grid connected building-integrated photovoltaic system. IEEE, *Compatibility and Power Electronics 2015*, ISBN: 978-1-4799-6301-0, pp. 332-336.

Ewen, M. T. (1991). A Critical Assessment of the U.S. Code for Lightning Protection of Boats. *IEEE, Transaction on Electromagnetic Compatibility*, vol. 33, no. 2, pp. 132-138.

Musznicki, P. (2013). Conducted EMI propagation paths in DC-AC hard switching converter. *International Review of Electrical Engineering*, vol. 8, no. 6.

González, V.H.J. (2013). Electromagnetic Compatibility in ships design and construction. *IEEE Electric Ship Technologies Symposium*. ISBN 978-1-4673-5245-1.

Kaiser, K.L. (2005). Electromagnetic compatibility handbook. *CRC Press, Taylor and Francis Group*. ISBN: 0-8493-2087-9.

Johns, D. (2016). Designing building structures for protection against EMP and lightning. *IEEE Electromagnetic Compatibility Magazine*, vol. 5, issue1, pp. 50 – 58, ISSN: 2162-2272.

Venskauskas, K.K & Krestianinov, V.V. (1984). The Ship Electromagnetic Environment and Methods of Improving Electromagnetic Compatibility Characteristics of Ship Radioelectronics Equipment. *International Symposium on Electromagnetic Compatibility*, ISBN: 978-1-5090-3171-9.

Nicolopoulou, E.P. et al. (2016). Electromagnetic simulation of a lightning strike on a ship. *IEEE-High Voltage Engineering and Application (ICHVE)*, ISBN: 978-1-5090-0496-6.

Marin, G. et al. (2014). Assessment of electromagnetic radiation exposure of embarked personnel on Romanian naval ships. *IEEE-Electrical and Power Engineering*, ISBN: 978-1-4799-5849-8.

NFPA 780: Standard for the Installation of Lightning Protection Systems

Coté, J. (2016). Modern Lightning Protection On Recreational Watercraft. *BoatU.S. Magazine.*

Proceedings of 12th International Conference on Marine Navigation and Safety of Sea Transportation, TransNav 2017
21-23 June 2017, Gdynia, Poland

Universal Recuperation System of Electricity from the Exhaust System of an Internal Combustion Engine as the Engine of Small Capacity

P. Olszowiec, M. Luft & Z. Łukasik
University of Technology and Humanities in Radom, Poland

ABSTRACT: This article presents a universal system of recovery of electric energy based on the flow of exhaust gases in the exhaust system. The system allows the implementation engines with a very large discrepancy capacity. Trials were carried out even in the supercharged spark-ignition engines with a capacity of 1200 cm3. The proposed solution enables the use of it in the existing turbocharged engine without significant design changes. The system works in both modern engines based on electronic systems and pumping units.

1 INTRODUCTION

In recent years, the pace of development of internal combustion engines does not go hand in hand with the vehicle increasing demand for electricity. Despite the implementation of a series of environmentally friendly solutions to reduce harmful emissions, the efficiency of the power unit has not improved significantly in the last decade. Modern turbocharged spark unit that uses high-pressure injection system, in fact achieves efficiency of 0.4 - 0.45[7,8].

This article presents the project implementation of the high-speed turbogenerator, implemented into the exhaust system of an internal combustion engine in order to recuperate energy. The nominal specifitations of the generator used in the project are the following: the power of 1kW, 187V three-phase voltage and variable speed of 100,000 rev / min. The conducted analysis of the use of the turbogenerator in the process of recovery of electric energy from the exhaust system, show an improvement in the energy balance of the vehicle[2]. For the power system with a turbo-generator, we developed an appropriate energy management system from two sources: a classic alternator and a high-speed turbogenerator. They are used depending on the load of the drive unit, operating conditions, the demand for energy and concern for environmental standards (Fig.1.).

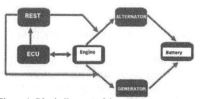

Figure 1. Block diagram of the vehicle energy management

2 SYSTEM REST (RECOVERY ENERGY SYSTEM OF TURBOGENERATOR

Developed within the project, the system REST (Recovery Energy System of Turbogenerator), cooperates with the drive unit, fitting into the contemporary automotive trend of "Downsizing" (Figure 2).

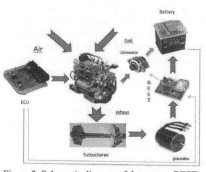

Figure 2. Schematic diagram of the system REST

The cooperation with the above type of exhaust unit allows the implementation of REST in internal combustion engines with low capacity (less than 1600cm3). In addition, the REST system uses the operating turbocharger to drive a high speed electric generator. This solution significantly reduces the cost of implementation of the system, since it is enough to connect the turbocharger to the generator, and thus, the system does not require upgrading of the exhaust system.

The purpose of a high speed generator is a partial or complete replacement of a classical electrical source, which is an alternating current generator. The main objective of the system is the use of the exhaust flow energy in the exhaust system, to drive a high speed generator to produce electricity. A high speed gas flow and low torque occurring in the exhaust system of the engine with spark ignition, prevent the use of any gear combined with the turbocharger shaft and the generator shaft. Therefore, the turbocharger and the generator and are driven by a common shaft. Given the high speeds of the turbine set (approx. 100,000 rev / min), it was necessary to introduce into the system a stabilizing voltage system of 14 V, necessary to work with the lead - acid battery.

3 THE DESIGN OF THE EXHAUST SYSTEM

As a result of the delivery and combustion of fuel-air mixture in an internal combustion piston engine, the energy generated in the combustion process can be divided into three parts: a vehicle power; heat transfer from the cooling system - constituting a loss; heat emission and energy flow of the exhaust gases[6].

The exhaust gas stream carries with it a certain amount of energy in various forms, namely:
- exhaust gas temperature - a measure of heat exhaust; can reach the order of 1000 °C;
- exhaust backpressure - at idle and at low and medium engine load, excess pressure appears alternately with a negative pressure.

Exhaust backpressure is necessary to overcome the resistance of the exhaust gas flow in the exhaust system. The exhaust gas flow takes place due to the pressure difference between the pressure prevailing in the combustion chamber when the exhaust valve is opened, and atmospheric pressure or slightly lower, prevailing at the end of the exhaust system. The difference between these pressures is necessary to:
- obtain the flue gas velocity at which the combustion chamber is left by the required amount of exhaust gas;
- overcome the flow resistance, associated with the flow of exhaust gases.

The flow resistance of the exhaust gas depends on the gas velocity. The higher the flow rate of the exhaust gas, the higher the flow resistance. It is not preferred to remove all the exhaust gases from the combustion chamber in all engine operating conditions. When the engine is running at low or medium load, it is appropriate to leave a certain amount of exhaust gases in the combustion chamber. The resistance of the exhaust gas flow is called exhaust backpressure. If the value of the exhaust backpressure is known, required by the engine manufacturer, its measurement should be made while assessing the exhaust system. It is possible then to determine whether the test system will allow the engine to achieve its maximum power or not. However, taking this measurement into account, it is not possible to assess the torque value and power of the engine running in the conditions of small and medium loads. For this reason, the exhaust system coming from the manufacturer and constructor of the exhaust unit has an advantage because its characteristics are then adjusted to the characteristics of the engine. For a typical exhaust system with a catalytic converter, exhaust backpressure is from 30 to 40 kPa. A typical muffler produces exhaust backpressure around 10 kPa, and in its special design, egr. for a sports version, this value can be lowered to 5 kPa[4].

4 THE TURBOCHARGER

The turbocharger, used for research on the possibility of functioning of the system of energy recuperation from of the exhaust system, is a model TD015 Mitsubishi(Fig.3). It is a radial turbocharger nominally dedicated to the exhaust unit Renault 1.2 Tce which was used for the study.

Figure 3 The view of the turbocharger of the engine 1.2 Tce

As part of the analysis of the output parameters of the turbine, it has been tested on the test stand where it was accelerated gradually to the speed of 100,000 rev / min to measure the resulting pressure and mass flow rate of air. At maximum speed, the turbine obtained a pressure of 0.8 bar. Although the flue turbine is characterized by a relatively high power, the obtained torque reaches a low value. Torque fluctuations are caused by pulsating flow of exhaust gases through the turbine at the time of opening and closing the outlet valve. The frequency of the pulse is in the range of 20 - 80 Hz. For the calculated flow and the measured exhaust gas temperature of the drive unit installed on the dynamometer, the power gas turbine was calculated[5]. To calculate the power of the turbine the following equation was used:

$$N_t = \eta_{izt}\eta_{mt}\dot{m}_t \frac{k_t}{k_t-1} R_t T_3 \left[\left(\frac{1}{\pi_t} \right)^{\frac{K_t-1}{K_t}} - 1 \right]$$

where:
N_t = W power of the gas turbine, W
η_{izt} = isentropic efficiency of the gas turbine
η_{mt} = mechanical efficiency of the gas turbine
$\dot{m}_t$ = flow rate through the compressor
R_t = exhaust gas constant, $J \cdot kg^{-1}K^{-1}$
T_3 = temperature at the compressor inlet
π_t = static pressure of the turbine
K_t = heat capacity ratio

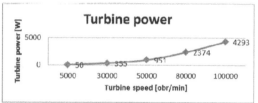

Figure 4. Speed turbine use in REST

Due to the fact that the turbine and the compressor are a single unit connected to each other by the rigid shaft, so isentropic and mechanical efficiency is defined cumulatively(fig.4).

The calculated power of the presented turbo set for the rotation in the range of the technical capabilities of the generator, significantly exceeds the power declared by the manufacturer of high speed electrical power source. During the study, the turbine at a speed of 100,000 rev / min gained power 4293W, while the maximum power of the proposed test generator was 1000W. This allows for the implementation of research in the future with the use of a generator with more power.

5 GENERATOR

The basis for the system REST is a high-speed brushless electric power generator with permanent magnets. Due to the number of windings, these machines can be divided into 2-phase and 3-phase ones. Mechanical parameters do not differ from brush electric machines, and their significant advantages include: high durability, precise speed control, possibility of use in adverse operating conditions (high temperature). It consists of a stator, on whose surface drains of the heat were milled, and a rotor with a permanent magnet. As a

protection against damage of the magnet, the flange of a titanium alloy was applied to the rotor, and its task is to carry the partial stress at high operating speeds (Tab.1). As the material for the rotor shaft, structural steel S235 JRG2 was used.. Taking into account the specific conditions of the generator work, especially in terms of temperature transferred from the exhaust system to integrate the power source to the turbine, a housing centering the shafts of both elements was created. The housing contains channels which use the flow of air entering the compressor, as a refrigerant for the generator.

Table 1. Parameters of a high-speed electric generator

Parameters					
U [V]	I [A]	F [Hz]	N [rev/min]	P [W]	M [Nm]
187	5	1667	100000	1000	0,1

Generator parameters for single phase
R_{Ph}: 2,8Ohm (20^0C)
L_d: 830μH
K_T: 0,05Nm/A
K_v: 1,79 V_ϱ/1000min^{-1}
Combination: triangle

The manufacturer of the components for the generator is the Austrian company ATE Antriebstechnik und Entwicklungs GmbH, specializing in high-speed electric drives reaching even 1000000 rev/m. According to the data provided by the manufacturer (Table 1), the theoretical power of the generator is sufficient to meet a demand for electricity in the vehicle.

An important issue in the use of the generator of electricity in the engine exhaust system with spark ignition, is the knowledge of the torque of the above the generator.

6 RESEARCH RESULTS

As part of the metering operation of the turbogenerator, the following parameters were recorded: voltage and current so as to calculate the value of the power. During the study, these parameters were changed to determine the generator power for different operating states of the internal combustion engine. This experiment was repeated for each case of the studied rotation speeds and for all the assumed load points of the engine. The graph shows the power waveforms as a function of the torque (Figure 5). For speed of 1500 rev / min, measurement of the power system for all the torques was carried out, while, at each measuring point the generator was loaded with three different current values, in which different power values were obtained. Similarly for the moment, these were the powers of 0.6W, 0,9W and 1,3W. With a torque 30Nm, the currents transferred to the turbogenerator had a value of 0.115A, 0.13A and 0.045A for the power 4W, 3,18W, and 2,5W. With a torque 45Nm, the obtained voltage is: 45V, 66V, and 79,5V for the power 14,5W, 13,2W, and 7,95W. For the above speed of 1500 rev / min, the highest values of the observed parameters were recorded at the engine load with a torque 75NM.

Characteristics of electric power

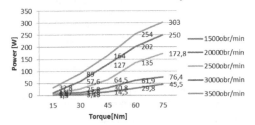

Figure 5. The graphs of power of the system REST for different rotation speeds of the internal combustion engine

At that point, the following values of power were obtained: 41W, 46,5W and 41,2W, analogously to the current 0.51, 0.4A and 28.3A. For the speed 2000 rpm / min, the highest parameters were recorded at the moment of 75Nm for the 61,7V, 127,3V and 143,2V for power values 70,3W, 76,4W, and 28,6W.

1500 obr/min

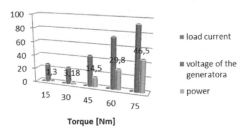

Figure 6. Parameters of the system REST for the selected loads with the rotation speed 1500 rev/ min

2000 obr/min

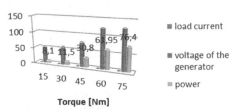

Figure 7. Parameters of the system REST for the selected loads with the rotation speed 2000 rev/ min

The speed of 2500 rev / min and a engine load of 75Nm were accompanied by the following electrical parameters of the turbogenerator (Fig.8): the required current 0.73A, 0.5A and 0.2A. Powers obtained with these currents are analogously 53,6W, 57,6W and 29,8W. Another investigated value of rotational speed is 3000 rev / min which is, as in previous studies, the highest power of the system REST which was recorded with a load of the order of 75Nm. The electric power obtained by the generator with such a torque is 223,3W; 250W and 208W for the current 2,9A, 1.79A and 1.2A.

2500 obr/min

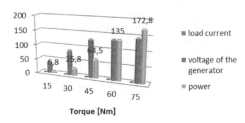

Figure 8. Parameters of the system REST for the selected loads with the rotation speed

3000 obr/min

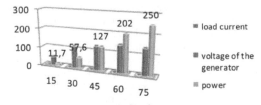

Figure 9. Parameters of the system REST for the selected loads with the rotation speed 3000 rev/ min

3500 obr/min

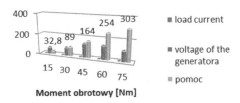

Figure 10. Parameters of the system REST for the selected loads with the rotation speed 3500 rev/ min

The highest electric power gained during the experiment, generated by the system REST, was obtained at a speed of 3500 rev/ min, and with a torque 75Nm (Fig. 10). The generator had values: 104V, 2,9A and power of 303W. In addition, the required current was at 2,1A with 140V and the obtained power was 294W, and also the required current was at 1.6A with 180V and the obtained power was 288W. However, the resulting maximum electrical power is not sufficient enough to replace the traditional alternating current alternator. On the other hand, these studies show that it is possible in certain operating conditions of the engine and the vehicle, disconnect the alternator. This solution also allows to use smaller nominal generator power, which enables the reduction of electricity.

7 CONCLUSIONS

The analysis of the conducted test results showed that the achieved results directly depend on the measuring operating points of the drive unit. These points were determined by the constant rotational speed and the degree of engine load expressed in Nm. The research and analysis of the overall efficiency of the internal combustion engine, in juxtaposition to its cooperation with the classic alternator and the system REST, generating the same electrical power, showed:
- For the speed of 1500 rev / min over the whole range of investigated loads, there was an average improvement of 3.6%.
- For the speed of 2000 rev / min the efficiency improvement is proceeding very evenly for all load points resulting in an average score of 5.8%.
- A similar course of the efficiency improvement is drawn on the graph for the speed of 2,500 rev / min. The average improvement in the overall efficiency of this speed is 3.0%.
- A steady efficiency improvement in all the investigated load points was also observed for the speed of 3000 rev / min. On average, the efficiency increases by 3.9%.
- The last tested speed range was 3500 rev / min. In this series of tests, the increase in efficiency reached 4.2%.
- The results of the conducted research of the system, which uses the turbogenerator in the

process of energy recuperation, show a significant improvement in the energy balance in the vehicle. This also creates a possibility to use a low-power alternator, which will cover the demand for electricity only in the case of inefficient working of the turbogenerator. The choice of the studied flue gas unit of small capacity shows that the system is universal, and it is possible to apply the system REST not only in urban vehicles but also as a system that supports the process of recovery of electric energy in hybrid vehicles.

REFERENCES

[1] Bourhis G., Leduc P., "Energy and Exergy Balance for Modern Diesel and Gasoline Engines", Oil&Gas Science and Technology – Rev. IFP, 2009,
[2] Dziubański S., Jantos J., "Improving thermal balance in internal combustion engine using the turbogeneratora", Machine Design, Novi Sad, 2010,
[3] Kuszewski H., Ustrzycki A.: Laboratorium spalinowych środków transportu Oficyna wydawnicza Politechniki Rzeszowskiej, Rzeszów 2011,
[4] Luft.S: Podstawy budowy silników, Wydawnictwa Komunikacji i Łączności WKŁ Warszawa, 3, 2012,
[5] Mitaniec W., "Metoda obliczania parametrów termo dynamicznych gazów w przewodach rozgałęzionych", Instytut Pojazdów Samochodowych i Silników Spalinowych, Czasopismo Techniczne, Wydawnictwo Politechnik Krakowskiej, 2008
[6] Wojciechowski K., Merkisz J., Puć P., Lijewski P., Schmidt M., "Study of Recovery of Waste Heat From the Exhaust of Automotive Engine", AGH University of Science and Technology, Institute of Combustion Engines and Transportation, Poznan University of Technology, 2007,
[7] Yang J., Stabler F., "AAutomotive Applications of Thermoelectroc Materials",Journal of Materials, 2009
[8] Zwyssig C., Duerr M., Hassler D., Kolar J.W., "An Ultra-Hihg-Speed, 500000 rpm,1 kW Electrical Drive System", Power Electronic Systems Laboratory ETH Zurich, Switzerland, 2007
[9] Zwyssig C., Round S.D., Kolar J.W., "Power Electronics Interface for a 100 W, 500000 rpm Gas Turbine Portable Power Unit", Power Electronic Systems Laboratory Swiss Federal Institute of Technology Zurich, Switzerland, 2006,
[10] http://www.magnetimarelli.com/excellence/technological-excellences/kers - z dnia 07.01.2016

A New Innovative Turbocharger Concept Numerically Tested and Optimized with CFD

L.C. Stan
Constanta Maritime University, Constanta, Romania

ABSTRACT: Turbocharging is perhaps the best way to increase the performance of a reciprocating internal combustion engine, gasoline or Diesel. While the basic concept is simple, the subtleties of a turbocharger system are actually very complex. The purpose of this article is to investigate the functioning of new innovative two stages turbochargers under the aspects of exploring the limits of this new design and make an optimisation study in order to pinpoint the influences of the two pressure stages upon the overall turbocharger behaviour. The software to be used across the numerical experiment is ANSYS 16 (CFX and Design Explorer. The main conclusion is that this new innovative arrangement is depending for its final working parameters on the speed of the LP Rotor as main contributor to the useful delivered pressure and the speed of the HP Rotor up-on the reaction time of the device, this being massively shortened by the presence of this stage.

1 INTRODUCTION

Turbocharging is perhaps the best way to increase the performance of a reciprocating internal combustion engine, gasoline or Diesel. While the basic concept is simple, the subtleties of a turbocharger system are actually very complex (www.turbokart.com).

Acting as a pump, the internal combustion engine piston needs air to burn the fuel injected inside. By burning the fuel the energy needed to move the crankshaft is developed transforming it into useful mechanical energy.

The amount of fuel to be burnt is in close chemical report to the proportion of air inside the combustion chamber and the released heat is proportional to the heat developed during the combustion process. The more air inside, the more fuel to be burnt and more energy will then be available to put in motion the engine.

In order to increase the amount of air inside the combustion chamber the superchargers and turbochargers are to be put in place. The superchargers are driven by the motor itself via a mechanical transmission and the main element is an air compressor rotating at high speeds and blowing the air under pressure inside the combustion chamber providing so an extra-amount of available air for burning (Baines, 2005).

The other way around is the so called turbo-supercharger, also known as a turbocharger, or just a turbo for short. Here the kinetical energy of exhaust gases from the cylinder is used to propel a turbine rotor which in turn is rotating an air compressor to blow the air under pressure inside the cylinders (Kraus, 2009). The turbocharger is driven by the exhaust gas leaving the cylinders of the diesel engine it serves. The gas has sufficient pressure and heat when released from the cylinder at exhaust opening, to drive the turbocharger. It is directed on to turbine blades by nozzles which are built into a nozzle ring in the axial flow type or into a radial turbine form the peripheral volute casing of smaller turbochargers.

A small turbocharger for a generator diesel prime mover may be driven by a radial flow gas turbine, which closely resembles the impeller it drives. This type of machine costs less to manufacture than the axial flow turbine and has a simpler construction.

In this context, it is proposed a new concept of Two Compression Stages Turbocharger with compressor rotors placed in opposition installed inside a common housing, solution which is simplifying the construction and is making it in the same time more compact. The construction solution itself is solving the problem of the reaction time (Lag) for the High Pressure compression stage when the engine demand for extra pressure is suddenly.

Some of the main elements are to be seen in the figures below as follows: 3-Turbine rotor for the Low pressure stage (LP); 4-Turbine Housing; 5-Exhaust Gases Low Pressure turbine duct; 9-Exhaust gases High Pressure (HP) turbine duct; 10-Turbine rotor for the High pressure stage; 14-Compressor housing; 15- High pressure compressor rotor; 18-Low pressure compressor rotor; 22-Inlet for exhaust gases for Low Pressure turbine; 23-Inlet for exhaust gases for high Pressure turbine.

The exhaust burned gases coming from the engine are directed to the 22-Inlet for exhaust gases for Low Pressure turbine and 23-Inlet for exhaust gases for high Pressure turbine. They are putting in motion 3-Turbine rotor for the Low pressure stage (LP) and 10- Turbine rotor for the High pressure stage. 3 and 10 are placed on the same shafts with 15- High pressure compressor rotor; and 18-Low pressure compressor rotor which are delivering compressed air to the engine.

For normal functioning of the engine the HP compression stage is not working, and that is via a throttling valve for instance, the exhaust gases coming from the engine is feeding only 22-Inlet for exhaust gases for Low Pressure turbine and only 18-Low pressure compressor rotor is working whereas the 15- High pressure compressor rotor is still playing just a role of guiding and transforming the kinetic energy of the air in static pressure.

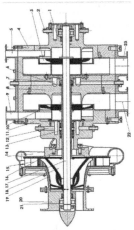

Figure 1. Cross-section of the proposed Turbocharger.

Once the engine demand for air is increasing the 23-Inlet for exhaust gases for high Pressure turbine receives exhaust gases and 10- Turbine rotor for the High pressure stage is delivering extra kinetic energy to the air and thus increasing the pressure of the air to feed the engine.

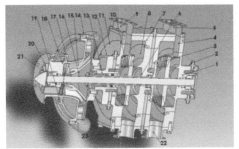

Figure 2. 3D View of Turbocharger.

Since the constructive solution is very compact and between compression stages the distance/gap is extremely small, the reaction time is almost instantaneous and depends entirely only of the reaction time of the throttling valve.

The purpose of this article is to investigate the functioning of new innovative two stages turbochargers under the aspects of exploring the limits of this new design and make an optimisation study in order to pinpoint the influences of the two pressure stages upon the overall turbocharger behaviour. The software to be used across the numerical experiment is ANSYS 16 (CFX and Design Explorer).

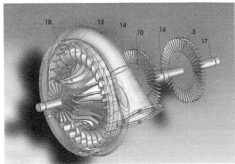

Figure 3. Turbines rotors and compressor rotors.

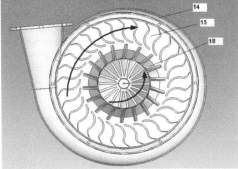

Figure 4. Compressor housing and compressor's HP and LP stages rotors.

2 MATERIALS AND METHODS

The departing point of numerical simulation is a new design of turbocharger with dimensions taken using the engineering common sense as in the figure 5.

In the following figures, the dimensions of the Low Pressure (LP) Rotor (ext. diameter 129 mm) and the High Pressure (HP) rotor (ex. Diameter 490 mm) are to be seen. The width of the rotor assembly is 174 mm

In order to establish the fluid domain after some CAD manoeuvres the following final geometry was generated (Fig.6).

One might be seen the inlet area where the air with its properties at 250C and at normal pressure-temperature conditions, is entering, and the outer region where the pressured air is leaving the turbocharger toward the engine cylinder.

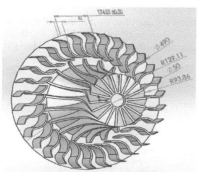

Figure 5. Compressor rotors dimensions.

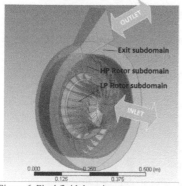

Figure 6. Final fluid domain.

Three subdomains were established: the LP, HP Rotors subdomains which are supposed to rotate in counter directions and the exit subdomain which is fixed. For the first two simulations the speed pf these moving domains will be ±1570.8 [rad/sec]=±15000 rpm, the plus/minus sign is describing that the rotor domains are moving in opposite directions. The rotors themselves will be kept fixed and only the domains are moving.

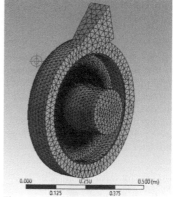

Figure 7. The fluid domain meshing.

The finite volumes mesh comprises 831869 elements with 173208 nodes and the quality of meshing is given in the figure below:

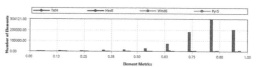

Figure 8. Finite volume elements mesh quality.

The boundary conditions are the Opening type with normal conditions for the Inlet region and for the Outlet region we'll have three types of conditions following the scenarios set for study:

- For the first scenario the Outlet condition is Wall and that is for exploring the limits of turbocharger functioning (i.e. the maximum delivered pressure) knowing that every turbocharger has different performances for different geometries. For our case size and geometry, we have therefore to explore these limits.
- For the second scenario the Outlet region will have the Opening condition meaning that this time the turbocharger is delivering the charge in atmosphere and that is for exploring the maximum velocity of the delivered air.
- The last condition will be deployed for the optimisation study and it is supposing a condition of Opening type, and certain counter-pressure to exist here to mimic the air access inside the engine cylinder. Tis pressure will be variate when the optimisation process will be started.

3 RESULTS AND DISCUSSION

3.1 *CFD analysis of the first scenario to investigate the maximum delivered pressure*

The model was run until the convergence was achieved, somewhere around 130 iterations. Due to the model complexity this is taking more or less an hour using a very performant computer.

– Total pressure profile

In the following figure, in the inlet region a negative relative pressure is calculated underlying here the suction effect of the device functioning. Gradually the total pressure is building up toward the exit domain where the peaks pressures of 1.21 e5 Pa are to be locally recorded inside the exit domain. Calculating the average pressure inside the device, this is 50142.9 Pa. This average pressure will be later on used as optimisation output parameter.

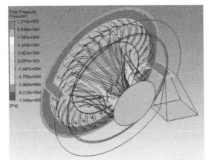

Figure 9. Total pressure distribution field for the first scenario.

– The velocity profile

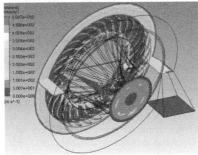

Figure 10. Total velocity distribution field for the first scenario.

For the outlet area the velocity is zero since there was placed a wall. But the velocity fields are big inside the exit domain since here the air is circulated in circular trajectories whereas inside the rotors it is almost stagnant less the zone in between the two rotors rotating in opposite directions. Locally peak velocities of 500 m/sec are recorded.

– LP Rotor total pressure

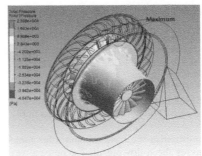

Figure 11. Total pressure distribution field LP Rotor for the first scenario.

The figure above is better pinpointing the total pressure distribution across the LP Rotor with 2.39 e4 peak pressures locate at the end of the blades.

– HP Rotor total pressure

The total pressure distribution for the HP rotor is shown below, with peak values of 9e4 Pa, three times bigger than those calculated for the LP Rotor.

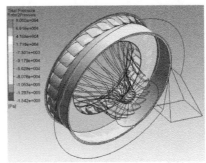

Figure 12. Total pressure distribution field HP Rotor for the first scenario.

– Turbulence kinetic energy

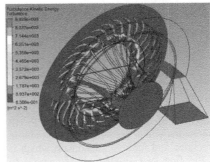

Figure 13. Turbulence kinetic energy distribution field for the first scenario.

Since the rotors are rotating on opposite direction, as expected the maximum turbulence is recorded inside the region in between the two rotors and it have a peak value of 8.92e3 [m2/s2].

3.2 *Optimisation of the design and Response Surfaces*

The design optimisation is done with another Ansys 16 Workbench module, namely the Response Surface Optimisation module.

This module is employing the Response Surface technology in order to determine a set of design points based on the input and output design parameters.

This time the goal of the optimisation study is not to geometrically optimise the design but to derive the most important parameters to impact the functioning of the device.

Outline of Schematic B2: Design of Experiments

	A	B
1		Enabled
2	⊟ ✓ Design of Experiments	
3	⊟ Input Parameters	
4	⊟ 15000rpm (A1)	
5	P2 - AngVelR2	☑
6	P1 - AngVelR1	☑
7	P6 - InputOutletPressure	☑
8	⊟ Output Parameters	
9	⊟ 15000rpm (A1)	
10	P4 - VolumePressure	
11	P7 - AveVelocityOutlet	

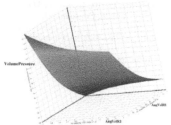

Figure 19. Input and Output parameters for the Optimisation.

The input parameters to be considered are the AngVelR1 meaning the velocity of the LP Rotor spanning from 10000 to 20000 rpm, AngVelR2 meaning the velocity of the HP Rotor spanning from 10000 to 20000 rpm, and the InputOutletPressure which is the pressure imposed at the Outlet region to mimic the air access inside the engine cylinder, spanning from 30000 to 70000 [Pa].

The Output design parameters are the VolumePressure standing for the entire total pressure as average inside the fluid domain and the AveVelocityOutlet standing for the averaged velocity on the outlet area.

- Response surface AngVelR1- AngVelR2- VolumePressure

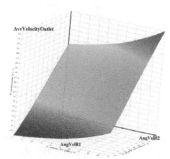

Figure 20. Response surface AngVelR1- AngVelR2- VolumePressure.

As seen in the figure above the most influential parameter upon the Total pressure averaged for the entire fluid domain is the velocity of the LP Rotor whereas the HP Rotor has a certain influence but several times more modest.

- Response surface AngVelR1- InputOutletPressure – VolumePressure

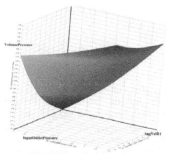

Figure 21. Response surface AngVelR1- InputOutletPressure – VolumePressure.

Now, comparing the influence of the Outlet working pressure and the velocity of the LP Rotor over the total pressure averaged on the fluid domain, it is obvious from the response surface above that both have comparable influence, the working pressure being slightly more influential.

- Response surface AngVelR1- AngVelR2- AveVelocityOutlet

Figure 22. Response surface AngVelR1- AngVelR2- AveVelocityOutlet.

From the above response surface, it is obviously now that the most influential parameter upon the outlet averaged air velocity is, this time, the speed of the HP Rotor. This means that the second compression stage is merely accelerating the air massively decreasing the response lag of the device making thus the device more responsive when the air parameters to be pumped inside the cylinder are quickly changing.

- Response surface AngVelR2- InputOutletPressure – AveVelocity Outlet

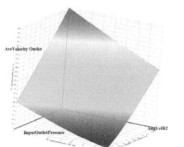

Figure 23. Response surface AngVelR2- InputOutletPressure – AveVelocity Outlet.

Now, comparing the influence of the Outlet working pressure and the velocity of the HP Rotor over the total pressure averaged on the fluid domain, it is obviously from the response surface above that the speed of the HP Rotor is more influential that the working pressure upon the average velocity of the air on the outlet surface.

– Sensitivity diagram

The overall design parameters sensitivity is shown in the diagram below:

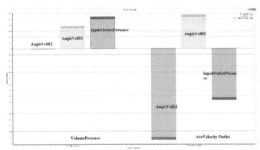

Figure 24. Design sensitivities.

– All the conclusions above are synthetically presented in the above diagram making obviously that the LP Rotor speed is mostly important for the pressure inside the device and the HP Rotor speed is responsible for the air velocity leaving the device.

– Optimum candidate

Whether we need a trade-off between the maximum pressure to be delivered and a minimum velocity of the air delivered, the optimum input design parameters are:

Table 1 The optimum candidate

	AngVelR2 (radian s^-1)	AngVelR1 (radian s^-1)	Input Outlet Pressure (Pa)
Optimum Candidate	-1016.5	1921.4	58241

From the Table 1, it can be noticed that if we need a maximum pressure along with a minimum air velocity, then the LP Rotor has to have a velocity almost twice that of the HP Rotor and the working pressure to be half of the investigated range.

4 CONCLUSIONS

This study is not meant to deliver the perfect shape and size of the proposed device which of course may be heavily improved mostly as rotors blades geometrical shapes which for this study were taken more using the engineering commons sense.

The study was meant to investigate the influence of the speed of the Rotors and the working pressure upon the capacity of the device to deliver air with certain velocities and pressures. The main conclusion is that this new innovative arrangement is depending for its final working parameters on the speed of the LP Rotor as main contributor to the useful delivered pressure and the speed of the HP Rotor upon the reaction time of the device, this being massively shortened by the presence of this stage.

REFERENCES

Baines, N. C. 2005. Fundamentals of Turbocharging. Concepts ETI. ISBN 0-933283-14-8.

Kraus, J. 2009. A Look Back: Genesis of the Automotive Turbocharger. Auto Universum.

Martinas, G., Cupsa, O.-S., Stan, L.-C., Arsenie, A. 2015. „Cold Flow Simulation of an Internal Combustion Engine with Vertical Valves Using Layering Approach" *Modern technologies in industrial engineering Modtech*, ISSN 2286-4369

Martinas, G., Stan, L.-C., Arsenie, A., Lamba, M.-D. 2014. „The Influence of a Wake Equalizing Duct over the Cavitation of a Maritime Ship Propeller" *Revista Analele Universitatii Maritime din Constanta*, vol 21, ISSN 1582-3601

McGeorge, H.D. 2002. *Marine auxiliary machinery*, Butterworth Heineman

Munteanu, A.-C., Buzbuchi, N., Stan, L.-C. 2014. Rotary internal combustion engines, pp 93-101, *Constanta Maritime University Annals* Year XV, Vol.21, Constanta Maritime University, Romania.

http://autouniversum.wordpress.com/a-look-back-genesis-of-the-automotive-turbocharger/.

http://www.turbokart.com/turbochargedengines. html

Study of Green Shipping Hybrid Diesel-Electric New Generation Marine Propulsion Technologies

G. Rutkowski
Gdynia Maritime University, Gdynia, Poland

ABSTRACT: The purpose and scope of this paper is to describe the factors to consider when determining the new solutions and technologies implemented in the shipping industry in the new generation marine propulsion system. Such changes are to promote a green ships concept and change the view of sea transportation to a more ecological and environment-friendly systems. Unfortunately, harnessing the wind, waves and solar power is not always the best solution for the shipping industry - especially in the frequently shadow or cloud-wrapped areas without the adequate winds or waves. In such specific situations, the best solution can be to use the different green shipping technologies such as: hybrid diesel-electric, waste heat recovery system (WHR) or LNG & duel fuel (DF) eco-power plants. In this paper, we are going to concentrate only on the new generation hybrid electric marine propulsion technologies.

1 INTRODUCTION

The pollution and waste created in the shipping processes have imposed the environmental burdens and accelerated resource depletion. The situation is set to worsen in the face of the intensifying trade globalization, which has contributed to the sustained growth in the international shipping activities [1], [2]. Overall, there are a lot of environmental hazards/aspects that can impact the air, seas or land. The common one is oil spill, but more and more air emissions have been in focus due to for instance, the global impact of global warming, and some of the impacts with acid rain.

Furthermore, it is common knowledge that ships are responsible for the release of a range of noxious gases and carbon dioxide. Therefore, the International Maritime Organization (IMO) has been working with the shipping industry, different stakeholders ranging from shippers and carriers to government bodies and international communities to reduce these emissions and make shipping more environment-friendly.

On the other hand, when one compares transporting the same cargo on the same distance by different means of transportation such as rail, road and marine it must be noted that the sea transportation is one of the most environment-friendly means (see table 1).

Anyway, to help protect the environment, many shipping firms have taken the initiative to find ways to lessen the environmental damage of their operations while enhancing their performance and conclude with managerial and policy implications of the conceptual framework to promote green shipping practices (i.e. [5], [7], [8], [9]).

Currently we have a wide range of modern marine green technologies available on the market used to enhance the performance and sustainability for the oceangoing vessels. Although harnessing wind, waves and solar power can be a very good solution for the oceangoing vessel [2] not always the same solution is the best option for other ships, especially in frequently shadow or cloud-wrapped areas without adequate winds, waves or solar power.

However, in case of the small river ships, tugboats or ferries which frequently pass low bridges and whose voyages are short, using e.g. huge wind sails installations or rotors systems makes no sense. In case of the DP (dynamic positioning) ships we always expect to have an immediate access to a full available power on propulsion system with intention to stay in position or have good manoeuvrability in bad weather conditions; hence the power available on thrusters should always be independent from the current solar or wind activities. In such specific cases, the best solution in fact can be not solar or wind sails but another type of green shipping technologies such as hybrid diesel-electric, waste heat recovery system, variable speed generators or i.e. LNG & duel fuel eco-power plant. Anyway, in this paper we are going to concentrate only on the hybrid electric new generation marine propulsion technologies.

2 HYBRID ELECTRIC MARINE PROPULSION SYSTEM

A hybrid electric marine propulsion system can achieve propulsion using a fueled power source (e.g. petrol or diesel engine or optionally steam turbine) or through a stored energy sources (e.g. a battery bank and electric motor) [6], [7], [8], [9].

This technology is not new and was implemented in the early 20th century [10]. The first diesel-electric ship was the Russian tanker 'Vandal' from Branobel launched in 1903. Steam turbine-electric propulsion system has been used on Tennessee-class battleships since the 1920's. Early submarines also used similar technology switching between the diesel engines for surface running, and electric motors for submerged propulsion. The Finnish coastal defense ships 'Ilmarinen' and 'Väinämöinen', laid down in 1928 and 1929, were among the first surface ships to use diesel-electric transmission.

Petrol-electric transmission (UK English) or Gasoline-electric or Gas-electric transmission (US English) is a transmission system for road, rail and marine transport which avoids the need for a gearbox [10]. The petrol engine drives a dynamo which supplies electricity to traction motors which propel the vehicle or boat. The traction motors may be driven directly or, in the case of a submarine, via a rechargeable battery.

In a diesel-electric transmission arrangement, as used on 1930s and later US Navy, German, Russian and other nations' diesel submarines, the propellers are driven directly or through reduction gears by an electric motor, while two or more diesel generators provide electric energy for charging the batteries and/or driving the electric motors. This mechanically isolates the noisy engine compartment from the outer pressure hull and reduces the acoustic signature of the submarine when surfaced [10]. Nowadays some nuclear submarines also use a similar turbo-electric propulsion system, with propulsion turbo generators driven by reactor plant steam. Gas turbines are also used for electrical power generation and some ships use a combination.

Nowadays, there are many companies that provide fully tailor-made new generation hybrid propulsion solutions for each individual ship (i.e. [5], [7], [8], [9]). The 'Skandi Marstein', a supply boat for drilling platforms, was the first vessel with a new generation diesel-electric drive launched in 1996 for North Sea sector. That ship was a milestone for Siemens group. On a three-day cruise, the 'Skandi Marstein' used 35 percent less energy than a diesel vessel [9].

Table 1. Comparison of the different means of transportation (ship, train, truck) efficiency with the indexed comparison to marine transportation. Sources: St. Lawrence Seaway website with RTG analysis for Combined Great Lakes-Seaway Fleet, which includes all categories of vessels operating within the Great Lakes-Seaway System (domestic carriers and international vessels) until October 2016.

Comparison of efficiency the different means of transportation (ship, train, truck) with percentage and/or indexed* comparison to marine transportation.	Base year 2010			Nowadays post renewal of all modes		
	Marine (Ship)	Rail (Train)	Road (Truck)	Marine (Ship)	Rail (Train)	Road (Truck)
Fuel efficiency [distance in kilometers to move 1tonne of cargo with 1 liter of fuel]	243 (100%)*	213 (88%)*	35 (14%)*	358 (100%)*	225 (63%)*	41 (11%)*
Energy Efficiency [tonne of cargo to be move on 1kilometer per 1liter of fuel]	243 (100%)*	213 (88%)*	35 (14%)*	358 (100%)*	225 (63%)*	41 (11%)*
Energy Consumption indexed to marine transportation (where @ 1 unit ≈ 130 kjoule/t-km)	1.0	1.14	6.94	1.0	1.59	8.73
Existing Modal Noise Footprint Comparisons (Severe Ldn Footprints (sq.km)= [km²])	2 (1.0)*	1687 (844)*	3933 (1967)*	2 (1.0)*	2354 (1177)*	5484 (2742)*
Greenhouse Gas (GHG) Emissions Intensity [g/t-km= grams emitted per cargo-tonne-kilometer]	11.9 (1.0)*	14.2 (1.19)*	75.5 (6.33)*	8.1 (1.0)*	13.3 (1.64)*	65.5 (8.07)*
Criteria Air Contaminants (CAC) Emissions of NOx [g/kCTK = grams emitted per thousand-cargo-tonne-kilometer]	233.4 (1.0)*	245.9 (1.05)*	392.0 (1.68)*	32.3 (1.0)*	35.2 (1.09)*	54.5 (1.69)*
Criteria Air Contaminants (CAC) Emissions of SOx [g/kCTK = grams emitted per thousand-cargo-tonne-kilometer]	82.9 (1.0)*	1.5 (0.002)*	0.7 (0.008)*	0.07 (1.0)*	0.10 (1.43)*	0.61 (8.71)*
Criteria Air Contaminants (CAC) Emissions of PM [g/kCTK = grams emitted per thousand-cargo-tonne-kilometer]	13.7 (1.0)*	7.0 (0.51)*	13.3 (0.97)*	2.1 (1.0)*	0.53 (0.25)*	2.7 (1.28)*
Indexed comparison to move 30000 tonnes of cargo with a Seaway-size vessel	1 ship	301 rail cars	963 trucks	1 ship	301 rail cars	963 trucks
Indexed comparison to move 62000 tonnes of cargo with a Great Lake 1000-foot vessel (L≈305 m)	1 ship	564 rail cars	2340 trucks	1 ship	564 rail cars	2340 trucks

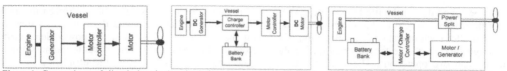

Figure 1. Comparison of diesel-electric (on the left) with serial hybrid (in the center) and parallel hybrid (on the right) electric marine propulsion systems. Source: Hybrid-UK website [6], Nov.2016.

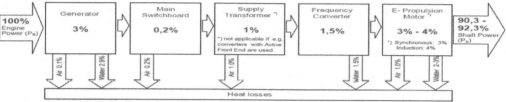

Figure 2. The typical losses observed on electrical components in standard Man diesel-electric propulsion plant. Source: MAN group guideline [3] and website [5], Nov. 2016.

New generation modern diesel-electric ships [5], [7], [8], [9], including DP class vessels, cruise ships, warships and icebreakers, usually use electric motors in pods called azimuth thrusters underneath to allow for 360° rotation, making the ships far more maneuverable. An example of this is the 'Harmony of the Seas', the largest passenger ship as of 2016.

Regarding hybrid electric marine propulsion system nowadays there are three basic hybrid configurations: diesel/electric, serial hybrid & parallel hybrid, and there are all available with many variations depends on the type of craft and how we are going to use it (i.e. [3], [6], [8]).

In diesel-electric configuration (see Figure 1 on the left) the engine (e.g. petrol or diesel) is connected directly to an electrical generator. From this point on the power in the system is transferred electrically to the propeller shaft via a motor controller and electric motor. The system may have multiple generators and multiple motors connected to a common electrical bus. This technology is used in diesel/electric trains and many large ships such as the 'Queen Mary 2'.

The 'Queen Mary 2' has a set of diesel engines in the bottom of the ship plus two gas turbines mounted near the main funnel, all are used for generating electrical power, including those used to drive the propellers. This provides a relatively simple way to use the high-speed, low-torque output of a turbine to drive a low-speed propeller, without the need for excessive reduction gearing. However, using the strict definition this configuration is not a hybrid yet as there is no electric storage of energy.

The serial hybrid system (see Figure 1 in the center) is very like the diesel/electric configuration in that it breaks the mechanical connection between the engine and propeller shaft. However, in this system a battery bank is also connected to the common electric power buss and finally because of such configuration, you can stop the engine if

needed and use the stored energy in the battery bank. With large batteries, you can have long periods of electric propulsion (and/or driving onboard electrical appliances) without resorting to the generator.

In such a case, a hybrid system seems to be very effective in operational costs and enables us to reduce fuel consumption and emissions. However, at this point it must be also noted, that there are always some losses of power in this configuration in the electric generator, batteries, motor controller and motor [3], [6]. In this situation if we require the same maximum power available at the shaft the engine needs to be bigger than nominated engine planned for such a ship to account for these losses (see Figure 2).

Additionally, the generator, motor controller and motor need to transmit the full power what makes them large and expensive items. In case of full power, the engine operates with relatively high efficiency so the hybrid does not provide any improvements. However, to get the same maximum operating power, we have sized the electrical components to operate at maximum engine power where the hybrid does not give any benefits. There is literally a very high price to pay for this action and explains why serial hybrid systems can be very expensive.

For example, in 2013, in Bremerhaven University of Applied Sciences the different hybrid drive concepts for a harbor tug and for a motor ferry was investigated with the following conclusions [4]: 'The use of hybrid drive systems on ships, in certain circumstances makes sense.' A hybrid drive concept offers the chance to achieve lower specific fuel consumption by adjusting power and speed. This action can have a big impact, especially in the partial load range.

'For the harbor tug the investment costs for the hybrid propulsion system are about € 820.000 higher compared to the diesel-electric system. These costs can only be saved during the lifetime of the system if the batteries are charged by the electricity of the public grid after each operation.' But there must be also enough time to charge the batteries during the break between two operations. The second type of hybrid ship that was investigated is a motor ferry, which operates every 20 min between Bremerhaven and Blexen/Nordenham. 'Compared to the harbor tug the time of full load operation is much higher. This is the main reason, that the hybrid propulsion system in such a situation does not save energy compared to the gas-mechanical system.'

A parallel hybrid configuration (see Figure 1 on the right) maintains the mechanical connection between the engine and propeller shaft. As its name implies, the electric motor acts on the drive shaft in parallel with the engine. The Power split is a mechanical device that allows transfer of power between its connections. In such a configuration, we can drive the propeller directly from the engine or from the electric motor or from both. We can also disconnect the propeller for a stand-alone generator function. During re-generation, the engine is disconnected.

In a parallel hybrid system, the motor & generator functions can be combined in one unit saving weight and cost ([5], [7], [8], [9]). The electric drive components are sized to the operating power where they can provide the most benefit, that is: low to mid speed cruising. Maximum power is supplied directly by the engine and the hybrid keeps out of the way.

A parallel hybrid thus gives improved efficiency at a high power compared to a serial hybrid. A parallel hybrid is also intrinsically more reliable than a serial system. If an electronic component breaks down, a serial system does not function. If a parallel hybrid breaks down, the engine can continue providing propulsion in the usual way. As the electric components are smaller and lighter, a parallel hybrid can be considerably lower in cost than a serial system. This provides a good cost/feature trade off and is why number of companies have decided to concentrate on developing parallel hybrid diesel electric systems [6], [2].

Another quite important question is where a hybrid system can help us to operate the engine with higher efficiency. There are two fundamental methods used to improve overall efficiency: by engine cycling and energy buffering [6].

To explain engine cycling we must imagine a situation where the engine is revving fast under low load. This equates to our ship operating at cruising speed. We are burning lots of fuel with very low fuel efficiency. In a car, in a similar situation we would instinctively go up a few gears up. For the same road speed the marine engine would then be turning at a slower RPM to provide power reduces with its speed

of rotation. So, changing gear has narrowed the gap between the amount of power the engine can supply (at that particular RPM) and the power demanded from it. This loads the engine closer to the maximum power curve and thus increased its efficiency.

So, in such a case all we need to do when cruising ship is to change up a few gears to improve fuel economy. The problem is that in our standard diesel marine engine craft we usually have only one forward gear. However, in a serial hybrid or diesel/electric system we have mechanically disconnected the engine from the propeller. We can now vary the engine speed to match the power required by the craft; the electronics converts this power to the voltage/current required by the motor to drive the shaft. Our electric transmission of power from engine to shaft performs the function of a continuously variable gear box. So, our hybrid gives us the ability to change gear and provides the same improvements in fuel economy as we can observe for the car. Engine cycling (that is varying its speed per load) is one way a hybrid can improve engine efficiency. This cycling is normally performed automatically by an intelligent control system.

Fuel (engine) efficiency can be defined as a distance in kilometers to move 1 tonne of cargo with 1 liter of fuel or optionally in shipping industry distance in nautical mile to move the ship (in ballast or in loaded condition) by burning 1 tonne of fuel. Adequately energy efficiency can be defined as the number of tonne of cargo to be move on 1kilometer per 1liter of fuel (see table 1).

Another method used to improve overall efficiency in hybrid electric propulsion system is energy buffering. Energy buffering can help us in the situation where the power requirement is low and regardless of the engine speed it cannot be operated efficiently. In such a specific situation, a hybrid can extract from the engine a higher power than required by propulsion. The extra energy supplied can then be stored (buffered) in the battery bank [3], [6]. Once the batteries are full we can turn the engine off and provide the low propulsion power demand with pure electric drive. When the batteries become depleted then we turn the engine back on again. The engine is thus operated in stop/start cycles and when it is running it is operating under a substantial load, this results in higher overall efficiency.

Offshore Support Vessels (OSV) or Anchor Handling Tug Supply (AHTS) vessels are good examples of offshore ships with high flexible power demand and, per consequence, different operation modes and sailing speeds. The operation profile is often split into two parts: one with a high propulsion power demand for transit and free sailing and another with a significantly lower power demand when operating the vessel in dynamic positioning mode and station keeping.

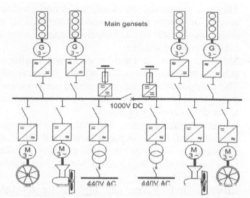

Figure 3. Hybrid Diesel-Electric configuration (redundant) of a PSV, with an energy-saving electric propulsion system with variable speed Gensets and energy storage sources offered by MAN Diesel and Turbo group. Source: MAN Diesel and Turbo group website [5], Nov. 2016.

In hybrid mode (see Figure 4 on the left), the electric machine is used as alternator as well as propulsion motor (PTO/PTI). This opens the way for flexible use of the main engine and the gensets. In electric/PTI-mode the propeller is driven with variable speed by a frequency converter. In PTO-mode the converter supplies a fixed voltage and frequency to the mains. The main engine and the shaft generator can operate in a range of 70% to 100% rpm. Doing so maximizes both propeller and engine efficiencies and also helps to reduce exhaust emissions [3].

In PTO-mode (see Figure 4 in the center) the main engine provides not only the power for ship propulsion, it also supplies the electric power needed for the ship's consumers. This mode is e.g. selected for transit sailing. It allows a high loading of the main engine, running with low specific fuel oil consumption and therefore minimal emissions. This often prolongs the maintenance period of the gensets as they can be switched off when not needed.

The PTI booster mode is mainly selected for maximum speed. Together with the main engine the electric machine works as an auxiliary motor, which delivers support to the propeller. The gensets deliver the electric power, both for propulsion and the vessel's consumers. The PTI booster mode increases mainly the flexibility of the propulsion system for peak loads. In addition, if extra power is needed, for bollard pull operation for example, this can be provided by the electric motor, using it for PTI boosting. The booster PTI can also be well-applied in ice operations.

Depending on the type of DP class ship, a diesel-electric propulsion system consists of four to six diesel generators, two main propulsions and separate engine rooms to ensure system redundancy and flexibility. In Siemens [9], MAN group [5] or Rolls Royce [8] diesel-electric propulsion systems the ships' propellers (screws) are turned by inverter-fed electric motors that get their energy from diesel-powered generators. A frequency converter controls the speed of the electric motors and propellers and because of this, the propeller can turn much more slowly. This configuration saves fuel because electric motors operate at high efficiency even at low speeds.

Depending on how much thrust is required, the converter adjusts the frequency and amplitude of the alternating current from the diesel generator. The ship's power demand determines also how many of the generators are running. Thus, since they produce only the energy that is needed and are not directly coupled to the speed of the screws, diesel engines can operate at a very high level of efficiency. For ships with a wide range of activities where the load on the propulsion system changes frequently, the hybrid system offers significantly lower fuel consumption, improved maneuverability and lower emissions compared to the other existing systems. Noise and vibration levels are also reduced, providing comfort to the crew [3]. Compared to purely diesel-powered vessels, diesel-electric ships are quieter and can use up to a third less fuel [8].

3 CONCLUSION

Economics constitutes the main reason why shipping is more efficient when it transports a comparatively huge amount of cargo. Based on research work presented in Table 1 we also know that marine transport is one of the most environment-friendly means of transportation. The Combined Great Lakes-Seaway Fleet for the example can move its cargo 59% farther (or is 59% more fuel-efficient) than rail and 873% farther (or is 873% more efficient) than trucks [1].

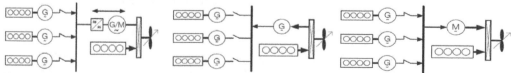

Figure 4. Comparison of hybrid diesel-electric mode (on the left) with PTO-mode (in the center) and PTI booster mode (on the right). Source: MAN Diesel & Turbo Group Guideline [3], Nov.2016.

In addition, when we take into consideration atmospheric pollutants - shipping emits 8.1 g/t-km of pollutants, trains emit 1.64 times the amount and trucks emissions are 8.07 times higher. These results also reflect the fact that the magnitude of technological change will be much greater for the marine mode than for the two ground modes such rail and truck. Therefore, shipping as a means of transport is projected to increase significantly in the future [1].

Regarding the advantages of new generation hybrid diesel-electric propulsion system the main ideas can be summarized as follows:

− Hybrids can help improve efficiency at cruising speed by engine cycling or buffering the energy from the engine and returning this to the drive shaft later [6].
− There is lower fuel consumption and emissions due to the possibility to optimize the loading of diesel engines / gensets [3]. The gensets in operation can run on high loads with high engine efficiency [5], [8]. This applies especially to vessels which have a large variation in power demand, for example for an offshore supply vessel, which divides its time between transit and station-keeping (DP) operation.
− We observe a better hydrodynamic efficiency of the propeller [5], [7], [8], [9]. Usually Diesel-electric propulsion plants operate a FP-propeller via a variable speed drive. As the propeller operates always on design pitch, in low speed sailing its efficiency is increased when running at lower revolution compared to a constant speed driven CP-propeller. This also contributes to a lower fuel consumption and less emission for a Diesel-electric propulsion plant [6].
− There is improved maneuverability and station-keeping ability, by deploying special propulsions such as azimuth thrusters or pods. Precise control of the electrical propulsion motors controlled by frequency converters enables accurate positioning accuracies.
− High reliability is noted, due to the multiple engine redundancy [3]. Even if an engine / genset malfunction, there will be sufficient power to operate the vessel safely. Reduced vulnerability to a single point of failure provides the basis to fulfill high redundancy requirements.
− We observe reduced life cycle cost, resulting from lower operational and maintenance costs [8].

− There is increased payload, as diesel-electric propulsion plants take less space compared to a diesel mechanical plant. Especially engine rooms can be designed shorter.
− More flexibility in location of diesel engine / gensets and propulsions is widely appreciated. The propulsions are supplied with electric power through cables. They do not need to be adjacent to the diesel engines / gensets.
− Lower propulsion noise and reduced vibrations are produced. For example, a slow speed E-motor allows to avoid the gearbox and propulsions like pods keep most of the structure bore noise outside of the hull [3].
− Efficient performance and high motor torques gives advantages for example in icy conditions [3], [8] as the electrical system can provide maximum torque also at low speeds.
− A hybrid can only improve overall efficiency if the losses incurred in the storage and return of energy is less than the improvements made in engine operating efficiency. This is a very important point as not every hybrid design can achieve this and some can even end up being less efficient than a standard diesel!

BIBLIOGRAPHY

[1] Environmental and Social Impacts of Marine Transport in the Great Lakes-St. Lawrence Seaway Region, Executive Summary prepared by Research and Traffic Group, January 2013, http://www.greatlakes-seaway.com, website dated November 2016.
[2] Global trends in renewable energy investment 2012, by REN21 (Renewable Energy Policy Network for the 21st Century); www.ren21.net/gsr, website dated November 2016.
[3] MAN Diesel & Turbo Group Guideline, Hybrid Propulsion Flexibility and maximum efficiency optimally combined; https://marine.man.eu, November 2016.
[4] Thorsten Völker, Hybrid Propulsion Concepts on Ships, University of Applied Sciences, Bremerhaven, Germany, 2013.
[5] https://marine.man.eu - MAN Diesel & Turbo Group website, accessed 15 Nov. 2016.
[6] http://www.hybrid-marine.co.uk, Hybrid Marine UK website, accessed 15 Nov. 2016.
[7] http://www.mhi.co.jp - Mitsubishi Heavy Industries website, accessed 15 Nov. 2016.
[8] http://www.rolls-royce.com – Roll- Royce website, accessed 15 Nov. 2016.
[9] http://www.siemens.com – Siemens group website, accessed 15 Nov. 2016.
[10] http://worldmaritimenews.com/ – World Maritime News website, accessed 15 Nov. 2016.

Ships Stability and Loading Strength

Assessment of the Realistic Range of Variation of Ship Equivalent Metacentric Height Governing Synchronous Roll Frequency

P. Krata & W. Wawrzyński
Gdynia Maritime University, Gdynia, Poland

ABSTRACT: The paper deals with the characteristics of the equivalent metacentric height of ships which governs the frequency of the resonance mode of ship motion called the synchronous roll. The frequency of synchronous roll can be effectively predicted on the basis of the equivalent metacentric height which is amplitude of roll depended. This frequency can be considerably different than usually estimated with the use of the initial metacentric height. The analyzed problem covers the need for realistic assessment of the range of variation of this equivalent metacentric height. Ship operators should be aware of significance of the expected GM_{eq} changes, since consequently the frequency of potentially dangerous synchronous roll occurs at different wave encounter frequency than predicted on the basis of the initial metacentric height GM_0 recommended nowadays by IMO for the routine on-board use. Numerous ships in their typical loading conditions are studied to obtain the basis for statistical analysis of the equivalent metacentric height variation.

1 INTRODUCTION

The ship buoyancy and her stability are two main features determining the feasibility of any operation when in a port or at seaway. The buoyancy-related problems are relatively simple since the shape of the submerged part of a hull does not play any role and only the volume of displaced water needs to be sufficient to support the weight of the ship. Moreover, the system is self-regulating and the natural draft adjusting is an inherent feature of the fluid – floating body interaction.

In the contrast to the buoyancy, the ship stability is a more complex matter and it has to be considered both at the design stage and during the routine operation. The main potential threat is related to excessive roll motion and the large amplitudes of roll. Although the large amplitudes of ship roll are highly unlikely during cargo handling in a port (Krata *et al.* 2013), the dangerous excessive rolling may happen at seaway (Kobylinski & Kastner 2003), thus then proper stability assessment should be applied as a routine part of ship management. Historically, the first approach towards ship stability evaluation was based on an experiment only. Till the time of boat launching one could not be sure the design correctness so the design process had to be based on architect's and seamen's experience and on the earlier projects similitude. Since the mid-

eighteenth century the development of researches in the field of ship stability is noticed. The studies of numerous scientists laid the groundwork for deeper understanding of the stability notion and characteristics. Primarily, the works of Bouguer from 1746, Euler from 1749, Bernoulli from 1757 and Attwood from 1796 are recognized as the foundations of the scientific approach towards ship stability related issues. Bouguer introduced the metacentric height notion which remains the only measure utilized to assess ship stability till the end of the nineteenth century (Kobylinski & Kastner 2003). In the mid-nineteenth century the righting arm curve was already known and thanks to Moseley's publication from 1850 also the concept of dynamical stability was introduced. Although, those theoretical achievements were not fully applied in practice and none stability standards were elaborated at that time (Kobylinski & Kastner 2003).

The significant development was noticed in the middle of the twentieth century thanks to Jaakko Rahola's dissertation from 1939 (Rahola 1939). He statistically obtained the margins between the sufficient and insufficient stability characteristics. He analyzed the initial metacentric height and the shape of the righting arm curve including some areas under this curve (Womack 2002). In the 60's under the auspices of the International Maritime Organization the contemporary ship stability standards were developed, mainly covering Rahola's results. In the 80's they were supplemented by the weather criterion inspired by the Japanese proposal and researches (Francescutto 2009).

Regrettably, the decades of experience revealed numerous stability-related incidents resulted from dynamic phenomena taking place in rough sea which are not covered by the existing stability standards. One of the response was the publication in 1995 the IMO MSC/Circ.707 which was then significantly enhanced and issued in 2007 as IMO MSC.1/Circ.1228 (IMO 2007). This *Revised guidance to the master for avoiding dangerous situations in adverse weather and sea conditions* is intended to give some help to ship masters when sailing in stormy conditions. The relatively up to date publication (in comparison to Rahola's findings from 1939 being the foundation of the *Intact Stability Code*) contains a set of remarks and advices regarding the avoidance of following dangerous dynamical phenomena at sea like surf-riding and broaching-to; reduction of intact stability when riding a wave crest amidships; synchronous rolling; parametric roll motions (IMO 2007). Some of them refer to the natural frequency of ship roll which is essential in the light of this paper.

Another reaction of the marine society was a new concept of the second generation ship stability criteria and then the risk-based ship stability approach (Kobylinski 2014). The works on both of these approaches are still ongoing (Hinz *et al.* 2016) and the final date of completion is not known, so the most reasonable approach towards ship stability nowadays seems to be complying with the *Intact Stability Code* requirements (which are obligatory) and also following the MSC.1/Circ.1228 recommendations. There are numerous software tools described in the literature and some of them are also available on the market, designed to help navigators to calculate the ship courses prone to arouse the synchronous roll according to the *Revised guidance to the master for avoiding dangerous situations in adverse weather and sea conditions* (Krata & Szlapczynska 2011, Ami seaware, Amarcon).

The application of the MSC.1/Circ.1228 guidance is strictly related to the need for the natural frequency of ship roll estimation which should be carried out with the use of any available and reliable method. One of such method is elaborated in 2016 and it applies the equivalent metacentric height of the ship which depends on the actual amplitude of roll (Wawrzynski & Krata 2016a). The pending paper deals with characteristics of this equivalent metacentric height which governs the possible range of frequency of the synchronous roll.

The rest of the paper is organized as follows. Section 2 introduces the background of the ship roll motion. Section 3 discusses results of the research,

e.g. variations of the equivalent metacentric height. Finally, Section 4 presents summary and conclusions.

2 NONLINEAR DYNAMICAL SYSTEM DESCRIBING ROLL MOTION OF THE SHIP

The ship rolling is described as a nonlinear dynamical system. The most essential parameter determining the nonlinearity of rolling is the restoring moment which is generally described by the righting arm curve (the *GZ* curve). The roll equation applied to solve the ship rolling problem is formulated as follows:

$$\ddot{\phi} + 2\mu \cdot \dot{\phi} + \frac{g}{r_x^2} GZ(\phi) = \xi_w cos(\omega_e t). \qquad (1)$$

where μ is the damping coefficient, g is the gravity acceleration, r_x is the gyration radius of a ship and added masses (which is assumed to be constant for the sake of simplicity), GZ is the righting arm, ξ_w means the exciting moment coefficient and ω_e denotes external moment frequency.

The roll equation (1) was thoroughly studied in (Wawrzynski & Krata 2016b) in the context of the natural frequency of the ship roll prediction. The Runge-Kutta method was applied to solve it with the initial conditions set to $\phi_{initial}=0$ and $\omega_{initial}=0$ and a range of the harmonic external exciting moment. The Matlab script was prepared to carry out the calculations and the results were compared to the exemplary ones obtained with the use of the build-in solver of the differential equations available in the Mathematica package. That research clearly revealed that the nonlinearities of roll motion are attributed not only to damping but the nonlinearity of *GZ* curve could have significant effect on roll period. This result may find the reference in the literature, for instance in (Falzarano & Taz Ul Mulk, 1994), (Contento *et al.*, 1996) and (Francescutto & Contento, 1999).

In nonlinear dynamical systems, including the rolling of the ship, the nonlinear oscillations usually appear. The typical phenomenon related to nonlinear oscillations is the resonance frequency dependence on the amplitude of the oscillations. The exemplary cases can be seen in figure 1 where the roll spectra are determined on the basis of the equation (1) with the use of two approaches: first, the linear righting arm which is written in the form $GZ(\phi)=GM_0\cdot\phi$ (left plot) and second, the actual nonlinear characteristics of the righting arm. The resonance frequencies trace the maxima of the roll spectra obtained for the increasing value of the excitation moment. It can be clearly seen that the simplified approach (left plot) produces the maxima appearing for one constant value of the roll frequency, while the more accurate approach, taking into account the actual nonlinearity of the *GZ* curve, results in more complex pattern of the roll spectra maxima track (right plot). The value of the ship natural roll frequency strictly depends on the amplitude of roll which is omitted nowadays in the IMO recommendations provided in the MSC.1/Circ.1228 guidance for ship masters.

When the linear righting arm characteristic is considered the maxima of roll spectra reflect the so called natural frequency of ship roll which can be easy calculated according to the well known formula provided in the *Intact Stability Code* (IMO 2008):

$$\omega_r = \frac{2\pi}{\tau}. \tag{2}$$

where

$$\tau = \frac{2 \cdot c \cdot B}{\sqrt{GM_0}}. \text{ with the value of } c \text{ coefficient:}$$

$$c = 0.373 + 0.023 \cdot \frac{B}{T} - 0.00043L. \tag{3}$$

where τ is the natural roll period, GM_0 is the initial metacentric height, c is the coefficient describing transverse gyration radius r_x ($r_x=c \cdot B$), B is the breadth, L is the length at the waterline and T is the mean draft of the ship.

Although the formulae (2) and (3) are approximate ones they still remain in common use

on-board for the purpose of IMO MSC.1/Circ.1228 guidance application. The problem is that the application of the formulae (2) and (3) yields a rough estimation of roll period rather than calculation of the precise values, however deck officers facing the lack of alternative simple and reliable formulae still follow the approximate ones and they use the results as exact periods.

In fig. 1 the line of resonance frequency calculated with the use of the GM_0 is marked „GM0" and it can be seen that the maxima of all presented curves coincide with the „GM0" line presented in the left plot while they significantly differs in right plot of figure 1.

The formula (3) is based on the GM_0, which is the parameter of the initial stability and it is valid only up to about 0.1 rad. The calculation of the natural roll frequency with the use of IMO method assumes that the righting arms are linearized according to the formula: $GM_0 \cdot \phi$. Despite the fact that in some cases, such a simplification can be used up to 0.3-0.4 rad (Fig. 2, left plot) and rarely even up to 1.0 rad (Fig. 2, right plot), it is a far-reaching simplification in majority of cases. Very often clear discrepancies occur already at 0.2 rad (Fig. 3 left plot) and sometimes they start even from the onset of the *GZ* curve (Fig. 3 right plot).

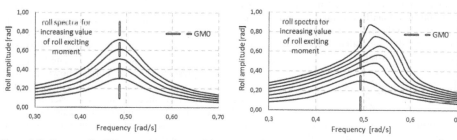

Figure 1. Roll spectra for different values of an exciting moment for the LNG carrier (T=12.00 m, GM_0=5.00 m) calculated with the linear restoring arm (left plot) and actual nonlinear shape of the *GZ* curve (right plot)

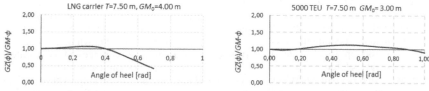

Figure 2. Cases of good convergence between the *GZ* curve and the righting arms linearized according to the formula: $GM_0 \cdot \phi$

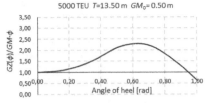

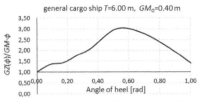

Figure 3. Cases of significant discrepancy between the GZ curve and the righting arms linearized according to the formula: $GM_0 \cdot \phi$

The roll period and the resonance frequency, taking the nonlinearity of the GZ curve and the roll amplitude into account, can be calculated according to the method proposed by Wawrzynski and Krata, (Wawrzynski & Krata 2016a). The method is based on the area under the GZ curve and the average inclination of the tangent line to the GZ curve, both calculated from zero up to the roll amplitude. The main formula looks very similar to the IMO's one:

$$\tau(\phi_A) = \frac{2 \cdot c \cdot B}{\sqrt{GMeq(\phi_A)}} . \tag{4}$$

where the value of c coefficient remains the same as given in the formula (3) and $GMeq(\phi_A)$ denotes the equivalent metacentric height for a specified roll amplitude ϕ_A.

The formula for the equivalent metacentric height calculation is thoroughly discussed in (Wawrzynski & Krata, 2016b). Generally, the approach is based on the energy balance method suggested in (Kruger & Kluwe, 2008) and additionally on the averaging of the GZ nonlinearities in the angle of heel domain. The final formula for the equivalent metacentric height calculation is the following:

$$GMeq(\phi_A) = \frac{\int_0^{\phi_A} GZ(\phi) d\phi}{\phi_A^2} + \frac{GZ_{\phi_A}}{2 \cdot \phi_A} \tag{5}$$

According to formula (5) both the value of the equivalent metacentric height GM_{eq} and as a consequence the roll period, depend on the roll amplitude. This makes up a noteworthy difference comparing to the original IMO recommended formula.

The verification of this method was performed for seven ships in different loading conditions (total number of analyzed cases add up to 30), and it revealed very good consistency of the roll period predicted by the formulae (4) and (5) with the results of numerical simulations for a wide range of roll amplitudes (Wawrzynski & Krata, 2016b).

Due to the combination with formulae (4) and (5), the roll resonance frequency also depends on the amplitude:

$$\omega_r(\phi_A) = \frac{2\pi}{\tau(\phi_A)} . \tag{6}$$

Formula (6) combined with formulae (4) and (5) was extensively analyzed (Wawrzynski & Krata 2016b). In the course of those investigations, the final form of the formula (6) was determined:

$$\omega_r^2(\phi_A) = \left(\frac{2\pi}{\tau(\phi_A)}\right)^2 - 2\mu^2 \tag{7}$$

The elaborated formulae allow calculating the roll resonance frequency as a function of amplitude and further they are called the "GM_{eq} method". The influence of the GZ curve nonlinearity affecting the roll characteristics with the increase in roll amplitude can be conveniently presented in the form of the resonance curve which is also called the backbone curve. Graphically, the resonance frequencies for consecutive amplitudes are presented in the form of the resonance curve which is marked "GMeq" in the following figures. In the research (Wawrzynski & Krata, 2016b) the accuracy of the GM_{eq} method was examined for different stability characteristics. In most cases, the method reveals very good agreement with the results of rolling simulations. In Fig. 4, the roll spectra for different values of an exciting moment, calculated for two ships, are presented. Each curve shows the roll amplitude versus roll frequency with the maximum reflecting the resonance frequency of oscillations. It can be seen, that the resonance curve calculated with the use of the GM_{eq} method accurately traces the maxima of all curves.

However, for the loading conditions where the simulation results reveal roll amplitude bifurcations the resonance curve obtained from GM_{eq} method is not very accurate (Fig. 5) whereas it still performs far better than the IMO recommended formula (3). Although, the bifurcation is known as the relatively difficult problem to model (Francescutto & Contento 1999). The observed inaccuracy is found in regions with significant jumps of the amplitude of roll. In Fig. 5, in both cases a significant shift of the resonance curve to the left hand side, e.g. towards lower values of the roll frequency, is noticed and it can be seen that there is an inaccuracy of the resonance curve which rises together with the increase in jump value.

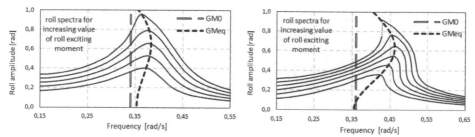

Figure 4. Roll spectra and resonance curves for the 5000 TEU Panamax container ship (T=7.50 m, GM_0=1.50 m) (left plot) and for the general cargo ship (T=6.00 m, GM_0=1.00 m) (right plot)

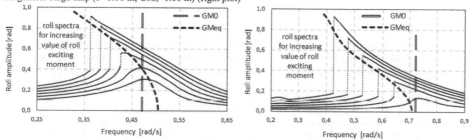

Figure 5. Roll spectra and resonance curves for the LNG carrier (T=7.50 m, GM_0=6.00 m) (left plot) and the offshore support vessel (T=6.10 m, GM_0=2.50 m) (right plot)

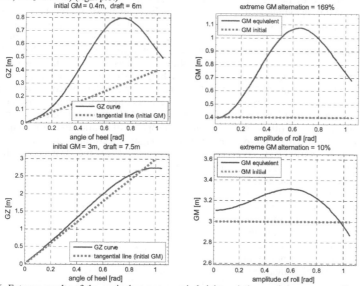

Figure 6. Extreme results of the equivalent metacentric height variations and the corresponding GZ curves for the general cargo ship (upper plots) and for the 5000 TEU Panamax container ship (lower plots)

Table 1. Main particulars of ships considered in the course of the research.

| Type of vessel | length [m] | breadth [m] | considered draft T | | metacentric height GM_0 | |
			min [m]	max [m]	min [m]	max [m]
1 general cargo ship	140.00	22.00	6.00	9.00	0.40	1.00
2 bulk carrier	156.10	25.90	6.50	10.50	1.00	3.50
3 LNG carrier	278.80	42.60	7.50	12.00	1.00	5.00
4 5000 TEU Panamax container ship	283.20	32.20	7.50	13.50	0.50	3.00
5 7500 TEU container ship	285.00	45.60	7.50	12.50	2.00	5.00
6 motor tanker	320.00	58.00	10.00	22.00	5.00	10.00

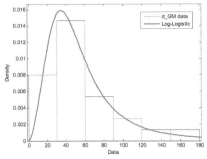 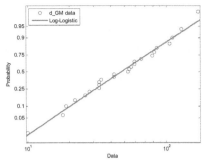

Figure 7. PDF distribution fitting over the histogram of the equivalent metacentric height (left plot) and the probability plot for Log-Logistic distribution (right plot)

3 VARIATION OF THE EQUIVALENT METACENTRIC HEIGHT WITH THE INCREASE IN SHIP ROLL AMPLITUDE

As the roll amplitude depended equivalent metacentric height governs the natural frequency of ship roll, the accuracy of synchronous roll prediction is affected. Thus, the fully justified question is related to the realistic extent of the equivalent metacentric height variations. In case of negligible change in GM_{eq} value the IMO recommended formula (3) might be still in use, otherwise the formula (3) should be replaced by the formula (4) or (7) to address the problem of the accurate prediction of possible synchronous roll of the ship.

To assess this crucial matter of GM_{eq} variation, the series of calculations was carried out aiming at the typical loading conditions of numerous cargo vessels. Six different ships were taken into account to cover a variety of ship purpose, their different particulars and loading condition from ballast up to fully loaded ones. The main particulars of considered ships are given in table 1.

The equivalent metacentric height was calculated according to formula (5) for all the vessels listed in the table 1. The wide range of possible amplitudes of roll was taken into account. The performed series of calculations revealed that the equivalent value of the metacentric height may vary significantly according to the amplitude of roll in many cases but not in all of them. Two extreme examples of calculation results are shown in Fig. 6.

It can be clearly seen that the variation of the equivalent metacentric height firmly depends on the shape of the GZ curve. This variation is rather typical for the vessels in considered typical loading conditions, however the alternation of the GM_{eq} value comparing to the initial GM ranges from unimportant 10% up to massive 169%. Thus, the further analysis is carried out.

All the obtained variations of the equivalent metacentric height for the considered cases are gathered as one data set and the distribution of the probability density function (PDF) is fitted (Fig. 7

left plot). The best performing distribution is Log-Logistic with the mean value equals 61.4% which is significant in terms of synchronous roll prediction. The satisfactory goodness of fit is presented in Fig. 7 in the right plot.

On the basis of the Log-Logistic distribution of PDF the more convenient cumulative distribution function is calculated (Fig. 8). This enables easy calculation of probability of exceeding a predefined level GM_{eq} variation.

The predefined levels of the GM_{eq} variation are set up to 50% and 100% (see Fig. 8). According to the obtained data set in about 46% of analyzed cases the equivalent metacentric height reaches for some roll amplitudes at least 1.5 the initial value of the metacentric height. Moreover, in about 13% of cases the GM_{eq} reaches the value twice as large as the GM_0. Such results confirm the significance of proper modeling of the GZ curve nonlinearity and they support the proposal of application of the equivalent metacentric height (according to the elaborated formula (5)) instead of the IMO recommended formula (3).

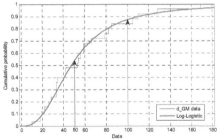

Figure 8. Cumulative distribution function for the equivalent metacentric height variation

4 SUMMARY AND CONCLUSION

The problem of the ship synchronous roll prediction is discussed in the paper. It is emphasized that the nonlinearity of the GZ curve plays a vital role in roll modeling while the contemporary recommended simplified formula for the natural roll frequency estimation omits this nonlinearity at all. To address the problem the equivalent metacentric height is proposed to be used instead of the initial GM. It is revealed that such modification of the natural roll frequency evaluation method makes this frequency depended on the actual roll amplitude of the ship. Such an approach induce the question how significant may be the variation of the equivalent metacentric height and whether it may be neglected or not. Thus, the study is focused on the assessment of range of variation of the equivalent metacentric height for numerous different types' ships sailing in realistic loading conditions.

The research reveals that in almost half of examined cases the equivalent metacentric height may significantly vary comparing to the initial GM even by 50%. Moreover, in one eighth of cases the GM_{eq} is even twice as large as the initial GM. In the light of the obtained results the conclusion is clear that the variation of the equivalent metacentric height governing the natural frequency of roll is significant enough to take it serious into account when applying the recommendations of the IMO MSC.1/Circ.1228 *Revised guidance to the master for avoiding dangerous situations in adverse weather and sea conditions*. This publication is intended to provide help to ship masters although it can be really helpful and accurate in case of accurate estimation of the natural frequency of roll. Therefore, the contemporary use of the initial metacentric height should be replaced in formula (3) by the equivalent metacentric height calculated according to formula (5).

The described adoption of the equivalent metacentric height to the formula enabling the ship natural period of roll estimation, facilitates the application of trustworthy set of dynamic constraints to the route planning algorithm. It should be also capable to make the on-board assessment of synchronous roll likelihood more accurate which would make the marine navigation just safer.

REFERENCES

AMI seaWARE EnRoute Live Onboard real-time seakeeping guidance, Aerospace and Marine International, http://www.amiwx.com/pdf/AMI_seaware_EnRoute_Live.pdf

Contento G., Francescutto A., Piciullo M., *On the Effectiveness of Constant Coefficients Roll Motion Equation*, Ocean Eng. 23, 1996, pp. 597-618

Falzarano J., Taz Ul Mulk M., *Large Amplitude Rolling Motion of an Ocean Survey Vessel*, Marine Technology, Vo. 31, 1994, pp. 278-285

Francescutto, A., Contento, G., *Bifurcations in ship rolling: experimental results and parameter identification technique*, Ocean Eng. 26, 1999, pp. 1095-1123

Francescutto A., *The intact ship stability code: present status and future developments*, Marine Technology 2009

Hinz T., Krata P., Montewka J., *A Meta-model for Risk Assessment of RoPax Capsizing as an Alternative Way of Ship Safety Evaluation*, Proceedings of the 6th International Maritime Conference on Design for Safety – DNV GL, November 28 - November 30 2016, Hamburg, Germany

IMO, *Intact Stability Code*, 2008

IMO MSC.1/Circ.1228, *Revised guidance to the master for avoiding dangerous situations in adverse weather and sea conditions*, 2007

Kobyliński L., Kastner S., Stability and Safety of Ships, Vol. 1: Regulation and Operation, Elsevier Ocean Engineering Books, Vol. 9, 2003

Kobylinski L., *Stability Criteria - Present Status and Perspectives of Improvement*, TransNav, the International Journal on Marine Navigation and Safety of Sea Transportation 8, No. 2 (2014), pp. 281–86

Krata P., Szlapczynska J., *Weather Hazard Avoidance in Modeling Safety of Motor-driven Ship for Multicriteria Weather Routing*, In A. Weintrit A, T. Neumann (eds.), Methods and Algorithms in Navigation - Marine Navigation and Safety of Sea Transportation, 2011, pp. 165-172

Krata P., Szpytko J., Weintrit A., *Modelling of Ship's Heeling and Rolling for the Purpose of Gantry Control Improvement in the Course of Cargo Handling Operations in Sea Ports*, Mechatronic Systems And Materials IV, Solid State Phenomena, Volume 198 (2013), pp. 539-546

Kruger S., F. Kluwe, *A simplified method for the estimation of the natural roll frequency of ships in heavy weather*, Technical University Hamburg, 2008

OCTOPUS-ONBOARD The new generation decision-making support system to optimize ship performance in waves, AMARCON B.V., www.amarcon.com

Rahola J., *The Judging of the Stability of Ships and the Determination of the Minimum Amount of Stability - Especially Considering the Vessels Navigating Finnish Waters*, Technical University of Finland, 1939

Wawrzynski, W., Krata, P. [a], *Method for ship's rolling period prediction with regard to non-linearity of GZ curve*, Journal of Theoretical and Applied Mechanics, Vol 54, no 4 (2016), pp. 1329-1343

Wawrzynski, W., Krata, P. [b], *On ship roll resonance frequency*, Ocean Eng. 126, 2016, pp. 92-114

Womack J., *Small commercial fishing vessel stability analysis, Where are we now? Where are we going?*, Proceedings of the 6th International Ship Stability Workshop, Webb Institute, New York, 2002

Buckling Strength of Rectangular Plates with Elastically Restrained Edges Subjected to In-plane Impact Loading

B. Yang & D.Y. Wang

Shanghai Jiao Tong University, Shanghai, China

ABSTRACT: The dynamic buckling of rectangular plates with the elastically restrained edges subjected to in-plane impact loading is investigated. Budiansky–Hutchinson criterion is employed for calculation of dynamic buckling loads. The displacement function concluding the elastically restrained boundary condition is expressed as Navier's double Fourier series. In order to solve the large deformation equations of plate, Galerkin method is applied. Also, the nonlinear coupled time integration of the governing equation of plate is solved by using fourth-order Runge–Kutta method. The correctness of the method presented in the paper has been validated by comparing the results with the published literatures. It is proved that the rotational restraint stiffness that is usually ignored by previous researchers plays an important role in dynamic response and dynamic buckling of the rectangular plates subjected to in-plane impact loading. Furthermore, the influence of the other parameters (initial imperfections, impact duration and geometric dimensions) on the dynamic response and dynamic buckling are studied in detail.

1 INTRODUCTION

Since ship plate are usually thin-walled, therefore they are susceptible to buckling if the applied compressive loads reach a dynamic value which is influenced by a variety of parameters including the plate dimensions, initial imperfections, impact duration and boundary conditions. Stiffened plating in ships is supported by various types of members along the edges, which have a finite value of the rotational restraint stiffness. Depending on the rotational restraint stiffness of support members, the rotation along the plate edges will to some extent be restrained. When the rotational restraint stiffness equal to zero, the boundary condition can be regarded as simply supported case; When the rotational restraint stiffness are infinite, the boundary condition can be regarded as clamped case. Most current practical design criteria from classification societies for the buckling and ultimate strength of ship plate are based on boundary conditions in which all (four) edges are simply supported. In the real engineering field, however, the idealized simply supported or clamped boundary conditions never exist because of finite rotational restraints. Therefore, it is of great importance and significance to study buckling behavior of rectangular plates with elastically restrained edges.

As for ultimate strength of stiffened plate, lots of researchers have proposed predicting formulas and buckling modes, such as (Zhang, 2009). In these formulas, influence factors are slenderness of plate between stiffeners and slenderness of combination of stiffener and attached plating. Buckling modes include stiffener buckling, plate buckling and overall stiffened plate buckling. These buckling modes are neither excluding nor independent; some of them may occur simultaneously. Applying enforced displacement with very slow velocity, which can be considered as quasi-static process without inertia effect, ultimate load of stiffened plate can be obtained when reacting load reaches largest value, followed by a decline process of the reacting load.

In terms of dynamic buckling or collapse strength of the rectangular plates, due to the load is the dependence of time, the subject is more complicated than static ultimate strength. In this condition, there are some additional issues required to be considered, such as inertia effect and dynamic constitutive equation of the material. Some research works have been investigated. Cui et al. (1999) pointed that dynamic buckling always took place elastically for the types of rectangular plates tested under fluid-solid slamming. They also found that enhancing boundary constraint was a useful method to strengthen the plates' capability to resist dynamic buckling. A series of dynamic collapse tests were carried out on rectangular plates under axial compressive loads by Paik (2003), with different loading speeds. The experiment results shows that the ultimate compressive strength and in-plane stiffness of flat plates both increase with increases of the loading speed, this due to the fact that the strain rate of the material increases with increasing loading speed. So Ji and Wang (2014) studied influence of load shape on dynamic deflection response of across-stiffened plate subjected to in-plane impact loading by using Abaqus/explicit FEM code, considering strain rate effect and strain hardening. Four parameters of load shape such as peak load, impact duration, the type of decaying after reaching the peak load, and ratio between rise time of the load and total impact duration are considered. Axial residual displacement of the impacted end is chosen as main detected factor of dynamic response of the plate. Representative failure modes after impact are also presented in the paper. However, initial imperfection of the panel is not considered in the paper. Kubiak (2013) investigated the buckling and post buckling response of thin plates and thin-walled structures with flat walls subjected to static and dynamic loading. It can be clearly known that the boundary conditions adopted in above literatures are simply-support or clamped-support. However, it is well known that there are no absolute simply supported edges and clamped edges in engineering field. Therefore, the evaluated dynamic buckling loads are underestimated in the case of simply supported boundary conditions and overestimated in the case of clamped boundary conditions. In design process, if regard the current results as design criteria, it may make the design values be conservative and lead to the waste of resource. It is very valuable to study the buckling behavior of rectangular plats with elastically restrained edges.

Since stiffeners provide a certain amount of rotational restraint stiffness, they can be represented as De Saint Venant torsion bars. It has been proved that the approximation is correct if the stiffeners don't buckle or distort before the local buckling happened in the panel. This method is presented by Paik and Thayamballi (2000) and it has the advantages of fast estimates of buckling loads and good accuracy. But this method applies only to isotropic materials. Tarján and Kollár (2010) proposed approximate expressions to determine the lowest buckling load of short and long composite plates which are elastically restrained at the four edges, but this method can only compute the critical buckling loads. Ahmad (2013) compared the theoretical expression in published literature and rule-based expressions to evaluate the elastic buckling load of T-bar stiffened plates and to assess the interactions of different buckling modes. From above published literatures, it can be clearly seen

that the buckling behavior of plate with elastically restraint edges are studied by some scholars. But these methods only apply to analyze the buckling behavior of plate under static loads.

To the authors' knowledge, there are no systematical studies on dynamic buckling of rectangular plate with elastically restraint boundary conditions under dynamic loads. All results obtained from the published literatures are based on simplified boundary conditions. So there are some errors between the current results and real results. In order to obtain more accurate results of dynamic response and dynamic buckling, it is necessary to investigate the dynamic buckling behavior of rectangular plate with elastically restraint edges under in-plane impact loading.

The objective of the present work is to utilize the large deformation theory of thin plate with Galerkin method for the dynamic response and dynamic buckling analyses of rectangular plates with elastically restraint edges subjected to in-plane impact loading. Comparison studies are carried out in order to validate the accuracy and correctness of the method for solving the aforementioned problems. Extensive new numerical results are presented to investigate the effect of various parameters (rotational restraint stiffness, initial imperfections, impact durations and geometric dimensions) on the dynamic response and dynamic buckling behavior of the plates.

2 MATHEMATICAL MODEL

The dynamic buckling of rectangular plates with elastically restrained edges subjected to impact loading is investigated. The rectangular plate is modeled as a plate elastically restrained along unloaded edges and simply supported along loaded edges, as shown in Figure 1.

2.1 Governing equations

Using the straight-line method and considering the initial defects, the strain at any point in the plate is expressed as follow

$$
\left.\begin{aligned}
\varepsilon_x &= \varepsilon_x^0 - z\frac{\partial^2 (w - w_0)}{\partial x^2} \\
\varepsilon_y &= \varepsilon_y^0 - z\frac{\partial^2 (w - w_0)}{\partial y^2} \\
\gamma_{xy} &= \gamma_{xy}^0 - 2z\frac{\partial^2 (w - w_0)}{\partial x \partial y}
\end{aligned}\right\} \quad \left(-\frac{h_p}{2} \le z \le \frac{h_p}{2}\right) \quad (1)
$$

where, $\varepsilon_x^0, \varepsilon_y^0, \gamma_{xy}^0$ are the strains in the middle surface of the plate; w is the total deflection along z-direction in the middle surface of the plate; w_0 is the initial imperfections of the plate; h_p is the thickness of the plate.

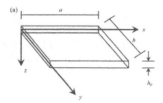

(a)

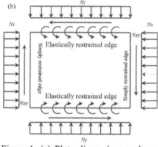

(b)

Figure 1. (a) Plate dimensions and coordinates; (b) rotational-restrained plates under uniaxial compression.

Considering the initial imperfections the following equations can be obtained:

$$\begin{cases} \varepsilon_x^0 = \dfrac{\partial u}{\partial x} + \dfrac{1}{2}\left(\dfrac{\partial w}{\partial x}\right)^2 - \dfrac{1}{2}\left(\dfrac{\partial w_0}{\partial x}\right)^2 \\[2mm] \varepsilon_x^0 = \dfrac{\partial v}{\partial y} + \dfrac{1}{2}\left(\dfrac{\partial w}{\partial y}\right)^2 - \dfrac{1}{2}\left(\dfrac{\partial w_0}{\partial y}\right)^2 \\[2mm] \gamma_{xy}^0 = \dfrac{\partial u}{\partial y} + \dfrac{\partial v}{\partial x} + \dfrac{\partial w}{\partial x}\dfrac{\partial w}{\partial y} - \dfrac{\partial w_0}{\partial x}\dfrac{\partial w_0}{\partial y} \end{cases} \tag{2}$$

The force and torque in the middle-surface of the plate can be represented as:

$$\begin{cases} N_x = \dfrac{Eh_p}{1-v^2}\left(\varepsilon_x^0 + v\varepsilon_y^0\right), M_x = -D\left[\dfrac{\partial^2(w-w_0)}{\partial x^2} + v\dfrac{\partial^2(w-w_0)}{\partial y^2}\right] \\[2mm] N_y = \dfrac{Eh_p}{1-v^2}\left(\varepsilon_y^0 + v\varepsilon_x^0\right), M_y = -D\left[\dfrac{\partial^2(w-w_0)}{\partial y^2} + v\dfrac{\partial^2(w-w_0)}{\partial x^2}\right] \\[2mm] N_{xy} = \dfrac{Eh_p}{2(1+v)}\gamma_{xy}^0, \qquad M_{xy} = -D(1-v)\dfrac{\partial^2(w-w_0)}{\partial x\partial y} \end{cases} \tag{3}$$

where, $D = \dfrac{Eh_p^3}{12(1+v)}$, E is the elastic modulus of the plate and v is the Poisson's ratio. The relationship between the middle-surface strains and internal force derived from the Eq. (3) are as follows:

$$\begin{aligned} \varepsilon_x^0 &= \dfrac{1}{Eh_p}\left(N_x - vN_y\right) \\[2mm] \varepsilon_y^0 &= \dfrac{1}{Eh_p}\left(N_y - vN_x\right) \\[2mm] \gamma_{xy}^0 &= \dfrac{2(1+v)}{Eh_p}N_{xy} \end{aligned} \tag{4}$$

According to the large deformation theory of plate, the membrane forces can't be ignored. Introduce Airy's stress function, F, which satisfies

$$N_x = \dfrac{\partial^2 F}{\partial y^2}, \ N_y = \dfrac{\partial^2 F}{\partial x^2}, \ N_{xy} = -\dfrac{\partial^2 F}{\partial x\partial y} \tag{5}$$

Substituting Eq. (5) into Eq. (4) yields the middle-surface strains' representation expressed by stress function.

$$\begin{aligned} \varepsilon_x^0 &= \dfrac{1}{Eh_p}\left(\dfrac{\partial^2 F}{\partial y^2} - v\dfrac{\partial^2 F}{\partial x^2}\right) \\[2mm] \varepsilon_y^0 &= \dfrac{1}{Eh_p}\left(\dfrac{\partial^2 F}{\partial x^2} - v\dfrac{\partial^2 F}{\partial y^2}\right) \\[2mm] \gamma_{xy}^0 &= \dfrac{2(1+v)}{Eh_p}\dfrac{\partial^2 F}{\partial x\partial y} \end{aligned} \tag{6}$$

Neglecting the terms of rotary inertia of the plate the following equation of motion can be obtained.

$$\begin{cases} \dfrac{\partial N_x}{\partial x} + \dfrac{\partial N_{xy}}{\partial y} = \rho h_p \dfrac{\partial^2 u}{\partial t^2} \\[2mm] \dfrac{\partial N_y}{\partial y} + \dfrac{\partial N_{xy}}{\partial x} = \rho h_p \dfrac{\partial^2 v}{\partial t^2} \\[2mm] \dfrac{\partial^2 M_x}{\partial x^2} + 2\dfrac{\partial^2 M_{xy}}{\partial x\partial y} + \dfrac{\partial^2 M_y}{\partial y^2} + \dfrac{\partial}{\partial x}\left(N_x\dfrac{\partial w}{\partial x} + N_{xy}\dfrac{\partial w}{\partial y}\right) \\[2mm] + \dfrac{\partial}{\partial y}\left(N_y\dfrac{\partial w}{\partial y} + N_{xy}\dfrac{\partial w}{\partial x}\right) = \rho h_p \dfrac{\partial^2 w}{\partial t^2} \end{cases} \tag{7}$$

Neglecting the terms of plane inertia, Eq. (7) can be simplified as follow:

$$\begin{aligned} &\dfrac{\partial^2 M_x}{\partial x^2} + 2\dfrac{\partial^2 M_{xy}}{\partial x\partial y} + \dfrac{\partial^2 M_y}{\partial y^2} + \dfrac{\partial}{\partial x}\left(N_x\dfrac{\partial w}{\partial x} + N_{xy}\dfrac{\partial w}{\partial y}\right) \\[2mm] &+ \dfrac{\partial}{\partial y}\left(N_y\dfrac{\partial w}{\partial y} + N_{xy}\dfrac{\partial w}{\partial x}\right) = \rho h_p \dfrac{\partial^2 w}{\partial t^2} \end{aligned} \tag{8}$$

Substituting Eq. (3) and (5) into Eq. (8) lead to the following equation of motion.

$$\begin{aligned} &\rho h_p \dfrac{\partial^2 w}{\partial t^2} - D\nabla^4(w - w_0) + \\[2mm] &\left(\dfrac{\partial^2 F}{\partial y^2}\dfrac{\partial^2 w}{\partial x^2} - 2\dfrac{\partial^2 F}{\partial x\partial y}\dfrac{\partial^2 w}{\partial x\partial y} + \dfrac{\partial^2 F}{\partial x^2}\dfrac{\partial^2 w}{\partial y^2}\right) = 0 \end{aligned} \tag{9}$$

According to the Marguerre's theory, by introducing Airy's stress function the compatibility equation can be found.

$$\nabla^4 F = E\left[\left(\dfrac{\partial^2 w}{\partial x\partial y}\right)^2 - \dfrac{\partial^2 w}{\partial x^2}\dfrac{\partial^2 w}{\partial y^2} - \left(\dfrac{\partial^2 w_0}{\partial x\partial y}\right)^2 + \dfrac{\partial^2 w_0}{\partial x^2}\dfrac{\partial^2 w_0}{\partial y^2}\right] \tag{10}$$

In the paper, the boundary conditions which are elastically restrained at unloading edges and simply supported at loading edges are adopted and assume the rotational restraint stiffness of the two elastically restrained edges are the same.

$$\begin{cases} x=0: & w=0, \quad \dfrac{\partial^2 w}{\partial x^2}=0 \\[4pt] x=a: & w=0, \quad \dfrac{\partial^2 w}{\partial x^2}=0 \\[4pt] y=0: & w=0, \quad \dfrac{\partial^2 w}{\partial y^2}+K\dfrac{\partial w}{\partial y}=0 \\[4pt] y=b: & w=0, \quad \dfrac{\partial^2 w}{\partial y^2}+K\dfrac{\partial w}{\partial y}=0 \end{cases} \tag{11}$$

where, the parameter $K=\zeta/D$ is the uniform elastic rotational stiffness of the edge support. The values of K range from 0 to infinity. The special cases of $K=0$ and $K=\infty$ in Eq. (11) correspond to simply supported edge and clamped edge, respectively. ζ is the rotational restraint stiffness of the constrained edges.

2.2 Solution methodology

As shown in Figure 1, longitudinal edges of rectangular plate are elastically restrained. Elastic supported boundary condition can be recognized as an intermediate condition between the simply and the clamped boundary condition. For the longitudinal edges of rectangular plate, in the case of the simply supported boundary conditions, out of the plane displacement (displacement in z-direction) and bending moment are zero; in the case of clamped conditions, out of the plane displacement and rotational angles are zero. For the elastically restrained conditions, out of the plane displacement is zero along longitudinal edges, but neither rotations nor bending moments are zero. For this reason the out of the plane displacement can be represented as superposition of a sine term and cosine term (Bisagni, 2009). The boundary conditions of rectangular plate are simply support along longitudinal direction (loaded edges) and elastic support along lateral direction (unloaded edges). According to the method presented in literature 16, the out of the plane displacement is written as follow:

$$w(x,y,t)=\sum_{m=1}^{k}\sum_{n=1}^{l}{}^{t}W_{mn}\sin\frac{m\pi x}{a}\left(\sin\frac{n\pi y}{b}+R\left(1-\cos\frac{2n\pi y}{b}\right)\right) \tag{12}$$

Geometrical imperfections, which are always present in real applications, are also introduced in the model. From a numerical point of view, they can be used to avoid the divergence problems. Geometrical imperfections are introduced through a double sines series and the values of each ${}^{t}W_{mn}$ are known numbers. They will be more accurate if determined through a Fourier analysis of experimental data which can't be obtained in this paper, so it is suitable to use the first buckling mode of the rectangular plate which are simply supported along all four edges. Consequently the expression of the initial imperfections $w_0(x, y)$ is taken equal to

$$w_0(x,y)=\sum_{m=1}^{k}\sum_{n=1}^{l}{}^{0}W_{mn}\sin\frac{m\pi x}{a}\sin\frac{n\pi y}{b} \tag{13}$$

Substituting Eq. (12) into Eq. (11), the dimensionless rotational restraint stiffness can be written as

$$R=\frac{bK}{4\pi}\frac{\displaystyle\sum_{n=1}^{l}n}{\displaystyle\sum_{n=1}^{l}n^2}=\frac{b\zeta}{4\pi D}\frac{\displaystyle\sum_{n=1}^{l}n}{\displaystyle\sum_{n=1}^{l}n^2} \tag{14}$$

Substituting Eqs. (12) and (13) into Eq. (10) the following equation is obtained:

$$\begin{aligned} \nabla^4 F=Eh\sum_{i=1}^{k}\sum_{j=1}^{l}\sum_{m=1}^{k}\sum_{n=1}^{l}&\left\{\left(\frac{ijmn\pi^4+j^2m^2\pi^4}{4a^2b^2}\right)\left(\cos\frac{m+i}{a}\pi x\cdot\cos\frac{j+n}{b}\pi y\right.\right.\\ &\left.+\cos\frac{m-i}{a}\pi x\cdot\cos\frac{j-n}{b}\pi y\right)+\left(\frac{ijmn\pi^4-j^2m^2\pi^4}{4a^2b^2}\right)\\ &\left(\cos\frac{m+i}{a}\pi x\cdot\cos\frac{j-n}{b}\pi y+\cos\frac{m-i}{a}\pi x\cdot\cos\frac{j+n}{b}\pi y\right)\\ &+\left(\frac{Rijmn\pi^4+2j^2m^2\pi^4}{2a^2b^2}\right)\left(\cos\frac{m+i}{a}\pi x\cdot\sin\frac{2j+n}{b}\pi y\right.\\ &+\cos\frac{m+i}{a}\pi x\cdot\sin\frac{n-2j}{b}\pi y+\cos\frac{m-i}{a}\pi x\cdot\\ &\left.\sin\frac{2j+n}{b}\pi y+\cos\frac{m-i}{a}\pi x\cdot\sin\frac{n-2j}{b}\pi y\right)\\ &+\left(\frac{2Rijmn\pi^4+R^2j^2m^2\pi^4}{4a^2b^2}\right)\left(\cos\frac{i+m}{a}\pi x\cdot\sin\frac{2n+j}{b}\pi y\right.\\ &\left.+\cos\frac{i+m}{a}\pi x\cdot\sin\frac{j-2n}{b}\pi y\right)+\left(\frac{2Rijmn\pi^4-R^2j^2m^2\pi^4}{4a^2b^2}\right)\\ &\left(\cos\frac{i-m}{a}\pi x\cdot\sin\frac{2n+j}{b}\pi y+\cos\frac{i-m}{a}\pi x\cdot\sin\frac{j-2n}{b}\pi y\right)\\ &+\left(\frac{Rijmn\pi^4-R^2j^2m^2\pi^4}{a^2b^2}\right)\left(\cos\frac{i+m}{a}\pi x\cdot\sin\frac{2n+2j}{b}\pi y\right.\\ &+\cos\frac{i+m}{a}\pi x\cdot\sin\frac{2n-2j}{b}\pi y+\cos\frac{i-m}{a}\pi x\cdot\sin\frac{2n+2j}{b}\pi y\\ &\left.\left.+\cos\frac{i+m}{a}\pi x\cdot\sin\frac{2n-2j}{b}\pi y\right)\right\} \end{aligned} \tag{15}$$

Airy's stress function F consists of two components, one is the general solution and the other is particular solution. Using different trigonometric relations, the particular solution component can be derived as:

$$\begin{aligned} F_1=Eh\sum_{i=1}^{k}\sum_{j=1}^{l}\sum_{m=1}^{k}\sum_{n=1}^{l}&{}^{(i,j)\ne(m,n)}\left\{A_1\cos\frac{i+m}{a}\pi x\cdot\cos\frac{j+n}{b}\pi y+A_2\cos\frac{i+m}{a}\pi x\cdot\cos\frac{j-n}{b}\pi y\right.\\ &+A_3\cos\frac{i-m}{a}\pi x\cdot\cos\frac{j+n}{b}\pi y+A_4\cos\frac{i-m}{a}\pi x\cdot\cos\frac{j+n}{b}\pi y+A_5\cos\frac{i+m}{a}\pi x\cdot\\ &\sin\frac{2j+2n}{b}\pi y+A_6\cos\frac{i+m}{a}\pi x\cdot\sin\frac{2j-2n}{b}\pi y+A_7\cos\frac{i-m}{a}\pi x\cdot\sin\frac{2j+2n}{b}\pi y\\ &+A_8\cos\frac{i-m}{a}\pi x\cdot\sin\frac{2j-2n}{b}\pi y+A_9\cos\frac{i+m}{a}\pi x\cdot\sin\frac{j+2n}{b}\pi y+\\ &A_{10}\cos\frac{i+m}{a}\pi x\cdot\sin\frac{j-2n}{b}\pi y+A_{11}\cos\frac{i-m}{a}\pi x\cdot\sin\frac{j-2n}{b}\pi y\\ &+A_{12}\cos\frac{i-m}{a}\pi x\cdot\sin\frac{j-2n}{b}\pi y+A_{13}\cos\frac{i+m}{a}\pi x\cdot\sin\frac{n}{b}\pi y\\ &+A_{14}\cos\frac{i-m}{a}\pi x\cdot\sin\frac{n}{b}\pi y+A_{15}\cos\frac{i+m}{a}\pi x\cdot\cos\frac{2n}{b}\pi y\\ &\left.+A_{16}\cos\frac{i-m}{a}\pi x\cdot\cos\frac{2n}{b}\pi y\right\}\left[{}^{t}W_{ij}{}^{t}W_{mn}-{}^{0}W_{ij}{}^{0}W_{mn}\right]+\\ Eh\sum_{m=1}^{k}\sum_{n=1}^{l}&\left\{A_{17}\cos\frac{2m}{a}\pi x+A_{18}\cos\frac{2n}{b}\pi y+A_{19}\cos\frac{2m}{a}\pi x\cdot\cos\frac{4n}{b}\pi y\right.\\ &+A_{20}\cos\frac{4n}{b}\pi y+A_{21}\cos\frac{2m}{a}\pi x\cdot\sin\frac{3n}{b}\pi y+A_{22}\cos\frac{2m}{a}\pi x\cdot\sin\frac{n}{b}\pi y\\ &\left.+A_{23}\sin\frac{3n}{b}\pi y+A_{24}\sin\frac{n}{b}\pi y+A_{25}\cos\frac{2m}{a}\pi x\cdot\cos\frac{2n}{b}\pi y\right\}\left[{}^{t}W_{mn}^2-{}^{0}W_{mn}^2\right] \end{aligned} \tag{16}$$

The abbreviations A1 to A25 are defined in Appendix A.

The general solution component can be derived as:

$$F_2 = \frac{1}{2}N_x x^2 + \frac{1}{2}N_y y^2 - N_{xy}xy \quad (17)$$

At last, the Airy's stress function F is expressed as follow:

$$F = F_1 + F_2 \quad (18)$$

Substituting Eqs. (12) (13) and (18) into Eq. (9), then a new equation is obtained and the two sides of it is multiplied by

$$\sin\frac{r\pi x}{a}\sin\frac{s\pi y}{b} + R\left(1-\cos\frac{2s\pi y}{b}\right)\sin\frac{r\pi x}{a}$$

$$r=1, 3,\ldots k; s=1,3\ldots l \quad (19)$$

and then integrated over the whole plate area. All anti-symmetric terms of the Fourier series become zero because of the symmetric initial imperfections shape (Petry, 2000).The summation over odd indices of Navier's double series can meet the criteria of accuracy and save a lot of CPU time. The equations may become unmanageable as the CPU time increases nearly quadratically when r and s increase. The calculations are performed by using a $r\times s$ series which lead to convergent solutions. The computer code has been developed to solve the dynamic response and buckling of rectangular plate under in-plane impact loading and the initial imperfections are also included. For time integration a fourth-order Runge–Kutta method has been used. The initial conditions are set as:

$$'\dot{W}_{mn}(t=0) = 0 \quad (20)$$

$$'W_{mn}(t=0) = {}^0W_{mn} \quad (21)$$

3 NUMERICAL EXAMPLES AND DISCUSSION

3.1 Numerical verification

It is necessary to validate the convergence of the numerical results with respect to the number of the terms selected form the series. The results of three different numbers of the terms are shown in Figure 2. It was observed that the models with r=s=5 have produced good results and therefore this terms was used in the following analysis.

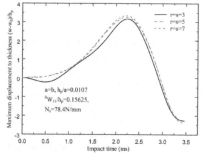

Figure 2. Effect of the number of the terms in series on the dynamic response of the rectangular plate.

In order to verify the method developed in the present study, the example in the literature (Zhang, 2004) was evaluated by the presented method, and a comparison of the present theoretical results with the results of the published literature is as shown in Figure 3. Vertical coordinates is the mid-point transverse deflection which has been normalized to the corresponding static deflection w_{stat}. Therefore, the presented method can be used to study dynamic buckling of rectangular plate with elastically restraint edges under impact loading.

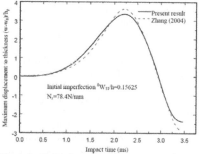

Figure 3. Deflection response curves at the center of plate ($N_{max}=3N_0$).

3.2 Calculation of the rectangular plates

The relationships of geometric dimensions are $a=b$, $h_p/a=0.0104$. The properties of rectangular plate are the elastic modulus $E=2.1\times10^5$MPa, Poisson's ratio $v=0.3$ and material density $\rho=2.7\times10^{-6}$ kg/mm^3. The impact loading is represented by polynomial function as follow:

$$N_x(t) = \begin{cases} \eta_1 t^2 + \eta_2 t & 0 \le t \le T_d \\ 0 & t > T_d \end{cases} \quad (22)$$

where $\eta_1=-4N_{max}/T_d^2$, $\eta_2=4N_{max}/T_d$, N_{max} is the amplitude of impact loading, T_d is the impact duration. The value of N_{max} is equal to N_0 which is the static critical buckling load of rectangular plate. N_0 is represented as follow:

$$N_0 = \frac{\pi^2 D}{b^2}\left(\frac{b}{a} + \frac{a}{b}\right)^2 \qquad (23)$$

As for a const aspect ratio a/b=1.0 (i.e., a square plate), the influence of rotational restraint stiffness on the dynamic response of the rectangular plate under impact loading are shown in Figure 4. Seven rectangular plates with different computing conditions (a-g) which are indicated as Table 1 are discussed. As shown in Figure 4 the larger rotational restraint stiffness obviously lead to smaller response amplitude. T_0 is the natural period of the rectangular plate with four simply supported edges. With the increase of rotational restraint stiffness, the time of displacement response reaching its peak decreases. When the R changes between 0.01 and 10, the effects of R on dynamic response are huge.

The explicit computing results, including the maximum displacement response, dynamic buckling loads of rectangular plates for various initial imperfections, impact durations and plate thickness, of all caseses (a-g) are summarized in Table 2. It shows that the difference value between maximum deflections and minimum deflections are $0.3940h_p$, $0.7575h_p$, $0.9545h_p$, $0.6491h_p$, $1.0585h_p$, $0.5877h_p$ and $0.3900h_p$ for the case a-g respectively. Obviously, the rotational restraint stiffness of boundary edges has a great influence on dynamic response of rectangular plate. And it is evident that the larger initial imperfections, impact duration and thinner plate thickness corresponds to higher dynamic deflections response. Moreover, the larger R is, the earlier transversal deflection response at center of rectangular plate reached maximum.

Dimensionless transversal deflections $(w-w_0)/h$ versus dimensionless peak loads N_{max}/N_0 curves are shown in Figure 5. As shown in Figure 5, lower value of R (rotational restraint stiffness) results in larger dynamic failure load with same impact duration Td and initial imperfections $^0W_{11}$, and vice versa. There are apparent bifurcation points to evaluate initial buckling load, which means there are apparent index to measure the dynamic buckling strength. The values dynamic buckling loads evaluated according to Budiansky–Hutchinson criterion for the case R=0.01 and R=10 are equal to the values for the case R=0 and R=∞, respectively. So it is reasonable to define when the R is smaller than 0.01 or larger than 10, rotational restraint edges can be handled as simply support edges and clamped support edges, respectively.

It can be found that larger initial imperfections lead to higher maximum dynamic deflection response and smaller dynamic buckling loads by comparing the cases (a)-(c) in Table 2. Moreover, decreasing velocity of dynamic buckling loads slows down with the increasing initial imperfections. Cases (c)-(e) show that dynamic deflection response increase with increasing the impact duration, in contrast to dynamic buckling load. The dynamic buckling loads change inconspicuously when the impact duration is greater than natural period of rectangular plate. Cases (c), (f) and (g) show that dynamic deflection response and dynamic buckling loads are very sensitive to plate thickness. With the increase of the plate thickness, the dynamic deflection response and dynamic buckling load will increase. In terms of dynamic buckling load, the difference between R=0 and R=∞ for the case (a), (b), (c), (d) and (e) are $1.0N_0$, $1.1N_0$, $1.1N_0$, $1.0N_0$ and $1.1N_0$, respectively. So it can be concluded that the influence degree of rotational restraint stiffness on dynamic buckling load changes slightly when initial imperfections and impact duration change largely. But there is a different results when the plate thickness changes. For the case of (c), (f), (g) (h_p/a=0.0107, 0.0120, 0.0130), the different results are $1.1N_0$, $1.5N_0$, and $1.8N_0$, respectively. It implies that the thicker of plate thickness is, the larger influence degree of rotational restraint stiffness on dynamic buckling load will become.

As is shown in Table 2, the value of the dynamic critical buckling load is larger than that of static buckling load, which suggests that inertial term can strength the carrying capacity of the plate. It also indicate that either increasing initial imperfection and impact duration or the decreasing rotational restraint stiffness and plate thickness result in the larger value of dynamic critical buckling load. The dynamic critical buckling load has a minim value (N_x=1.9N_0) for the case (e). There will exist waste of the resource and increasing cost when the static buckling load is adopted as allowable load criterion during the design process. Therefore, it is necessary to consider inertial term on the load carrying capacity of the plate.

Table 1. Parameter value

Case	Initial imperfection	Impact duration	Plate thickness
a	$^0W_{11}/h_p$=0.0625	T_d/T_0=1	h_p/a=0.0107
b	$^0W_{11}/h_p$=0.1563	T_d/T_0=1	h_p/a=0.0107
c	$^0W_{11}/h_p$=0.3125	T_d/T_0=1	h_p/a=0.0107
d	$^0W_{11}/h_p$=0.3125	T_d/T_0=2/3	h_p/a=0.0107
e	$^0W_{11}/h_p$=0.3125	T_d/T_0=4/3	h_p/a=0.0107
f	$^0W_{11}/h_p$=0.3125	T_d/T_0=1	h_p/a=0.0120
g	$^0W_{11}/h_p$=0.3125	T_d/T_0=1	h_p/a=0.0130

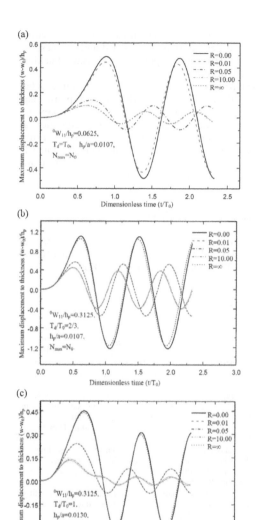

(a)

(b)

(c)

Figure 4. Response curves at the center of the plate for different rotational restraint stiffness.

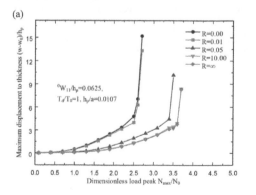

(a)

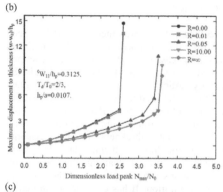

(b)

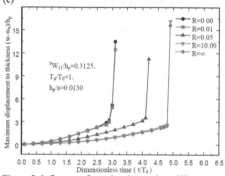

(c)

Figure 5. Influence of rotational restraint stiffness on dynamic buckling.

Table 2. Dynamic deflection response and dynamic buckling loads response

Case		Rotational restraint stiffness				
		$R=0$	$R=2$	$R=5$	$R=10$	$R=\infty$
a	Maximum response	$0.489h_p$	$0.445h_p$	$0.139h_p$	$0.097h_p$	$0.095h_p$
	Dynamic buckling load	$2.6N_0$	$2.6N_0$	$3.4N_0$	$3.6N_0$	$3.6N_0$
b	Maximum response	$0.987h_p$	$0.915h_p$	$0.329h_p$	$0.235h_p$	$0.230h_p$
	Dynamic buckling load	$2.2N_0$	$2.2N_0$	$3.1N_0$	$3.3N_0$	$3.3N_0$
c	Maximum response	$1.364h_p$	$1.287h_p$	$0.561h_p$	$0.417h_p$	$0.410h_p$
	Dynamic buckling load	$2.0N_0$	$2.0N_0$	$2.9N_0$	$3.1N_0$	$3.1N_0$
d	Maximum response	$1.092h_p$	$1.045h_p$	$0.565h_p$	$0.449h_p$	$0.442h_p$
	Dynamic buckling load	$2.5N_0$	$2.5N_0$	$3.4N_0$	$3.5N_0$	$3.5N_0$
e	Maximum response	$1.407h_p$	$1.316h_p$	$0.498h_p$	$0.356h_p$	$0.349h_p$
	Dynamic buckling load	$1.9N_0$	$1.9N_0$	$2.8N_0$	$3.0N_0$	$3.0N_0$
f	Maximum response	$0.817h_p$	$0.787h_p$	$0.374h_p$	$0.238h_p$	$0.229h_p$
	Dynamic buckling load	$2.5N_0$	$2.5N_0$	$3.5N_0$	$4.0N_0$	$4.0N_0$
g	Maximum dynamic response	$0.548h_p$	$0.534h_p$	$0.291h_p$	$0.167h_p$	$0.158h_p$
	Dynamic buckling load	$3.0N_0$	$3.0N_0$	$4.1N_0$	$4.8N_0$	$4.8N_0$

4 CONCLUSIONS

The correctness of the presented method has been validated through comparison with the published literatures in the case of simply support and clamped support, and then the method was used to study the nonlinear dynamic buckling of rectangular plate subjected to impact loading with elastically restraint boundary conditions. The effects of rotational restraint stiffness, initial imperfections, impact durations and plate thickness on the dynamic response and buckling of rectangular plate were investigated. For impact loading, the maximum deflection response of rectangular plate occurs before impact duration. It has been demonstrated in this paper that rotational restraint stiffness that is usually ignored by previous researchers plays an important role in dynamic response and buckling of the rectangular plates subjected to in-plane impact loading.

With increasing the rotation restraint stiffness, the dynamic buckling load of rectangular plate increases but dynamic displacement response of rectangular plate and the time of dynamic displacement response reaches its peak decreases. As mentioned in the above figures, when the value of the dimensionless rotational restraint stiffness (R) is more than 10, the boundary condition can be regarded as clamped support; and simply support when the value of the rotational restraint stiffness is less than 0.01.

For the different initial imperfections, with the increase of magnitude of initial imperfections, the dynamic deflection response increases and the time of dynamic deflection response reaching its peak decreases. With the increase of impact duration, the maximum displacement of plate increases and the time of dynamic deflection response reaching its peak decrease. It is evident that the longer impact duration corresponds to smaller dynamic buckling loads. The influence degree of rotational restraint stiffness on dynamic buckling load changes a little when initial imperfections and impact duration have changed.

For the different plate thickness, the time of dynamic response reach its peak are same, but dynamic displacement response amplitude and dynamic buckling loads are different. With increase of the plate thickness, the dynamic deflection response and dynamic buckling load will increase.

And the thicker of plate thickness is, the larger influence degree of rotational restraint stiffness on dynamic deflection response and dynamic buckling load will become.

The results shows rotational restraint stiffness has a significant impact on dynamic response and buckling of rectangular plate. Since the previous research usually regard the boundary condition as simply support or clamped support, so the results can not reflect the reality, the method presented in this paper overcomes this question successfully and saves a large amount of resources and money.

ACKNOWLEDGEMENTS

This work is co-supported by the Ministry of Education and Ministry of Finance of China (Grant No. 201335), and by NSFC(51239007).

REFERENCES

Ahmad R R. 2013. Elastic coupled buckling analysis in stiffened plates with T-bar stiffeners. *Proc IMechE, Part C: J Mechanical Engineering Science* 227(6): 1135-1149.

Cui S, Cheong HK and Hong H. 1999. Experimental study of dynamic buckling of plates under fluid-solid slamming. *Int J Impact Eng* 22(7): 675–691.

Ji Z H and Wang D Y. 2014. Influence of load shape on dynamic response of cross-stiffened deck subjected to in-plane impact. *Thin-Walled Struct* 82: 212–220.

Kubiak T. 2013. Static and Dynamic Buckling of Thin-Walled Plate Structures. *Springer*.

Lai S K and Xiang Y. 2009. DSC analysis for buckling and vibration of rectangular plates with elastically restrained edges and linearly varying in-plane loading, *Int J Struct Stab Dyn* 9(3): 511-531.

Paik J K and Thayamballi A. K. 2000. Buckling strength of steel plating with elastically restrained edges. *Thin-Walled Struct* 37(37): 27–55.

Paik J K and Thayamballi AK. 2003. An experimental investigation on the dynamic ultimate compressive strength of ship plating. *Int J Impact Eng* 28(7): 803–811.

Tarján G and Kollár L. P. 2010. Buckling of axially loaded composite plates with restrained edges, *J Reinf Plast Compos* 29(23): 3521-3529.

Zhang S and Khan I. 2009. Buckling and ultimate capability of plates and stiffened panels in axial compression. *Mar Struct* 22(4):791–808.

Zhang T, Liu T G and Zhao Y. 2004. Nonlinear dynamic buckling of stiffened plates under in-plane impact load. *J. Zhejiang Univ Sci A* 5(5): 609-617.

Cargo Loading and Port Operations

A Comparison of Loading Conditions Effects on the Vertical Motions of Turret-Moored FPSO

S.A. Erkurtulmus, E. Peşman & H. Copuroglu
Karadeniz Technical University, Trabzon, Turkey

ABSTRACT: The tanker based floating, production, storage and offloading units (FPSOs) have six degrees of freedom motion during operation in the sea. The loading conditions can be changed with using ballast tanks. This has an impact on the amplitude of motions. In this study, the loading effects on the movements are investigated. Polyester mooring lines are commonly used at deep-water locations for a FPSO. Four polyester mooring lines are used. Internal turret is chosen and modeled as a single point. The first- and second- order wave forces calculated from the hydrodynamics program ANSYS-AQWA for multiple bodies. For this, the vessel and line dynamics are solved simultaneously for the environmental and boundary conditions. Gulf of Mexico is chosen for the non-parallel wind-wave-current 100 year hurricane conditions for the analysis. For comparison, 50%, 80% and fully loaded conditions are chosen. The results are presented and discussed.

1 INTRODUCTION

A floating, production, storage and offloading units (FPSO) system for offshore oil and gas production employs a custom-built ship or tanker with structural modifications. It is equipped with hydrocarbon processing equipment for separation and treatment of crude oil, water and gases, arriving on board from sub-sea oil wells via flexible pipelines. FPSOs were found to be viable production systems that can withstand the harsh environment offshore Gulf of Mexico (Wichers et al. 2001).

Model tests were conducted at the Offshore Technology Research Center (OTRC) multi-directional wave basin to examine the behavior of generic FPSOs in wave, wind and current conditions typical of the passage of severe hurricane (Ward et al. 2001).

The Dynamic Loading Approach (DLA) provides an enhanced structural analysis basis to assess the capabilities and sufficiency of a structural design.

An important design consideration for an FPSO is its mooring system. A turret-based single-point mooring system is relatively new, and has proven to be good design feature of several FPSOs today (Thiagarajan et al. 1998). Turrets can either be internal to the structure or external based on specific requirements. Polyester mooring lines are commonly used at deep-water locations for a FPSO. Recommended way to describe dynamics of FPSO is to set a coupled model that also contains dynamics of mooring system. Breaking point of a polyester rope can be found at 15% elongation (Kim et al. 2001). The vessel and mooring line dynamics are solved simultaneously. They examined acceptance tests performed with actual full scale cables. As a result they suggested a formula for specific modulus of polyester ropes in terms of dynamic analysis. A coupled dynamic analysis of floating structures with polyester mooring lines is studied and the mathematical model allowed relatively large elongation of polyester rope and nonlinear stress-strain relationship (Tahar et al. 2003). The mooring line dynamics was based on elastic rod theory. Numerical calculations are done utilizing nonlinear FEM. The vessel and stiffness polyester mooring lines are solved simultaneously (Catipovic et al. 2012).

In this study, effects of operational loading conditions on the vertical motion of the turret-moored FPSO is discussed and compared.

2 DESCRIPTION OF SYSTEM

In this study, floating, production, storage and offloading units (FPSOs) encountered during the operation, and response to the coupled dynamic analysis made in the loading conditions that occur due to environmental conditions. For this, diffraction and radiation theory to analyses the ship motions is explained and the analysis was carried out on FPSO hull forms using ANSYS AQWA program which is based on this theory and carries out calculations via panel method using green functions.

The analysis in time domain is developed for the global vertical motion (heave, pitch and roll motion) simulation of a turret-moored tanker based FPSO designed for 1829 m water depth and non-parallel wind wave current 100 years hurricane conditions in Gulf of Mexico in different operational loading conditions.

2.1 Designed parameters of turret-moored FPSO and mooring lines

The turret-moored FPSO is designed for 1829 meters water depth. It has 317 m length, 47.17 m of the breadth. Full draft is 18.9 m and 80% loaded draft is 15.121 m. The design parameters are given in Table 1.

Table 1. Designed parameter of the FPSO

Description	Unit	50% loaded	80% loaded	Full loaded
LBP	m	317.0	317.0	317.0
B	m	47.17	47.17	47.17
D	m	22.02	22.02	22.02
T	m	9.45	15.121	18.90
Δ	MT	123112	188829	243221
C_b		0.85	0.85	0.85
X_{turret}*	m	-46.33	-46.33	-46.33

* Turret in centerline behind F_{pp}

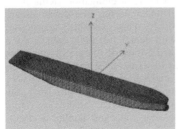

Figure 1. Basic model of FPSO

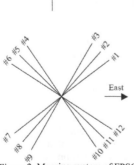

Figure 2. Mooring system of FPSO

The mooring lines and risers are spread from the turret. There are 12 combined mooring lines with chain, wire and chain, and 13 steel wire risers. Table 2 shows the main particulars of mooring lines. The schematic plot of the arrangement for mooring lines is shown in Figure 3. There are 4 groups of mooring lines, each of which is normal to the other group. Each group is composed of 3 mooring lines 5 degree apart from each mooring line in the group. The center of the first group is heading the true East, and so the second group is toward the true North.

Table 2. Main particulars of mooring systems

Description	Unit	Quantity
Pretension	kN	1424
Number of line		4x3
Degrees between 3 lines	degree	5°
Length of mooring lines	m	2652
Segment 1 (ground position): chain		
Length at anchor point	m	121.9
Diameter	cm	9.52
Weight in air	N/m	1856
Weight in water	N/m	1615
Stiffness, AE	kN	820900
Mean breaking load	kN	7553
Segment 2: polyester		
Length at anchor point	m	2438
Diameter	cm	16.0
Weight in air	N/m	168.7
Weight in water	N/m	44.1
Stiffness,, AE	kN	168120
Mean breaking load (MBL)	kN	7429

Table 2. Main particulars of mooring systems (cont.)

Segment 3* (hang-off position): chain		
Length at anchor point	m	91.4

* Other parameters are the same as for the Segment 1.

Table 3. Environmental conditions for the simulation

Description	Unit	Quantities
Wave		
Significant wave height, HS	m	7.1
Peak period, TP	s	14
Wave spectrum		JONSWAP (γ=2.5)
Wave direction	degree	180
Wind		
Velocity at 10 m above MWL	m/s	41.12
Wind spectrum	NPD	

Table 3. Environmental conditions for the simulation (cont.)

Wind direction	degree	150
Current		
at free surface 0 m	m/s	0.9144
at 60.96 m	m/s	0.9144
at 91.44 m	m/s	0.0914
on the sea bottom	m/s	0.0914
Direction	degree	210

2.2 Environmental conditions

For this study, the 100 year hurricane conditions for the Gulf of Mexico (GoM) is chosen. The wave condition is given by JONSWAP spectrum with a significant wave height of 12.1 m, a peak period of 14 s and peak enhancement factor of 2.5. NPD wind spectrum is used for the simulation and wind speed 41.12 m/s at 10 m high above the sea level.

3 RESULTS AND DISCUSSIONS

The tanker based turret-moored FPSO analysis is described in this section. Sea depth is 1829 m. The hydrodynamic coefficients are calculated by AQWA-LINE and these coefficients are directly used as input values in the AQWA-DRIFT for time domain simulation.

3.1 Hull hydrodynamics

The design data L, B, D, T, KG, the turret position and the top tension of mooring lines are taken from other study (Ward et al. 2001).

The methodology for hull/mooring/riser coupled static/dynamic is similar to that of experimental works (Kim et al. 2002). The mooring lines are assumed hinged at the turret and anchor position. The wave force quadratic transfer functions are computed for 81 wave frequencies, ranging from 0.015 to 1.215 rad/s. The hydrodynamic coefficients and wave forces are expected to vary appreciably with large yaw angles and effects should be taken into consideration for reliable prediction of FPSO global motions (Kim et al. 2003). Therefore, they are calculated in advance for various yaw angles with 5° interval and the data are then tabulated as inputs (Arcandra et al. 2002).

From the AQWA output, the water-plane area, the displacement volume, the center of buoyancy and the restoring coefficients were obtained. Based on these data, vertical static equilibrium of the FPSO can be checked, i.e. the sum of the vertical line top tensions and weight is to be equal to the buoyancy. The relations between the natural frequency, and the restoring coefficients/masses are defined as follows:

$$f = \frac{1}{2\pi} \sqrt{\frac{C_{ij}}{M_{vij}}} \quad \text{(1/s or Hz)} \quad (i, j = 1, 2, .., 6) \qquad (1)$$

where f is the natural frequency, C_{ij} is the restoring coefficients (hydrostatic + mooring), and M_{vij} $(M_{aij}+m_{ij})$ is the virtual mass in which M_{aij} is the added mass near natural frequencies and m_{ij} is the mass/inertia of the body (Kim et al. 2004).

3.2 Slender member dynamics

For the static/dynamic analysis of mooring lines and risers, an extension of the theory developed for slender rods was used (Garrett 1982). Assuming that there is no torque or twisting moment, one can derive a linear momentum conservation equation with respect to a position vector $r(s,t)$ which is a function or arc length s and time t:

$$-(B\vec{r}'')'' + (\lambda\vec{r}')' + \vec{q} = m\ddot{\vec{r}} \qquad (2)$$

$$\lambda = T - B\kappa^2 \qquad (3)$$

where primes and dots denote spatial s-derivative and time derivative, respectively, B is the bending stiffness, T the local effective tension, k the local curvature, m the mass per unit length, and q the distributed force on the rod per unit length. The scalar variable λ can be regarded as a Lagrange multiplier.

The normal component of the distributed external force on the rod per unit length, q_n, is given by a generalized Morison equation:

$$q_n = C_I \rho A_e \dot{v}_n + \frac{1}{2} \rho D |v_{nr}| v_{nr} + C_m \rho A_e \ddot{r}_n \qquad (4)$$

where C_I, C_D and C_m are inertia, drag, and added mass coefficients and $\dot{v}_n$, v_{nr} and $\ddot{r}_n$ are normal fluid acceleration, normal relative velocity and normal structure acceleration respectively and the symbols ρ and D are fluid density and local diameter (Kim et al. 2004).

3.3 Vertical motions

In this study, standard floating body conventions are followed. Ship-based right-handed coordinate system, with x from stern to bow, y towards port, z positive upwards. The origin is located at the intersection of mid-ships and the waterline. Heave motions (ζ_3) is positive upwards and pitch motion (ζ_5) is positive bow downward (Thiagarajan et al. 1998).

The vertical motion of a point distant x_0 from the origin with reference to a geo-stationary observer is;

$$z = \zeta_3 - (x_0 - x_g)\zeta_5 \qquad (5)$$

where x_g is the distance of center of gravity from the origin. Expressing all variables as sinusoids with frequency equal to exciting wave frequency, the following equation can be obtained for the magnitude of vertical motion at a point:

$$|z|^2 = |\zeta_3|^2 + (x_0 - x_g)^2 |\zeta_5|^2 - 2(x_0 - x_g)|\zeta_3||\zeta_5|\cos(\delta) \qquad (6)$$

where δ is the phase difference between the heave and pitch motions. The velocities and accelerations are obtained by successively differentiating Equation (1) with respect to time. The surge velocity and acceleration can be ignored, as they are with second order in magnitude compared to heave and pitch velocities and accelerations (Thiagarajan et al. 1998). The vertical acceleration of a point is given to first order by:

$$\ddot{z} = \ddot{\zeta}_3 - (x_0 - x_g)\ddot{\zeta}_5 \qquad (7)$$

The motion results are presented in this section in frequency domain.

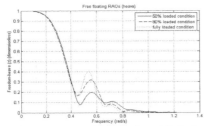

Figure 3. A comparison of free floating RAOs for heave

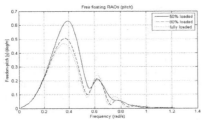

Figure 4. A comparison of free floating RAOs for pitch

3.4 *Time domain analysis*

In this section, time domain analysis is done. For the simulation, the Gulf of Mexico has been chosen for 100 year of hurricane conditions. The current is assumed to be continuous, irregular waves are unidirectional and come from the beginning. The JONSWAP spectrum was chosen as the wave spectrum. The results are given and compared for 50%, 80% and fully loading condition.

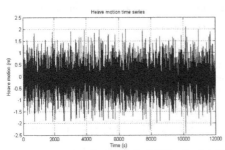

Figure 5. Heave motion time series for 50% loading condition

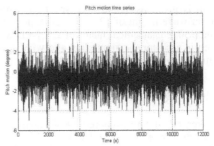

Figure 6. Pitch motion time series for 50% loading condition

Figure 5. and Figure 6. show heave and pitch motion time series for 50% loading condition respectively.

Figure 7. and Figure 8. show heave and pitch motion time series for 80% loading condition respectively. Table 4 shows a comparison of 80% loading condition and full loading condition.

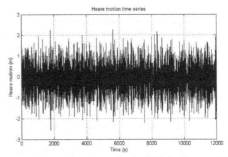

Figure 7. Heave motion time series for 80% loading condition

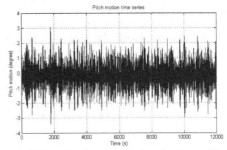

Figure 8. Pitch motion time series for 80% loading condition

Figure 9. and Figure 10. show heave and pitch motion time series for fully loading condition respectively. Table 4 shows a comparison of 50%, 80% and fully loading condition.

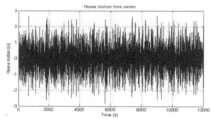

Figure 9. Heave motion time series for fully loading condition

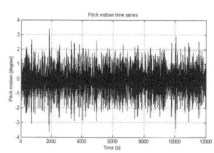

Figure 10. Pitch motion time series for fully loading condition

Table 4. A comparison of the 100-year hurricane condition results

		50% loaded	80% loaded	Full loaded
Heave	Mean	-0.06	0.12	-0.0014
	Std.	0.65	1.76	0.82
Pitch	Mean	-0.68	-0.37	-0.14
	Std.	1.25	0.99	0.88

Table 5. A comparison of mooring tensions (3 mooring lines)

		50%	80%	Fully
Mooring no 1	Mean	4242	6917	4239
	Std.	0.009	823	39
Mooring no 2	Mean	1224	2338	1444
	Std.	267	314	455
Mooring no 3	Mean	1109	2909	1203
	Std.	271	350	263
Mooring no 4	Mean	4242	7273	4239
	Std.	0.022	883	39

4 CONCLUSIONS

In this study, the global and mooring dynamics of a deep-water (1829 m) tanker based turret-moored FPSO in non-parallel 100-year hurricane conditions for the Gulf of Mexico (GoM) are numerically simulated and results show comparison of loading conditions effects on the vertical motions of turret-moored FPSO.

A turret-moored FPSO model with three different operational loading conditions which are 50% loaded, 80% loaded and fully loaded. Results indicate that the optional loading conditions are quite important in determining the vertical motion response of the model.

When the global motion simulations in hurricane conditions are examined, it can be observed that there are differences between various loading conditions. As indicated in the Table 4. For the heave motions, 80% loaded condition has the highest value. But for pitch motion, we can see that the vessel has such a big pitch amplitude when vessel 50% loaded. The results show that vessel look more stable when the FPSO is fully loaded.

The argument can be clearly seen in the mooring tensions in Table 5. that mooring line no. 2 and no. 3 are slacker than the others. This reason can be clarified by the anchor position on the sea bed. Also it can be clearly seen that the differences between mooring tensions for 80% and fully loaded conditions, is almost 40%. This is an expected result. Because the position of the center of gravity is more below when the FPSO operating fully loaded condition and the mooring tensions are less.

REFERENCES

Arcandra, T & Nurtjahyo, P. & Kim, M. H. 2002. Hull/mooring/riser coupled analysis of a turret-moored FPSO 6000 ft: comparison between polyester and buoys-steel mooring lines. *11th Offshore Symposium the Texas Section of the Society of Naval Architects and Marine Engineers, SNAME*:1-8.

Catipovic, I. & Coric, V. & Vukcevic, V. 2012. Dynamic of FPSO with Polyester Mooring Lines. *22nd International Offshore and Polar Engineering Conference (ISOPE),Rhodes, Greece, 17-22 June 2012.*

Garret, D. L. 1982. Dynamic Analysis of Slender Rods. *J. Energy Resources Technol. Trans. ASME 104*:302-307.

Kim, Y. B. 2003. Dynamic Analysis of Multiple-Body Floating Platforms Coupled with Mooring Lines and Risers. *PhD Dissertation, Civil Engineering Department, Texas A&M University, College Station*, TX.

Kim, Y. B. & Kim, M. H. 2002. Hull/mooring/riser Coupled Dynamic Analysis of a Tanker-based Turret-moored FPSO in Deep Water. Proc. *12th Offshore Polar Engineering Conference, (ISOPE), Kitakyushu, Japan, 26-31 May 2002*:169-173.

Kim, M. H. & Ran, Z. & Zheng, W. 2001. Hull/mooring Coupled Dynamic Analysis of a Truss Spar in Time Domain. *International Journal Offshore Polar Engineering. 1(1)*: 42-54.

Tahar, A. & Kim, M. H. 2003. Hull/mooring/riser Coupled Dynamic Analysis and Sensitivity Study of Tanker-based FPSO. *Applied Ocean Research 25*: 367-382.

Thiagarajan, K. P. & Finch, S. 1998. An Investigation Into the Effect of Turret Mooring Location on the Vertical Motions of an FPSO Vessel. *Journal of Offshore Mechanics and Arctic Engineering (OMAE). 20 April 1998*: M. M. Bernitsas.

Ward, E. G. & Irani, M. B. & Johnson, R. P. 2001. Responses of a Tanker-based FPSO to Hurricanes. *Proc. Offshore Technology Conference, OTC 13214 [CD-ROM], Houston, TX.*

Wichers, J. E. W. & Develin, P. V. 2001. Effect of Coupling of Mooring Lines and Risers on the Design Values for a turret moored FPSO in Deep Water of the Gulf of Mexico. *Proc. 11th International Offshore Polar Engineering Conference, Stavenger, Norway, 17-22 Junw 2001*: 480-48.

The Analysis of Container Vessel Service Efficiency in the Aspect of Berth and Handling Equipment Usage in Polish Ports

A. Kaizer, L. Smolarek & E. Ziajka
Gdynia Maritime University, Gdynia, Poland

K. Krośnicka
Gdansk University of Technology, Gdańsk, Poland

ABSTRACT: Nowadays, container terminals face a lot of challenges because of constantly growing container throughput. The ship owners choose only the terminals that are able to serve them efficiently. Due to minimising the cost of shipping one unit, the ship owners decide to order bigger ships that are economically favorable. The changes in the world fleet of container vessels have an impact on the functioning of container terminals. One of the most crucial aspects is the issue of appropriate use of the length of the berth and optimizing the number of mooring posts. The length of the berth and its proper use determines also the number of technical equipment, especially Ship-to-Shore Cranes used for loading and unloading operations. The article presents the analysis of the influence of crane cycle times effectiveness and estimation of the indicator of quays usage in polish container terminals.

1 INTRODUCTION

Seaports are very important elements of transportation networks, which strongly influence the efficiency of freight handling. Port productivity greatly depends on proper and high speed operations. Nowadays, the vessel owners prefer ports which provide them with enough effective service.

Although the representatives of container terminals, in order to strengthen its market position and become more competitive are trying to take projects allowing shortening the time of ship's mooring in port [10]. Therefore main aim of ports strategies include investments in automated handling equipment, large storage area, upgraded infrastructure, access to the port, as well as adapt to all the requirements of currently floating units. Currently, polish ports constitute an important point on the route of Corridor Trans-European Transport Network TEN-T. Therefore, the development trends of container terminals are focused heavily on expanding the port infrastructure, as well as projects aimed at shortening service of containerships [11].

The goal of published study is to valorize the automation reloading processes in the port for a short stay of the vessel in port. Authors hypothesized: automation of reloading is a key factor in the retention time of the ship in port.

2 CHARACTERISTICS AND TURNOVER OF CONTAINERIZED CARGO IN POLISH SEAPORTS

Container transport market is under dynamic changes. Sea transport, which is the most frequently used form of transport has lead the containerization to become one of the most popular ways to transport goods.

Polish seaports, where the container terminals are located are free of ice and tides what makes good conditions for development of containerization. In the Port of Gdynia there are two container terminals: Baltic Container Terminal (BCT) and Gdynia Container Terminal (GCT), in Gdańsk: Deepwater Container Terminal (DCT) and Gdańsk Container Terminal (GTK) and also located in port of Szczecin-Świnoujście: DB Port Szczecin and OT Świnoujście [6].

Due to large potential of Baltic's deep-sea trading routes in three years from the moment of launch the Deepwater Container Terminal received first container vessel departing directly from Far East. In 2011 when DCT started handling E-class container vessels the terminal handled 634 871 TEU, what makes the terminal one of the fastest growing terminals in Poland and also a hub for Baltic Sea Region. From 2010 to 2014 terminals located in Gdynia as well as DCT Gdansk recorded an upward trend. A breakthrough year for DCT was 2013 when the terminal exceeded one million TEU in container handling. In April 2013 OT Świnoujście was launched because of growing demand for container transport. Port of Szczecin-Świnoujście Authority reported in total 87 784 TEU in 2015. Since 2010 the container handling in Port of Szczecin-Świnoujście is continuously rising [10], [11], [12].

Unfortunately in 2015 DCT Gdańsk recorded a drop in container handling which was caused by decrease in transit to Russia. Entering into force of the alliance M2 and G6 has lead to a decline in container handling in Gdynia. Actually GTK has an annual decline in transshipment. In 2016 the container handling decreased by about 40% in comparison to 2015.

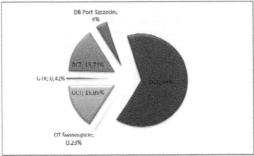

Figure 1. The share of container transshipments in container terminals in Poland in 2016

Last year polish container terminals handled in total about 2029718 TEU (Fig. 1). The largest share in handling had DCT Gdańsk, which achieved a score of 1289842 TEU. Gdynia Container Terminal achieved 321626 TEU. That result is better than achieved by BCT (318 871 TEU). Also DB Port Szczecin and OT Świnoujście terminals are becoming to have a greater impact on the amount of container transshipment in Poland.

3 ORGANISATION AND HANDLING TECHNOLOGY IN POLISH CONTAINER TERMINALS

Port authorities and the administration of container terminals in Gdańsk, Gdynia, Szczecin and Świnoujście support the development of containerization by long-term investments. The process of increasing the efficiency of operating vessels cause shorter staying in ports, and thus the attractiveness of containerization in maritime trade. Development trends of container terminals are aimed at purchasing new ship to shore (STS) gantry cranes and automated handling equipment [5].

Polish container terminals mostly operate with: STS; trailers (quay area); rubber tyred gantry cranes (RTG) or reachstackers (storage yards); rail mounted

gantry cranes (RMG) or tracks (gates area) in the process of handing [7], [8].

Technical specifications and a brief description of the operation of individual polish container terminals are as follows:

3.1 Container terminals in Port of Gdynia

Baltic Container Terminal has a quay for container handling (Helskie I) with length of 800 m and maximum depth 12.7 m. Terminal offers five positions to handle container ships in the system lift on-lift off, including one roll on- roll off system as well. Terminal handle vessels with 8 ship to shore cranes: 2 x 13 rows; 2 x 17 rows; 4 x 19 rows [15].

BCT container turnover in 2016 amounted to 318871 TEU.

Gdynia Container Terminal is located at Bułgarskie Wharf with a length of 450 meters, including 366 meters for container lift on-lift off operations. Terminal operates container feeder services, that connect Poland with the base ports of Western Europe. GCT owns and operates the following cargo handling equipment:

- two Super-Post-Panamax ship-to-shore gantry cranes with twin-lift 65 T capacity and 61 m outreach,
- one post-panamax ship-to-shore gantry crane with twin-lift 60 T capacity and 54 m outreach,
- one post-panamax ship-to-shore gantry crane with twin-lift 41 T capacity and 41 m outreach,
- two panamax ship-to-shore gantry cranes with 31,5 T capacity and 35 m outreach [14].

3.2 Container terminals in Port of Gdańsk

The Deepwater Container Terminal Gdansk became operational on the 1st of June 2007 with the arrival of the first commercial container ship at T1 berth. In the first years of operation, the terminal specialised in handling feeder ships. The direct connection with Asia contributed to the development of the DCT, which became a Baltic hub and one of the fastest developing terminals in the world. In 2011, the terminal began to handle E class container ships with a capacity of 15500 TEU, and in 2013, Triple-E class container ships with a capacity of 18000 TEU, the world's largest container ships at the time, operated by the Maersk Line.

In January 2015, the construction of new terminal quay began. The second berth T2 in DCT Gdańsk started operation in 2016. The new 650 m long quay with five new ship to shore gantry cranes and 17 m depth will increase DCT's annual handling capacity up to 3 million TEU. Actually terminal owns 71 hectares operational area (where storage capacity is 55000 TEU) and operates container vessels with eleven STS cranes [13].

Gdańsk Container Terminal has been operating at the Port of Gdańsk since September 1998. The terminal is placed at the estuary of the Visla River in the Inner Port of Gdańsk. The terminal can handle ships of a maximum capacity of 20000 DWT. The terminal is equipped with one ship-to-shore gantry crane, two shore cranes, one 100 t mobile harbour crane and two gantry cranes. The quay allows to handle ships up to 225 m in lift on-lift off and roll on-roll off systems.

3.3 Container terminals in Port of Szczecin – Świnoujście Complex

The Container Terminal DB Port Szczecin started operation in 2009 as a part of Deutsche Bahn group. In the following years, there has been an intensification of investment activities in supra - and infrastructure of the container terminal. Nowadays terminal offers reloading capacity of 120 000 TEU and the storage capacity of 4 000 TEU. DB terminal owns 300 m long berth for container vessels (max draft 9.15 m) equipped with two STS cranes [17].

The OT Port Świnoujście in the middle of 2013 commenced first container handling in Hutników quay and Górników quay. The quays have lengths respectively 328 m and 330 m. The draught 13.2 meters allows to enter post panamax container vessels with the capacity of 4500 TEU. Container terminal in Świnoujście is equipped with two Liebherr LHM 500 cranes with a spreader that ensures handling of containers weighing 40 tonnes (for 17 rows) and after the adaption also handing of over high containers as well as 140 tonnes heavy. What is more, terminal owns storage yard with an area of 9300 square meters that allows storage of full containers up to 4-5 layers – about 2000 TEU at the same time [16].

4 ANALYSIS OF THE EFFICIENCY OF HANDLING CONTAINER CARRIERS BASED ON THE INDICATOR OF THE USAGE OF CONTAINER TERMINAL QUAYS IN POLISH PORTS

4.1 The Indicator of quay usage

The indicator of quay usage is defined as the volume of cargo (for example TEUs or tones) per one unit length of the berth (1 meter long) in a period of time (usually one year). Calculation of that indicator for container terminals consists of dividing the annual throughput by a length of the berth. The analyzing of the indicator of quay usage can help to optimize the utilization of terminals' work [1]. Therefore observation of the results of indicator is important in long-time investments planning to notice needed and prepare new investments.

4.2 The Indicator of quay usage in polish container terminals

Polish container terminals are common to have a total of approximately 4045 linear meters of quays where operating container ships takes place. Taking into account the annual turnover in the year 2016 equal to 2029718 TEUs it can be calculated that the average value of the use of quays is 502 TEUs / 1 meter of a quay. The average value of the work throughput in the individual terminals presents table 1, table 2 and table 3.

Table 1. The average value of the work throughput in the Gdynia container terminals [15], [14].

Container terminals	BCT	GCT
Indicator of berth usage [TEU/ meter of quay]	399	519
1 STS average handle work in 2016 [TEU/ year]	39859	53604
Turnover in 2016 [TEU]	318871	321626
Berth length (meters)	800	620
Terminal STS handle equipment [amount]	8	6

Table 2. The average value of the work throughput in the Gdańsk container terminals [13], [18].

Container terminals	DCT	GTK
Indicator of berth usage [TEU/ meter of quay]	992	23
1 STS average handle work in 2016 [TEU/ year]	117258	2128
Turnover in 2016 [TEU]	1 289 842	8510
Berth length [meters]	1300	365
Terminal STS handle equipment [amount]	11	4

Table 3. The average value of the work throughput in the Szczecin- Świnoujście container terminals [16], [17].

Container terminals	DB	OT
Indicator of berth usage [TEU/ meter of quay]	287	7
1 STS average handle work in 2016 [TEU/ year]	43088	1564
Turnover in 2016 [TEU]	86176	4693
Berth length [meters]	300	660
Terminal STS handle equipment [amount]	2	3

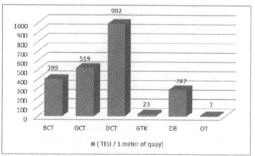

Figure 2. The indicator of berth usage in polish container Terminals

In addition, by analyzing the annual turnover of cargo and number of cranes working on the quay in the fast developing polish container terminal - DCT, we can see that the average handle work in 2014 reached 237676 TEUs for individual STS gantry crane by year (Fig.3). Such a high value can be the incentive to buy new crane to maintain an appropriate level of handling rate. Especially with a vision to operate the biggest container vessels in future.

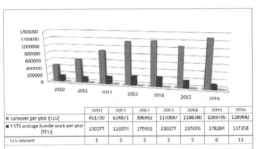

	2010	2011	2012	2013	2014	2015	2016
Turnover per year (TEU)	451730	634871	896962	1150088/	1188380	1069705	1289842
1 STS average handle work per year (TEU)	150577	126974	179392	230177	237676	178284	117258
STS amount	3	5	5	5	5	6	11

Figure 3. The DCT Terminal statistics turnover and STS cranes work effectiveness [13]

5 MODELING OF HANDLING WORK AT THE EXEMPLARY CONTAINER TERMINAL

Investigation on the effects of different modeling hypotheses on container terminal performances is very important. The focus on the effects of different hypotheses on handling equipment models may have an impact on the container terminal performances [9]. Such effects could not be negligible and should be investigated with respect to different planning horizons, such as tactical. The aim is to find a sort of a guideline useful to point out the strengths or weaknesses of different approaches [3],[4].

5.1 Basic elements of model

Quay modeled has a length of 800 m with theoretically designed 4 mooring positions with a length of 200 m each. Researched berth is a place for container handling in the lo-lo system (3 positions) and one position lo-lo and ro-ro. The terminal is equipped with 8 post panamax STS cranes with moving rate from 25 to 32 movements per hour. In the proposed model, the initial four scenarios assumed the manual configuration of container ships of different lengths and capacities. The authors assume that the number of containers loaded and unloaded is the same (import = export). Additionally the model consists of the limit for the time spent by a vessel at the quayside (demurrage / despatch) and the limit of the amount of cranes handling the ship at the time: ship no. 1 max 6 STS at one time, ship no. 2 max 3 STS at one time and ship no. 3 max 2 STS at one time. Additional details of the model shows the Figure 4. The capacity of each vessel and the amount of containers for handling represent table 4.

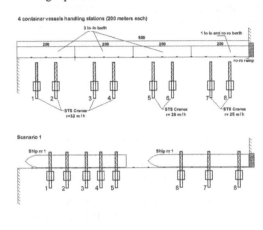

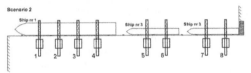

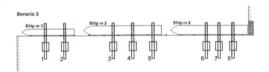

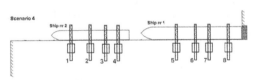

STS efficiency [r] (avarage rate):

number 1- 4= 32 m / h
number 5, 6= 28 m / h
number 7, 8= 25 m / h

m / h- moves per hour

Vessel's berth time limit and demurrage/ despatch specification:

ship nr 1 max 48 h; demurrage: 100 mu / h; despatch: 50 mu / h
ship nr 2 max 36 h; demurrage: 70 mu / h; despatch: 35 mu / h
ship nr 3 max 20 h; demurrage: 50 mu / h; despatch: 25 mu / h

mu / h- monetary unit per hour
h- hour

Figure 4. General layout of berths and details of the model

Table 4. The number of boxes handled at the terminal depending on the capacity of an operated ship

	Ship no. 1	Ship no. 2	Ship no. 3
Ship capacity [TEU]	7.500	4.100	2.100
% Of ship capacity [TEU] served in a terminal [%]	65%	100%	100%
Part of a ship capacity served in a terminal [TEU]	4875	4100	2100
Estimated* number of FEU on a ship served in a terminal	1950	1640	840
Estimated* number of TEU on a ship served in a terminal	975	820	420
Number of containers [boxes] on a ship served in a terminal (import)	**2925**	**2460**	**1260**
Number of containers [boxes] on a ship served in a terminal (export)	**2925**	**2460**	**1260**

* Assumed distribution of containers' types (sizes) in the general ship's capacity is: percentage of containers 40 feet and larger - 80%, percentage of containers 20 feet and less - 20%.

5.2 *Results of modeling*

The program written in Visual Basic was used to count the economic characteristic of different container terminal performances.

For different configurations of STS post panamax gantry cranes, ships' locations at quay and ships' capacities the economic effects were calculated (Fig. 7). First introduced data is related to ships (Fig. 5).

DATA

	STS TYP1	STS TYP2	STS TYP3	
	4	2	2	units
	32	28	25	m/h

	TEU	LOA	STS	TIME	DEMURRAGE	DESPATCH	STATE/NUMBER
SHIP1	5850	350	6	48	100	50	2
SHIP2	4920	268	3	36	70	35	0
SHIP3	2520	180	2	20	50	25	0

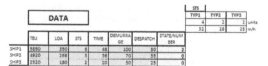

Figure 5. Initial data

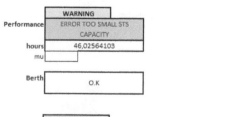

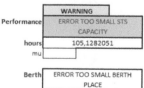

Figure 6. Control module

If any of the limits, the performance of STS cranes or a berth is exceeded, it turns on a warning (Fig 5).

In case of lack of space at the quay calculations are stopped.

Scenario1									
phase-A									
Ship1	5850	1	1	1	1	0	0	0	0
Ship1	5850	0	0	0	0	1	1	1	1
phase-B									
Ship1	0	0	0	0	0	0	0	0	0
Ship1	1005,5	0	0	1	1	1	1	1	1
									-246,92
Scenario1									
phase-A									
Ship1	5850	1	1	1	1	1	0	0	0
Ship1	5850	0	0	0	0	0	1	1	1
phase B									
Ship1	0	0	0	0	0	0	0	0	0
Ship1	2925	0	0	1	1	1	1	1	1
									-145,59
Scenario1									
phase-A									
Ship1	5850	1	1	1	1	1	1	0	0
Ship1	5850	0	0	0	0	0	0	1	1
phase-B									
Ship1	0	0	0	0	0	0	0	0	0
Ship1	4280,5	0	0	1	1	1	1	1	1
									-75,096

Figure 7. Economic results for different scenarios

Costs and benefits are calculated for different scenarios of STS cranes locations per ships Fig. 7,8. Respecting the order of arrangement of STS cranes various scenarios differ in their allocation to individual vessels.

Output per hour	The dwell time at the quay	demurrag	despatch	limit	Result
128	45,703125	0	114,84	48	
106	45,703125	0	22,969	48	
0	45,703125	0	114,84	48	
170	51,61764706	361,76	0	48	
		-361,76	114,84		-246,92
156	37,5	0	525	48	
78	37,5	0	0	48	
0	37,5	0	525	48	
170	54,70588235	670,59	0	48	
		-670,59	525		-145,59
184	31,79347826	0	810,33	48	
50	31,79347826	0	810,33	48	
0	31,79347826	0	810,33	48	
170	56,85421995	885,42	0	48	
		-885,42	810,33		-75,096

Figure 8. Calculation module

The existing articles on managing a container terminal as a system are trying to simulate all elements or a subset of activities as a followed predefined hierarchy. The main contributions are to maximize overall terminal efficiency or the efficiency of a sub-area of the terminal (Fig. 9)[2].

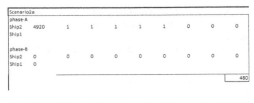

Figure 9. The influence of the number of STS used to unload a ship on the economic results

6 CONCLUSIONS

The analysis of the turnover of containerized goods in polish seaports as well as proposed model enable to notice the dependence of handling equipment, especially number of STS cranes, their cycle times and the length of the berth on vessel service efficiency in container terminals. The indicator of quay usage is essential for planning long term terminals' investments in infra- and supra-structure. That could be used to assume the terminals needs at the time of dynamic changes on container transport market. Created model helps to optimize the configuration of ships served in container terminal by taking into account different configurations of ship's lengths and capacities and also different crane cycle times. According with model results and the author's hypothesis the automation reloading processes have influence on a efficiency of port operation.

REFERENCES

[1] Adi Budipriyanto, Budisantoso Wirjodirdjo, Nyoman Pujawan, Saut Gurning, *Berth Allocation Problem Under Uncertainty: aconceptual Model using Collaborative Approach,* Industrial Engineering and Service Science, 2015

[2] Aravindan S., Thiruvenkatasamy K., *An Analysis on the Modeling of Container Terminal Operations,* Indian Journal of Science and Technology, Vol 9(39), 2016.

[3] Bradley Skinner, Shuai Yuan, Shoudong Huang, Dikai Liu, Binghuang Cai, Gamini Dissanayake, Haye Lau, Andrew Bott, Daniel Pagac, *Optimisation for job scheduling at automated container terminals using genetic algorithm,* Computers & Industrial Engineering, 2012

[4] Elentably A., *Simulation of a Container Terminal and it's Reflect on Port Economy,* TransNav, the International Journal on Marine Navigation and Safety of Sea Transportation, 2016, Tom Vol. 10 nr 2, pages 331—337.

[5] Kaizer Adam, Leszek Smolarek., *The analysis of dredging project's effectiveness in the Port of Gdynia, based on the interference with vessel traffic,* In: A. Weintrit, T. Neumann (eds.), Safety of Marine Transport - Marine Navigation and Safety of Sea Transportation, CRC Press/ Balkema, Taylor & Francis Group, 2015, page 167-172

[6] Kaizer Adam, Ziajka Ewelina, Truszczyński Mateusz., *Ocena założeń rozwojowych trójmiejskich terminali kontenerowych,* Inżynieria Morska i Geotechnika, 2016.

[7] Kaup Magdalena, Chmielewska-Przybysz Maja, *Analiza i ocena technologii wykorzystywanych do obsługi kontenerów w portach morskich,* Logistyka, vol 5, 2012

[8] Krośnicka Karolina, *Rozwiązania przestrzenno-technologiczne zwiększające płynność dostaw ładunków do portowych terminali kontenerowych,* Logistyka, vol. 3, 2014

[9] Lubiński K , Cirocki K., Bargieł K., Meyer M., Trella P., Miszke W., *Conveyor- different approaches based on student - made model,* 20TH IEEE International Symposium on Industrial Electronics, ISIE 2011, Gdańsk, Poland. page. 1393 - 1396

[10] Martin Jeffrey, Martin Sally, Pettit Stephen, *Container ship size and the implications on port call workload,* International Journal of Shipping and Transport Logistics, vol. 7, 2015.

[11] Salomon Adam, *Organizacja i funkcjonowanie portowych terminali kontenerowych oraz perspektywy ich rozwoju,* Zeszyty Naukowe Akademii Morskiej w Gdyni, 2013, pages 70-80.

[10] http://www.port.gdynia.pl 15.02.2017
[11] https://www.portgdansk.pl 20.02.2017
[12] http://www.port.szczecin.pl 23.02.2017
[13] http://dctgdansk.pl 23.02.2017
[14] http://www.gct.pl 15.02.2017
[15] http://www.bct.gdynia.pl 24.02.2017
[16] http://otport.swinoujscie.pl 18.02.2017
[17] http://portszczecin.deutschebahn.com 05.01.2017
[18] http://www.gtk-sa.pl 13.02.2017

Development Investments at Container Terminals in the Case of Cargo Congestion

M. Truszczyński & A. Pezała
Gdynia Maritime University, Gdynia, Poland

ABSTRACT: The increase in the volume of container traffic is a constant tendency. The growth of container ships is a chance to optimize the operating costs of ship owners. Effect of this is push the market for smaller units. The container terminals is under pressure to expand the infrastructure and technology that will handle this volume of cargo. Long investment period makes that congestion may be a result of using fully handling potential of the terminal. Large parts of the cargo handled in a short time as in the case of operating ULCV by terminals may also contribute to the causing of congestion. Congestion is undesirable and causes negative effects in economic, environmental and social. The aim of the article is the recognition and classification of congestion occurring at container terminals, to develop a model decision-making in case of congestion and the analysis of solutions to reduce the risk of congestion.

1 INTRODUCTION

1.1 *Current situation on container shipping market*

Year 2016 and beginning of 2017 was continuation of difficult situation within container shipping market. Over the years the size of container vessels were growing up to accommodate increasing volume of containerized cargo. The gaps between accessing generations of container ships were appearing earlier and were bigger then between former generations. (Fig 1.) From first container vessels (generation of early containerships with the example of Ideal X) it took almost 15 years to appear next generation (fully cellular containerships –second generation). From second generation it took next 15 years to reach new size of vessels. New generation was Panamax, appeared on 1985 and had capacity of about 4000 TEUs (Twenty-feet Equivalent Unit). The next generation appeared in 1996 with Post Panamax I with size of 6600 TEUs. When also barrier of 6600 TEUs was breached it won't take long time to reach vessels with capacity reaching 8000 TEUs. This generation started to oppress ports in dredging of port canals. Ten years took to see alive next generation of vessels – Post Panamax III, called also E-class from the beginner of this class Emma Maersk. The range of this class war from 11000 to 14500 TEUs. The last existing for this moment class (Triple-E) appeared in 2013. Vessels

of this class have capacity from 18000 to 19200 TEUs. This class characterize with 3 major features: "economy of scale", "energy efficiency" and "environmentally improved". Key factors for the current situation on shipping market have "economy of scale" and "energy efficient". Economy of scale had to be an advantage that due to the large capacity of vessels, ship-owners had to minimalize incurring transport cost per one TEU. Energy efficiency of new builded vessels had to ensure that in spite of the grooving size of vessels they characteristic of fuel consumption was similar.

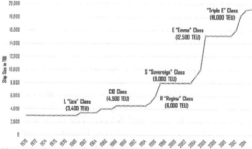

Figure 1. Growth of container ships through years (source: https://people.hofstra.edu/geotrans/eng/ch3en/conc3en/largestc ontainerships.html 15.01.2017)

Larger size of Triple-E class ensure without magnifical change of vessels sizes (only 3 m longer,

4 meters wider, and have only immersion only 0.5 m bigger than Class-E), despite negible changes in sizes, new vessels were able to carry 2,500 TEUs more. Decision to built Triple-E class was introduced after economical crisis in 2008, moreover launching of new ULCC (Ultra Large Container Carrier) were

just after economic slowdown in 2012. (Rodrigue 2012). Maersk ability to cut down freight charges caused that other ship-owners followed Maersk and also ordered new ULCC. The number of that vessels affected in chain reaction. New ones increased the existing overcapacity, which caused the further drop in freight charges so that next ship-owners were following and ordered new vessels to cut down operational costs.(Kyunghee 2017)

1.2 *Merges, alliances and acquisition in shipping market*

Ship-owners to deliver to customer frequent services from diversified ports are forced to cooperate with their opponent. One of the most significant way of cooperation is to be formed in alliance. Operating within alliances enable ship-owners further development of port network without investing in new vessels to cover new services. They can serve new routes by sharing slots with their partners from alliance. Ship-owners operating within alliance are able to further investments in larger and larger vessels. So far (since second quarter of 2015) four of them was present in container shipping market:
– M2 (Maersk Line and MSC)
– Ocean Three (CMA CGM, China Shipping, UASC)
– CKY(H)E (Hanjin, K-Line, Yang Ming, COSCO and Evergreen)
– G6 (APL (NOL), MOL, HMM, Hapag-Lloyd, NYK and OOCL).

Anyway the situation within industry are tough, 2016 shocked container carriers business with bankrupt of Hanjin. Further operating on dumped freight rates are unable for ship-owner. The only possible solution is consolidation of market by merges. In 2016 shipping business were witness of merge of China Shipping and COSCO. Takeover UASC by Hapag-Lloyd or APL by CMA CGM, which all off them were belonging to competing alliances, oppressed industry to reorganize commitments and agreement. The next merge will be forming of joint venture by Japanese ship-owners (MOL Liner, K-Line and NYK). The next possible consolidation is acquisition of Hamburg Sud by Maersk Line. Marges and acquisitions meaningly effect on alliances performance so that they are going to transformate.

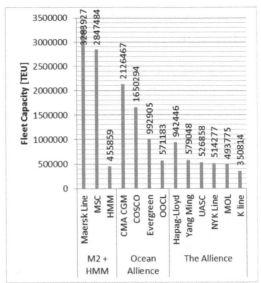

Figure 2. Fleet capacity according to alliances (source: author study on the basis: Alphaliner, TOP 100 - Existing fleet on February 2017)

From the beginning of April 2017 two new alliances will enter into force (Fig 2.) as:
– The Alliance
– Ocean Alliance
These two alliances will replace: Ocean Three, CKY(H)E and G6. Changes also affect on M2 to which join HMM. (Notteboom 2016)

2 CONGESTION PROBLEM

Weak position of seaport and sea terminals pointed A. Przybyłowski. Plans for development and investing policy on containers terminals are significally connected with the policy of ship-owners. Although terminals mainly operates on their own, they have difficulties with adjustment their plans to the changes in connecting shipping with rail, road and inland modes of transport. Ports are not longer controlling the cargo flow in inland market, due to fact that control of sea transport chains are mainly under control of ship-owners. Ports and terminals have to invest enormous sums of money in development of infrastructure and superstructure. In spite of investment process containers terminal are in danger of cargo congestion. (Przybyłowski 2009)

As James Kiritsis from Drewry pointed while "Transport Week" 2017, alliances encourage ship-owners to invest in big vessels. Bigger ships affects in changes in patterns of volume exchange. Numbers of services provided by ship-owners were decreasing in line with uprising average vessel size. This patters due to cascade effect affect on almost each port which have to struggle with peaks in cargo flow.

Table 1. Comperation of terminals served by 7500 TEUs vessel and 15 000 TEUs vessel

Vessel size TEU	15000	7500
No. of calls a week	2	4
No. of moves pre call	5000	2500
Times [h] of operation with 6 operating cranes with efficiency 25 moves/hour	33	16,5

Terminal have to handle more container in cumulated time. (Tab. 1) Peaks in cargo flow affects in threat of congestion. Time operation on the ship is connected with the crane productivity (Fig.1) (Kiritsis, unpubl.). High density of containers unloaded in short period affect in higher demand in use of superstructure and higher peak manning.

We can divide three main of terminal where congestion can occur: Berth, Yard and Gate.

Berth congestion can affect in:
− Ship berth congestion − which forces vessels to wait in queue to call. It is result of extension of occupating berth by other vessel. It can be also result on accumulation of vessel to be operated in terminal after severe weather condition.
− Ship handling congestion − this congestion affect in extension in unloading/loading of vessel. It could be effect in inadvertence in process of loading and unloading as well as a problem in transportation of containers to yard.

Yard congestion could affect in:
− Cargo stack congestion − which is a result of extending landfill of containers within the yard. Terminals which are close to fulfill their storage capacity have problems with container handling.
− Vehicle work congestion − happens when road trucks which entered to terminal have to wait unusually long time to be un/loaded. Usually this is a result of low efficient operations within landside or lack of required equipment to serve whole landside flow of cargo.

Gate congestions are usually an effect of unsuitable way of trucks notification to terminal. It results in extending of waiting time in queue before entering truck into terminal. This congestion can be caused by accumulation of truck arrivals to terminal in certain hours or by low terminal operation efficiency due to which additional trucks are not allowed on the terminal. (Gidago 2015)

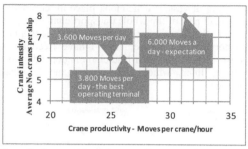

Figure 3. Crane productivity according to number of cranes (source: Kyritsis, J. 2016 Big ships, big alliances An impact on regional markets)

2.1 Ways investigate and eliminate cargo congestions

First of all congestion is nonmeasurable but can be treated as things generating extra, unreasonable costs for terminals. The action that is needed to be taken in every terminal is to control average quantity of stored containers separately for export and import. This gives an opportunity to predict cargo congestion. The capacity utilization which should warn about possible delays in cargo flow through terminal is volume exceeding 90% of terminal storage capacity. Despite that piece of advice, for some terminals even exceeding their storage capacity does not affect in generation of cargo congestion. Whenever it could result in higher operational cost and lower effectivity of terminal is going down. In case of congestion appearance the first step which should be done by terminal operators is identification of existing bottlenecks inside the terminal. The largest possibility of congestion is in the landside of terminal, the least possible is to observe congestion in seaside.

There are a lot of different solutions of solving that problem that have already become. In article we will focus on strategic and tactical investments which can solve congestion problem. Firstly, cargo congestion can be minimalized by better cooperation between different parts of transport chain. Secondly, cargo congestion can be solved by increasing terminal area. This is quite expensive and often impossible because of space lack. Last but not least, some changes on exploitationally-technical areas could be made so it does not cause any need to increasing terminal area.

2.2 Cooperation and integration inside the transport chain

Integration in shipping industry should take place not only between ship-owners. Containers allowing to transport goods in one transport unit by various transport modes (ships, trucks, railroad cars, barges) help to organize door to door transport. Loading and unloading of vessels which took place in terminal is only small part of the whole transport chain. In spite of fact that there are many parts of container transport chain, shipper doesn't have freedom to choose each of part. Sea-land chain is depending on shipping lines, which choose port and terminal coverage.

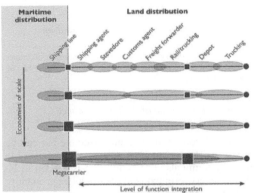

Figure 4. Integration of container chain distribution parts (source: Rodrigue, J. The Sage Handbook of Transport Studies, page 75)

Ship-owners during last years gives their attention into horizontal integration by merges, acquisitions and forming alliances, however now they should focus on vertical integration (Fig. 4).Vertical integration will give ship-owners not only incomes from most proficient part of supply chain (from logistic service providing) but it also improve transparency of supply chain. More effective flow of information between involved parties will affect in better managing of container flow (Rodrigue & Notteboom & Shaw 2013). For example in situation when both shipping line, terminal, trucking or rail company and logistic operator have access to loading plan of container vessel, export of containers could be organized at the beginning for containers unloaded from the vessel as first, excluding the period of stacking container in terminal. McKinsey admit that in spite of some barriers the future of alliances is integration in landside operation which will be horizontal and vertical integration at one moment. That operating alliances could exchange information about terminal operation a live. Then few terminals operating in one port could be treated as one terminal and then call of ships could be menaged in most efficient manner and ships would unloaded by the least occupied quey. (Joerss, Murnane, Saxon & Widdows 2015).

2.3 Improving of efficiency in berth and yard operations.

The easiest way to improve berth operations is multi-pick lifting. Twin-lift enables the STS to lift two 20-in containers, tandem lift enable to lift two 40-in containers. (Song & Panayides 2012). The next possibility is to implement fully automated STS cranes. They are remote operated and equipped with two trolleys. One which is operated manually unload containers from vessel, and move them to elevated platform on landside of the crane, the second one, operated automatically transfer the containers from platform straight onto on AGV (Automated Guided Vehicle). (Gharehgozli, Roy & Koster 2014)

The most important way to improve yard operations is implementation of automated yard marshalling. In such kind of terminals internal transport are performed usually by AGV which is driverless and run inside the terminal following standard track which are pointed by system of wires or transponders. Containers transported by AGV are stored on stacking area by ASC (Automated Stacking Crane). ASC is characterized by better productivity and higher yard utilization than traditional RTG (Rubber Tire Gantry) (Sharif Mohseni 2011).

As far as ships handling is concerned, also mentioned by Malchow Port Feeder Barge can be used. Due to its size, capacity, and technical possibilities it can be used as proper solution for cargo congestion. According to Malchow "Other major and even minor ports could benefit from the operation of this innovative type of vessel as well as it improves the efficiency and at the same time reduces the ecological footprint of intra-port container haulage. Additionally it can even facilitate container handling at places which are not suited at all for this kind of operation." (Malchow 2014) As it is not widely used in Ports of Europe it could be considered as solution helping to reduce berth congestion.

2.4 Dry Ports and extended gateways

According to F. Lannone, in European Union quarter of inland freight is related with port. Furthermore, huge demand on truck transportation still dominating in container distribution. Because of its characteristics trucking could be one of factors improving cargo congestions. Another unfavorable one is customs dwell time caused by customs activities. Visible view into documents, essential information and also exchange of data is crucial for hinterland transportation. (Iannone 2013) Due to that fact, effective hinterland processes provide effective cargo handling on terminal. According to fact, that demand on transportation is still very high, information exchange is very important to optimalise service of containers. Due to fact that ships that need to be serviced is getting bigger and terminals needs to be fastly released from containers, conception of extended gateways and dry ports was created. In general the clue of extended gateways is to push containers from sea terminals to inland terminals with customs clearance and connected procedures. According to "Collaborative Information Services for Container Management" project run by COMCIS, extended gateways was introduced in such terminals as Europe Container Terminal in Rotterdam. As organization claims, extended gateways are favorable both to environment and customers. They gives an opportunity to use multimodal transportation in better and more effective way, so not only emission of CO_2 is lower but also it makes transportation cheaper and more effective than transportation only by road. By European Gateway Services (EGS) ECT offer more efficient, cheaper and easiest transportation of container to customer. COMCIS project was to develop that services to make it more predictable and connected with ECT in better way. (Transport Research & Innovation Portal) Before implementation of COMCIS project, actual position of container on vessel was unknow, so customers usually added 24 hours to Actual Time of Arrival (ATA). To make organizing of inland transportation more effective, COMCIS wanted to combine all conformations about container to support EGS and to make possible to release containers to inland terminals faster. That implementation caused with removing bottlenecks form terminal so possibility of cargo congestion was reduced. (COMCIS 2017)

Considering extended gateways dry ports cannot be not mentioned. They can be a part of extended gateways like on ECT, but can only be a part of sea terminal. Basically purposes of dry ports are quite similar to other container terminals, but dry port is not only a container terminal. Firstly they are directly connected with port mostly by rail but also by road. Dry ports are the most advanced type of inland terminals, so they can be said to be a part of port, but moved into hinterland. They are a point where different modes of transport meets and cargo can be transshipped, cleared through customs and shipped directly to port. They can be discussed as extended gateways that shipper can see as adequate terminal to port what is the main difference between dry port and inland terminal. Their main clue is to be a terminal, which release sea terminal from cargo congestion. Dry port is terminal directly connected by rail with one or several ports and serve all logistics facilities that are needed to forwarding and shipping container such as custom clearance, transshipment, containers handling and reparations, storage of containers and many more (Trainaviciute, Bentzen, Caruso, Laugesen, 2009).

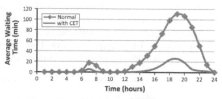

Figure 5. Time of service of truck according to part of the day, (source: Dekker, Heide, Asperen, Ypsilantis, 2015)

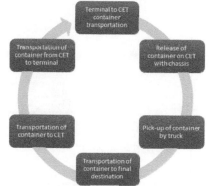

Figure 6. Flow of containers via CET, (source: authors study according to Dekker, Heide, Asperen, Ypsilantis, 2015)

As a dry ports can be seen as quite similar to extended gateway conception, and in general the main difference is difference in ownership. This two solutions in literature are usually presented together as two similar, so developing dry port description in this article is not necessary, because of fact, that it can be seen as one in general.

2.5 Chassis Exchange Terminals (Dekker, Heide, Asperen, Ypsilantis, 2015)

Another bottleneck in container terminal are gates for truck service. Because of big demand on road transportation, fast and effective service of trucks on terminal are very difficult so it easily lead to congestion. As studies on Los Angeles and Long Beach ports show, more than 40% of truck are waiting for release of container more than 2 hours. It is caused by time-consuming truck handling process. Operation scheme is simple. Firstly, there is a pre-announcement, check of documents, clearance, then trucker is permitted to drop off or pick up of container. After that, goes to exchange point, arrives at gate, trucker and container are checked, container is loaded or unloaded. Finally, trucker goes to the gate one more time, customs is checked and after final check he can leave the terminal. What's more, number of trucks on terminal is not constant in time. Quantity of trucks per hour depends on day of the week and hour of the day. There are several points in a day where number of truck is sublime. Also number of trucks in Friday are smaller than in Monday in the same hours. Time that trucker spend on terminal is unfavourable for both sides: a terminal and trucking company. Due to that problem that is needed to be solved conception of Chassis Exchange Terminals was created. This kind of terminals are dedicated to road transportation and their main aim is to reduce time needed to service of trucks to minimum.

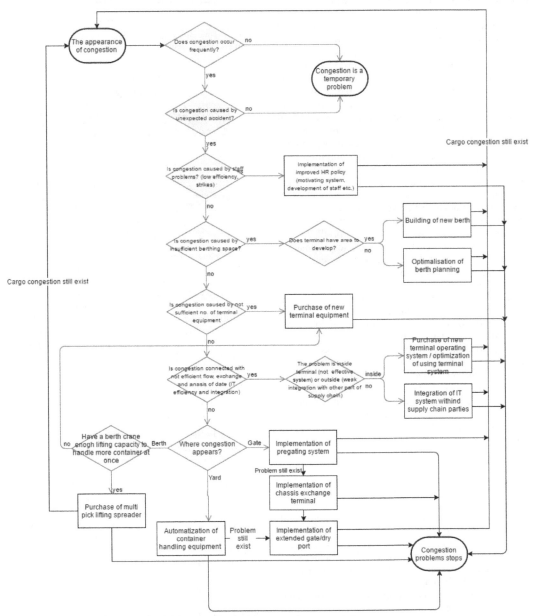

Figure 7. Cargo congestion preventing investment. flowchart, source: authors own study

Containers operated by CET have to cover following way. (Fig. 6) Containers are transported by terminal trucks from sea terminal to CET, where they are be picked up by other trucks to transport them to final destination. Then, chassis with container is turned back to CET, where terminal truck pick up it and transport to terminal. This flow of containers can result with smaller number of trucks waiting on terminal gates, and minimalization empty truck running. As a main advantage shortage of service time must be mentioned. According to Fig. 5, time of service can be meaningfully reduced. As it can be seen (Fig. 5) on terminal in rush hours truck could be serviced up to 2 hours. In CET time is reduced to nearly half an hour. It is essential according to regulations about drivers work time. As a result of faster pickup or release of container, driver don't lose his work time for waiting, what is essential if he have to drive under restrictive regulations about his work time.

However, this solution has its drawbacks. Firstly, this type of terminals needs a lot of space. Containers can not be stacked and the space for save service of trucks is needed. What's more, to effective work of CET road transportation in very good condition is needed. CET must be localized nearly to highways and must enable fast and effective transportation of containers between CET and sea terminal, so that distance between then shouldn't be too big. Due to that facts, building CET could be impossible because of lack of space, but despite that fact, CET are worth considering as solution for cargo congestion at gates.

2.6 *Pre-gates*

As trucks arriving to terminal have to be announced and checked, due to cargo congestion on terminals it could be shifted to prior steps of transportation. As an example of pre-gating system "e.Brama" in DCT Gdańsk (Deepwater Container Terminal Gdańsk) can be shown. This year this due to opening of second berth on this terminal, to prevent cargo congestion and to make service if trucks more efficient, test of pre-gating system started. In general, pickup of container is possible after announcement of its avaibility and custom clearance. (DCT Gdańsk 20.01.2017) Pre-gate system enables checking how much time does it take to service one truck on average, how much trucks are waiting or what ships are already on berth. System is in last stage on introducement, so data about difference in truck service is not fully avaible now. (Deepwater Container Terminal 10.01.17) Similar system "eModal" is running in Port of Virginia. Single or multi-visit can be done on different terminals. User also can choose if trucker will pick up a container or release empty one on terminal. (Port of Virginia 02.01.2017) Pre-gating system enables to announce

trucker before he arrive to terminal, due to that fact time of service is shorter so efficiency is higher. According to information from Port of Botany in Australia in 2006 DP Word terminal noted that after introducing pre-gate system percentage of trucks services in 1 hour increased from 73% to 85%. (Davies 2009) That numbers can show, that pre-gates system are change, that will help to provide effective, fast and professional service of trucks on terminal what can prevent cargo congestion.

2.7 *Solutions summary*

On basis of literature overview general model of decisions in case of cargo congestion was created. (Fig. 7) Its aim is to help in choosing solutions to eliminate arised problem. Model could operate in strategic and tactical wide time horizon, it does not have application for temporal problems. Presented graphic is based only on general overview, due to that fact it may does not be only appropriate source of information. Every decision about solving cargo congestion must be based on actual geographical, economical, social or environmental factors.

3 CONCLUSION

Congestion can occur in terminal due to various causes. Some circumstances are inevitable, like the delay caused by severe weather conditions, arriving of vessel which operation in previous terminal was extended or by accident. However terminal operators should gather information from various sources to match to expectation of ship-owners. There are some investments which are possible to implement without involving large amount of money notwithstanding we should remember that investment must be analyzed in term of the possibilities to eliminate bottlenecks. Investments must me designed to suit to anticipated throughput of containers. We should also remember that each investment must be analyzed in respect of specific case because each terminal have their own unique specification.

REFERENCES

Alphaliner, TOP 100 - Existing fleet on February 2017, source http://www.alphaliner.com/top100/ access 22.01.2017
COMCIS official website. Access: 14.01.17 source: http://www.comcis.eu/ect.html
Davies, P. (2009) Container Terminal Reservation Systems, Paper presented at the 3rd Annual METRANS National Urban Freight Conference
Deepwater Container Terminal, Podręcznik użytkownika systemu awizacji samochodów ciężarowych Access: 10.01.17 source: http://dctgdansk.pl/wp-content/uploads/2017/01/e-brama1.8.pdf

Dekker, R. Heide, S. Asperen, E. Ypsilantis, P. (2015) A chassis exchange terminal to reduce cargo congestion at container terminals Access: 07.01.17 source: https://www.researchgate.net/publication/257562698_A_ch assis_exchange_terminal_to_reduce_truck_congestion_at_c ontainer_terminals

Gharehgozli, A.H., Roy, D., Koster, R. 2014 Sea Container Terminals: New Technologies, OR models, and Emerging Research Areas, SSRN Electronic Journal,

Gidago, U. 2015. Consequences of Port Congestion on Logistics and Supply Chain in African Ports, *Developing Country Studies* Vol 5, No 6 (2015) 160-167

Iannone, F. (2013) Dry Ports and the Extended Gateway Concept: Port-Hinterland Container Network Design Considerations and Models Under the Shipper Perspective. Access: 18.01.2017 source https://ssrn.com/abstract=2320394

Joerss, M., Murnane, J., Saxon, S. Widdows, R., 2015 Landside operations: The next frontier for container-shipping alliance, access: 20.01.2017, source: http://www.mckinsey.com/industries/travel-transport-and-logistics/our-insights/landside-operations-the-next-frontier-for-container-shipping-alliances

Kyritsis, J. 2016 Big ships, big alliances An impact on regional markets, Paper presented onTransport Week 2016

Kyunghee, P. 2017, Asia's Shipping Lines Are Facing More Mergers, access 12.01.2017 source: https://www.bloomberg.com/news/articles/2017-01-03/swim-or-sink-outlook-prompts-asia-shipping-lines-to-face-mergers

Malchow, U. 2014 Port Feeder Barge: Advanced Waterborne Container Logistics for Ports, *TransNav,* vol 8. No 3, 411-416

Notteboom, T. 2016, PortGraphic: alliances in container shipping-impact of Korean/Japanese carrier dynamics, access 01.01.2017 source: http://www.porteconomics.eu/2016/11/01/portgraphic-alliances-in-container-shipping-impact-of-koreanjapanese-carrier-dynamics/

Port of Virginia, Proffesional Trucker Orientation Brochure. Access: 02.01.17 source: http://www.portofvirginia.com/pdfs/tools/APMTVA_OTRDriver.pdf

Przybyłowski, A. 2009. A. Challenges for Polish Seaports' Development in the Light of Globalisation Processes in Maritime Transport, *TransNav*, vol 3. No 4. 2019, 457-462

Rodrigue, J.P, Evolution of Containerships, 2012 access: 10.01.2017, source: https://people.hofstra.edu/geotrans/eng/ch3en/conc3en/containerships.html

Rodrigue, J.P, Notteboom, T, Shaw, J. 2013, *The SAGE Handbook of Transport Studies,* Londonm The SAGE Publications Ltd.

Sharif Mohseni, N. 2011, Developing a Tool for Designing a Container Terminal Yard, Access: 20.01.2017 source: http://repository.tudelft.nl/islandora/object/uuid:020efc36-c130-4429-a1b6-7028235400ab?collection=education

Song D.W, Panayides P.M. 2012, *Maritime Logistics: Contemporary Issues*, Bingley, Emerald Group Publishing

Trainaviciute, L. Bentzen, K. Caruso, A. Laugesen, M. (2009) The Dry Port – Concept and Perspectives. Access: 05.01.2017 source: http://archive.northsearegion.eu/files/repository/20130301142236_WPC-TheDryPortConcept.pdf

Transport Research & Innovation Portal, Collaborative Information Services for Container Management. Access: 15.01.17 source: http://www.transport-research.info/project/collaborative-information-services-container-management

Maritime Education and Training (MET)

Innovation Methods of Assessment and Examination System for Universities Engaged in Bologna Process

I. Sharabidze
Batumi State Maritime Academy, Batumi, Georgia

ABSTRACT: Generally, the teachers and professors of the universities engaged in Bologna process make assessment according to syllabuses and at their discretion. The frame for the assessment system should be created for the purpose to evaluate students' skills and competence from different angles – in particular: written, oral, presentation, etc. Assessment enables students to demonstrate that they have fulfilled the objectives of their course and achieved the required standard. Assessment also helps students to reflect on their learning, and to recognize and enhance their achievements.

Therefore, we elaborated unified system, which is divided into 4 components per semester. Semester is subdivided into 3 exams on a five (five) week bases. I and III five (five) week bases exams are held by the subject teacher, who is provided 7 types of assessment prior to exams, at the beginning of the semester. II five (five) week bases exam and the final exams are held by Examination Centre.

The Assessment System:

I on a five (five) week bases – 20 credits

II on a five (five) week bases – 20 credits

III on a five (five) week bases – 20 credits

Final Exam – 40 Credits

Above mentioned assessment and examination system provides us with full information regarding students' learning, skills, and understanding; besides writing, oral and presentation skills. Taking into account all above mentioned at the end we have professional graduates who duly correspond with STCW Convention (Manila amendments) requirements.

1 INTRODUCTION

World maritime educational process is regulated by STCW Convention including 2010 Manila amendments and IMO Model Courses as recommendations, which promotes designing the syllabuses, distribution of hours, etc.

Let us see the vital importance of STCW Convention on maritime educational system discussing below mentioned tables:

Part A. Chapter II. Table A-II/1for Navigation at the operational level, Table A-II/2 for Navigation at the management level, Table A-II/3 for officers in charge of a navigational watch, Table A-II/4 for Navigation at the support level, Table A-II/5 for ratings as able seafarer deck; A-III/1 for officers in charge of an engineering watch, A-III/2 for chief engineer officers and second engineer officers, A-III/3 for chief engineer officers and second engineer officers, A-III/4 Marine engineering at the support level, A-III/5 Marine engineering at the support level, A-III/6 Electrical, electronic and control engineering at the operational level, A-III/7 for electro-technical ratings, A-IV/1 for certification of GMDSS radio operators.

These tables give us clear view what are the main criteria for seamen to meet all requirements: 1. Competence, 2. Knowledge, understanding and proficiency, 3. Methods for demonstrating competence, 4. Criteria for evaluating competence.

Therefore, it's mandatory for the students of all higher educational institutions to meet all the requirements mentioned above. Coming out of these requirements students should be able to demonstrate and prove their competence both verbally as well as in written form.

Practice over the years showed that a full-time student would need to complete 60 ECTS per academic year, which represents 240 ECTS for 4 academic years. During this period evaluation process should be fulfilled with different means, coming out of the specification of the subject and additionally if the subject gives us the opportunity we should break down the methods of learning per subject.

Researches showed that it's really compulsory to use different methods of teaching for high educational institutions where the student of bachelor level completes 240 ETCS for 4 academic years.

We present you the special scheme worked out by the specialists of our academy:

LEPL - Batumi State Maritime Academy

Faculty			Semester		
Academic Department			Group		
Professor/Teacher			**Teaching Load**		
Subject			Lecture	Practical	Laboratory
Academic Year					

	Ist 5 Week Bases Assesment							IInd 5 Week Bases Assesment	IIIrd 5 Week Bases Assesment						
	1	2	3	4	5	6	7		1	2	3	4	5	6	7
	TEST	Presentation	Verbal	Written	role playing	Group Work	Questions/ answers	Computer base Test	TEST	Presentation	Verbal	Written	role playing	Group Work	Questions/ answers
Date															
Auditorium															

Note:	Date	the date and hour of assesment, according to the methods of assesments chosen by teacher
	Activity	90 % and more - 5 | 90 % - 70 % -4 |70 % - 50 % - 3|50 % - 30 % - 2
	Assesment	If during the five week period a student has 30% or more absence the assesment of a teacher will be reset.
		The Student will have right to pass the final exam

	Name/Surname of a Student	Ist 5 Week Bases Assesment																														
		1	2	3	4	5	Σ	1	2	3	4	5	Σ	1	2	3	4	5	Σ	1	2	3	4	5	Σ	1	2	3	4	5	Σ	
1																																1
2																																2
3																																3
4																																4
5																																5
6																																6
7																																7
8																																8
9																																9
10																																10
11																																11
12																																12
13																																13
14																																14
15																																15
16																																16
17																																17
18																																18
19																																19
20																																20

IInd 5 Week Bases Assesment

Name/Surname of a Student	1	2	3	4	5	Σ	1	2	3	4	5	Σ	1	2	3	4	5	Σ	1	2	3	4	5	Σ	1	2	3	4	5	Σ	
1																															1
2																															2
3																															3
4																															4
5																															5
6																															6
7																															7
8																															8
9																															9
10																															10
11																															11
12																															12
13																															13
14																															14
15																															15
16																															16
17																															17
18																															18
19																															19
20																															20

IIInd 5 Week Bases Assesment

Name/Surname of a Student	1	2	3	4	5	Σ	1	2	3	4	5	Σ	1	2	3	4	5	Σ	1	2	3	4	5	Σ	1	2	3	4	5	Σ	
1																															1
2																															2
3																															3
4																															4
5																															5
6																															6
7																															7
8																															8
9																															9
10																															10
11																															11
12																															12
13																															13
14																															14
15																															15
16																															16
17																															17
18																															18
19																															19
20																															20

Grading system in Georgia is based on 100 scores. According to the requirements of "Law of Georgia on Higher Education" the score on final exam should be 40 and in written form. Which means that we have 60 scores which should be distributed in the following way: one semester is divided into three 5 (five) week's bases exams. The Ist 5 (five) week's bases exam has 15 scores, the IInd (five) week's bases exam has 25 scores, IIIrd (five) week's bases exam has 20 scores. 5 scores in each exam is given to students' activities, presence on lectures and discipline, the second 5 scores is given to teachers' assessment of a student during 5 weeks, as for the rest of the scores see the following: 5 scores in the Ist 5 (five) week's bases exam is given to a teacher, 15 scores in IInd (five) week's bases exam – computer bases test, 10 scores – teachers' assessment. For your information, the IInd (five) week's bases exam, which is computer based test and the final exam is carried out by examination and assessment center of the academy. Professors have right to choose the method of assessment from 1-7 (mentioned in the table above).

Above mentioned assessment and examination system provides us with full information regarding students' learning, skills, and understanding; besides writing, oral and presentation skills. The students show us the knowledge and competence they achieved during studies from different angles, like verbal, written, presentation, group work, etc. As for monitoring of the system, it is monitored by the subject professor, as well as by examination and assessment center, which gives us the full picture of students' skills and competency.

REFERENCES:

http://www.mes.gov.ge/?lang=eng
http://www.classbase.com/Countries/Georgia/Grading-System
STCW Convention (Manila Amendments)
"Law of Georgia on Higher Education"

Effects of Deck Cadets' Working Conditions on Quantity and Perceived Quality of Sleep Among Marine Science Students

H. Yılmaz, E. Başar & A. Ayar
Karadeniz Technical University, Trabzon, Turkey

ABSTRACT: Deck officers are among to member of crew with important responsibilities which includes cooperation with the Master during voyage, being involved in navigation and port watching as well as maintenance of the ship and its safety equipment. Students studying to become a deck officer aboard in commercial vessels, perform their practical training on board. They can be subjected to unusual working hours and inadequate rest periods during the mission on the ship. Considering the importance of adequate sleep for productivity, vigilance, sustaining attention and even over-all health and well-being, it is not only sleep quantity but sleep quality is also critical. The aim of this study was to investigate possible effects of the long-term onboard training (aprox. 7 months) on the sleep quantity and quality of the maritime students. Data regarding sleep quantity and perceptive quality among the subjects were obtained by using Pittsburg Sleep Quality Index (PSQI). PSQI is a scale providing information on type and violence of sleep disorders and the quality of sleep during the past month. A total of 60 maritime students were asked to fill PSQI before and after onboard training. Demographic information such as age, gender, weight, height and information regarding the internship were also obtained. Of the students 43.3% rated poor sleep quality experience before the internship and this ratio increased to 73.3% after the internship. A prominent decrease in sleep quality was determined. Sleep duration did not change significantly. This study showed a significant negative effect of onboard training on sleep quality among participants and they did not recover within a month after returning from the sea which indicates potential long term consequencies.

1 INTRODUCTION

Deck cadets are trained to become a watchkeeping officer in commercial vessels. According to the STCW 78 (amended in 2010) training standards, students studying in maritime schools are obliged to undertake 12 months of sea training. In Turkey, the three-part internship system is widely applied in the undergraduate programs that provide the qualification of oceangoing watchkeeping officer. Students studying in related programs perform their first two internships between June-September at the end of each academic year and their third internship between February and September during the third academic year (Turkish Seafarers Regulation 2016).

Maritime is a profession that involves a lot of physical and mental effort and many risks. Officers play a role in ensuring the safety of life and in the development of the healthy working attitude of the crew. Deck cadets are also marine students trained on board to be officers. Jeżewska et al. (2006) stated that students at sea are more vulnerable to the stress compared to experienced officers, due to reasons such as social relationships in the marine environment, physical burdens, lack of control and lack of support. The World Health Organization sees work-related stress sources as determinants of health and illness.

It is known that the shift system and working hours at the ship also affect the health of the seafarers.

Many studies in the literature show that working hours cause sleeping problems in seafarers (Hystad & Eid 2016, Jepsen et al. 2015, Yılmaz 2012). An individual should begin sleeping simultaneously with the biological clock for maximum benefit from the sleep. If sleep time is not synchronized with the biological clock, proper sleep becomes difficult.

The schedules of watchkeeping in vessels may conflict with the individual's biological clocks (Hystad & Eid 2016, Yılmaz et al. 2013, Calhoun 1999, Colquhoun et al. 1988). This situation can disrupt sleeping and sleep quality at a significant level, and adversely affect both the general health and the professional performance of the person due to lack of sleep.

Maritime students have to learn a wide range of tasks and procedures during the limited training period at the ship. They have to work with professional seafarers because they do job based learning. It is well known in the industry that deck cadets are used as "joker" member in reducing the workload of the officers and the crew. They are subjected to unusual working and resting hours for these reasons (Yıldız et al. 2016, Magramo & Gellada 2009).

When adaptation problems, stress, workload and working time factors are combined, it can be predicted that maritime students continue to their educational life on board and in the academy as "unhealthy".

In this study, it was aimed to investigate the effect of the long-term onboard training (aprox. 7 months) on the sleep quantitiy and quality of the maritime students. The Pittsburg Sleep Quality Index (PSQI) was administered to 60 students before and after the onboard training period.

2 METHODS

2.1 Participants

In this study, PSQI was applied to Turkish students who will take a long term internship, and their sleeping qualities during school life were determined firstly. This questionnaire has been applied in relation to the last month that does not include the exam process. Secondly, the PSQI was repeated a month later the students' disembarkation. Data collection was completed in 11 months (December 2015-November 2016).

2.2 Demographic Variables

The individuals involved in the study are required to have a seafarer's medical report, have not passed a disease within the past month, have to be taken to a long-term internship and have personal rations.

The participant's age, height, weight, gender data and the duration of internship, ship type, average daily work / rest / sleep / night sleep hours on board were obtained with the help of student information form.

2.3 Data Collection and Items

The Pittsburgh Sleep Quality Index (PSQI) is a self-rated questionnaire which assesses sleep quality and disturbances over a 1-month time interval. Nineteen individual items generate seven "component" scores: subjective sleep quality, sleep latency, sleep duration, habitual sleep efficiency, sleep disturbances, use of sleeping medication, and daytime dysfunction. Each component takes a score between 0-3 ("0" is very good, "3" is very bad) and the sum of 7 component scores is the global PSQI score. The global score has a value between 0-21. The higher the global score, the worse the sleep quality. A global PSQI score of ≤5 indicates "good sleep quality" and a score of >5 indicates "poor sleep quality". The poor sleep quality suggests that at least two of the above mentioned components are severely impaired or three of the components are moderate impaired. The index is known to be valid and reliable in Turkish society (Buysse ve ark.,1989; Ağargün ve ark., 1996).

2.4 Information of Participants' Internship

Participants in the study were trained in bulk carriers (15%), dry/general cargo vessels (23.3%), Ro-Ro (6.7%), container (11.7%) and tankers (43.3%). The average duration of internship is 6.9 months (SD 0.73).

The average daily study reported by students was between 10 and 21 hours (mean 15.4±2.4); mean daily rest between 3 and 14 hours (mean 8.2±2.1); mean daily sleep was between 3 and 10 hours (mean 6.2±1.4) and mean daily nighttime sleep varied between 0 and 8 hours (mean 5.2±1.8). "Daily sleep" refers to all the sleep processes that occur within 24 hours, the "night sleep" expression at midnight and the sleep processes that occur around it.

2.5 Analysis of Data

Participants' global PSQI scores, subscores, gender, age, type of ship, daily average work/rest/sleep/night sleep hours, watchkeeping system and duration of internship were assessed using excel data analysis tools. Frequency distribution, mean and standard deviations are taken into account in the evaluations.

As the values of the four women who participated in the study did not cause any significant differences in the average PSQI scores, they were evaluated in 60 people without discrimination between men and women in the calculations.

2.6 Limitations of the Study

The recurrence of PSQI questionnaire on board was considered while studying the effect of students' working conditions on sleep quality. However, the working method was limited to the comparison of pre- and post-internship interrogations. This limitation caused from the distances and communication difficulties stemming from the nature of the profession, and the thought of students hesitate from give accurate information in the working environment.

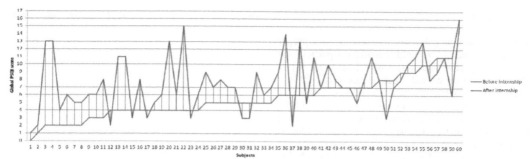

Figure 1. Subjects' global PSQI scores before and after internship

Table 1. Scores of global PSQI and subscales before internship (mean, standard deviation)

									Before Internship Scores of PSQI subscales							
		Global PSQI		Sleep quality		Sleep latency		Sleep duration		Sleep efficiency		Sleep disturbance		Use of sleeping medication		Daytime dysfunction	
	N	Mean	SD	Mean	SD	Mean	SD	Mean	SD	Mean	SD	Mean	SD	Mean	SD	Mean	SD
Gender																	
Female	4	3.5	1.0	1.0	0.0	0.5	1.0	0.5	0.6	0.0	0.0	0.8	0.5	0.0	0.0	0.8	1.0
Male	56	5.8	2.9	1.1	0.8	1.4	1.0	1.1	1.0	0.4	0.7	1.1	0.4	0.0	0.0	0.7	0.8
All	60	5.6	2.9	1.1	0.7	1.3	1.0	1.1	1.0	0.4	0.7	1.0	0.4	0.0	0.0	0.7	0.8
Age (Year)																	
21-24	51	5.6	2.9	1.2	0.8	1.3	1.0	1.1	1.0	0.4	0.7	1.0	0.4	0.0	0.0	0.7	0.9
24-27	9	5.4	2.9	1.0	0.7	1.2	1.1	1.1	1.2	0.2	0.4	1.2	0.4	0.0	0.0	0.7	0.7
Type of Vessels																	
Bulk carrier	9	4.8	2.5	0.7	0.7	1.3	1.1	1.3	1.0	0.0	0.0	1.0	0.5	0.0	0.0	0.4	0.5
Dry/General Cargo	14	6.4	3.8	1.4	0.8	1.7	1.1	1.0	1.1	0.6	0.9	1.1	0.3	0.0	0.0	0.6	0.9
Container	7	5.3	2.0	1.3	0.8	1.1	0.9	1.0	1.0	0.1	0.4	1.0	0.6	0.0	0.0	0.7	0.8
Ro-Ro	4	6.5	2.4	1.5	1.0	1.8	0.5	0.8	1.0	0.8	0.5	1.0	0.8	0.0	0.0	0.8	1.0
Tanker (Chemical/Cruide)	26	5.4	2.8	1.1	0.6	1.0	1.0	1.2	1.0	0.3	0.6	1.0	0.3	0.0	0.0	0.8	0.9
Daily Working Hours																	
10-12	9	4.6	1.7	0.9	0.6	1.2	0.8	0.7	0.9	0.2	0.4	0.8	0.4	0.0	0.0	0.8	1.0
13-15	13	6.0	2.8	1.1	0.6	1.6	1.0	1.2	1.1	0.4	0.8	1.1	0.5	0.0	0.0	0.6	0.8
16-18	35	5.8	3.2	1.2	0.8	1.2	1.0	1.2	1.0	0.4	0.7	1.1	0.4	0.0	0.0	0.7	0.9
19-21	3	4.3	2.5	1.0	0.0	1.0	1.0	1.0	1.0	0.3	0.6	1.0	0.0	0.0	0.0	0.0	0.0
Daily Resting Hours																	
3-5	4	5.3	2.8	1.0	0.0	1.0	0.8	1.3	1.0	0.5	0.6	1.0	0.0	0.0	0.0	0.5	1.0
6-8	39	5.9	3.1	1.3	0.8	1.3	1.0	1.1	1.0	0.4	0.7	1.1	0.4	0.0	0.0	0.7	0.8
9-11	11	5.3	2.7	1.0	0.6	1.5	1.1	1.2	1.1	0.3	0.6	1.0	0.6	0.0	0.0	0.3	0.5
12-14	6	4.5	1.4	0.7	0.5	1.2	0.8	0.8	1.0	0.2	0.4	0.8	0.4	0.0	0.0	0.8	1.0
Daily Sleeping Hours																	
3-4	7	5.6	3.2	1.1	0.7	1.3	1.0	1.3	1.1	0.3	0.5	1.1	0.4	0.0	0.0	0.4	0.8
5-6	31	6.1	3.2	1.2	0.8	1.2	1.1	1.2	1.0	0.5	0.8	1.1	0.4	0.0	0.0	0.8	0.9
7-8	20	5.1	2.3	1.0	0.8	1.5	0.9	0.9	0.8	0.3	0.4	1.0	0.4	0.0	0.0	0.6	0.8
9-10	2	4.0	2.8	1.0	0.0	1.0	1.4	1.5	0.7	0.0	0.0	0.5	0.7	0.0	0.0	0.0	0.0
Daily Night Sleeping Hours																	
0-2	4	7.5	3.5	1.5	0.6	1.3	1.5	1.0	0.0	1.5	1.0	0.0	0.0	0.0	0.0	0.0	0.0
3-5	24	5.8	3.2	1.3	0.7	1.3	1.1	1.2	1.0	0.3	0.7	1.0	0.5	0.0	0.0	0.7	0.9
6-8	32	5.3	2.6	1.0	0.8	1.3	0.9	1.0	1.0	0.3	0.6	1.1	0.4	0.0	0.0	0.6	0.8
Schedule of Wathckeeping (Time)																	
00-04/12-16	14	6.1	3.1	1.4	0.6	1.3	1.0	1.1	1.0	0.5	0.7	1.0	0.4	0.0	0.0	0.8	0.9
00-06/12-18	1	5.0	-	1.0	-	1.0	-	0.0	-	0.0	-	1.0	-	0.0	-	2.0	-
04-08/16-20	32	5.6	3.2	1.1	0.8	1.3	1.0	1.2	1.1	0.4	0.8	1.1	0.5	0.0	0.0	0.7	0.9
08-12/20-24	8	4.9	1.9	1.0	0.5	1.4	0.9	1.1	0.6	0.3	0.5	0.9	0.4	0.0	0.0	0.3	0.5
MIXED	5	5.4	2.1	1.2	1.1	1.6	1.5	0.6	0.9	0.2	0.4	1.0	0.0	0.0	0.0	0.8	0.8
Duration of Internship (Month)																	
5.1-6.0	12	5.6	3.5	1.1	0.8	1.3	1.1	1.3	1.0	0.4	0.7	1.0	0.4	0.0	0.0	0.5	0.8
6.1-7.0	21	5.5	2.3	1.0	0.8	1.3	0.8	0.9	1.1	0.2	0.5	1.2	0.5	0.0	0.0	0.8	0.8
7.1-8.0	25	5.8	3.2	1.2	0.7	1.3	1.1	1.2	0.9	0.5	0.8	0.9	0.3	0.0	0.0	0.6	0.9
8.1-9.0	2	4.5	2.1	1.0	0.0	1.0	1.4	1.0	0.0	0.0	0.0	1.0	0.0	0.0	0.0	0.5	0.7

3 RESULTS

Sixty individuals (56 men, 4 women) with proper suitability (seafarer's health certificate) participated in the study. The mean age was 23.1 years (21.3-26.8) and the mean body mass index was 24.0 kg / m2 (18.5-32.4).The average global PSQI score of the students was 5.6 ± 2.9 before and 7.6 ± 3.5 after the internship, respectively (confidence interval 0.74-0.90, P<0.001). In addition, number of subjects with poor sleep quality was also increased after the internship activity. Before the training, 34 subjects (56.7%) had a global PSQI score of ≤ 5 while the remaining 26 subjects (43.3%) had a score > 5. After the internship, 16 subjects (26.7%) had a global score ≤ 5, and 44 individuals (73.3%) had a score >5. For each participant, global PSQI scores before and after the internship are shown in Figure 1. When the global PSQI score is ≤ 5, it is evaluated as "good sleep quality" and when it is> 5, it is considered as "poor sleep quality" (Buysse et al. 1989).

The frequency distributions of the global PSQI scores before and after the internship are shown in Figures 2 and 3.When PSQI subscores were compared before and after the internship, it was determined that the total subscores for all the individuals changed as provided in Figure 4. The mean and standard deviation (SD) values of the participants' global PSQI scores and component scores before and after the internship, are given in Tables 1 and 2. Mean and SD values of PSQI scores were calculated separately according to gender, age, type of ship, daily average work / rest / sleep / night sleep hours, shift system and internship period. If the post-internship scores is more than pre-training scores that means sleep quality was negatively affected.

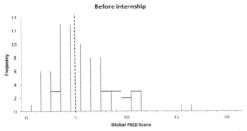

Figure 2. Frequency distrubation of global PSQI scores before internship

4 DISCUSSIONS

The students studying at the Maritime Transportation and Management Engineering Undergraduate Program are participated in this study. The third-year students' sleep quality was examined before and after internship by considering onboard working conditions.

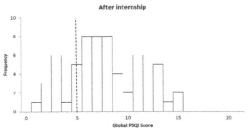

Figure 3. Frequency distrubation of global PSQI scores after internship

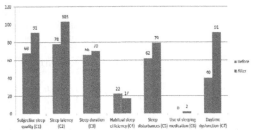

Figure 4. Total scores of PSQI subscales before and after onboard training

Aysan et al. (2014) found that the mean PSQI score of the students was 6.15 ± 1.9 and ratio of poor sleep quality (PSQI> 5, poor sleepers) was 59%, in a study they conducted with Turkish university students studying in the field of health sciences. Saygılı et al. (2011) found a general PSQI average of 6.9 ± 2.4 in their study with Turkish university students at different levels of education and in different disciplines (from health to social sciences). 30.5% of these students were considered to have poor sleep quality. Baert et al. (2014) found that 30.4% of first-year college students were poor sleepers before the first exam period in Economics and Commercial Sciences in Belgium.

Andruskiene et al. (2016) found that 45.0% of students were poor sleepers in the study they conducted with university students in various maritime departments in Lithuania. In the same study, it was seen that marine students and also third / fourth year students had worse sleep quality than the others. In this study, it was seen that the marine students were "poor sleepers" at a higher rate (43.3%) than the other departments before the internship. However, the overall PSQI average of 5.6 ± 2.9 indicates that the maritime students had lower sleep complaints before the internship for the Turkish population.

Table 2. Scores of global PSQI and subscales after internship (mean, standard deviation)

| | | Before Internship Scores of PSQI subscales | | | | | | | |
| | Global PSQI | Sleep quality | Sleep latency | Sleep duration | Sleep efficiency | Sleep disturbance | Use of sleeping medication | Daytime dysfunction |
N	Mean SD	Mean SD	Mean SD	Mean SD	Mean SD	Mean SD	Mean SD	Mean SD
Gender								
Female	4	7.3 5.5	1.3 1.3	1.5 1.3	1.3 1.3	0.3 0.5	1.3 0.5	0.0 0.0	1.8 1.5
Male	56	7.6 3.4	1.5 0.8	1.7 1.0	1.2 1.0	0.3 0.6	1.3 0.6	0.0 0.2	1.5 1.0
All	60	7.6 3.5	1.5 0.8	1.7 1.0	1.2 1.0	0.3 0.6	1.3 0.6	0.0 0.2	1.5 1.0
Age (Year)								
21-24	51	7.6 3.4	1.5 0.8	1.8 1.1	1.2 1.0	0.3 0.6	1.3 0.5	0.0 0.0	1.5 1.0
24-27	9	7.2 4.3	1.6 0.9	1.3 0.9	1.0 1.0	0.3 0.7	1.4 0.7	0.2 0.4	1.3 1.0
Type of Vessels								
Bulk carrier	9	7.8 3.5	1.7 0.7	2.0 1.0	1.1 0.6	0.4 0.7	1.3 0.7	0.0 0.0	1.2 0.8
Dry/General Cargo	14	7.6 4.3	1.4 0.8	1.5 1.2	1.4 1.3	0.5 0.8	1.3 0.6	0.0 0.0	1.5 1.2
Container	7	7.4 3.8	1.3 1.0	1.6 0.8	1.1 1.2	0.0 0.0	1.4 0.8	0.0 0.0	2.0 1.3
Ro-Ro	4	7.5 4.5	1.5 1.3	2.5 1.0	0.5 0.6	0.3 0.5	1.3 0.5	0.0 0.0	1.5 1.3
Tanker (Chemical/Cruide)	26	7.5 3.1	1.6 0.8	1.7 1.0	1.2 1.0	0.2 0.5	1.3 0.5	0.1 0.3	1.5 0.8
Daily Working Hours								
10-12	9	6.1 2.2	1.0 0.9	1.7 1.0	0.6 0.7	0.0 0.0	1.0 0.0	0.0 0.0	1.9 1.2
13-15	13	7.8 3.8	1.5 0.8	1.8 0.9	1.5 1.1	0.2 0.6	1.5 0.7	0.0 0.0	1.4 1.0
16-18	35	7.6 3.6	1.7 0.8	1.6 1.1	1.2 1.0	0.3 0.6	1.3 0.6	0.0 0.2	1.5 1.0
19-21	3	10.0 3.6	1.7 0.6	2.7 0.6	1.3 1.5	0.7 1.2	1.7 0.6	0.3 0.6	1.7 0.6
Daily Resting Hours								
3-5	4	9.5 3.1	1.8 0.5	2.8 0.5	1.3 1.3	0.5 1.0	1.5 0.6	0.3 0.5	1.5 0.6
6-8	39	7.8 3.6	1.7 0.8	1.7 1.1	1.2 1.1	0.3 0.6	1.3 0.6	0.0 0.2	1.6 1.0
9-11	11	7.1 3.4	1.3 0.8	1.5 0.9	1.3 1.1	0.2 0.6	1.5 0.7	0.0 0.0	1.4 1.1
12-14	6	5.7 2.4	0.8 0.8	1.5 1.0	0.8 0.8	0.0 0.0	1.0 0.0	0.0 0.0	1.5 1.2
Daily Sleeping Hours								
3-4	7	10.7 2.1	2.3 0.5	2.1 0.9	1.9 0.9	0.6 0.8	1.7 0.5	0.3 0.5	1.9 0.7
5-6	31	7.6 3.9	1.6 0.8	1.7 1.1	1.3 1.1	0.4 0.7	1.3 0.6	0.0 0.0	1.4 1.1
7-8	20	6.6 2.6	1.3 0.8	1.6 1.0	0.9 0.9	0.1 0.2	1.3 0.5	0.0 0.0	1.5 0.9
9-10	2	5.0 0.0	0.5 0.7	1.0 0.0	0.5 0.7	0.0 0.0	1.0 0.0	0.0 0.0	2.0 1.4
Daily Night Sleeping Hours								
0-2	4	10.8 3.1	2.3 1.0	2.0 1.2	2.0 1.4	0.8 1.0	1.8 0.5	0.0 0.0	2.0 1.4
3-5	24	8.7 3.8	1.8 1.1	2.0 1.1	1.3 1.1	0.5 0.7	1.4 0.6	0.1 0.3	1.6 1.0
6-8	32	6.3 2.8	1.2 0.7	1.5 1.0	0.9 0.8	0.1 0.3	1.2 0.5	0.0 0.0	1.4 1.0
Schedule of Wathckeeping (Time)								
00-04/12-16	14	7.2 3.1	1.4 0.9	1.4 1.0	0.9 1.1	0.2 0.6	1.4 0.5	0.1 0.3	1.8 1.1
00-06/12-18	1	9.0 -	2.0 -	1.0 -	2.0 -	0.0 -	1.0 -	0.0 -	3.0 -
04-08/16-20	32	7.8 3.7	1.5 0.8	1.9 1.0	1.3 1.0	0.3 0.6	1.3 0.7	0.0 0.2	1.4 0.9
08-12/20-24	8	6.9 2.5	1.4 0.5	1.8 1.0	1.4 0.7	0.3 0.5	1.1 0.4	0.0 0.0	1.0 0.8
MIXED	5	7.8 5.2	1.8 0.8	1.6 1.3	0.8 1.3	0.4 0.5	1.4 0.5	0.0 0.0	1.8 1.3
Duration of Internship (Months)								
5.1-6.0	12	6.2 3.1	1.4 0.8	1.3 0.9	1.0 1.0	0.1 0.3	1.1 0.5	0.1 0.3	1.2 0.9
6.1-7.0	21	7.9 3.4	1.7 0.9	1.7 1.2	1.1 0.9	0.2 0.5	1.4 0.6	0.0 0.0	1.8 0.9
7.1-8.0	25	7.8 3.8	1.4 0.8	2.0 1.0	1.2 1.1	0.4 0.7	1.3 0.6	0.0 0.2	1.4 1.1
8.1-9.0	2	8.5 3.5	1.5 0.7	1.5 0.7	2.0 0.0	0.5 0.7	1.5 0.7	0.0 0.0	1.5 0.7

After the internship, there was a 2-point increase (deterioration) in overall sleep quality (interval 0-21). While the average sleep quality was close to the "good" limit with a score of 5.6 ± 2.9 before the internship, the average sleep quality was accumulated in the "poor" area with a score of 7.6 ± 3.5 after the internship. 43.3% of the pre-training students were "poor sleepers", but this ratio increased to 73.3% after the internship.

It was observed that the students were subject to shift work procedures during onboard training but continued to work outside shifts. As discussed in the STCW 78 (amended in 2010) and MLC 2006; normal working hours should not exceed 8 hours per day in port and sea in terms of remuneration; from the point of view of fatigue, the maximum working hours shall not exceed 14 hours in a period of 24 hours and 72 hours in a period of 7 days. The students reported that the average daily work-up was between 10 and 21 hours (mean 15.4±2.4); the average daily rest between 3 and 14 hours (mean 8.2±2.1); the average daily sleep was between 3 and 10 hours (mean 6.2±1.4) and the average daily nighttime sleep was between 0 and 8 hours (mean 5.2±1.8) in the internship. In general, they have performed against the regulations of STCW 78 (amended in 2010) and Maritime Labor Convention 2006 (MLC 2006).

Karakoç (2009) reported a mean PSQI score of 6.96±3.19 in the study of health workers' sleep quality. The percentage of those with a PSQI score of 5 or higher was 76.5%. Çalıyurt (1998) reported average PSQI scores for shift health workers, nurses, doctors and jet lag were as 8.26±3.60, 6.60±2.69, 6.33±2.69 and 7.00±2.63 respectively. The mean PSQI score of the control group without any work and shift irregularities was 3.3±1.49. The mean PSQI score of 7.6±3.5 determined by this study shows that the sleep quality of the students returning to land after onboard training was as bad as the health personnel and jet-lag long flight staff.

Table 1 and 2 show the differences in PSQI scores before and after the internship. According to the tables; there is no significant difference in PSQI changes between individuals aged 21-24 years and 24-27 years.

When ship types are examined, the greatest increase (deterioration) in the average PSQI score was seen in those who practiced on bulk carriers. This is followed by cadets on tankers and containers, dry/general cargo and Ro-Ro vessels. However, when the n and *standard deviations* are taken into consideration, it can be concluded that the worst deterioration in sleep quality occurs in individuals performing internships in tankers (n:26, the amount of the global change was mean 2.1±0.3).

When the daily average working hours are examined, it is seen that the PSQI score increases as working hours increase. The highest increase in the PSQI score was seen in 19-21 hours of employees. This is followed by 16-18 hours, 13-15 hours and 10-12 hours respectively. As the average daily rest time decreased, the PSQI score increased. The highest score increase was seen in those who rested 3-5 hours a day. This is followed by those who rest 6-8 hours, 9-11 hours and 12-14 hours, respectively.

The PSQI score increased as the daily total sleep duration decreased. The highest score increase occurred during 3-4 hours of sleep. This is followed by 5-6 hours, 7-8 hours and 9-10 hours respectively. Similarly, when nightly sleep decreased, the PSQI score increased. The highest score increase occurred in the 0-2 hours night sleepers. This is followed by 3-5 hours and 6-8 hours night sleepers.

Participants reported that they were subject to variable shift systems throughout their onboard training duties. In order to be able to make the classification, the shift hours that have been covered in the last month of the internship are taken into consideration. Accordingly, the individuals were subject to the 00-04/12-16, 00-06/12-18, 04-08/16-20, 08-12/20-24 and MIXED shift systems. The MIXED statement indicates that individuals in that group do not have a shift schedule during their working hours. The increase in PSQI peaked on a person subject to the 00-06 / 12-18 shift system. If this data for one person is ignored, the maximum increase in score has occurred in the MIXED group, which does not have a certain working order. This is followed by 04-08 / 16-20, 08-12 / 20-24 and 00-04 / 12-16 shift systems, respectively.

When the effect of the internship period on PSQI score increase was investigated, it was seen that the increase in the score was most prominent among the students who worked for 8-9 months. This is followed by 6-7 months, 7-8 months and 5-6 month interns, respectively. As the training period increased, the PSQI score increased.

Figures 4 shows the changes in sleep quality index components for all participants. The greatest impairment occurred in the field of daytime dysfunction (C7). This is followed by sleep latency, subjective sleep quality and sleep disturbance components, respectively. Individuals experiencing daytime dysfunction are forced to remain awake during driving, eating, or other social activities, and can not have enough desire to do anything. It is an undeniable fact that the educational life and social life of the students in this situation will be adversely affected. In the habitual sleep efficiency (C4) of the PSQI components, the total score fell unlike other components. Sleep efficiency is roughly calculated as the ratio of sleep time to lying time. The score increase in the sleep duration component (C3) suggests that individuals do not have an increase in sleep times. In this case, it can be mentioned that there is a decrease in lying time after the internship. It is understood that there are individuals who are experiencing sleep latency in the sample population (C2), as well as those who tend to fall asleep as soon as they reach the bed. However, when the total score changes of the components are examined, it is seen that there are no significant changes in the duration and efficiency of sleep.

The information in the literature attracts to attention that an individual with a normal nighttime sleep of 8 hours will experience a sleep deprivation with a 5-hour nighttime sleep. The only way to compensate for lack of sleep is to sleep more. For 1 hour of daily sleep loss, sleep deprivation can be corrected by sleeping one or two hours more each night. It is generally said that for short-term deprivations it can take up to 7 days, and for long-term deprivations it may take several weeks. Sleep deprivation is known to affect health negatively. For example, it increases the amount of cortisol in the blood. Long-term increases in cortisone can also trigger disorders such as depression, obesity, and heart problems (Epstein&Mardon 2006, URL-1, URL-2, URL-3). The components of sleep duration in Figure 4 showed that the students did not tend to sleep too much to compensate for sleep deprivation after they had passed from sea life to land life. This may be thought to be related to the component of sleep disturbances (C5). In addition, the necessity of continuing daily life and educational activities may also be related to this situation.

Figures 4 shows a significant increase (deterioration) in the sleep latency (C2) component. Individuals have more than 15 minutes to fall asleep suggesting that have sleep latency. In addition, the frequency of do not fall asleep in 30 minutes determines the severity of sleep latency.

Sleep disturbance component (C5) indicates that individuals experience problems include waking in the midnight or early morning hours, get up feeling a need for bath/toilet, cannot breathing comfortably, coughing or loud snoring, feeling too hot or cold, have bad dreams and feeling pain. A significant increase was determined in the number of people experiencing toilet/bathroom needs, nightmares and waking in nighttime or early morning respectively, after the internship. And that the most severe complaint is related to waking up at midnight or early in the morning.

Studies in the literature show that quality and adequate night's sleep are important in terms of healthy life and cognitive performance and may have an impact on academic achievement (Zeek et al. 2015, Baert et al.2014, Koçoğlu et al. 2010, Eliasson&Lettieri 2010, Gilbert & Weaver 2010, Perez-Chada et al. 2007, Howell et al. 2004, Wagner et al. 2004). Students in the sample and in the widespread system in Turkey need to continue their academic life at the end of the internship without interruption. The findings of this study show that sleep quality of marine students is negatively affected by the process they are working as deck cadets. It has been observed that the students did not recover in next a month of returning to land and they continued their education life by sleeping problems.

5 CONCLUSION

When the data from this study and the data in the literature were considered together, it was found that the marine university students had better sleep quality (PSQI global score: 5.6±2.9) than the students in other university departments before the internship; but they had worse sleep quality after the internship (PSQI global score: 7.6±3.5).

Third-year students who were trained as a oceangoing watchkeeping officer were found to have a significant deterioration in sleep quality after onboard training (P<0.001, mean 6.9 months).

It was observed that the students were subject to shift work procedures during onboard training but continued to work outside the shift (mean 15.4±2.4). Students were generally in violation of STCW 78 (amended in 2010) and MLC (2006) regulations as deck cadets. Increased working hours have resulted in decreased rest and sleep hours and impaired sleep quality in the individual. As the training period increased, the deterioration of sleep quality increased.

After the internship, daytime dysfunctions were observed to increase significantly indicating the seriousness of the situation.

Additionally, the finding that the students did not recover from the compromised sleep pattern within a month after returning from the sea is also of importance since they have continue to the academic life with this disturbed sleep pattern.

In further studies, it would be useful to review the attendance status and academic achievement of marine students after the internship. In addition, the effects of ship types, shift systems and training periods can be examined in more detail by increasing the number of samples.

It would be useful to reorganize the programs of the maritime schools so that they allow the students to recover physically before the transition to academic life after the internship. In addition, firms employing deck cadets should observe their working hours; it will be useful to do work arrangements in such a way that night rests will not be divided and enough sleep can be attained for at least 1 month before students return to land.

REFERENCES

Ağargün, M.Y., Kara, II., Anlar, O. 1996. Pittsburgh uyku kalitesi indeksinin geçerliği ve güvenirliği (in Turkish), Türk Psikiyatri Dergisi, 7(2), 107-115

Andruskiene, J., Barseviciene, S., Varoneckas, G. 2016. Poor Sleep, Anxiety, Depression and Other Occupational Health Risks in Seafaring Population, Transnav, the International Journal on Marine Navigation and Safety of Sea Transportation, Volume 10, Number 1, DOI: 10.12716/1001.10.01.01

Aysan, E., Karaköse, S., Zaybak, A., Günay-İsmailoğlu, E. 2014. Üniversite Öğrencilerinde Uyku Kalitesi ve Etkileyen Faktörler, Dokuz Eylül Üniversitesi Hemşirelik Fakültesi Elektronik Dergisi, 7(3),193-198

Baert, S., Omey, E., Verhaest, D., Vermeir, A. 2014. Mister Sandman, Bring Me Good Marks! On the Relationship Between Sleep Quality and Academic Achievement, Institute for the Study of Labor, Germany, IZA DP No. 8232, Available at: http://ftp.iza.org/dp8232.pdf, 30.01.2017

Buysse, D.J., Reynolds, C.F.3rd, Monk, T.H., Berman, S.R., Kupfer, D.J., 1989. The Pittsburgh Sleep Quality Index: A new instrument for psychiatric practice and research, Psychiatry Research, May:28(2), 193-213

Calhoun, R. 1999. Human Factors in Ship Design: Preventing and Reducing Shipboard Operator Fatigue, University of Michigan, Available at: http://citeseerx.ist.psu.edu/viewdoc/download?doi=10.1.1.6 04.286&rep=rep1&type=pdf Accessed: 29 January 2017

Colquhoun, W.P., Rutenfranz, J., Goethe, H. 1988. Work at sea: a study of sleep, and of circadian rhythms in physiological and psychological functions, in watchkeepers on merchant vessels, International Archives of Occupational and Environmental Health, 60:321, doi:10.1007/BF00405665

Çalıyurt, O. 1998. Sirkadiyen uyku uyanıklık düzenini etkileyen iş ve çalışma gruplarında uyku kalitesinin değerlendirilmesi (in Turkish), Master Thesis (Supervisor: Abay E.), Trakya University, School of Medicine, Department of Psychiatry, Edirne

Eliasson, A.H., Lettieri, C.J. 2010. Early to bed, early to rise! Sleep habits and academic performance in college students, Sleep Breath, 14:71, doi:10.1007/s11325-009-0282-2

Epstein, J.E., Mardon, S. 2006. The Harvard Medical School Guide To A Good Night's Sleep, McGraw-Hill E-books, Harvard College, USA, DOI: 10.1036/0071467432

Gilbert, S.P., Weaver, C.C. 2010. Sleep Quality and Academic Performance in University Students: A Wake-Up Call for College Psychologists, Journal of College Student Psychotherapy, 24:295–306, DOI: 10.1080/87568225.2010.509245

Howell, A.J., Jahrig, J.C., Powell, R.A. 2004. Sleep quality, sleep propensity and academic performance, Perceptual and Motor Skills, 99,525-535.

Hystad, S.W., Eid, J. 2016. Sleep and Fatigue Among Seafarers: The Role of Environmental Stressors, Duration at Sea and Psychological Capital, Safety and Health at Work, 7, 363-371

Jepsen, J.R., Zhao1, Z., Leeuwe, W.M.A. 2015. Seafarer fatigue: a review of risk factors, consequences for seafarers' health and safety and options for mitigation, International Maritime Health, 66, 2:106–117

Jezewska, M, Leszczyńska, I, Jaremin, B. Work-related stress at sea self estimation by maritime students and officers. Int Marit Health 2006; 57(1-4):66-75.

Karakoç, B. 2009. Uyku kalitesi üzerine bir çalışma: Özel Dal Hastanesi sağlık çalışanları örneği (in Turkish), Master Thesis (Supervisor: Alpar Ş.E.), Marmara University, Institute of Health Sciences, İstanbul

Koçoğlu, D., Tokur-Kesgin, M., Kulakçı, H. 2010. İlköğretim 2.Kademe Öğrencilerinin Uyku Alışkanlıkları ve Uyku Sorunlarının Bazı Okul Fonksiyonlarına Etkisi (in Turkish) / The Influence of Sleep Habits and Sleep, Problems on Some School Functions of Elementary School 2nd Level Students, Hacettepe University Faculty of Health Sciences Nursing Journal, 24–32

Magramo, M., Gellada, L., 2009. A Noble Profession Called Seafaring: the Making of an Officer, TransNav, the International Journal on Marine Navigation and Safety of Sea Transportation, Volume 3 Number 4

MLC, 2006. Maritime Labour Convention 2006, as amended, ILO

Perez-Chada, D., Perez-Lloret, S., Videla, A.J., Cardinali, D., Bergna, M.A., Fernández-Acquier, M., Larrateguy, L., Zabert, G.E., Drake, C. 2007. Sleep Disordered Breathing And Daytime Sleepiness Are Associated With Poor Academic Performance In Teenagers. A Study Using The Pediatric Daytime Sleepiness Scale (PDSS), SLEEP, Vol. 30, No. 12

Saygılı, S., Çil-Akıncı, A., Arıkan, H., Dereli, E. 2011. Üniversite Öğrencilerinde Uyku Kalitesi ve Yorgunluk (in Turkish), Electronic Journal of Vocational Colleges, December 2011, pg.88-94

STCW, 2010. International Convention on Standards of Training, Certification and Watchkeeping for Seafarers 1978, with 2010 Manila amendments, IMO

Turkish Seafarers Regulation 2016. Gemiadamları Yönetmeliği (in Turkish)

URL-1 https://www.scientificamerican.com/article/fact-or-fiction-can-you-catch-up-on-sleep/, 30.01.2017

URL-2 (https://www.quora.com/Is-sleep-deprivation-irreversible#), 30.01.2017

URL-3 https://www.scientificamerican.com/article/fact-or-fiction-can-you-catch-up-on-sleep/, 30.01.2017

Wagner, U, Gais, S, Haider, H, Verleger, R, Born, J. 2004. Sleep inspires insight, Nature, 427(6972):352-355.

Yıldız, S., Uğurlu, Ö., Yüksekyıldız, E. 2016. Occupational Issues and Expectations of Turkish Deck Cadets, the International Journal on Marine Navigation and Safety of Sea Transportation, Volume 10, Number 3

Yılmaz, H. 2012. Determination of fatigue and sleep states of watchkeeping officers with help EEG and bridge simülatör / Vardiya zabitlerinin yorgunluk ve uykusuzluk hallerinin EEG ve köprüüstü simülatör yardımı ile belirlenmesi (in Turkish), Master Thesis (Supervisor: Başar E.), Karadeniz Technical University, The Graduate School of Natural and Applied Sciences

Yılmaz, H., Başar, E., Yüksekyıldız, E., 2013. Investigation of Watchkeeping Officers' Watches Under The Working Hours Ineligible to STCW Regulation, TransNav, the International Journal on Marine Navigation and Safety of Sea Transportation, Vol.7, pp.493-500

Zeek, M.L., Savoie, M.J., Song, M., Kennemur, L.M., Qian, J., Jungnickel, P.W., Westrick, S.C. 2015. Sleep Duration and Academic Performance Among Student Pharmacists, American Journal of Pharmaceutical Education, 79 (5), Article 63

Online Learning Technology in the Academic Educational Process

L.C. Stan, R. Hanzu-Pazara & A. Duse
Constanta Maritime University, Constanta, Romania

ABSTRACT: The modern society needs citizens more trained and specialized for its evolution and development. These requirements can be covered through a better opening of the education system, at all levels, to the civil society. Many top universities from Romania validated this trend offering online courses to their students. The paper aims to present the online teaching system that Constanta Maritime University (CMU) introduced, designed to offer easy access to information for teachers and students, but also available for former students, in order to be able to update their knowledge to the latest information about technical development in maritime field necessary in their duty activities. The focus of the present case study is on the particularization of an important European project that CMU developed and implemented together with other 4 partners, project with an important online component.

1 INTRODUCTION

The philosophy of education system is now changing globally and rapidly towards a continuous learning process. The modern society needs citizens more trained and specialized for its evolution and development. These requirements can be covered through a better opening of the education system, at all levels, to the civil society.

It is an important mission of the system, mainly of the academic level training system, to ensure the necessary techniques and information volume for a more educated society, with a higher level of knowledge, in a continuous contact with the latest researches and technical development. The teachers must determine the most appropriate pedagogical approaches for their market of students and the types of courses that they teach.

In the present, inside of the training system, there are used two concepts of teaching and training, in generally, the traditional concept, based on the paper text and courses audition, on one hand and on the other hand, the modern concept, using computerized technologies, as simulators, virtual reality and online courses. Both concepts are useful, because not all types of information can be communicated using exclusively the type of education traditional or modern.

There is knowledge, as fundamentals, which are better developed using traditional concept, where the teachers express clearly the terms, definitions, formulas and interact with the students for a higher understanding.

On the other side, courses developed especially for specialization or for upgrading of knowledge, after finishing of the academic training, can be more easily communicated through the modern technologies, as online or distant learning. In this case, it is considered that the receivers of information already have the fundamental knowledge and this new information come to complete it.

In the recent years, many universities offered to their students another educational option next to the traditional one: the online courses. The online learning and online schools revolutionized the education. They appeared as a direct result of the new needs for education of the persons involved in the social and economic life of civil society and do not have enough time to attend to the classes or of the persons that from other different reasons cannot attend the courses. But this new educational technology could not have been practical without the computerized technologies. In the present days, the computers and computerized programs are part of the life for many people, becoming indispensable for many activity fields. The online teaching technology combines successfully the new needs of learning and teaching with the computer performance.

There are many discussions on the definitions about the online learning methods and the difference among e-learning, distance learning and online learning methods.

It is not uncommon that researchers face difficulties when performing meaningful cross-study comparisons for research (Moore et al., 2011).

Online learning is described by most authors as access to learning experiences via the use of some technology (Benson, 2002; Carliner, 2004; Conrad, 2002). Both Benson (2002) and Conrad (2002) identify online learning as a more recent version of distance learning which improves access to educational opportunities for learners described as both nontraditional and disenfranchised. Other authors discuss not only the accessibility of online learning but also its connectivity, flexibility and ability to promote varied interactions (Ally, 2004; Hiltz & Turoff, 2005; Oblinger & Oblinger, 2005). Hiltz and Turoff (2005) in particular not only elude to online learnings' relationship with distance learning and traditional delivery systems but then, like Benson (2002) makes a clear statement that online learning is a newer version or, and improved version of distance learning.

2 ONLINE LEARNING BENEFITS

The involvement of the new technologies in the academic educational process has benefits for the students and lecturers, increasing competencies for both and approaching the second by the latest revolutionary educational trends. Significant changes appear to be emerging in higher education and in many components of school education.

The online teaching means to conduct a course partially or entirely through the Internet; thus the Internet is the primary way of communication.

With the Internet, the online courses have become a viable and valuable option for the student who may not be able to enroll full-time in a traditional educational institution. They could be reasons of accessibility, flexibility or quality, all compelling and contributing to the attractiveness of this mode of learning (Boettcher and Conrad, 2010), (Ko and Rossen, 2010). Through the online courses:

- The student can attend a course at anytime from anywhere in the world if he has internet access, according with his own schedule. This option is valuable for those persons who have a heavy life program, for who the regular teaching schedule is difficult to be keep. This is one from the possibilities offered by the online learning with a great importance in the maritime activity, where the program is made under the local time; ship hours, mostly different by the school local time.
- The course material is accessible 24 hours a day 7 days a week. The students have the ability to read and reread lectures, discussions, explanations and comments. Using the online teaching techniques, the student or trainee have possibility to access more courses on the same time, option to take all the information one time and to cover the curricula in a shorter period than can be done during regular classes.
- In an online environment, attendance to class is only evident if the student actually participates in classroom discussion. This increases student interaction and the diversity of opinion, because everyone can express an opinion, not just the most talkative one.
- Online instructors come with practical knowledge and may be from any location across the globe. This allows students to be exposed to knowledge that cannot be learned from books and to see how class concepts are applied in real business situations.
- Participating online is much less intimidating than in the classroom. Anonymity provides students a level playing field undisturbed by seating arrangement, gender, race and age. The students can also think longer about what they want to say and add their comments when they are ready. In a traditional class room, the conversation could have gone way past the point where the student wants to comment. In many cases, the interaction between person and computer is more benefic than an interaction person to person related to the process of information and knowledge transfer. The computerized information is taking as impersonal by the receiver and can be interpreted and adapted to the own perception. When the same information is expressed by a person, it could become personalized, the sender putting belong the information, his own remarks or opinions.
- The online environment makes instructors more approachable. Students can talk openly with their teachers through online chats, email and in newsgroup discussions, an open forum for general impressions and opinions, without waiting for office hours that may not be convenient. These communication procedures can help to the improvement of the present data and to the development of additional subjects with role of covering of missing date or useful information's for the main course. Also, this option for communication provides enhanced contact between instructors and students.
- Online course development allows for a broad spectrum of content. Students can access the school's library from their PC's for research articles, e-book content and other material without worries that the material is not available.
- Critical to the explosion in the online field is the increasing number of programs and courses now available.

Thus, the major differences between the online and the campus courses can be summarized in five characteristics:
- The faculty role shifts to coaching and mentoring;
- Meetings are asynchronous;
- Learners are more active;
- Learning resources and spaces are more flexible;
- Assessment is continuous.

Today, the computers and computerized programs are part of the life for many people. The present paper, a case study, aims to reveal a piece of the new learning technology, online teaching technology, that combines successfully the new needs of learning and teaching with the computer performance. Putting together the social and economic conditions, economical being prior, and the opportunities created by the online teaching procedures, the development possibilities of this lifelong learning concept increased considerable (Dumitrache et al., 2010).

3 ONLINE LEARNING PROCESS IN CONSTANTA MARITIME UNIVERSITY

Because there are no geographic barriers to online learning, students can find diversity of course material that may not be available to them where they live or work. One of the activity field intensive interested in this learning opportunity, the online procedure, is the maritime field. Here, the characteristics of activity do not allow, all the time, the direct presence in the classes, for improving the existing knowledge and skills. In the last decades, the changes, especially technically, has been produced, making necessary a periodically improvement and update. The online teaching techniques represent the better solution for this case, the internet being accessible from the middle of the ocean now.

In this area of activity, Constanta Maritime University developed an important project for distant simulation and tutorial systems on board. The project is under the Leonardo da Vinci program, the cooperation partners being: University of Bremen, Germany as promoter, Kongsberg Maritime, Norway with simulation technology, Elsfleth University of Maritime Studies, Oldenburg, Germany, Constanta Maritime University, Romania, Escola Nautica Infante D. Henrique, Lisbon, Portugal and also seven shipping companies from Romania, Germany and Portugal (Chiotoroiu et al., 2006).

The premise from which the project started was the lack of practical experience of the young officers while the complexity of the work on a ship increased in the last time due to several factors, including here the automation of the ships functions, multinational crew format and reduction of the crew number. In the recent years, a problem in international maritime training became obvious: the lack of experiential learning of entry-level officers or "lost apprenticeship". Like in many other technical work systems, the work processes on board have been an object of extensive automation. People on board modern ships are need predominantly for planning, control and supervision. However, in critical and unusual situations, they have to step in actively. Such situations require flexible problem solving, improvisation and intuition. Furthermore, decreasing number of (experienced) crewmembers on board and the pressure of fast promotion into responsible positions have increased the experiential learning gap of the junior-officers. This problem is most obvious in the tanker shipping, due to the demands on junior officers which are higher in this branch of the maritime industry as compared, for example, in the field of container shipping.

Based on these important and complex premises, the aim of the project is the development of a common strategy to find a solution to cover the lack of experience of the entry level maritime officers, those ones must acquire the necessary abilities, capabilities for the on board work. The attention was focused on the tanker operations where the young officers must handle dangerous cargo. The solution found through the project includes three based components:
- experiential learning (on board of the ships),
- advanced distant learning technology (on board and ashore, online learning with special attention on the simulation technology)
- formal learning in the classroom and simulator (ashore).

The experiential learning of the young cadets could be achieved through an exchange of experience between the junior officers and specialists, experts with good experience and maritime knowhow.

An important part of the project was the implementation of the internet infrastructure for distant learning used for courses release, guidance for the investigative learning. This was done based on the cooperation between educational institutions and shipping companies. Also, Kongsberg Maritime, Norway investigated how maritime simulators could be used and included in the online learning, the focus being on the implementation of the liquid-cargo-handling-simulator into the learning-management system.

Through this project, a virtual e-learning platform was established. Kongsberg Maritime, the simulation software developer, always has a server connected between the four educational institutions through which the students during the cadet's practice can communicate, work and perform exercises, tests, simulations, online exams. Thus, they have the possibility to visualize the simulated application regarding different operation necessary to be known on a tanker ship and to be familiar with the particular installations and operational procedures characteristically to oil, chemical and gas carrier ships. Also, the platform enables communication and knowledge cooperation between educational institutions and shipping companies (Stan et al., 2010).

Moreover, on this server, the junior officers can even log on board merchant vessels and can create scenarios for the vessels developed by Kongsberg and develop solutions for them.

The communication between trainers and students can be done through different ways, using the electronic correspondence or an open forum for general impressions and opinions. The internet forum gives flexibility and fluency to the exchange of the communication and short-term problem solutions. These communication procedures help the improvement of the present data and generate the development of additional subjects with role of covering of missing date or useful information's for the main course.

The training developed under such online platform is in line with the STCW convention rules for training and certification and derives from the IMO model courses.

The following pictures present an example of an exercise developed online through this project. This exercise will demonstrate the normal ballasting procedure based on the product carrier simulator.

The software program allows two modes of operation, the instructor mode, where all the options and settings are available, and the operator mode, or the student's interface, with limited access.

From the instructor mode, different preset conditions can be loaded, on which the simulation process can start and the own condition can be created and also intervene at any time. Also, it can act as the bridge or as a shore terminal.

In the operator mode, the previous mentioned facilities are not present.

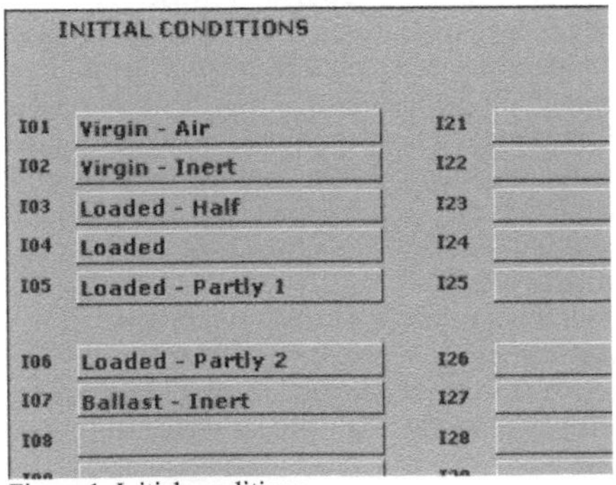

Figure 1. Initial conditions.

When starting the simulator, the Picture Directory screen is displayed.

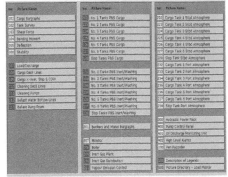

Figure 2. Picture directory screen.

This interface can be thought as a browser which allows the user to navigate between installations or other indicating screens related to the status of the ongoing operation. (load diagrams, shear forces, bending moment etc.)

The Load Master screen offers stability and force calculation in advanced, to avoid any excessive stress that could damage the ship.

The stresses vary with the cargo distribution throughout the length of the ship. Incorrect loading can damage the ship and so, the cargo/ballast must be placed according to a carefully calculated plan.

The cargo barograph will give the operator a total view of the cargo and ballast tanks with information about tank level, flow rate, cargo density and quantity in each tank.

Figure 3. Cargo barograph.

Also, it displays information about heel angle, metacentric height and trim.

Tank survey screen displays the cargo, ballast and slop tank configuration along the ship, giving information about tank level and ullage. Also, ship conditions will be dynamically updated based on tank ullage.

Figure 4. Tank survey screen displays.

The shear forces diagram calculates its values from the load distribution of the ship including the steel weights of the different hull sections and the corresponding buoyancy forms.

The bending moments are calculated from the shear forces. The actual bending moment is drawn in blue, the red and the yellow curves give maximum limits for respectively harbor and seagoing condition.

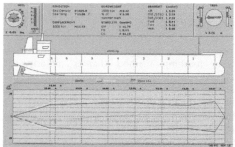

Figure 5. Picture of bending moments and forces (a).

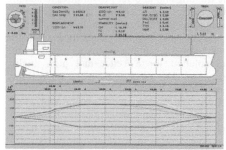

Figure 6. Picture of bending moments and forces (b).

The stability curve in the form of righting arm values is calculated for heel angles ranging from 0 to 60 degrees.

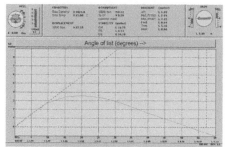

Figure 7. The stability curve.

All righting arm values are corrected (reduced) for possible "free surface" effects. The reduction in metacentric height is specifically given. The area under the stability curve represents the heel resistance or dynamic stability.

The following pictures were designed as an introduction to the simulator, and starts by presenting the initial condition used and the keys used for operation.

The initial condition used in this exercise is "Virgin – Air" which means the ship is empty, with no cargo, ballast or supplies onboard.

After selecting the work condition, the instructor must engage the simulation, switching the simulator from FREEZE mode to RUNNING mode, by pressing the F1 key on the keyboard.

In the operator mode, giving the specific exercise, the student must check the load and stability conditions prior to any ballasting planning. A special consideration must be given to the actual bending moment. The diagram shows that in this case, the best option is to fill tanks no. 3 port and starboard, otherwise the existing bending moment could exceed the permitted value and the ship could suffer serious damage.

Figure 8. Initial conditions – empty ship.

After planning the ballasting chart, the student must start the operational procedure for filling tanks 3 S/P.

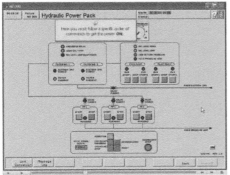

Figure 9. Hydraulic Power Pack (a).

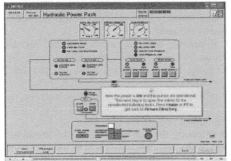

Figure 9. Hydraulic Power Pack (b).

Firstly, he must start the power generators for the electrical driven pumps. This is done in the "Hydraulic Power Pack" screen, where he must follow a specific algorithm to get the power ON. This is explained in the demonstration.

The next step consists of opening all the valves towards the designated tanks, and starting the ballast pumps.

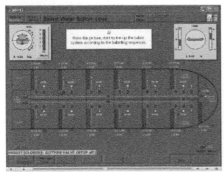

Figure 10. Ballast Water Bottom Lines screen.

This is done from the "Ballast Water Bottom Lines" screen and "Ballast Pump Room" screen.

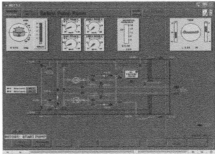

Figure 11. Ballast Pump Room screen.

While the process is underway, there are flow, pressure and temperature indicators that give precise information feedback, helping the student for a better control.

Finalizing the ballasting procedure, the student must find that the bending moment was reduced and of course, as a consequence of weight loading, the metacentric height was reduced.

4 CONCLUSIONS

The paper reveals through a rigorous presentation the new teaching method that a majority of educational institutions has implemented in the educational system, with good results for the students' instruction and not only for them. Also, it presents as example an exercise developed online through the project presented in the paper.

In nowadays, it is compulsory for the educational institutions to develop systems able to provide adequate training and to offer possibilities to acquire new competencies and knowledge during life.

The usage of the newest teaching techniques as online and distant learning, combined with the traditional forms of training seems to represent the optimum solution for a better and high quality learning system. The paper certifies the applicability of this new learning method in our university with good results. The "experiential learning gap" of the junior-officers, the main objective of the project, was covered in a successfully way. Thus, it was noticed a considerable improvement of the practical skills level of the junior officers on board of the ships.

Also, the constant cooperation and communication give to the cadets the confidence in their professional capabilities and abilities.

The method makes learning and training more effective and attractive for students, it is a valuable step in the student training. The paper comes to develop the picture of this new learning method, with good replications in the educational system.

The contribution of the paper refers to a higher understanding of the role of the new training method through a successful particular case presentation.

REFERENCES

Ally, M. 2004. Foundations of educational theory for online learning. In Terry (Ed.), *The theory and practice of online learning* (pp. 3–31). (2nd ed). Athabasca, AB: Athabasca University

Benson, A. 2002. Using online learning to meet workforce demand: A case study of stakeholder influence. *Quarterly Review of Distance Education*, 3(4), 443–452.

Boettcher, V. J., Conrad, R.-M. 2010. *The online teaching survival guide: simple and practical pedagogical tips*, published by Jossey-Bass, San Francisco, USA.

Carliner, S. 2004. *An overview of online learning* (2nd ed.). Armherst, MA: Human Resource Development Press.

Chiotoroiu, L., Dinu, D., Hanzu-Pazara, R., Pana, I. 2006. Simulation Models in Maritime Distant Learning: Tankers Topping-Off, *Conference "TECHNONAV '2006"*, Constanta, Romania

Conrad, D. 2002. *Deep in the hearts of learners: Insights into the nature of online community*. Journal of Distance Education, 17(1), 1–19.

Dumitrache, R., Stan, L., Dumitrache, C., Hanzu Pazara, R., (2010). E-Learning and simulation in maritime learning process, *Proceedings of the 6th International Seminar on the Quality Management in Higher Education*, pag.423 – 427, Tulcea.

Hiltz, S. R., & Turoff, M. 2005. *Education goes digital: The evolution of online learning and the revolution in higher education*. Communications of the ACM, 48(10), 59–64,

Ko, S., Rossen, S. 2010. *Teaching online: a practical guide*, Third edition, published by Routledge, New York, USA.

Moore Joi L., Dickson – Deane, C., Galyen, K., 2011 E-learning, online learning, and distance learning environments: Are they the same, *Internet and Higher Education* 14 (2011) 129–135

Oblinger, D. G., & Oblinger, J. L. 2005. *Educating the next generation*. EDUCAUSE.

Stan, L., Memet, F., Buzbuchi, N. 2010. Engine Room Simulator, a new teaching method for maritime education, *Proceedings of the 6th International Seminar on the Quality Management in Higher Education*, pag. 647 – 651, Tulcea.

International Convention on Standards of Training, Certification and Watch keeping for Seafarers, STCW95, IMO Sales Number: IMO-938E

User's manual Liquid Cargo Handling – Product Carrier Cargo Handling, Kongsberg Maritime Ship Systems AS

IMO Model Course 2002, Engine Room Simulator, UK. Intype libra Ltd.

Human Factor, Crew Manning and Seafarers Problems

Underlying Causes of and Potential Measures to Reduce Long-term Sick Leave Among Employees in the Service Department on Board Swedish Passenger Vessels

G. Praetorius, C. Österman & C. Hult
Kalmar Maritime Academy, Linnaeus University, Kalmar, Sweden

ABSTRACT: This paper presents findings from a workshop focused on the physical, social and organizational work environment in the service department on board Swedish passenger vessels. Twenty-seven maritime professionals participated to provide input to potential causes and measures for long-term sick leave. During the workshop, an affinity diagram was created to systematically order the input of the respondents. The results show a wide range of causes and potential measures across multiple organizational levels. Unclear leadership, ambiguous or high demands with limited decision latitude, as well as aspects of work organization, i.e. manning, pressure for effectiveness, and extensive working hours were identified as important contributors to long-term sick leave. Although the findings are based on a workshop with Swedish maritime professionals, the results may be transferable to the international maritime sector to gain a deeper understanding of how to create, organize and promote a safe and efficient work environment.

1 INTRODUCTION

From a transportation perspective, Sweden is an island, relying heavily on sea transport services. Since the early 20th century, passenger traffic has developed into a segment of specialized cargo vessels with a large diversity ranging from harbor and commuter ferries to purpose-designed cruise liners (Stopford, 2009) offering a floating vacation resort. In 2015, 26 million passengers were transported to and from Swedish ports, which represents 74% of all port calls. Furthermore, despite the global economic crises, passenger traffic to and from Sweden, especially the numbers of cruise liners, has steadily increased over the past decade (Trafikanalys, 2016).

While an integral part of the maritime industry and making up the lion share of the crew in the Swedish Register of Seafarers, the crew in the catering department has not received much attention in research. The large body of the research on maritime work environment has focused mainly on the deck and engineering departments (e.g. Allen, Wadsworth, & Smith, 2007; Kataria, Holder, Praetorius, Baldauf, & Schröder-Hinrichs, 2015; Nielsen, Bergheim, & Eid, 2013). A wider perspective is needed when investigating causes of ill-health and possible measures towards improved work and living conditions for seafarers.

1.1 *The project*

This paper presents findings from a workshop held as part of a research project focusing on the work environment and explores work related health issues in the service department on board passenger vessels.

The overall aim of the research project was to analyze work-related experiences based on interviews, observations, survey data and social insurance statistics concerning sick leave longer than 60 days. It has been a cooperation between researchers at Kalmar Maritime Academy and the Swedish Social Insurance Agency for Seafarers (Försäkringskassan Sjöfart). The project was conducted during January 2015 to January 2017 and was funded by the Swedish Mercantile Marine Foundation.

Specifically, the aim of the workshop was to present initial results and triangulate data from social insurance statistics, a survey, interviews and observations in an open brainstorm in order to identify potential causes and measures for long-term sick leave within the service department onboard of Swedish passenger vessels.

2 BACKGROUND

Through the past 60 years, passenger shipping has seen a constant growth, which can mainly be associated with the increasing number of passenger operations. This development has led to a wide variety of specialized vessels reaching from commuter ferries to cruise liners that provide a full vacation experience including hotel services, entertainment, shops etc. In comparison to other segments of the maritime domain, cruise liners are slightly different as the business is focused on a direct interaction with customers and competitors are not only to be found in shipping, but also in the hotel and tourism business ashore (Stopford, 2009).

Within the variety of passenger vessels, the so-called Baltic Ferries represent a specialized type of vessel common within Europe and especially around the Baltic Sea. Baltic ferries are characterized by providing full hotel service and entertainment, but, unlike cruise ships, they also offer roll-on roll-off cargo services – RoPax (Stopford, 2009). Compared to many international cruise ships, the working conditions on these Swedish controlled RoPax vessels differ. Commonly, the crew is onboard for 1-2 weeks, rather than several months, many employed on a permanent 1:1 contract that allows for one week payed leave for one week worked. Swedish flagged vessels are also covered by the Swedish Work Environment Act (AML), which, in many respects, is more stringent than IMO rules. For example, the AML demands that crew is given the possibility to influence their working conditions and co-operate in the systematic work environment management.

2.1 *Earlier research*

Earlier research with focus on the service department on board passenger vessels has been very limited. Most studies addressing work environment aspects in merchant shipping have focused on cargo vessels, and primarily on the bridge or engine-room personnel (t.ex. Hetherington, Flin, & Mearns, 2006; Ljung, 2010; Lützhöft, Grech, & Porathe, 2011). Furthermore, the need for improved design of human—machine interaction, user-centered design of workspaces and technology, as well as end-user involvement has been the focus of several recent publications (t.ex. Costa, Holder, & MacKinnon, 2017; Kataria et al., 2015; Österman, Berlin, & Bligård, 2016).

Publications with a specific focus on the service department's work environment are rare and mostly focus on cruise liners from a more business-oriented perspective addressing the impact of employee satisfaction and motivation on customer satisfaction (t.ex. Larsen, Marnburg, & Øgaard, 2012; Sehkaran & Sevcikova, 2011). Other research in the cruise liner domain has among others addressed occupational communities (Lee-Ross, 2006), work perceptions (Dennett, Cameron, Bamford, & Jenkins, 2014) and working and living conditions onboard (Gibson, 2008).

However, there are two recent studies that address the service department onboard of Swedish passenger vessels. Forsell et al. (2017) conducted a study on work environment for seafarers on board of Swedish vessels and found indications for work environment problems associated with noise exposure and heavy workload on neck, back and arms. Ljung and Oudhuis (2016) primarily focused on the safety perceptions of service department employees. However, they also discuss the increased need for flexibility among the service department's employees as a consequence of a constantly decreasing number of crew, as well as temporary work contracts and economic pressure on the business.

3 METHODOLOGY

3.1 *Affinity Diagram*

The Affinity Diagram, or KJ-Method, was developed by Kawakita Jiro as a method to systematically organize large amounts of data derived through brainstorming to reveal connections and underlying relationships between ideas (Beyer & Holtzblatt, 1998). The KJ-Method has been widely used in quality management (Babbar, Behara, & White, 2002), but has also been recognized as an effective tool within usability testing and for requirement developments (Kuniavsky, 2003).

An affinity diagram is built bottom-up, which means that relationships and connections are developed based on a wide range of ideas and concepts. The first step is a brainstorming activity, where all ideas are noted down. Each idea is put onto a single sticky note. In a second step the ideas are discussed and ordered into categories. In a third step the connections between ideas and categories are developed through discussion. This can also mean that new relationships emerge and new concepts are built based on the discussion (Beyer & Holtzblatt, 1998).

3.2 *Participants*

Twenty-seven representatives from the Swedish maritime cluster participated within the workshop. Table 1 presents an overview of the participants and their organizational affiliation.

Table 1. Overview of the workshop participants

Participants (n)	Function
Swedish Mercantile Marine Foundation (4)	Foundation promoting well-being and health of Swedish seafarers
Swedish Transport Agency (2)	Administration, responsible for oversight of work environment policy implementation
SEKO Seafarers Union (2)	Union representatives
Shipping company I (5)	HR representative
	Head Chef
	Senior safety delegate
	Vice-senior safety delegate
	Safety delegate
Shipping company II (3)	2 HR representatives
	Representative for local work environment council
Shipping company III (2)	Nurse
	Senior safety delegate
Shipping company IV	Director of HR
	HR representative
	Local union representative
Shipping Company V (2)	CEO
	Senior safety delegate
Shipping company V (2)	Head of HR
	HR administrator
Swedish Confederation of Transport Enterprises (2)	Representative for work environment initiatives
Maritime Officers' Association (1)	Managing director

3.3 Procedure

Prior to the workshop, the participants had been invited via email and had received a short summary of the day's schedule.

The workshop started by a short introduction of the research project, preliminary results and information about the day program. This also included information about anonymity, confidentiality and informed consent.

The workshop was split into four main sessions; one session for identifying and categorizing potential causes and measures, as well as three sessions for specifying these and their connection to the physical, organizational, or social work environment. Each session started out by a short presentation of the current topic and was followed by a discussion about causes and potential measures for long-term illness/ absences. The sessions lasted about 40 minutes to one hour each.

Before the first session, the participants were asked to brainstorm with the help of sticky notes. They were asked to individually write down their opinion of which underlying causes could be related to the increased number of long-term sick leave within the service department on passenger vessels. In conjunction with the causes, the participants were prompted to identify one to two potential measures for the causes they had identified. After 10 minutes, the participants put their sticky notes on a wall in common sight split between causes and measures.

The first session focused on discussing and grouping the potential causes for long-term sick leave. The second to fourth sessions were focused on countermeasures within the physical, social and organizational work environment.

The discussions were led by a moderator who modified the grouping of sticky notes and added new concepts, if mentioned in plenary. The participants were asked to confirm each modification to ensure that the developing affinity diagram still matched the intent uttered as part of the discussion.

As an outcome of the workshop, an affinity diagram grouping potential measures to long-term sick leave was created in plenary with the participants.

To increase the overall credibility and transferability (Fishman, 1999) of the workshop's outcome, the results were compiled and sent to the participants for as a means of expert auditing (Lützhöft, Nyce, & Styhr Petersen, 2009).

4 RESULTS

The following section presents the results of the workshop in three sections: *physical, organizational* and *social work environment*. Each section will first present the identified causes for long-term sick leave and then present the measures suggested by the participants.

4.1 Physical work environment

4.1.1 Causes

The participants of the workshop identified two main categories of causes for long-term sick leave relating to the physical work environment: *physical workload and disorders*, and *design and ergonomics*.

Many work tasks performed by the catering department imply a high *physical load*, leading to an increased risk of musculoskeletal disorders. This risk is further aggravated by the constant exposure to whole body vibrations induced by ship movements, propulsion and reduction gear. The restaurant workers, for instance, spend almost the entire workday on their feet, carrying food and drinks to and from the tables. Here, the weight and design of plates and glasses was mentioned as an often unnecessarily high load, especially during stressful conditions with many guests waiting to be served.

Design and ergonomics addresses the overall design of the vessel, especially concerning noise, vibrations and sea state, as well as the design of workspaces and equipment onboard. It was noted several times that workspaces are very narrow and hard to reach making it hard to use equipment, such as serving carts, that might ease the work.

Furthermore, it was also emphasized that the as the decreasing number of crew increases the physical workload on the remaining employees in the service department.

4.1.2 Measures for the physical work environment

Measures to improve the physical work environment can be split into three categories; *design of workspaces*, *techniques and work strategies*, and *health-promoting measures*.

It was raised that the design of workspaces and equipment could significantly be improved through end-user involvement in the (re-)design of the service department, or as one participant stated: *"We need to improve our relation to the naval architects"*

The quote above shows the experience of not having an extensive influence over the design of workspaces within the service department's work environment leading to difficulties as tasks are not supported by the physical design of the work environment.

Techniques and work strategies concern the overall way in which work is conducted. The participants highlighted the importance of a thorough workplace introduction, e.g. learning ways of lifting heavy weights to minimize the effects of the physical load on muscles and joints, and the need to conduct risk and consequence analyses before tasks and work processes are changed to ensure that the desired improvement does not lead to unanticipated negative consequences. Furthermore, the participants identified rotation of work stations and job tasks with the aim to vary the strains on the body, as well as the need to increase the employees' awareness for body postures and the effects of physical workload on the overall well-being.

Finally, the participants emphasized that workplace design and techniques need to be complemented by *health-promoting measures*. These measures are both an individual responsibility, such as taking care of oneself and using the right equipment, and the responsibility of the organization, such as making sure that the employees have the right prerequisites and equipment to perform a good job. Organizational measures to promote physical well-being concern offering access to training facilities and shaping awareness for certain values, such as a non-drinking policy or support for those that attempt to quit smoking, as well as having access to physical therapy and massage.

4.2 Organizational work environment

4.2.1 Causes

Within the organizational work environment, underlying causes for long-term absences were associated to *aspects of leadership* and *organization of work*.

Aspects of leadership concern the way in which leadership is exercised, as well as the knowledge on how to manage and lead personnel. Many of the participants emphasized that there is a lack of knowledge about how to lead and support staff in their work, and a lack of clear organizational structures and salient responsibilities assigned to team leaders.

Too little, or suboptimal manning, time pressure, long working hours and monotonous tasks were highlighted as causes related to the *organization of work* within the service department. Several respondents mentioned that many of the causes could be related to the overall pressure for economic effectiveness and therefore the slimmed down organization on board.

4.2.2 Measures for the organizational work environment

To counterbalance the identified causes, four different categories of measures were suggested: Owner-managers, *leadership, training and education; crew size and shift system, and tasks and organization of work.*

The first category *owner-managers* comprises managers focused on communication within the organization, cooperation between shore and ship, recruitment processes and possibilities for career advancements. It was emphasized that the shipping company sets the organizational framework that provides employees with the preconditions for a safe work environment. Communication between managers, employees and teams, as well as between ship and shore organization needs to be continuous and clear. Organizational changes should be communicated early in the process and offer the opportunity to provide input and feedback. It is further of outmost importance that surveys and other evaluations conducted with employees are followed up and results communicated to the service department. This is essential to build common values within the organization, or as one of the participants said; *"We need to speak the same language"*.

Furthermore, time should be spent on profiling new recruitments to ensure that *"the right people"* are hired for the work on board. This also addresses the importance of clearly communicating career opportunities before a person joins the company.

In the category *leadership, training and education,* participants raised aspects of how successful leadership can be exercised on board and what the preconditions to achieve this comprise. Many of the participants emphasized that it is quite common that team leaders are recruited internally, which means that there is no specific education provided by the shipping company to staff that is promoted. Furthermore, some of the participants felt that the hierarchical structure of work has led to "*too many leaders*" which creates a work environment where it is not always clear who oversees certain work processes and who has the right to make decisions.

Measures concerning *education and training* partially focused on leadership courses, but also emphasized the need of the service personnel in receiving more guidance on certain aspects of their work, such as on customer service, or training of ergonomic job practices. Furthermore, joint training seminars between leaders and subordinates were suggested to form a basis for an increased mutual understanding.

When discussing *crew size and shift system*, the consequences of economic pressures were highlighted by the participants that work as crew onboard. One of the core problems is the increased workload and long working hours for what is perceived by some as a continuously decreasing number of crew. Suggestions for counterbalancing negative effects were therefore mainly directed towards offering more flexible working hours, increased manning and better employment security.

Task and organization of work addressed the overall distribution of tasks over the employees in the service department, as well as the assignment of responsibilities among the leaders on various levels in the organization. Some of the participants felt that "*job titles can sometimes get in the way*" of effective cooperation among co-workers and departments. Another aspect raised within the category of *task and organization of work* is the ability of the organization to successfully offer the opportunity to rehabilitation at work. After an employee's long-term sick leave the organization should be able to facilitate the return to work.

4.3 *Social work environment*

With regards to long-term absence from work and underlying causes within the social work environment, *identity and recognition as seafarer* was identified as a crucial contributing factor. Although the service department on board represents the largest department in terms of number of employees, the employees are often not considered as seafarers by other staff. Consequently, there might be friction between the service department and other departments, such as the engine-room or deck crew, or as one of the participants highlighted a "*we and them*" –*feeling* among the crew members.

Measures to promote a collective identity focused on joint seminars or training, as well as activities across the departments on board to increase mutual understanding for each other's work and work identity, as well as a feeling of belonging to the same team/crew.

5 DISCUSSION

This workshop aimed to identify underlying causes to long-term sick leave in the service department on board Swedish passenger vessels. The workshop united a wide range of representatives for both employers and employees, as well as representatives for the legislator. The results clearly show that there is a wide range of causes that can be associated to the constant economic pressure within the maritime domain. However, many of the potential measures highlighted the need for more cooperation and end-user involvement in the design of equipment and workspaces to promote the well-being of the workforce aboard.

Further, many of the causes of long-term sick leave captured by the workshop participants are related to high levels of job demands, performed with limited decision latitude, a combination well known to cause stress related ill-health (Karasek & Theorell, 1990). The work and living environment on board a ship has a high incidence of physical and psychosocial stressors that can affect the individual seafarer, and especially so for the catering department that performs on an arena where their work is constantly evaluated, and sometimes commented upon, by the guests. The adaption to a continuous influx of new colleagues and the forming of interpersonal relationships can be a significant stressor (Carter, 2005).

As emphasized by several recent publications (t.ex. Cervai & Polo, 2017; Kataria et al., 2015; Rasmussen et al., 2017), participatory design and end-user involvement in the design of workspaces is offering an opportunity to positively impact on the employees' well-being. To be able to do a job well requires a good fit between an employee's physical capabilities, workspace and organizational environment. If already considered during the design stage, end-user requirements for increased workability and safety of operation can be incorporated, often even at a lower cost.

The workshop generated a wide array of potential measures to promote the well-being of employees and improve the overall system performance. However, there have been suggestions and comments, which were beyond the scope of the current research, but should be considered further.

Several participants raised questions about the impact of life at home onto the overall well-being of an employee. Many of the respondents felt that life at home and work on board are inseparable when talking about the impact on a person's health. Some of the respondents raised that aspects of life at home may have a negative impact on the well-being, but cannot and should not be the responsibility of the employer. Thus, more research to address and explore the topic of work-life balance for active seafarers in the merchant fleet should be conducted to get a deeper understanding for how life and lifestyle at home are connected to an employees' ability to work and well-being.

Furthermore, one of the core outputs of the workshop addressed the need for more training and education. However, it was never specified in, greater detail, what type of training is needed and if the needs for certain education, e.g. courses in leadership, addresses a more global concept, or requires to be adapted to the specific circumstances of life and work onboard. This should be explored further.

6 CONCLUSION

This article presented the findings from a workshop with maritime professionals with the aim to reveal underlying causes for long-term sick leave and measures that might counterbalance these.

The main findings of this research indicate that increased cooperation and end-user involvement has the potential to become an important part of promoting well-being of employees and overall system performance. End-users, such as employees in the service department, have a tremendous contextual knowledge of their work environment, regardless whether this concerns physical, social or organizational aspects of work. Engaging these employees in the design of workspaces and work task on board can therefore become a source of motivation, as well as it can promote a better fit between organizational frame, employees and physical workspaces.

ACKNOWLEDEGEMENT

The authors would like to express their deepest gratitude towards the financial support of the Swedish Mercantile Marine Foundation for the project "Indenturens arbetsmiljö". Furthermore, we would like to thank the participants in this workshop. Their input has been invaluable for the research presented here.

REFERENCES

Allen, P., Wadsworth, E., & Smith, A. (2007). The prevention and management of seafarers' fatigue: a review. *Int Marit Health, 58*(14), 167-177.

Babbar, S., Behara, R., & White, E. (2002). Mapping product usability. *International Journal of Operations & Production Management, 22*(10), 1071-1089. doi:10.1108/01443570210446315

Beyer, H., & Holtzblatt, K. (1998). Creating one view of the custormer *Contextual design : defining customer-centered systems*: San Francisco, Calif. : Morgan Kaufmann Publishers, cop. 1998.

Carter, T. (2005). Working at sea and psychosocial health problems - Report of an International Maritime Health Association Workshop. *Travel Medicine and Infectious Disease, 3*, 61-65.

Cervai, S., & Polo, F. (2017). The impact of a participatory ergonomics intervention: the value of involvement. *Theoretical Issues in Ergonomics Science*, 1-19. doi:10.1080/1463922X.2016.1274454

Costa, N. A., Holder, E., & MacKinnon, S. N. (2017). Implementing human centred design in the context of a graphical user interface redesign for ship manoeuvring. *International Journal of Human-Computer Studies, 100*, 55-65. doi:http://dx.doi.org/10.1016/j.ijhcs.2016.12.006

Dennett, A., Cameron, D., Bamford, C., & Jenkins, A. (2014). An investigation into hospitality cruise ship work through the exploration of metaphors. *Employee Relations, 36*(5), 480-495. doi:doi:10.1108/ER-08-2013-0111

Fishman, D. B. (1999). *The Case for Pragmatic Psychology*. New York: New York University Press.

Forsell, K., Eriksson, H., Järvholm, B., Lundh, M., Andersson, E., & Nilsson, R. (2017). Work environment and safety climate in the Swedish merchant fleet. *International Archives of Occupational and Environmental Health, 90*(2), 161-168. doi:10.1007/s00420-016-1180-0

Gibson, P. (2008). Cruising in the 21st century: Who works while others play? *International Journal of Hospitality Management, 27*(1), 42-52. doi:http://dx.doi.org/10.1016/j.ijhm.2007.07.005

Hetherington, C., Flin, R., & Mearns, K. (2006). Safety in shipping: The human element. *Journal of Safety Research, 37*(4), 401-411.

Karasek, R., & Theorell, T. (1990). *Healthy work: stress, productivity, and the reconstruction of working life*. New York: Basic Books.

Kataria, A., Praetorius, G., Schröder-Hinrichs, J. U., & Baldauf, M. (2015, 9-14 August 2015). *Making the case for Crew-Centered Design (CCD) in merchant shipping*. Paper presented at the 19th Triennial Congress of the IEA, Melbourne, Australia.

Kuniavsky, M. (2003). *Observing the user experience. A practitioner's guide to user research*. San Francisco: Elsevier Science.

Larsen, S., Marnburg, E., & Øgaard, T. (2012). Working onboard – Job perception, organizational commitment and job satisfaction in the cruise sector. *Tourism Management, 33*(3), 592-597. doi:http://dx.doi.org/10.1016/j.tourman.2011.06.014

Lee-Ross, D. (2006). Cruise Tourism and Organizational Culture: The Case for Occupational Communities. In R. K. Dowling (Ed.), *Cruise Ship Tourism*. Wallingford, UK: CAB International.

Ljung, M. (2010). Function Based Manning and Aspects of Flexibility. *WMU Journal of Maritime Affairs, 9*(1), 121-133.

Ljung, M., & Oudhuis, M. (2016). Safety on Passenger Ferries from Catering Staff's Perspective. *Social Sciences, 5*(3), 38.

Lützhöft, M., Grech, M., & Porathe, T. (2011). Information Environment, Fatigue, and Culture in the Maritime Domain *Reviews of Human Factors and Ergonomics* (Vol. 7, pp. 280-322): Human Factors and Ergonomics Society.

Lützhöft, M., Nyce, J., & Styhr Petersen, E. (2009). Epistemology in Ethnography: Assessing the quality of knowledge in Human Factors research.

Nielsen, M. B., Bergheim, K., & Eid, J. (2013). Relationships between work environment factors and workers' well-being in the maritime industry. *International maritime health, 64*(2), 80-88.

Rasmussen, C. D. N., Lindberg, N. K., Ravn, M. H., Jørgensen, M. B., Søgaard, K., & Holtermann, A. (2017).

Processes, barriers and facilitators to implementation of a participatory ergonomics program among eldercare workers. *Applied Ergonomics, 58,* 491-499. doi:10.1016/j.apergo.2016.08.009

Sehkaran, S. N., & Sevcikova, D. (2011). 'All Aboard': Motivating Service Employees on Cruise Ships. *Journal of Hospitality and Tourism Management, 18*(1), 70-78. doi:http://dx.doi.org/10.1375/jhtm.18.1.70

Stopford, M. (2009). *Maritime Economics*. New York: Routledge.

Trafikanalys. (2016). *Sjötrafik 2015* (Statistik 2016:17). Retrieved from http://www.trafa.se/globalassets/statistik/sjotrafik/rapport-2016_17-sjotrafik-2015.pdf

Österman, C., Berlin, C., & Bligård, L.-O. (2016). Involving users in a ship bridge re-design process using scenarios and mock-up models. *International Journal of Industrial Ergonomics, 53,* 236-244. doi:http://dx.doi.org/10.1016/j.ergon.2016.02.008

Human Reliability Analysis of a Pilotage Operation

J. Ernstsen, S. Nazir & B.K. Roed
University College of Southeast Norway, Borre, Norway

ABSTRACT: About 80-85 % of maritime accidents and incidents are attributed human error. To analyze the role of humans in maritime operations is of paramount importance to investigate and understand how tasks are carried out at the sharp end. In this work, a hierarchical task analysis is conducted to capture tasks required in a piloting operation. The data has been gathered using structured interview and transcribed verbatim prior to the analysis. The transcript was coded into categories with help of subject matter experts before structured into a hierarchy. The presented structure enlightens tasks and collaboration patterns that requires elevated attention for efficient and safe piloting operations using the Systematic Human Error Reduction and Prediction Approach, a human reliability analysis technique. Results indicate that omission of acting was the frequent most error type, and communication error second most frequent. The results are discussed considering pilot's comments from interviews and observation.

1 INTRODUCTION

A wide range of accidents in the maritime domain are attributed to human error, and researchers claim between 80-85 % for such errors in the maritime industry (Di Pasquale, Miranda, Iannone, & Riemma, 2015). It is challenging to predict circumstances in which humans perform errors. In general; and widely accepted, humans with expertise and knowledge of an operation tend to be less error prone than novices. Human errors commonly occur when the cognitive and physical resources overcome the environmental demands (Reason, 1990). Experts have more sophisticated mental schemata (which triggers perceptual understanding of the situation), and thus comprehend situations with less mental effort than their novice counterparts (Davis, 1982).

The concept of human error is complicated; established schemata are also prone to errors, often by relying on heuristics (e.g. to more easily perceive an item as items you see frequently or recently) to generate efficient and *good enough* understanding of a situation (Kahneman & Tversky, 1981). When the cues are misleading, experts can misinterpret the situation if she or he is not explicitly aware of the heuristic (Tversky & Kahneman, 1975). Whenever the situation deviate slightly from the acquired cognitive schemata, a novice's interpretation of the situation may, in fact, be more accurate, albeit using more mental resources to comprehend it.

A multitude of maritime operations are complex because of the high consequences if errors are done, e.g. Korean Sewol ferry accident and the capsizing of Costa Concordia (Kim & Nazir, 2016; Schröder-Hinrichs, Hollnagel, & Baldauf, 2012). Considering the dynamic nature of pilotage operations, i.e. that the safest situation often is to keep going, puts elevated pressure on the captain's and the pilot's continuous situational understanding, which defines much pilotage operations as complex.

Both experts and novices are prone to errors, critical situations within maritime operations need to be understood to implement measures that reduce the probability and consequences of human errors. Such understanding may help to combat this through precise training and experience (from novice to experts) and by controlling the situation directly (i.e. operational design) (Kluge, Nazir, & Manca, 2014).

Today, the maritime industry is evolving, and have already in large parts moved towards fewer operators to perform essential work tasks (Bloor, Thomas, & Lane, 2000; Lundh, Lützhöft, Rydstedt, & Dahlman, 2011), which puts more pressure on the operators, emphasizing the need to effectively train and design the work system e.g. assess for flexibility in the work organization and task execution, which is found to be important in a study on maritime engineer officers (Mallam & Lundh, 2016).

To improve operational performance, you can alter the work environment (e.g. change procedures, design layout, and how tasks are executed and staffing). Another way is to improve skills and knowledge of operators with training or staffing (Salman Nazir & Manca, 2015). Either approach, whether you alter the work environment or staffing to improve performance, demands explicit knowledge of which tasks are required and how they are executed. The higher the consequence and probability of errors in such tasks elevates the demand for a thorough review of how the tasks are carried out. Considering that much human errors in the maritime domain occurs in pilotage operations, e.g. Godafoss, Federal Kivalina and Crete Cement accidents, (Accident Investigation Board, 2010a, 2010b, 2012). This current research will investigate how tasks are carried out in such an operation, and use a human reliability analysis technique to find potential human errors and their probability of occurring and to asses if the human errors are safety critical (i.e. higher consequence for safety, environment and cost if error is done). In the current study, the well validated (Kirwan, 1992) Systematic Human Error Reduction and Prediction Approach (SHERPA) will be used, which to best of our knowledge, for the first time in the maritime domain.

A scenario is presented for interviewees. The scenario is based on a normal pilot operation at Slagentangen with no introduced accidents or incidents, a port in the Norwegian Oslo Fjord which is known to be challenging due to high underwater current and plenty of shallow skerries. To navigate an oil tanker into Slagentangen is challenging and requires careful collaboration between the pilot and the captain of the vessel, as well as making fast paced decisions. A shared understanding of the internal and external physical and social factors between the pilot and the captain is necessary to promote an efficient and safe operation, by ensuring a distributed situational awareness (S. Nazir, Sorensen, Overgard, & Manca, 2014; Stanton et al., 2006). This requires a comprehensive understanding of both technical and non-technical skills by the pilot and captain, as well as knowledge about the domain to solve novel situations which commonly arises in dynamic maritime operations, and which rightfully justifies the high level of experience required of both captains and pilots.

2 METHOD

2.1 Interview technique

The interviews were open-ended and unstructured. A standardized piloting scenario was presented to the pilots for a consistent understanding of the operation in topic. The interviewees were asked to define an undesirable incident or accident which they had prior experience or knew of. They were asked to define the accident to a level 4 on a 10-level scale, where 10 would be explosion and several casualties. This to ensure some consistency in the undesirable situations to be analyzed. The same interviewer was used and the probed questions during the analyzed were formulated to not be leading the interviewee, this to ensure that the interviewer would be consistent and not contaminating the gathered interview data. All interviews opened with a review of the informed consent. The interviews were around 1 hour 15 minutes on average, with the longest of 01:37 hours and shortest on 01:05 hours. The interviews were recorded using audio recorder or iPhone voice memos application. When enough data was collected to see structure in the work tasks, the interviews was focused to elaborate further to ensure a valid representation of the piloting operation. The data collection process and storing has been approved by Norwegian Centre for Research Data.

2.2 Sampling and response rate

Interviewees were gathered using the snowball-approach (i.e. ask interviewers to provide colleague/friends fitting the criteria of interviewee) and network of the research team.

Eight interviewees with piloting experience contributed to the analysis. To be a captain or pilot requires much sailing experience, thus all applicable interviewees were deemed experts considering their working positions. The interviewees were included in different parts of the research; in the development and validation testing of the Hierarchical Task Analysis (HTA) and SHERPA. The interviews were conducted in-person or using FaceTime®, which is an alternative to the renowned Skype™.

2.3 Tools used to structure and analyze results

The interview data were transcribed verbatim in the beginning, as data satiety was approaching, more efficient transcription methods was used, e.g. to transcribed only relevant sections of the interview. Tasks and functions are identified using a combination of a grounded (i.e. bottom-up) approach and theoretical/practical evaluation (i.e. top-down), where the information is evaluated by subject-matter experts (SMEs). The tasks and functions are structured hierarchically, following the steps of conducting hierarchical task analysis (Annett, Duncan, Stammers, & Gray, 1971). A task analysis is widely used as a tool to structure tasks prior of subsequent analyses, such as SHEPRA (Baber & Stanton, 1996).

SHERPA is used to identify the following three trends for pilotage operations: (1) Errors with a high probability of occurring. (2) Errors which are deemed critical, i.e. substantial damage to vessel, personnel or yield environmental hazards. (3) Findings with a high frequency of the same error type, e.g. multiple errors are categorized as an action error (please see Table 1 for an overview of error categories).

Table 1. Error categories SHERPA

Error categories	Action errors
	Checking errors
	Retrieval errors
	Communication errors
	Selection errors

A probability of occurring in SHERPA is defined in an ordinal manner, i.e. low probability is assigned errors which have never occurred, medium probability is assigned if it has occurred on previous occasions, and high probability are assigned if the error frequently occurs. The probability data will rely on historical trends and/or SMEs. An error with critical consequences is defined binary; that is, either yes or no.

3 RESULTS AND ANALYSIS

3.1 Result Hierarchical task analysis (HTA)

The task analysis was structured hierarchically in a timeline, please see Figure 1 for an overview of the task analysis. The main tasks (Figure 1) identified for pilotage operations are, in ascending order, (1) order and get pilot aboard, (2) development of bridge relation, (3) checklist pilot, (4) assessment of environmental conditions, (5) route planning, (6) give navigational inputs, (7) commanding and coordinating tugboats, (8) berthing of vessel. 28 sub-tasks were identified in total for all 8 main tasks during the hierarchical task analysis. A further breakdown of the 28 sub-tasks revealed a total of 55 operations. The analysis did not go into more detail as motoric, mechanical and cognitive operations would not contribute further information to conduct SHERPA.

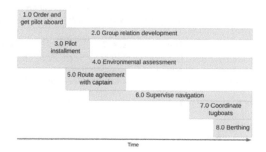

Figure 1. Timeline overview of tasks identified in a pilot operation

3.2 Result Systematic Human Error Reduction and Prediction Approach (SHERPA)

The tasks were analyzed in terms of human error prediction and probability subsequent of the task analysis. SHERPA was developed and validated in accordance with SMEs to ensure consistency of the findings.

3.3 Validity and reliability considerations

Reliability of the findings was assessed using two other researchers to analyze a sub-section of the data. The result was further validated by two SMEs subsequent of the analyses to ensure that the researchers have structured and analyzed the data consistently.

Ultimately, to further ensure validity and reliability of the results, a researcher tested the task analysis result and HRA result in a real-life piloting operation: Piloting a car cargo vessel from Oslo Port to Hvasser pilot station. The results provided evidence for reliable and valid findings from the human reliability analysis.

3.4 Analysis of the results from HTA and SHERPA

Findings suggests that pilotage operations are associated with tasks that have a low probability of errors which have critical consequences if they occur. We see in Table 1 that 47 of the identified errors have low (no known occurrence) and medium (seldom occurrence) ordinal probability. At the same time, 15 critical errors are identified. Only 8 of the 55 identified tasks have high probability of human error, determined from interviews with pilots and SMEs. Most of the high probability errors are within bridge team relationship development, function 2.0 (Figure 1).

Table 2. Most frequent error type, second most frequent error type, number of errors with low-, medium-, and high probability and number of errors with critical consequence distributed among eight main tasks discovered in HTA. P = Probability, C = Critical. Hyphen "/" indicates a tie.

	Task 1	Task 2	Task 3	Task 4	Task 5	Task 6	Task 7	Task 8	Total
1st most error type	Information/ Action	Action	Checking	None	Action	Information	Action	Action	Action
2nd most error type	Information/ Action	Information	Action	None	Retrieval	Action	Information	Retrieval/ Information	Information
Low P	1	0	3	0	0	0	7	6	20
Medium P	7	5	3	0	3	6	0	5	27
High P	0	4	2	0	1	0	0	0	8
No of Critical errors	2	3	2	0	2	6	2	0	15

SHERPA reveals that the most frequent human error in pilotage operations are related to a decision not to act (or action omission, following Hollnagel (2000)). A need to find possible causes for a decision not to act is thus identified.

Second most frequent type of errors are communication errors. Pilotage operations are heavily dependent on communication, both formal and informal communication, often with international crews with cultural and language barriers that put strain on communication.

Communication errors are often attributed to what interviewees commonly referred to as "cold bridge climate", as opposed to "warm". Bridges with cold climate had communication errors because pilots and captains only shared need-to-know information, instead of warm bridge climates where nice-to-have information was shared also, as communicated by the interviewees.

In Table 2 below, error types, frequency of the respective probabilities and tasks which were deemed critical, both for all 8 main tasks and in total are presented.

Subsequent investigation of the results reveals a possible connection between cold bridge climates and the frequency of omission errors (i.e. decision of not to act, perhaps due to the social climate (at the bridge).

An example of three potential effects emerge from this relationship. In Figure 2, effects (1) and (2) are when the pilots enter bridge climates which are cold, for whatever reason (e.g. personality differences, pressure from ship-owners). In effect (2), a pilot with skill in team communication can act as a moderating factor, which influence how much and if action errors occurring during the operation. In effect (3), the bridge climate is considered warm from when the pilot enters the bridge, and thus, no action omission errors are due to exist, which renders the skill of team communication less important regarding how much actions omission errors will occur.

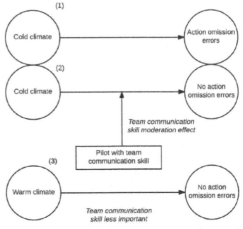

Figure 2. Identified relationship among bridge climate, action errors and team communication-skill of pilot. (1) In a cold bridge climate, action omission errors occur more frequently. (2) In a cold climate, a pilot with team communication skills is necessary to prevent action omission errors. Lastly, (3), a warm bridge climate does not require pilots with team commutation skills to ensure no action omission errors.

4 DISCUSSION

The research uncovers detailed specification and understanding of pilotage operation. Cognitive challenges are identified in the analysis, which put load on the operator. When the mental workload overcome the operator's cognitive resources, errors are more prone to occur (Reason, 1990). In the analysis, frequent and/or critical tasks are identified: From a cost-benefit perspective, such tasks should receive elevated attention as their impact (due to increased consequence or frequency) on operational performance is higher than other tasks.

4.1 Theoretical and practical implications of the research

Human reliability analyses are used to predict and explain why and how human errors occur, as well as the probability of occurrence. A plethora of methods are developed for this purpose, mostly because different domains and operations have own requirements in terms of human reliability assessments. The current research, however, attempts to use a well-underpinned HRA technique to assess its applicability within the maritime domain.

Pilotage operations are conducted in local waters, often where people are living, so that local communities have increased interest in ensuring that such operations does not have negative impacts on the environment, e.g. with inefficient operations or even accidents that generate oil spill in local waters.

As the results show a valid, precise depiction of the human reliability in a pilotage operation, the study's findings may have theoretical implications that it provides evidence to support that SHERPA technique is viable to assess human reliability in maritime operations, at least for pilotage operations. More research may provide evidence that SHERPA can be used in other maritime operations as well, e.g. dynamic positioning.

For omission of errors, found to be related to pilotage operations in the analysis, should be further investigated. A possible hypothesis, presented in the same section, is that action omission errors occur in cold bridge atmospheres. Controlled simulator experiments should be conducted to provide evidence or disprove this hypothesis, and to investigate effect sizes of the various effects presented in Figure 2.

At the same time, the findings in the current research provide some practical implications to understand pilotage operations. In the SHERPA, as mentioned above, we see that a common error is to omit actions. This information give insight in how pilots should be trained, an example could be to design exercises that increase pilot skill in issuing commands and actions in various team atmospheres (warm vs cold). This knowledge of the operation may be compared across cultures and nationalities as well as across industries (e.g. process, aviation, oil and gas).

4.2 Limitations

The qualitative nature of the research conducted share some common limitations with such research; that is, reflexivity considerations. Attitudes and prior knowledge and experience of the analysts and interviewers may shed systematic influence upon the findings and the gathered data itself. Nonetheless, although acknowledging such reflexivity in research on and with humans, measures have been taken to reduce its systematic effect on the results. E.g. the use of other researchers to ensure consistency – reliability – of the analysis has been conducted, the interviewer have wide training and preparing of the interview, to make sure that interviewees are not led

in any directions, and interviewers were ensured that their participation is utterly voluntary and that the interviews may have been discontinued without any explanation: these effects are commonly mentioned to reduce reflexivity in qualitative research (Willig, 2008).

5 CONCLUSION

This research shed some light and structure to maritime pilotage operations. Such operations are costly, and at times complex to carry out for both the captain and pilot, yet mandatory and necessary to ensure a safe voyage in close waters. There is potential for future work in the current state of this research. The results can be used to further develop training and assessment tools to train and measure the performance of pilot operations. This will make the operations more cost-efficient, environmentally friendly and safe, which will benefit ship-owners, captains, pilots, crews, passengers and the local communities.

A thorough analysis of pilot's tasks contribute to pinpoint staffing and which competencies are needed to ensure a safe and efficient operation. Considering the competitive nature of the maritime industry, which constantly need to find ways to increase safety, productivity and efficiency (Bhattacharya, 2015; Stopford, 2009), knowledge of where to focus regarding staffing and training is of great value.

REFERENCES

Accident Investigation Board, N. (2010a). Crete Cement IMO NO. 9037161, Grounding at Aspond Island in the Oslo Fjord, Norway, on 19 November 2008. Report Sjø, 1.

Accident Investigation Board, N. (2010b). Report on Marine Accident Federal Kivalina-IMO NO. 9205885 Grounding at Årsundøya, Norway 6 October 2008. Report Sjø, 1.

Accident Investigation Board, N. (2012). Report on Investigation Into Marine Accident M/V Godafoss V2PM7 Grounding in Løperen, Hvaler on 17 February 2011. Report Sjø, 1.

Annett, J., Duncan, K., Stammers, R., & Gray, M. (1971). Task analysis. Department of Employment Training Information Paper 6: HMSO, London.

Baber, C., & Stanton, N. A. (1996). Human error identification techniques applied to public technology: predictions compared with observed use. Applied ergonomics, 27(2), 119-131.

Bhattacharya, Y. (2015). Employee engagement in the shipping industry: a study of engagement among Indian officers. WMU Journal of Maritime Affairs, 14(2), 267-292. doi:10.1007/s13437-014-0065-x

Bloor, M., Thomas, M., & Lane, T. (2000). Health risks in the global shipping industry: An overview. Health, Risk & Society, 2(3), 329340. doi:10.1080/713670163

Davis, R. (1982). Expert Systems: Where are we? And where do we go from here? AI magazine, 3(2), 3.

Di Pasquale, V., Miranda, S., Iannone, R., & Riemma, S. (2015). A simulator for human error probability analysis (SHERPA). *Reliability Engineering & System Safety, 139*, 17-32.

Hollnagel, E. (2000). Looking for errors of omission and commission or The Hunting of the Snark revisited. *Reliability Engineering & System Safety, 68*(2), 135-145. doi:http://dx.doi.org/10.1016/S09518320(00)00004-1

Kahneman, D., & Tversky, A. (1981). *The simulation heuristic*. Retrieved from

Kim, T.-e., & Nazir, S. (2016). Exploring marine accident causation: A case study. *Occupational Safety and Hygiene IV*, 369-373.

Kirwan, B. (1992). Human error identification in human reliability assessment. Part 2: Detailed comparison of techniques. *Applied ergonomics, 23*(6), 371-381.

Kluge, A., Nazir, S., & Manca, D. (2014). Advanced Applications in Process Control and Training Needs of Field and Control Room Operators. *IIE Transactions on Occupational Ergonomics and Human Factors, 2*(3-4), 121-136. doi:10.1080/21577323.2014.920437

Lundh, M., Lützhöft, M., Rydstedt, L., & Dahlman, J. (2011). Working conditions in the engine department–A qualitative study among engine room personnel on board Swedish merchant ships. *Applied ergonomics, 42*(2), 384-390.

Mallam, S. C., & Lundh, M. (2016). The physical work environment and end-user requirements: Investigating marine engineering officers' operational demands and ship design. *Work, 54*(4), 989-1000.

Nazir, S., & Manca, D. (2015). How a plant simulator can improve industrial safety. *Process Safety Progress, 34*(3), 237-243. doi:10.1002/prs.11714

Nazir, S., Sorensen, L. J., Overgard, K. I., & Manca, D. (2014). How distributed situation awareness influences process safety. *Chemical Engineering Transactions, 36*, 409-414.

Reason, J. (1990). *Human error*: Cambridge university press.

Schröder-Hinrichs, J.-U., Hollnagel, E., & Baldauf, M. (2012). From Titanic to Costa Concordia—a century of lessons not learned. *WMU journal of maritime affairs, 11*(2), 151-167. doi:10.1007/s13437-012-0032-3

Stanton, N. A., Stewart, R., Harris, D., Houghton, R. J., Baber, C., McMaster, R., Green, D. (2006). Distributed situation awareness in dynamic systems: Theoretical development and application of an ergonomics methodology. *Ergonomics, 49*(12-13), 1288-1311.

Stopford, M. (2009). *Maritime economics 3e*: Routledge.

Tversky, A., & Kahneman, D. (1975). Judgment under uncertainty: Heuristics and biases *Utility, probability, and human decision making* (pp. 141-162): Springer.

Willig, C. (2008). Phenomenological psychology: Theory, research and method. *Existential Analysis, 19*(2), 429-433.

Supporting Seafarer and Family Well-being in the Face of Traumatic Events: A Before, During and After Model

A. Dimitrevich & S. Welch
Sailors' Society, Southampton, United Kingdom

V. Torskiy
National University "Odessa Maritime Academy", Ukraine

D.C. Seyle
One Earth Future Research, Broomfield, United States

ABSTRACT: A seagoing career is an attractive prospect in many countries, however, it is still one of the most dangerous careers and seafarers face a difficult working environment, social isolation and stress. Industry usually talks openly about some of the physical challenges of working at sea but mental health and wellbeing is a topic usually left out of the discussion. One in four people experience a mental health problem in their lifetime and seafarers are no different. In particular, the high rates of accidents and fatal incidents at sea suggest that seafarers are highly likely to face issues related to post-traumatic stress exposure. Seafarer studies have found that up to a 1/3 of survivors of fatal accidents at sea may suffer with PTSD and high rates of mental ill health have been found in the survivors of piracy and their families. (Prof. Neil Greenberg, Managing Director of March on Stress and Professor of Defence Mental Health at King's College London). There is good evidence that the risk of developing PTSD can be substantially diminished if organisations put in place measures to prevent the development of the condition and to detect the early signs of it early on and manage it actively to prevent it progressing. Trainings are one of the best tools to mitigate levels of distress in the future and to combat stigma which prevents many from addressing their mental health issues to specialists. Whether it's general stress and anxiety, or feelings of depression, it can be hard to know for seafarers what to do and difficult to know who to talk to in such circumstances without proper training and that is why a comprehensive approach is likely to require a coordinated program that includes more developed pre-departure training, engagement with broader group of seafarers' social surrounding including shipping companies HR & crewmanning agencies, maritime charities, ship visitors, state agencies and seafarers' family members. Such a system, particularly in the area of pre-departure training, can also be a platform for providing more information about coping and resilience as a way to boost recovery from maritime trauma including maritime piracy. Expanding the content of pre-departure training to include psychoeducation and coping may provide seafarers with tools to increase resiliency not just in the face of piracy, but the other stresses of a maritime career.

1 INTRODUCTION

A career at sea is an attractive prospect in many countries, however, it is still one of the most dangerous careers to pursue with many seafarers facing a difficult physical working environment, social isolation, fatigue and stress. The *industry openly talks about some of the physical challenges of working at sea, but mental health and well-being is a topic too often omitted from discussion. This is a common issue internationally: while mental health challenges are very common, they are frequently not openly discussed either by those who experience them or by programs focusing on workplace satisfaction. This is despite the fact that one study of the US population found a lifetime prevalence of some mental health issue approaching 50%* (Kessler et al., 1994). *Seafarers are not immune from these challenges. In particular, the high incidence of accidents and fatalities at sea suggest that seafarers are highly likely to face issues related to post-traumatic stress exposure. Seafarer* studies have found that up to a 1/3 of survivors of fatal accidents at sea may suffer from PTSD, and high rates of mental health issues have been found in the survivors of piracy and their families. (Prof. Neil Greenberg, Managing Director of March on Stress and Professor of Defence Mental Health at King's College London, Seyle et al., 2016)).

Mental health issues, and particularly challenges related to traumatic stress, can be impacted by training and support. In particular, pre-event training can develop good coping skills that can reduce the impact of trauma, and effective post-event care and support can reduce the risk of lasting PTSD. More broadly, training is one of the best tools to mitigate levels of distress in the future and to combat stigma which prevents many from addressing their mental health issues with specialists. *Whether it is general stress and anxiety or feelings of depression, it can be difficulet for seafarers to know what to do and who to talk to in such circumstances without proper training. That is why a* comprehensive approach requires a coordinated program which includes a more developed pre-departure training, engagement with broader groups of seafarers' social surroundings, which includes shipping companies, human resources and manning agencies, maritime charities (including ship visitors), state agencies and seafarers' families.

Such a system, particularly in the area of pre-departure training, can also be a platform for providing more information about coping and resilience, as a way to boost recovery from maritime trauma (including maritime piracy). Expanding the content of pre-departure training to include psycho-education and coping may provide seafarers with tools to increase resiliency not only in the face of piracy, but also other stresses of a career in the maritime industry.

One model of this is the Crisis-at-Sea Response Programme. The concept for this programme is built up within a Before, During and After paradigm.

2 BEFORE, DURING AND AFTER

"Before" – This stage includes soft skills training for seafarers and maritime industry's office staff (shipping companies, manning agencies, welfare and other organizations which provide a wide range of services for seafarers). Traumatic events and other crises at sea in part have lasting impact because the seafarer may be unprepared not just for the fact of the accident, but for how to react to his or her emotional response. Pre-event training can provide a basic framework to support recovery by giving the seafarer tools to understand his or her responses and mechanisms for better coping.

Based on existing research, pre-event training should include self and situational awareness tips, "psycho-education" (anatomy of stress, coping techniques (including breathing and stretching exercises), psychological first aid (survivor stress reactions, core actions, provide self-care, resources, guidance for use of the pfa app), breaking bad news techniques, conflict management, and suicide prevention. It is aimed at self-assessment and care, as well as supporting others during and after traumatic events at sea. This kind of pre-event educational training has some evidence suggesting that it does support recovery after traumatic events, but primarily through providing healthy coping mechanisms.(Wessely et al., 2008).

"During:" This stage is focused on providing support to seafarers and family during long-duration traumatic events. While some traumatic events a seafarer may experience will be over quickly, such as an accident, others can last for days or weeks. This is particularly true about kidnapping, which is an enduring threat to seafarers in some piracy hotspots internationally. The "during" stage of crisis response is aimed at providing seafarers and families better tools for managing negative emotions and supporting them where possible. Work on this stage includes coordination and facilitation of first response/crisis response teams and their activities. Supervision and intervision intervention for helpers from different organizations, including those from shipping companies and manning agencies, is highly important for avoiding secondary and vicarious traumatization of personnel. Breaking bad news and trauma-informed survivor interview techniques, as well as trauma risk management assessment should become part of best practice for the Maritime Industry.

"After" – is the stage when a number of supportive measures should be taken in the next port of call and/or at home The network Sailors' Society chaplains are trained to provide and use a holistic approach for trauma survivors and will:

- Provide for their basic physical needs
- Help with social re-adaptation and reintegration
- Facilitate contacts for medical and psychological rehabilitation
- Widen the "corridor of tolerance"
- Promote no blame policies
- Give continuum of care

The continuum of care is not only to look after the survivors' recovery, but also to encourage them into an active social role by means of education and other available resources. One of the main ideas is to introduce and involve survivors in the train-the-trainer concept allowing them to share their experience and teach others.

3 SEAFARER WELL-BEING IS A DATA-SCARCE ENVIRONMENT

This model was developed based on existing research on what kinds of interventions support recovery after traumatic events. In the case of seafarer well-being, though, there is a challenge with developing evidence-informed programs. There is a relatively severe lack of publicly available data about traumatic stress in seafarers. One study was a 2016 report by Oceans Beyond Piracy, examining the impact of piracy on seafarers. This study collected data on 465 seafarers including 101 former hostages, and found that more than 25% of former hostages showed symptoms consistent with PTSD (Seyle et al., 2016). Piracy is relatively rare, though, and most seafarers will not be impacted by it. This study also examined rates of non-piracy traumatic events, and found very high rates of exposure to other forms of traumatic events.

Table 1. Prior exposure to negative events aboard ships (Seyle et al., 2016).

Negative experience	Per cent reporting
Fire aboard ship	30.42%
Injured from an accident	26.48%
Experienced medical emergency	21.88%
Ship grounding	17.72%
Witnessed death from accident aboard ship	13.79%
Involved in serious fight	11.60%
Fell overboard	8.10%
Witnessed suicide aboard ship	4.81%

This study is consistent with other research showing that the maritime industry is a highly risky place to work. Assessments of the risks to seafarers health and wellbeing regularly identify issues that can contribute to injury or death, and persistent differences between different flags and countries in risk (Bloor, Thomas, & Lane, 2000; Jaremin, 2005; LI & WONHAM, 1999; O'Connor & O'Connor, 2006). Even in seafarers from tightly regulated countries, though, the maritime profession is riskier than others: a 2005 study of British merchant shipping found that the mortality rate among seafarers was almost 30 times higher than the general population, and significantly higher than the next most risky profession (Roberts & Marlow, 2005).

While these studies and others have shown that the rate of accident and injury among seafarers remains high, there is significantly less research available on seafarer mental health and in particular traumatic stress. Some research has explored the impact of seafarer workplace conditions such as social isolation and fatigue on seafarers wellbeing and mental health, or seafarer mental health overall (H & M, 2003; Iversen, 2012; Smith, Allen, & Wadsworth, 2006). There has also been some research on the impact of trauma on seafarers (GREENBERG AS ABOVE). In general, however, this field remains under-studied and the specific impact of a kind of psychoeducation-focused program like that described above has not been demonstrated in a seafarer population.

Additional research in this area is needed to identify exactly what kinds of supports are most of value to seafarers and their families. Until the time that a body of research can be built up that demonstrates the effects of programs such as these using good methods such as randomized trials or prospective studies, this kind of program must necessarily be considered more research-informed than evidence-based.

REFERENCES

Bloor, M., Thomas, M., & Lane, T. (2000). Health risks in the global shipping industry: An overview. *Health, Risk & Society, 2*(3), 329–340. https://doi.org/10.1080/713670163

H, S., & M, T. (2003). The social isolation of seafarers: causes, effects, and remedies. *International Maritime Health, 54*(1–4), 58–67.

Iversen, R. T. B. (2012). The Mental Health of Seafarers. *International Maritime Health, 63*(2), 78–89.

Jaremin, B. (2005). Work-site casualties and environmental risk assessment on board Polish vessels. *International Maritime Health, 56*(1–4), 17–27.

Kessler, R. C., McGonagle, K. A., Zhao, S., Nelson, C. B., Hughes, M., Eshleman, S., ... Kendler, K. S. (1994). Lifetime and 12-Month Prevalence of DSM-III-R Psychiatric Disorders in the United States: Results From the National Comorbidity Survey. *Archives of General Psychiatry, 51*(1), 8–19. https://doi.org/10.1001/archpsyc.1994.03950010008002

LI, K. X., & WONHAM, J. (1999). Who is safe and who is at risk: a study of 20-year-record on accident total loss in different flags. *Maritime Policy & Management, 26*(2), 137–144. https://doi.org/10.1080/030888399286961

O'Connor, P. J., & O'Connor, N. (2006). Work-related maritime fatalities. *Accident Analysis & Prevention, 38*(4), 737–741. https://doi.org/10.1016/j.aap.2006.01.004

Roberts, S. E., & Marlow, P. B. (2005). Traumatic work related mortality among seafarers employed in British merchant shipping, 1976–2002. *Occupational and Environmental Medicine, 62*(3), 172–180. https://doi.org/10.1136/oem.2003.012377

Seyle, D. C., Bahri, C., Brandt, K., Dimitrievich, A., Fernandez, K. T. G., Holmer, T., & Malhotra, N. (2016). *After the Release: The long-term behavioral impact of piracy on seafarers and families.* Broomfield, CO: One Earth Future Foundation.

Smith, A., Allen, P., & Wadsworth, E. (2006). *Seafarer fatigue: The Cardiff research programme.* Cardiff, UK: Centre for Occupational and Health Psychology , Cardiff University.

Wessely, S., Bryant, R. A., Greenberg, N., Earnshaw, M., Sharpley, J., & Hughes, J. H. (2008). Does Psychoeducation Help Prevent Post Traumatic Psychological Distress? *Psychiatry, 71*(4), 287–302. https://doi.org/10.1521/psyc.2008.71.4.287

Grey List Danger of Turkish Flagged Vessels

U. Yıldırım, Ö. Uğurlu & S. Yıldız
Karadeniz Technical University, Trabzon, Turkey

ABSTRACT: Turkish flag is at Paris MoU's white list since 2008; but according to the inspections conducted in 2016, Turkish flag is in risk of falling into the grey list. Flag state has initiated an intensive inspection campaign, due to unsatisfactory standards of Turkish flagged vessels in terms of seafarers' qualifications, safety drills and ISM applications. In this study, the statistical analysis has performed for the deficiencies identified for the Turkish Flagged vessels; also, the changes in Turkish vessels have put forth, and recommendations have presented to correct these deficiencies.

1 INTRODUCTION

Today the shipping industry has reached more than 1.2 million seafarers (ILO, 2016), 90.197 commercial vessels and 1.8 billion tons of cargo handling capacity (UNCTAD, 2016). In 1978, The Hague Memorandum was first developed to control the working and living standards in the maritime industry, which has reached its present size with global developments in industry and commerce in the last decades. However, the same year, the VLCC Amoco Cadiz accident occurred and following this disaster; a new port state control treaty was signed in Paris / France in 1982 with the cooperation of 14 countries (Paris MOU, 2016). After IMO recognized the Port State Control (PSC) program in 1982, Port State control played a crucial role in ensuring safety at sea (Li and Zheng, 2008). Port State Control (PSC) inspection conducted by port state authorities when the vessels are visiting foreign ports; in order to check the compliance of the ship's condition, equipment, personnel and operations with international regulations (Hänninen and Kujala, 2014). Audit report issued after each PSC audit; including the flag of the inspected ship, the IMO number, the type of ship, the year and date of construction, as well as the relevant ship details and deficiencies (including 0 if there are no deficiencies) recorded during the inspection (Cariou et al., 2008). Today, Paris MOU is an organization of 27 member countries with a combination of Belgium, Bulgaria, Canada, Croatia, Cyprus, Denmark, Estonia, Finland, France, Germany, Greece, Iceland, Ireland,

Italy, Latvia, Lithuania, Malta, the Netherlands, Norway, Poland, Portugal, Romania, the Russian Federation, Slovenia, Spain, Sweden and the United Kingdom (Paris MOU, 2017a). There are currently 8 more regional port state contracts;

- Asia and the Pacific (Tokyo MoU)
- Latin America (Acuerdo de Viña del Mar);
- Caribbean (Caribbean MoU);
- West and Central Africa (Abuja MoU);
- the Black Sea region (Black Sea MoU);
- the Mediterranean (Mediterranean MoU);
- the Indian Ocean (Indian Ocean MoU);
- and the Riyadh MoU.

The United States Coast Guard maintain the tenth PSC regime (Heij et.al. 2011; IMO, 2016). Port State Control (PSC) is the inspection of foreign ships in national ports to verify that the condition of the ship and its equipment comply with the requirements of international regulations and that the ship is manned and operated in compliance with these rules (IMO, 2016). The Paris and Tokyo MOUs use "The Black, Grey and White List" (WGB list) of flag states as one of the parameters for determining a ship's risk profile (Perepelkin et.al., 2010; Kara, 2016). The "White, Grey and Black (WGB) list" presents the full spectrum, from quality flags to flags with a poor performance that are considered high or very high risk. It is based on the total number of inspections and detentions over a 3-year rolling period for flags with at least 30 inspections in the period (Paris MOU, 2017b). Flags with an average performance are shown on the "Grey" list. Their appearance on this list may act as an incentive to improve and move to the "White" list (EMSA, 2017).

Turkey has been on the white list of the Paris MOU since 2008. Until 2015, Turkey has highly successful profile, however due to the increasing detention rates in the PSC audits, risk of dropping to the grey list has emerged. But towards the end of 2016; with the increase of the Republic of Turkey Ministry of Transport, Maritime Affairs and Communication initiated control and audit campaign, this danger was removed and Turkey continued to maintain its place in the white list. In this context, deficiencies and the detention deficiencies of the Turkish flagged vessels have been examined, which have been identified in the inspections since 2009. As a result, profile of deficiencies has been put forward and suggestions have been made about the flag state.

2 LITERATURE REVIEW

There are numerous academic studies in which data on deficiencies identified in port state controls are used. Li (1999), reviewed 20 years of data, revealed that there is high correlation between total loss of ship, ages and detention rates. Cariou et. al. (2008), used data on PSC reviews carried out on foreign vessels at Swedish ports between 1996 and 2001. As a result, it has been found that port state controls caused a reduction of 63% in reported deficiencies.

Cariou et. al. (2009), in other study, used 26.515 PSC inspection data from the Indian Ocean MOU region between 2002 and 2006. In the study; they investigated the determinants of the probability of detention and the number of deficiencies. The results showed that the effects of the main headlines that led to detention are; 40% for age of ship, 31% for recognized organization and 17% for the location of PSC inspection.

Baniela and Rios (2011), investigated the relationship between the level of compliance of cargo ships with international standards and the severity level of the events they are involved in. For this purpose, between 2005 and 2006, events were investigated which 2,584 cargo ships have been involved. They based on Paris Memorandum's criteria's to measure substandard vessels and the level of vessels. In order to measure seriousness of the incident, they based on the number of days the ship has been repaired. The results show that there is a meaningful relationship between the safety standards of the ships and the severity of the accidents, showing that the substandard vessels have an accidental inclination with more serious consequences.

In their study Knapp et al. (2011), have examined port state controls from point of view of savings. As a data set, 15.819 vessel's port state controls between

2002 and 2007 were examined. The dataset includes inspection of the vessels, the number of deficiencies after the inspection, and information about detentions. The study has shown that port state control and sectoral inspections are very cost effective for tankers, elderly vessels and large ships. Moreover, the results emphasize the importance of determining the vessels with high risk of total loss.

Port state controls are an important control mechanism for ensuring safety at sea (Cariou et al., 2008). As a result of port state controls, the danger of a country flag falling on the grey list from the white list makes it necessary to inquire the current problems on that country's ships. For this purpose, this study aimed to reveal the general profile of the deficiencies of Turkish flagged vessels that are in danger of falling to the grey list. Also in this study, the main deficiencies that led to the detention were investigated.

3 DATA AND METHODOLOGY

The study covers the statistical analysis of inspections made under the Paris MOU to Turkish flagged vessels between 2009 and 2016. Data in the study were obtained from the Paris MOU, Republic of Turkey Ministry of Transport Maritime Affairs and Communications and the Thetis system. On the official web page of the Paris MOU, under the Detention & Bannings tab, the list of detention is published monthly from 2009 onwards. The detention lists contain basic information such as name of the ship, IMO number, company, class organization, charterer, type of ship, flag, gross tonnage and year of built. In addition, the public can reach information about the port where the ship is detained, date of release, duration of days, total number of deficiencies, number of detentions for the last 36 months, shortcomings causing detentions and action taken.

Thetis, the central information system of the European Maritime Safety Agency and the Paris MOU, contains information on the inspections made under the Paris MOU, and information on the vessels being inspected (Paris MOU, 2017d). Thetis, which helps member countries in selecting the right ship for inspection by constantly profiling data about ships, also provides statistics on the results and performances of the inspections (Thetis, 2017a).

Under the Paris MOU, a total of 4407 inspections and 204 detentions were carried out on Turkish flagged vessels between 2009 and 2016 (MTMC, 2017). As can be seen in Figure 1, the maximum number of inspections were made in 2010, the least inspections were made in 2015. Detention rates were at least in 2010 and 2013, the highest in 2015 and 2016. The increasing detention rates since 2014 have revealed the risk of falling on the gray list for

Turkish flag, and this has been the trigger of this study.

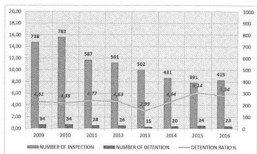

Figure 1. Detention rates of Turkish flagged vessels (MTMC, 2017)

The deficiencies identified in all inspections carried out between 2009 and 2016 on Turkish flagged vessels, and the numbers of detention deficiencies were presented in Table 1. The data for the years 2009-2015 presented in Table 1 were obtained from the Paris MOU (2017c) and MTMC (2017) institutions. For the 2016 data, Thetis (2017b) database was used. The classification of deficiencies was made according to the following headings, as in the Paris MOU annual reports.

4 RESULTS

13566 deficiencies were identified, between the years 2009 and 2016 belonging to Turkish flagged vessels. 1023 of these deficiencies have caused detention. The proportion of deficiencies causing detention within the total number of deficiencies is 7,54%. The years 2015, 2011, 2014 and 2016 stand out as the years with the highest rates of detention deficiencies respectively. This situation has revealed the gray list danger in the last 3 years' period.

The most important 7 deficiencies, which account for 73,21% (749 cases) of total detentions, were presented in Table 2. Deficiencies related to fire safety have reached the highest level in 2015 after 2009 and have been identified as the most important reason for detention. Other major deficiencies are the ISM, navigational safety, lifesaving appliances,

emergency systems, water/weather tightness and structural conditions. In Turkish flagged vessels, the deficiency types other than lifesaving equipment have increased in 2015. Between 2015 and 2016, deficiencies related to emergency systems increased compared to other years.

The Paris MOU system has a top 20 deficiencies section for deficiencies identified in all inspections made between 2014 and 2016 (Paris MOU, 2017e). Ten most important detention deficiencies are ISM, charts, nautical publications, fire doors/openings in fire-resisting divisions, fire detection and alarm systems, engine room, main engine, auxiliary engine, emergency systems and passage plan. The findings showed that, there is a great similarity between the deficiencies in Turkish flag and other flags. Likewise, the list of most frequent detainable deficiencies in the annual report of Tokyo MOU 2015 is similar to the Turkish flag. (Tokyo MOU, 2015).

Percentage rates of detention deficiencies on annual basis under subheadings were shown in Figure 2. For example; the rate of ISM deficiencies for 2010 is calculated from Table 1 as (33/129) * 100: 25,58. By the calculations made for the years 2009-2016, it was determined that the emergency systems showed the greatest increase in the detention rate. Other deficiencies that have tendency to increase were Fire safety, ISM, Safety of Navigation. Water / Weather tight conditions and Structural Conditions were determined as the deficiencies that have tendency to decrease. In 2015, other major deficiencies have increased, except for lifesaving equipment.

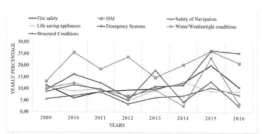

Figure 2. Percentage of most important detention deficiencies by year

Table 2. Number of most important detention deficiencies between 2009 and 2016

	2009	2010	2011	2012	2013	2014	2015	2016	Total
Fire Safety	26	12	19	22	16	18	27	19	159
ISM	18	33	12	18	8	12	16	13	130
Safety of Navigation	21	25	21	9	13	10	17	13	129
Lifesaving Appliances	24	16	18	7	9	17	9	9	109
Emergency Systems	8	16	14	7	8	9	18	18	98
Watertight Weathertight Conditions	13	15	12	7	9	2	10	2	70
Structural Conditions	9	11	11	4	9	2	7	1	54

Table 1. Deficiencies between 2009 and 2016

Deficiencies	2009 NR DEF	2009 Det DEF	2010 NR DEF	2010 Det DEF	2011 NR DEF	2011 Det DEF	2012 NR DEF	2012 Det DEF	2013 NR DEF	2013 Det DEF	2014 NR DEF	2014 Det DEF	2015 NR DEF	2015 Det DEF	2016 NR DEF	2016 Det DEF	Total NR DEF	Total Det DEF
Alarms	17	0	18	2	16	0	16	1	13	0	8	0	6	0	4	0	98	3
Cargo operations including equipment	13	1	14	1	15	2	10	1	13	0	3	1	11	0	6	0	85	6
Certificate & Documentation -Crew Certificates	31	3	35	4	22	6	19	3	13	1	29	3	22	1	30	2	201	23
Certificate & Documentation -Documents	206	4	197	1	166	2	150	2	103	0	121	2	100	6	120	1	1163	18
Certificate & Documentation -Ship Certificates	130	7	113	3	77	1	69	0	59	2	61	0	54	16	43	3	606	32
Dangerous goods	13	0	3	0	1	0	5	0	3	0	6	0	4	0	4	0	39	0
Emergency Systems	83	8	99	16	111	14	113	7	73	8	78	9	68	18	71	18	696	98
Fire safety	226	26	198	12	217	19	232	22	160	16	143	18	134	27	179	19	1489	159
ISM	136	18	129	33	65	12	76	18	54	8	59	12	61	16	62	13	642	130
ISPS	33	0	53	0	39	0	24	0	15	0	12	0	11	0	0	0	187	0
Labour conditions -Accommodation, recreational facilities, food and catering	0	0	0	0	0	0	0	0	8	0	29	0	74	2	44	0	155	2
Labour conditions-Conditions of employment	0	0	0	0	0	0	0	0	3	0	13	1	22	6	12	0	50	7
Labour conditions-Health protection, medical care, social security	10	0	8	0	8	0	2	0	18	0	73	0	70	2	69	1	258	3
Labour conditions-Minimum requirements for seafarers	0	0	0	0	0	0	0	0	0	0	0	0	2	0	2	0	4	0
Lifesaving appliances	226	24	178	16	179	18	145	7	95	9	132	17	100	9	101	9	1156	109
Living and Working Conditions-Living Conditions	132	4	104	10	107	5	86	3	72	1	37	0	8	0	26	1	572	24
Living and Working Conditions-Working Conditions	203	3	240	4	206	4	189	2	148	1	57	4	38	0	62	3	1143	21
Other	20	0	14	0	15	1	16	0	8	0	9	0	9	1	20	0	111	2
Pollution prevention-Anti Fouling	7	0	6	0	3	0	2	0	0	0	1	0	1	0	0	0	20	0
Pollution prevention-Marpol Annex I	57	10	53	9	48	4	44	10	23	0	23	3	9	0	13	2	270	38
Pollution prevention-Marpol Annex II	0	0	0	0	2	0	0	0	1	0	1	0	0	0	0	0	4	0
Pollution prevention-Marpol Annex III	0	0	0	0	0	0	0	0	0	0	0	0	0	0	1	0	1	0
Pollution prevention-Marpol Annex IV	16	1	5	1	13	1	17	3	12	0	7	0	10	1	3	0	83	7
Pollution prevention-Marpol Annex V	23	0	16	0	19	0	14	0	31	0	9	0	15	0	14	0	141	0
Pollution prevention-Marpol Annex VI	2	0	8	0	14	1	13	1	7	0	8	0	13	0	8	0	73	2
Propulsion and auxiliary machinery	116	5	153	2	115	11	89	3	68	9	46	9	43	3	47	5	677	47
Radio Communications	95	10	60	3	70	10	42	6	45	3	44	4	31	1	32	2	419	39
Safety of Navigation	370	21	354	25	240	21	256	9	206	13	143	10	161	17	175	13	1905	129
Structural Conditions	102	9	95	11	111	11	79	4	50	9	46	2	53	7	40	1	576	54
Water/Weathertight conditions	128	13	121	15	129	12	99	7	94	9	72	2	43	10	56	2	742	70
Total	2395	167	2274	168	2008	155	1807	109	1395	89	1270	97	1173	143	1244	95	13566	1023

5 CONCLUSION

Port State Control ensures that international maritime transport is carried out in accordance with environmental and safety standards (Sage, 2005). It provides detection and correction of deficiencies and helps to reduce loss of life and property, accident risk (Heij, 2011) and pollution. Publication of inspection reports and the causes of deficiencies also leads to the academic studies contribute to the maritime industry.

The performance of the flag states in the inspections and the rates of detention constitute the risk profile for that country. Flags with a high risk profile were found to have a strong relation with serious accidents (Baniela and Rios, 2011), and detention rates have a strong relation with accidents that resulting with total loss (Li, 1999). For this reason, it is extremely important to examine the deficiencies that cause detention.

When the average age of Turkish flagged vessels is examined (DWT); 318 vessels are in the 0-19 age range and 275 vessels are over 20 years old (TCS, 2016). As of the beginning of 2016, the average age of the world trade fleet is 20.3 years (UNCTAD, 2016). The excessive number of elderly vessels in the Turkish flag causes numerous deficiencies in the bridge, living quarters, fire safety and lifesaving equipment. Under navigational safety, nautical publications, charts, passage plans, magnetic compass, lights, shapes, sound signals, gyro compass deficiencies were highlighted. The deficiencies of the emergency systems are mostly; emergency lighting, batteries and switches, muster list, enclosed space entry and rescue drills, emergency fire pump and its pipes, and fire drills. It is thought that the renewal of Turkish ships, especially the coaster fleet, will lead to a decrease in detentions. Also, non-institutionalized companies and companies which do not have maritime culture, making limited investments in their vessels, because they cannot get high profit in international trade. This situation reduces the condition and quality of the crew.

Authors who have worked on Turkish flagged vessels for many years think that the deficiencies related to drills are mainly due to seafarer's point of view; because mostly seafarers think that drills are stealing time from their resting hours and drills are unnecessary jobs. Company managers and captains, instead of putting pressure on crew about drills, they need to change the point of view of the seafarers.

ISM is one of the most important reasons for detention on Turkish flagged vessels. ISM is a company and ship operating system that targets the safety of the environment and goods. In this context, the ISM requires that the land and ship personnel to have the necessary knowledge on safety and environment. Despite the fact that the ISM is known by crew of Turkish flagged vessels, in practice deficiencies often arise from the shortcomings in the safety culture. Therefore, it will be useful to conduct awareness trainings as integrated at the training institution, the company and onboard.

REFERENCES

Baniela, S., I. and Rios, J., V., 2011. Maritime Safety Standards and the Seriousness of Shipping Accidents, The Journal of Navigation, 64, 495–520.

Cariou, P., Mejia Jr., M.Q. and Wolff, F.C., 2008. On the effectiveness of port state control inspections, Transportation Research Part E, 44, 491–503.

Cariou, P., Mejia Jr., M.Q. and Wolff, F.C., 2009. Evidence on target factors used for port state control inspections, Marine Policy, 33, 847–859.

EMSA. 2017. Port State Control. Available: http://www.emsa.europa.eu/implementation-tasks/port-state-control/item/505.html [Accessed 07.02.2017].

Hanninen, M. and Kujala, P. 2014. Bayesian network modeling of Port State Control inspection findings and ship accident involvement. Expert Systems with Applications, 41, 1632-1646.

Heij, C., Bijwaard, G.E. and Knapp, S. 2011. Ship inspection strategies: Effects on maritime safety and environmental protection, Transportation Research Part D, 16, 42–48.

ILO, 2016. http://www.ilo.org/global/standards/subjects-covered-by-international-labour-standards/seafarers/lang--en/index.htm [Accessed 01.02.2017].

IMO, 2016. http://www.imo.org/en/OurWork/MSAS/Pages/PortStateControl.aspx [Accessed 05.01.2017].

Kara, E., G., E., 2016. Risk Assessment in the Istanbul Strait Using Black Sea MOU Port State Control Inspections, Sustainability 2016, 8(4), 390; doi:10.3390/su8040390

Li, K.X., 1999. The safety and quality of open registers and a new approach for classifying risky ships, Transportation Research Part E, 35, 135-143.

Li, K.X. and Zheng, H. 2008. Enforcement of law by the Port State Control (PSC). Maritime Policy & Management, 35, 61-71.

MTMC, 2017. Republic of Turkey Ministry of Transport Maritime Affairs and Communications, Circular Number 80368960-105.02.04/E.1619, 05.01.2017

Paris MOU, 2016. https://www.parismou.org/about-us/history [Accessed 25.12.2016].

Paris MOU, 2017a. https://www.parismou.org/about-us/organisation [Accessed 01.01.2017].

Paris MOU, 2017b. https://www.parismou.org/inspections-risk/white-grey-and-black-list [Accessed 15.01.2017].

Paris MOU, 2017c, https://www.parismou.org/detentions-banning/monthly-detentions [Accessed 07.02.2017].

Paris MOU, 2017d. https://www.parismou.org/inspection-search [Accessed 07.02.2017].

Paris MOU, 2017e. https://www.parismou.org/inspection-search/inspection-results-deficiencies [Accessed 07.02.2017].

Perepelkin, M., Knapp, S., Perepelkin, G. and Pooter, M. 2010. An improved methodology to measure flag performance for the shipping industry, Marine Policy, 34, 395–405

Sage, 2005. Identification of 'High Risk Vessels' in coastal waters, Marine Policy, 29, 349–355.

TCS, 2016. Turkish Chamber of Shipping, Maritime Sector Report 2015, http://www.denizticaretodasi.org.tr/Shared%20Documents/sektorraporu/2015_sektor_en.pdf [Accessed 27.012.2016].

Thetis, 2017a. https://portal.emsa.europa.eu/web/thetis [Accessed 04.02.2017].

Thetis, 2017b. https://portal.emsa.europa.eu/web/thetis/inspections [Accessed 03.02.2017].

Tokyo MOU, 2015. http://www.tokyo-mou.org/doc/ANN15.pdf [Accessed 01.02.2017].

UNCTAD, 2016. Review Marine Transport, 2016. United Nations Publication, ISBN:978-92-1-112904-5.

Finding a Balance: Companies and New Seafarers Generation Needs and Expectations

V. Senčila & G. Kalvaitienė
Lithuanian Maritime Academy, Klaipeda, Lithuania

ABSTRACT: MET institution's mission is not only preparing young people for maritime career at sea, forming their professional competences, but also providing help in their career management. For this reason, it is important to know new seafarers' generation needs and career ambitions, as well as expectations from employers' point of view. Mutual understanding would help to find a balance between expectations of both and adequately manage changes. Results of research, conducted by surveying 4th year full-time students studying at the Lithuanian Maritime Academy and experts from shipping and crewing companies in Lithuania and Latvia regarding desirable contract conditions and seafarers' personal and/or professional characteristics, important for successful career, are presented in the article.

1 INTRODUCTION

Generally, a labour market is understood as a place where two sides, workers and employees, interact with each other. In case of maritime labour market ship owners/managers and seafarers have changeable characteristics and expectations. MET institution's mission is not only preparing young people for maritime career at sea, forming their professional competences, but and providing help in their career management. For this reason, it is important to know new seafarers' generation needs and career ambitions, as well as expectations from side of maritime labour market. Mutual understanding would help to find a balance between expectations of both and adequately manage changes.

Shipping is, obviously, an extremely old industry and one that has established and maintained a strict hierarchical occupation system in which seafarers' career development follows a common route from ratings, to junior officers to senior officers, depending on the accumulation of knowledge, experience and skills (Wu & Morris, 2006).

The shipping industry is highly cyclical in nature and is characterised by extended periods of bust and boom (Sampson & Tang, 2016).

The maritime labour market currently became extremely global. Over 90 per cent of the seafarers from the advanced economies work on board their home fleets, while 80 per cent of seafarers from the developing economies and over 60 per cent of the seafarers from transitional economies work on board the fleets of advanced economies. World shipping industry has offered opportunities for seafarers from developing countries and, increasingly, from transitional economies to develop careers outside their home country (Wu & Morris, 2006).

Work on board a merchant ship can be stressful (Agterberg & Passchier, 1998). Today the seafarer often works alone, comrades have finished eating when he or she enters the mess, and the fun and excitement of port visits has disappeared (Ljung & Widell, 2014). Seafaring careers entail long-term separation between seafaring couples, which make seafarer-partners feel that they live different lifestyles and that nonseafaring people could not fully appreciate their experiences (Tang, 2010; Thomas at al., 2003).

The motives to become a seafarer have in recent decades changed. The myth of the exciting and free life at sea has largely been cracked. Research on how the major changes that permeated the work content and the organisation have influenced young people's motivation to choose a career in seafaring profession is hard to find (Ljung & Widell, 2014).

The hiring practices of shipping industry employers should be targeted at meeting the career expectations of recruits (Caesar & Cahoon, 2015): what future career ambitions do they have, including where do they hope to be employed in ten and twenty years' time, and what do they know about the working career of seafarers?

Lately MET institutions and shipping industry admit new young seafarers' generation, called generation Z, those who were born during the middle 1990s and late 2000s.

The main categories describing differences between generations are: social environment, technological environment and historical environment. Characteristics of the generation Z vary by cultural and regional environment but are most impacted by technological development and generation Z can be defined as "instant online" (Levickaite, 2010).

They are always online on any technical device virtually, with no stop. To solve problems, they try to find the solutions on the internet. Arising from their habit, they have different expectations in their workplaces. We can speak about a careerist, professionally ambitious generation, but their technical and language knowledge are on a high level (Bencsik at al., 2016).

Authors indicate different features of Z generation. Analysis that is more thorough allows determining some other featuring characteristics of this generation: growing hyperactivity, infantilism, multimedia literacy, loop reading, social autism, consumerism, lack of analytical evaluation of communication and a text as well as its critical evaluation and etc. (Targamadze, 2014).

Because of increasing the retirement age, often 3 or 4 generations are forced to cooperate and to work together. Based on the data, it is mainly the communication and the difference in the way of thinking that can cause difficulties in the process of working together. The research justified that the cooperation of different age groups could provide not only conflicts, but also positive results as well for the organisation (Bencsik at al., 2016).

Generation Z can bring us an additional advantage to the huge problem that we face in the industry: the overload of information. For example, think of the navigator; he has too much information through the many layers of ECDIS that simply cannot handle it. This generation can be trained at early stage sorting a lot of information quickly. Therefore, they will be much more able to process navigational information (Recruitment…, 2017).

Because of the global shortage of ship officers and the dynamic nature of the seafarer labour markets, shipping industry employers need to adopt appropriate measures to improve the future supply of maritime labour (Caesar & Cahoon, 2015). However, the new generation also should be ready to look for the balance of interests and be able for adaptation to the working environment and changing circumstances.

2 THE RESEARCH METHODOLOGY

2.1 *The sample of the research*

The research was conducted by surveying 4th year full-time students studying at the Lithuanian Maritime Academy, shipping and seafarer crewing companies in Lithuania and Latvia. The sampling was based on the principle of free will, i.e. all the final year students, who attended the Academy on the days of the survey, were included into the sample. Such survey sampling method is considered to be reliable.

In January 2017, 67 final year bachelor degree students were surveyed: out of which 38 percent were Marine Navigation study programme students, 62 percent were Marine Electrical and Marine Power Plant Operation study programmes students. Other respondents were the experts from five shipping and crewing companies in Lithuania and three shipping and crewing companies in Latvia, where students have had seagoing practice.

2.2 *Research instrument*

The data of the qualitative research was collected by using a written survey method. Original questionnaires made by authors were used for the survey.

The students of the Maritime Academy were asked about the planned duration of work at sea and what reasons would lead to the termination of a seafarer's career? The questions were aimed at identifying students' expectations and needs related to the work at sea: regarding the term of voyage and working conditions and their opinion about the personal and/or professional characteristics that determine successful employment of a seafarer and his career in a company. Two questions were aimed at identifying, what impact communication with older seafarers had during practice on board a ship and what means facilitated the adaptation, coping with the routine and other emotional issues relevant to the final year students.

Respondents of the qualitative research as experts include five shipping and crewing companies in Lithuania, Klaipeda, and three companies in Latvia. Meanwhile, a qualitative survey consists of four questions. These questions reveal main things that are important for settling the balance between the expectations and needs of shipping and crewing companies and young maritime sector specialists:
1 What general skills (foreign languages, teamwork, etc.) and personal characteristics (reliability, communication skills, etc.) are important for seafarer's career?
2 What most significant changes took place during the recent decade: regarding requirements for

seafarers and regarding seafarers' expectations and needs?

3 What work/contract conditions (in addition to salary and term of voyage) are important for young seafarers?

4 What means are decisive for seafarers' self-determination to work in the company for a long term?

Only survey question regarding the term of students' future seagoing career was closed-ended, required a short single-word answer. All other questions were open-ended, designed to encourage respondents' full, meaningful answers. The obtained data was processed using the content analysis technique, interpreting and coding textual material.

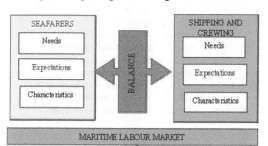

Figure 1. Model of investigation of balance between companies and seafarers needs and expectations

3 RESULTS OF INVESTIGATION

3.1 *Maritime students opinion regarding the term of seagoing career*

The research was aimed at identifying how long the final year students are planning to work at sea.

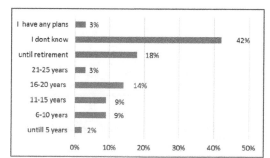

Figure 2. Percentage distribution of the respondents' opinions on the term of maritime career after the completion of the studies

A large part of the total number of respondents (42 percent) was uncertain about the term of marine career after the completion of the studies (Fig.2). 18 percent of respondents are planning to work at sea until retirement, 9 percent are planning to work at

sea for 6-10 years, 9 percent – for 11-15 years, 2 percent of respondents are planning to work at sea up to 5 years.

From 67 final year maritime students, who participated in the survey, most would terminate their career because of the following reasons:
- health problems, disease (N=44);
- family, small children (N=36);
- occupational injuries (N=28);
- low salary (N=27);
- long term of voyage (N=26);
- poor working conditions on board a ship (N=25);
- psychological issues (N=5);
- establishment of own business or another adequate alternative (N=1).

Three respondents did not identify a reason that would lead them to the termination of the seafarer's career.

3.2 *Maritime students' needs and expectations regarding desirable contract conditions*

Maritime students (67 respondents) were asked to define, in their opinion, the best job offer.

The most important factors defining a job offer as the best one were named *salary* (bigger, payable on time and adequate to the position held) (N=59) and *shorter employment contracts* (N=53). Most students, 41 percent expressed the opinion, that optimal voyage term would be 1-2 month, 33 percent named 2-3 month and 26 percent named 3-4 month voyage duration.

Maritime students were asked about other working/contract conditions (beside salary and term of voyage), which are important to them personally and would motivate them to work at sea.

Besides salary and term of voyage, the most important working/contract conditions named by maritime students were (Fig.3) the following:
- good relations between the members of the crew (N=48),
- good employment contract and accommodation conditions on board a ship (N=31);
- ship condition and type (N=27);
- navigation area (N=25);
- quality food (N=23);
- proper conditions for rest and sports (N=21);
- good reputation of the company and taking good care of employees (N=21).

Students named *good work tools and good supply* (N=18) as important working/contract conditions. *Internet connection* (N=17) and *career opportunities* (N=13) were indicated as important as well.

Only six students determined the opportunity to travel, *opportunity to see the world* as important working/contract condition to them personally and would motivate them to work at sea.

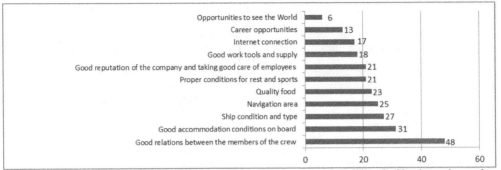

Figure 3. Maritime students' opinion about desirable working/employment contract conditions (beside salary and term of voyage)

3.3 Shipping and crewing companies experts opinion about working/employment contract conditions, most important for seafarers

The content analysis of the experts' opinions revealed that the following working/employment contract conditions are important to the seafarers and determine their decision to work in a company for a long period:

– salary;
– social security guarantees;
– good reputation and reliability of the company;
– policy of the company;
– term of employment contract;
– compliance of employment contract to the laws;
– career opportunities;
– composition of crew (national);
– navigation area;
– good microclimate on board a ship and professional assistance of the crew.

3.4 Maritime students and experts opinions about seafarers' desired personal and/or professional characteristics

Maritime students (67 respondents) and experts (8 respondents) were asked open-ended question about seafarers personal and/or professional characteristics, which determine their successful employment and career in a company.

Shipping and crewing companies' experts gave very similar responses, indicating their common approach (Fig. 4). Content analysis discovered that all eight experts described professional and personal features: *professional skills, good training, responsibility, communication skills, teamwork skills,* as most important for successful employment and career. Seven of them also named *foreign languages, reliability and tolerance.* Two of them also noted as important *the ability to adapt to the marine shipboard living and working monotony and physical health.*

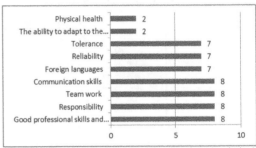

Figure 4. Experts' opinions about seafarers' desired characteristics

Students named as most important features for successful employment and career *assertiveness and diligence* (N=26) and *good professional skills and knowledge* (N=20). After that in descending order they identified *flexibility* (N=17) and *communicability* (N=13), *responsibility* (N=11), *leadership* (N=8), *professional experience* (N=6), *foreign languages and motivation* (N=5).

Figure 5. Students' opinions about seafarers' desired characteristics

Some overlaps and differences in opinions of maritime students and shipping and crewing companies' experts were identified. Both groups, students and companies' experts, highly rated professional skills, attitude (assertiveness, diligence, responsibility) and communicability (teamwork, flexibility). Thus, in general, the younger generation of seafarers demonstrated good adaptability for labour market needs, taking over the main companies' attitudes and approach to work. Here it is worth keeping in mind that students, who participated in the survey, were final fourth year maritime students, already having about one year practical experience.

On the other hand, students evaluated as less important foreign languages and did not name at all, as important, reliability and tolerance, which were highly rated by the companies' experts.

3.5 The impact of communication with older seafarers on the students' professional knowledge, skills and attitude

Maritime students were asked about the impact of communication with older seafarers during seagoing practice on board a ship on the students' professional knowledge, skills and attitude to towards seafarer's career (Tabl.1).

Table 1. Maritime students answers about the impact of communication with older seafarers during seagoing practice on board a ship on their professional knowledge, skills and attitude to a seafarer's career

Given the professional knowledge	*Understood the specifics of the job* (R7)
	Learnt things that helped to perform my assignments safer (R46)
	...how something functions and how to do something better (R53)
	...need Russian language skills (R57)
Given the motivation to seek a seafarer's career	*Promising career opportunities* (R17)
	Made a good impression, motivated to strive for career heights (R20)
	Motivated to study and go to work at sea, not to abandon seafarer's career (R9)
Given help, forming attitude on seafarer's career	*Changed my attitude to performed work (R34)*
	They changed my attitude to...(R43)
	I was advised to try to grasp a bigger picture (R27)
	Enabled to understand the real life of a seafarer (R23)
	I always tried to communicate with them not only about work, but also about life, they gave me good pieces of advice (R58)
	Helped me to make up my mind on future career (R19)
Revealed negative aspects of seafarer's career	*Hard and disadvantageous work, not worth doing due to low salaries* (R67)
	There are no friends, everybody wants to harm each other (R53)
	If I find a job on shore, work at sea is not worth choosing it (R5)
	Most seafarers advised to continue studies and to work on shore (R40)
	To look for as short employment contracts as possible....(R37)
	Dangerous work (R55)

Maritime students noted positive impact of communication with older seafarers' generations on their professional knowledge, on the motivation to seek a seafarer's career and help forming attitude towards seafarer's career. Besides, students during the communication revealed negative aspects of

seafarer's career, a phenomenon known as *"learned helplessness"* (Ljung & Widell, 2014).

3.6 Conclusions

Maritime labour market, ship owners/managers and seafarers, have changeable characteristics and it is important to know new seafarers' generation needs and career ambitions, as well as expectations from the employers' point of view.

Presented in the article qualitative research was conducted attracting experts from five shipping and crewing companies in Lithuania and three shipping and crewing companies in Latvia and 67 final year maritime students. The qualitative survey revealed main things that are important for settling the balance between the expectations and needs of shipping and crewing companies and young maritime sector specialists' generation.

Answering about the best job offer most maritime students identified salary and short-term (1-2 month) voyages, which offer on the market is not so big. Besides salary and term of voyage, students and companies' experts, as the most important working/contract conditions, named good relations between the members of the crew, good reputation and reliability of the company, navigation area. But only students named as important ship's type and condition, good accommodation conditions on board a ship, quality food and proper conditions for rest and sports, internet connection. As important working/contract conditions students also named good work tools and good supply, but did not mention social security guarantees, named by companies' experts.

Analysing seafarers' personal and/or professional characteristics, which determine their successful employment and career in a company, some overlaps and differences in opinions of maritime students and shipping and crewing companies' experts were identified. Both groups, students and companies' experts, highly rated professional skills, attitude (assertiveness, diligence, responsibility) and communicability (teamwork, flexibility). On the other hand, students evaluated as less important foreign languages and did not name at all, reliability and tolerance, which were highly rated by the companies' experts.

Maritime students noted positive impact of communication with older seafarers' generations on their professional knowledge, on the motivation to seek a seafarer's career and assistance in forming positive attitude towards seafarer's career. However, during the communication older seafarers conveyed to the students negative aspects of seafarer's career as well.

In general, the younger generation of seafarers demonstrated good adaptability for labour market needs, taking over the main companies' attitudes and approach to work.

Accordingly, shipping industry employers need to be ready to adopt appropriate measures to keep the balance between the needs and expectations of both, seeking to improve the future supply of maritime labour.

REFERENCES

Agterberg, G. & Passchier, J. 1998. Stress among seamen. *Psychological Reports*, 83: 708-710.

Bencsik, A., Horváth-Csikós, G., Juhász, T. 2016. Y and Z Generations at Workplaces. *Journal of Competitiveness*, Vol. 8, Issue 3: 90 – 106.

Caesar, L. & Cahoon, S. 2015. Training Seafarers for Tomorrow: The Need for a Paradigm Shift in Admission Policies. *Universal Journal of Management* 3(4): 160-167.

Levickaite, R. 2010. Generations x, y, z: How social networks form the concept of the world without borders (the case of Lithuania). *LIMES: Cultural Regionalistics*, Vol. 3, No. 2: 170-183.

Ljung, M. & Widell, G. 2014. *Seafarers' working career in a life cycle perspective - driving forces and turning points.* Gothenburg: Chalmers University of Technology.

Recruitment and Training of Generation Z. Retrieved from internet 2017-01-27. *http://www.safety4sea.com/recruitment-training-generation-z/.*

Sampson, H. & Tang, L. 2016. Strange things happen at sea: training and new technology in a multi-billion global industry. *Journal of Education and Work*, Vol. 29, No. 8: 980-994.

Tang, L. 2010. Development of online friendship in different social spaces. *Information, Communication & Society*, Vol. 13, No. 4, June: 615–633.

Targamadzė, V. 2014. Z karta: charakteristika ir ugdymo metodologinės linkmės įžvalga. *Tiltai* 4: 95-104.

Thomas, M., Sampson, H. & Zhao M. 2003. Finding a balance: companies, seafarers and family life. *Maritime Policy & Management*, Vol. 30, No. 1: 59–76.

Wu, B. & Morris, J. 2006. A life on the ocean wave: the postsocialist careers of Chinese, Russian and Eastern European seafarers. *The International Journal of Human Resource Management*, 17:1, January: 25–48.

On the Connection between Teamwork and Political Correctness Competence Provision for the Seafarers

Z. Bezhanovi, N. Vasadze, M. Abashidze & L. Khardina
Batumi State Maritime Academy, Batumi, Georgia

ABSTRACT: The aim of the presented paper is to discuss some directions of the English language teaching intended to meet the requirements of (Manila amended) STCW 78/95, the IMO Model Courses 3.17 and 1.39. The Maritime English teaching is mainly directed to provision of the Specification of minimum standard of competence for officers in charge of a navigational and engineering watches.

At the same time, it is not of less importance to focus the teaching directions on provision of performance of the officer's duties under requirements of the IMO model course 1.39 "Leadership and Teamwork" directly related to knowledge and ability to apply effective resource management through efficient cross-cultural communication on board and ashore; cultural awareness, allocation, assignment and prioritization of resources; decision making reflecting team experience; assertiveness and leadership, including motivation; obtaining and maintaining situational awareness; appraisal of work performance.

Thus, the empirical data of the paper presents a set of case studies (a real situations on board merchant ships) giving possibility on one hand to analyse the factors causing effectiveness of situational awareness and on the other hand the propose the principles of the possible conflicts prevention.

Taking into account a newly introduced term "a helmsperson" instead of a traditional "a helmsman" and the other terms gradually entering the maritime terminology, we'd like to mention that the future ship officers, provided with such conversational skills, will successfully perform their duties showing ability to apply effective communication on board and ashore at the same time providing the principles of leadership and team working.

In order to provide the above mentioned advantages, we propose to introduce politically corrected communicative technologies and cross-cultural communication principles into the Maritime Education and Training.

1 INTRODUCTION

The aim of the presented paper is to break the language barriers, possibly caused by political incorrectness between the seafarers. Nowadays, we live in the rapidly changing world and moreover in the society where new values and new ways of thinking are promoted. Since the language is a living phenomenon and ubiquitous it attracts and mirrors each individual's viewpoint and distinctly is reflected in the language. That is why new lexical units occur within the scopes of Political Correctness.

Taking into account seafarers work on the so-called "locked space", where they are obliged to be for months, away from their home countries, families etc. being a part of group where representatives are of different culture presenting their values and identities differently forming the group where are united people with diverse race, ethnicity, religion, country of origin, even sometimes gender inequality and etc are acquainted with some stressful condition, which is primarily revealed in the language. In such cases, the sender of the message and the receiver of the message are from different cultures and backgrounds. Of course, this introduces a certain amount of uncertainty, making communications even more complex. As it is known, widely spoken on board language is English which is basically the 2nd language for them. That is why, there occur problems in spoken English which can be foreseen in their grammar, pronunciation and so on.

When the speakers communicate cross-culturally, one should make particular effort to keep the communication process clear, simple and unambiguous. One communicates the way s/he does because any individual is raised in particular culture. The listener may interpret the speaker's behavior from an opposite standpoint. Sometimes this type of situation leads and creates misunderstanding and can be conflict producing. Whereas, Understanding the other's culture facilitates cross-cultural communication.Linguistic realization of political correctness has become one of the burning issues in the modern world. As is known, political correctness implies verbal behavior that excludes any form of verbal discrimination: racial, gender, religious or political: "The principle of avoiding language and behavior that may offend particular groups of people" (Oxford Dictionary 2010).

The term political correctness was coined and first used in 1793 when a judge, James Wilson, used it in the decision Chisholm v. Georgia (1793) to say it is not politically correct to speak of the United States instead of the people of the United States.

Political Correctness is linked to the face and politeness theories since the study has revealed that the choice of a particular strategy largely depends upon the context a particular term is used in.

2 GENDER INEQUALITY

In current English gender inequality used to be conveyed lexico-grammatically. As a result of the feminist movement certain changes were introduced into the English language in the second half of the 20th c. viz.: The use of the lexical unit man in a generic sense denoting a human being is avoided. The morpheme man as a component of compounds denoting different professions has been replaced by gender-neutral items. Accordingly, the terms related to occupations do not reflect gender. Below presented lexical units ending on "-man" is advised to be replaced "person" where the gender of a person is not indicated.

Politically Correct	Politically Incorrect
helmsperson	helmsman
pump person	pump man
mess person	mess man/boy
Sea person	seaman

If we discuss the lexical unit "seaman" which according to the Cambridge dictionary means the following: "a person skilled in seamanship", also we meet the following definition which is „a general-purpose for a man or a woman who works anywhere on board a modern ship, including in the engine spaces, which is the very opposite of sailing". Based on the definitions the term "seaman" is used both for male and female addressing form. But the protest can be raised due to the morpheme "man" because

there are women seafarers as well. For instance, a female Georgian graduate from Batumi State Maritime Academy Ms. Natia Labadze is employed in the crewing company Columbia ship management LTD. She started as a deck cadet and a novice sea person was promoted as 3rd Officer on board. That is why the addressing term "seaman" will be derogatory and offensive one since the morpheme "man" excludes women's rights and does not foresee that women can be sea persons as well. In the case of the term "sea person", the morpheme "person" has a general meaning, where is not stated a person is a female or a male.Thus, in accordance with the recent IMO researches, the safety and security of life at sea, protection of the marine environment and over 90% of the world's trade depends on the professionalism and competence of seafarers. The human element is a complex multi-dimensional issue that affects maritime safety, security and marine environmental protection involving the entire spectrum of human activities performed by ships' crews, shore based management, regulatory bodies and others. All need to co-operate to address human element issues effectively. Thus, the IMO constantly develops different activities aimed at provision of such an important issue. One of the modern approaches aimed at the same direction is the introduction of the IMO model course 1.39 Leadership and Teamwork intended to provide a person with the knowledge, skill and understanding of leadership and teamwork at the operational level on board a ship. The course is designed to meet STCW requirements for the application of leadership and teamworking skills, in accordance with the 2010 Manila Amendments, specifically as stated in table A-II/1, Function: Controlling the operation of the ship and care for persons on board at the operational level.

3 CORRELATION OF LEADERSHIP AND TEAMWORK WITH LANGUAGE

On completion of the course the learner/trainee should be able to demonstrate sufficient understanding and knowledge of leadership and teamworking and have the relevant skills to competently carry out the duties of officer in charge of a navigational watch on ships of 500 gross tonnage or more, or officer in charge of an engineering watch in a manned engine-room or designated duty engineer in a periodically unmanned engine-room. The knowledge, understanding and proficiency should include, but not be limited to, those listed in Column 2 of table A-II/1 and table A-III/1:

Working knowledge of shipboard personnel management and training include organization of crew, authority structure, responsibilities; cultural awareness, inherent traits, attitudes, behavior, cross-cultural communication; shipboard situation, informal social structures on board; human error, situation awareness, automation awareness, complacency, boredom; leadership and teamworking.

Knowledge of related international maritime conventions and recommendations: SOLAS, MARPOL, STCW, MLC, as well as national legislation.

Ability to apply task and workload management involves planning and coordination; personnel assignment; human limitations; personal abilities; time and resource constraints; prioritization workloads, rest and fatigue; management (leadership) styles; challenges and responses.

Knowledge and ability to apply decision-making techniques contain situation and risk assessment; identification and consideration generated options; evaluation of outcome effectiveness; decision making and problem solving techniques; authority and assertiveness; judgment; emergencies and crowd management).

Self-awareness, personal and professional development include knowledge of personal abilities and behavioral characteristics; opportunities for personal and professional development.

Knowledge and ability to apply effective resource management foresees effective communication on board and ashore; allocation, assignment and prioritization of resources; decision making reflecting team experience; assertiveness and leadership, including motivation; obtaining and maintaining situational awareness; appraisal of work performance; short and long term strategies. [1]

The whole set of competences and especially the last extract actually state new and high standards for the English language competence. Everything noted above should be implemented through the conversational ability of the officers, presenting a new challenge to the Maritime English teaching development. That is why, we want to propose political correctness as a potentially useful Maritime English teaching direction. Linguistic realization of political correctness has become one of the burning issues in the modern world. As it is known, political correctness implies verbal behavior that excludes any form of verbal discrimination: racial, gender, religious or political: ''The principle of avoiding language and behavior that may offend particular groups of people''[4]. The term political correctness was coined and first used in 1793 when a judge, James Wilson, used it in the decision Chisholm v. Georgia (1793) to say it is not politically correct to speak of the United States instead of the people of the United States. There are different domains of political correctness such as gender, that of sexual minorities, racial and disability one. The racial domain will be presented in our paper.

4 CONCLUSION

As the conclusive part of a paper could be proposed a brief politically corrected glossary which may be used within the frames of the English language competency for a future seafarers:

- Business person - a person in business or one who works at a commercial institution
- Chairperson - a chairman or chairwoman, someone who presides over a meeting, board, etc.
- Craftsperson - someone who is highly skilled at their trade.
- Differently able - disabled or handicapped. Fireperson - A fireman or firewoman. A firefighter
- Flight attendant - a member of the crew (staff) of an airplane who is responsible for the comfort and safety of its passengers.
- Fresh person - a freshman (male or female).
- Handicapable – Physically challenged.
- Helmsperson - a helmsman or helmswoman.
- Native American - an American Indian.
- Nurseryperson - a nurseryman or nurserywoman.
- police officer – a policeman/policewoman
- Salesperson - a salesman or saleswoman.
- Serviceperson - a serviceman or servicewoman.
- Undocumented immigrant – illegal immigrant.
- Waitperson - a waiter or waitress

5 CONCLUSION

Taking into account a newly introduced term "a helmsperson" instead of a traditional "a helmsman" and the other terms gradually entering the maritime terminology, we'd like to mention that the future ship officers provided with such conversational skills will successfully perform their duties showing ability to apply effective communication on board and ashore.

REFERENCES

I M O P u b l i s h i n g. Leadership and Teamwork (Model Course 1.39), IMO, 2014.
I M O P u b l i s h i n g. STCW including 2010 Manila Amendments, IMO, 2011.
I M O P u b l i s h i n g. IMO Standard Marine Communication Phrases, IMO, 2002.
http://www.oxforddictionaries.com/
www.imo.org
http://www.bbc.com/
www.urbandictionary.com
https://en.wiktionary.org/wiki/Category:English_politically_co rrect_terms
https://en.wiktionary.org/wiki/Category:English_politically_co rrect_terms
Duignan, Peter; Gann, L.H. (1995). Political correctness

Evaluation of Occupational Risk Factors for Cardiovascular Disease in Romanian Seafarers

L. Hanzu-Pazara
"Ovidius" University, Constanta, Romania

R. Hanzu-Pazara & G. Raicu
Constanta Maritime University, Constanta, Romania

ABSTRACT: Seafarer job implies a lot of occupational risk factors, along with psychosocial factors that may influence directly by damaging the endothelium or indirectly by aggravating the traditional risk factors. The study was analyzing data from 200 Romanian seafarers, comparatively in between different compartments, with respect of stress level given by occupational status. Results: Cardiovascular diseases were higher in deck operational crew (32.85%), compared with engine (21.05%) and other compartments (27.27%). Most prevalent independent risk factors in seafarers were high low-density lipoproteins (68.5%), differentiated on working compartments (deck 70.71% vs. engine 65.78% vs. else 59.09%). Overweight played an important addictive role. The measurement of ejection fraction in cardio – Doppler ultrasound exam revealed a decreased value in operational compartments compared with rating. Conclusions: The cardiovascular diseases seem to re-main important factors of distress on board vessels, despite precise medical standards of evaluation, and is necessary to overview all the elements that augment it.

1 INTRODUCTION

Cardiovascular diseases represent one important cause of morbidity and mortality worldwide. Seafarer job implies a lot of occupational risk factors like stressful situations on board vessel, modified diet or lack of exercise and sedentarism, along with psychosocial factors, that may influence directly by damaging the endothelium or indirectly by aggravating the traditional risk factors like smoking, dislipidemia or diabetes.

2 BACKGROUND

The most common cause of working disability and of major risk for sea safety are the cardiovascular diseases. In order to prevent conditions that may lead to an emergency situation onboard vessels, there have been implemented more strict inclusion and periodical medical examinations. Population-based epidemiologic data provide a good assessment of the evolution, progress and prevention of cardiovascular diseases, and different working areas analyses can implement a more precise evaluation of risk factors

Typically, the cardiovascular risk is influenced by some independent factors like dislipidemia (LDL cholesterol, HDL cholesterol), smoking, family history of cardiovascular diseases, hypertension.

3 OBJECTIVES

The objective of the study was to investigate the presence of cardiovascular changes in seafarers according to their position onboard vessel (operational, managerial, others), along with the evaluation of traditional cardiovascular risk factors.

4 METHODS

The study was analyzing data from 200 Romanian seafarers, with an age ranging between 20 to 65 years, evaluated during 12 months, at periodical pre-employment examination and periodical medical fitness tests in IOWEMED MEDICAL CENTER, Constanta, Romania. The programme was included into Loss Prevention evaluation criteria, with reference to human factor. There were collected demographic data, biological results (ex. lipid profile, glicaemia), BMI (body mass index) and cardiovascular investigations (electrocardiogram, Doppler heart ultrasound exam), that were comparatively analyzed in between different compartments, with respect of stress level given by occupational status. All data were included into a statistical analysis computerized program (SPSS 20.0).

Table 1. Evaluation of demographic and biologic features

Demographic & biologic features	No. = 200
Age, mean +/- SD	45,17 ± 10,103
Deck/engine	2,33:1
BMI mean +/- SD	26,27 ±2.972
Total Cholesterol mean +/-SD (mg/dL) (min/max)	217,85 ± 47,506 (142/320)
LDL mean +/-SD (mg/dL) (min/max)	142,35 ± 38,537 (63,44/230)
HDL mean +/-SD (mg/dL) (min/max)	51,13 ± 13,89 (22/95)
Triglycerides mean +/-SD (mg/dL) (min/max)	143,29 ± 96,439 (43/424)
Systolic BP mean +/-SD (mm/Hg) (min/max)	129,32 ± 10,605 (110/175)
Diastolic BP mean +/-SD (mm/Hg) (min/max)	78,88 ± 44,82 (60/100)
EF LV mean +/-SD (%) (min/max)	61,09 ±3,376 (38/72)

LDL – low density lipoproteins; HDL – high density lipoproteins; BP – blood pressure; EF – ejection fraction; LV – left ventricle

5 RESULTS AND DISCUSSION

In the total sample (n = 200), 60 seafarers (30%) were presenting cardiovascular changes including mostly arterial hypertension. Cardiovascular diseases were higher in deck operational crew (32.85%), compared with engine (21.05%) and other compartments (27.27%) (Table II). Most prevalent independent risk factors in seafarers were high low-density lipoproteins and dislipidemia in general (68.5%), differentiated on working compartments (deck 70.71% vs. engine 65.78% vs. else 59.09%).

Table 2. Pathological changes in seafarers according to working compartment

	Deck	Engine	Rating
Total (No.)	140	38	22
CVD (No.,%)	46 (32.85%)	8 (21.05%)	6 (27.27%)
DM (No.,%)	16 (11.42%)	2 (5.26%)	2 (9.09%)
Smokers (No.,%)	49 (35%)	11 (28.94%)	8 (36.36%)
Dislipidemia (No.,%)	99 (70.71%)	25 (65.78%)	13 (59.09%)

CVD – cardiovascular disease; DM – diabetes mellitus

Changes mostly present on ECG were conduction changes, compared with arrhythmias and ischemia (Fig.1)

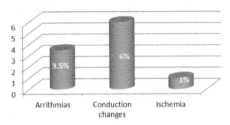

Figure 1. ECG changes in seafarers

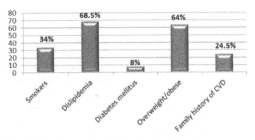

Figure 2. Cardiovascular risk factors in seafarers

Overweight played an important addictive role. Diabetes mellitus did not present a different scale compared to general population, even if in between compartments there were differences (deck 11.42% vs. engine 5.26% vs. else 9.09%). Blood pressure was higher in operational deck compartment compared with others, but the difference was not statistically significant. An interesting observation was given by the measurement of ejection fraction in cardio – Doppler ultrasound exam that revealed a decreased value in operational compartments (deck, engine) compared with rating. It is important to mention that the ideal ejection fraction for left ventricle is over 70% (Fig.3).

EF LV

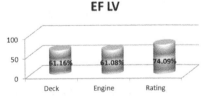

Figure 3. Values of ejection fraction measured by cardio Doppler ultrasound exam

Smoking was an independent risk factor that did not differ among compartments. The prevalence of smoking is important to be considered [Teo KK, Ounpuu S, Hawken S et al., 2006].

Correlations performed for entire seafarer group showed a statistically significant relation between dislipidemic status (DLP) and overweight/obese status and the presence of cardiovascular disease (p = 0,05, respectively p = 0,01) (Fig 4 and 5).

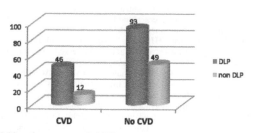

Odds ratio	2.0197
95 % CI:	0.9797 to 4.1636
z statistic	1.905
Significance level	P = 0.0568

Figure 4. Correlation between dislipidemic status and cardiovascular disease in seafarers

LDL – cholesterol was measured in two studies involving seafarers – Oldenburg et al. [Oldenburg M et al., 2010] showed 18% seafarers with high LDL-cholesterol (>160 mg/dL), and Purnawrma et al [Purnawarma I et al., 2011], even higher rate of 26,6%.

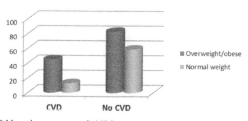

Odds ratio	2.4606
95 % CI:	1.2199 to 4.9631
z statistic	2.515
Significance level	P = 0.0119

Figure 5. Correlation between overweight/obese status and cardiovascular disease in seafarers

Example from literature are showing observations that were made across geographic regions: a dramatic increase in cardiovascular mortality in Beijing was attributed to greater cholesterol levels (mean cholesterol level of 166 mg/dL in 1984 and only 15 years later, 206 mg/dL [Critchley, J, Liu, J, Zhao, D, et al, 2004]. Less cigarette smoking along with favorable changes in blood pressure and lipids seemed to lead to decline in incidence of myocardial infarction [Hardoon, SL, Whincup, PH, Lennon, LT, et al, 2008].

Studies made on seafarers measured also other lipid parameters like HDL-cholesterol and triglycerides [Oldenburg M et al., 2010, Purnawarma I et al., 2011, Balanza Galindo S, 1996], showing low levels for HDL (< 35 mg/dL) and high levels of triglycerides (> 150 mg/dL).

Surprisingly, family history of cardiovascular disease correlated with the presence of cardiovascular changes was not statistically

significant compared with data from medical literature (p=0,20) (Fig. 6). Another studies showed this as a risk factor for 8,7% of the sailors [Oldenburg M et al., 2010, Oldenburg M et al., 2008].

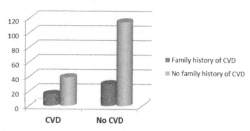

Odds ratio	1.5726
95 % CI:	0.7754 to 3.1897
z statistic	1.255
Significance level	P = 0.2095

Figure 6. Correlation between family history of cardiovascular disease and cardiovascular disease in seafarers

More prone to correlate to cardiovascular risk, and statistically significant (p = 0.008), smoking represented an independent risk factor for developing cardiovascular changes (Fig.7).

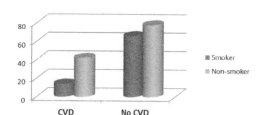

Odds ratio	0.3939
95 % CI:	0.1980 to 0.7838
z statistic	2.654
Significance level	P = 0.0080

Figure 7. Correlation between smoking status and cardiovascular disease in seafarers

Ideal mean BMI (≈25) was reached only in rating, compared with superior values in deck and engine compartment (25,6 vs. 26,4 vs. 27). Value of systolic blood pressure did not differ between compartments.

6 CONCLUSIONS

Even if the medical standards are tending to become more precise in evaluating the health status and the cardiovascular risk, with the reduction of the possibility of allowing unfit crew to go to sea, the cardiovascular diseases seem to remain important factors of distress on board vessels and is necessary to overview all the elements that augment it.

REFERENCES

Balanza Galindo S. (1996) Consumo de alcohol y factores deriesgo cardiovascular en una poblacion laboral maritima. *Med Maritima*, 1: 2.

Critchley, J, Liu, J, Zhao, D, et al. (2004) Explaining the increase in coronary heart disease mortality in Beijing between 1984 and 1999. *Circulation*, 110:1236.

Hardoon, SL, Whincup, PH, Lennon, LT, et al.(2008) How much of the recent decline in the incidence of myocardial infarction in British men can be explained by changes in cardiovascular risk factors? Evidence from a prospective population-based study. *Circulation*, 117:598.

Oldenburg M, Jensen HJ, Latza U, Baur X. (2008) Coronary risks among seafarers aboard German-flagged ships. *Int Arch Occup EnvironHealth*, 81, pp.735–741.

Oldenburg M, Jensen HJ, Latza U, Baur X. (2010) The risk of coronary heart disease of seafarers on vessels sailing under a German flag. *IntMarit Health*, 62, pp.123–128.

Purnawarma I, Jensen OC, Canals ML, Urkullu AC, Bercedo RG. (2011) Prevalence of cardiovascular risks factors and 10 year predictions of coronary heart disease in seafarers of Pertamina shipping (Indonesia). *Med Maritima*, 1: 11.

Teo KK, Ounpuu S, Hawken S et al.(2006) INTERHEART Study Investigators. Tobacco use and risk of myocardial infarction in 52 countries in the INTERHEART study: a case-control study. *Lancet*; 368, pp. 647–658.

Economic Analysis

Economic Analysis of Introducing Free Public Transport

J. Mikulski

Higher Technical School, Katowice, Poland

ABSTRACT: Municipal bodies responsible for transport policy more and more emphasis on measures to reduce the negative effects of growth of the automotive industry in these cities. One of the increasingly willing to use the tools in order to encourage car drivers to change the main means of transport for public transport is the introduction of the possibility of free use of its services.

The paper describes examples of cities that have implemented at home free public transportation, and the effects (benefits and costs), what solution it brought both to the bodies that organize public transport as well as for users. Then described an example of a city that could be an area where it could introduce such organizational solutions, and the article ends with presentation of the survey population.

1 INTRODUCTION

Public transport plays an enormous role in providing an efficient way to move around the city, giving the ability to travel to people, who are not motivated or for some other reasons do not use the privet means of transport. It may also be a tool used by local governments to more efficient development of the city. This is due to a range of factors public transport influences. Many of such factors are indirect, which means that they do not only influence the people using public transport, but also all other city traffic users. Some of them result from the existence of public transport itself, others are the result of using that means of transport. The limitation of the privet vehicle travels and many other significant aspects that change the level of the city functioning are of great importance.

The traditional approach to the development of public transport functioning has the tendency to omit or underappreciate many advantages, which are brought by the evolution of this branch of the economy. Current development of the transport is mainly directed at expansion of the infrastructural facilities and is focused on the speed of the journey and lowering its costs. Thus approach does not include the benefits of the investments into development of public transport, which would allow to decrease the costs for both its users and the local governments organising such transport.

Public transport plays a very important role in maintaining the balance between the quickly progressing technological growth, care for the environment and prevention of social exclusion. The necessary resources to create public transport an attractive and competing alternative to car traffic are mainly in hands of politicians, to some extent at the national level and especially at the local level. The best effects for this important area of public transport activity can be achieved by making available the latest fleet and simultaneous setting or increasing the fares for using the transport infrastructure (e.g. car parks). Of course, ensuring the properly functioning transport is necessary and essential to obtain the assumed effects and implementation of the additional charges for the car drivers without providing them with convenient alternatives most certainly would not result in nothing good. Current analysis indicate four factors that have the largest influence on the making decisions by the commuters regarding the means of transport they select. These are:

– frequency of running
– punctuality
– price
– speed of travelling

Among the mentioned factors the level of prices and punctuality have the worst opinion.

Until early 1990's public transport in Poland was managed by the state. Later, those duties were passed to the local governments. Since then the car market has also developed very quickly, and thus the road congestion has increased and become more troublesome. Its highest increase took place in the 1990's, just after the market liberalisation. At that time the number of registered vehicles nearly doubled. At the same time the number of public transport users decreased by 50%.

2 PUBLIC TRANSPORT AND ITS TYPES

The most common journeys of people using the urban transport are home - work and work - home. Thus, the more effective the transport system, with wider network of connections, shorter times of journeys and more frequent courses, the more attractive the city becomes. Other destinations are connected with entertainment, arranging some matters in the offices or shopping. Due to greater mobility, people are able to deal with their matters quickly, and as a result they have more time. The result of that is increased activity of the residents.

Other issue that is resolved by the efficient functioning of public transport is avoiding the problem of social exclusion. Thanks to the ability to travel provided by the city government people who cannot travel by means of privet transport due to different circumstances remain mobile and can easily function in the city. This group of people often includes students, elderly people, somewhat disabled people and those with very low income. For them, public transport is not a matter of choice but necessity for normal functioning in the present world.

A very large issue is also transport congestion, which in direct proportion concerns the developed cities, and so the tendency is that the more developed the agglomeration, the more congested it becomes. Jamming of the cities is directly correlated with its developmental status, hence during the rush hours there will always be a congestion. For that reason the target clients of public transport are also car owners. The more people give up the privet means of transport for public transport, the lower the issue of transport congestion becomes and the centres of cities become passable.

3 EXAMPLES OF CITIES THAT TAKE ADVANTAGE OF THE FREE OF CHARGE PUBLIC TRANSPORT

It is worth to take a look at the examples of cities in order to assess what has changed after implementation of the free of charge public transport and whether it has brought the intended effect [1, 3]. Worldwide such examples become increasingly common and more local governments discuss the decision of implementation of the free of charge

public transport. In Poland there are already several of such cities, e.g. Żory, Bełchatów, Świerklany or Nysa [9]. Not only small cities discuss free public transport for their residents. Manchester, Melbourne, Sydney, Calgary, Bangkok, Boston, Miami, Seattle, Baltimore, Tallinn are only some of the numerous group of worldwide agglomerations that do not charge their residents for public transport [4].

Currently, the most popular example of a city that decided to implement the free of charge public transport system is Tallinn [5, 6]. In 2013 it become the first European capitol city allowing its resident to use public transport without any charges. The main reason for that was to allow the unemployed people searching for a job further from the place of residence, and to allow change jobs for people who had the lowest earnings and who give up searching for jobs in more distant placed due to very high transport charges. The full evaluation of the decision made by the Tallinn local government is not jest ready, however it is worth to pay attention to conducted research, which indicated that the number of journeys in the studied period of time increased by 3% and at the same time the travelled distance increased by 2.5%. According to the researchers only a half of the change is caused by charging no fares for transport. The remaining part of those changes that increased the demand for public transport services is, in their opinion, determined by sectioning separate bus lines, significantly increasing the speed of buses and introducing other technological novelties such as: efficiently working passenger information system. Studies also showed that people who decided to use public transport used to be pedestrians and cyclists. Of course, it is not the expected result as the implementation of the free of charge public transport is mainly aimed at car drivers.

The passengers authorised for free of charge travels are all people who have proof of permanent residence in Tallinn. They have to have the card done, which authorises them to have free journeys, and which they are obligate to swipe through the reader for conducting the statistics.

Apart from Tallinn, the most popular examples of the European cities with free public transport are Belgian city of Hasselt and German city of Templin. Hasselt (70 thousand residents) is probably the most often pointed example of a city, where the implementation of the free of charge public transport system achieved a great success [2, 4]. It is also a great example showing the effects of making available free travels to their residents in a longer period of time. Free public transport was introduced there at the same time as other decisions aiming at the limitation of the investment outlays resulting from suddenly increasing car traffic. It needs to be noticed that the city's public transport network had already been very modern before the solution was implemented. After a decade since the implementation of the free public transport, the number of the served passengers increased more than ten times. It was a result obtained by on the one hand, the residents switching from privet vehicles to means of public transport, on the other hand, the influx of the new residents to the city, which suddenly became very attractive (lower costs of living).

In 2013 the free solution was ended due to increasing costs. The rising number of passengers requires constant increase in traffic capacity. The local authorities decided to reintroduce the fares for transport hoping that the percentage of people using public transport does not decrease. So far Hasselt still remains the city where many people use the collective transport solution. This proves that people, who are convinced to the positive aspects resulting from properly functioning public transport network are willing to use it even if it is not that less expensive than using the means of privet transport. Thus, one can state that public transport may also be a great advertisement of the collective transport, which is supposed to encourage people to use public transport.

German city of Templin introduce the free of charge public transport in a similar time [4]. The main purpose was to reduce the external costs generated by the privet means of transport. The town (14 thousand residents) recorded a 12 time increase in the number of passengers in only a couple of years.

The French example of a city with free public transport is the city of and the surrounding urban area [4]. The total number of people covered with the option of free travelling in that area is over 100 thousand. This solution was introduced due to gradual reduction of the incomes from the ticket sale to a place where before the introduction of the free transport they covered only 10% of the costs generated by public transport operator.

Norwegian city of Stavanger introduce the free of charge bus transport in the city centre for a short, trial period of time [4]. The purpose was to encourage people to use the newly implemented ecological vehicles. In this case the free public transport was a marketing campaign. In other Norwegian city - Bergen, the local authority decided for a certain period of time to pay for its clients travel to the city centre. It is then, not a typical example of a city deciding to cover its clients transport costs in order to discourage them to use the parking facilities.

On the premises of the Polish city of Żory there is a free of charge bus transport, which was introduced in 2014 [10, 11]. Such decision was caused by the decrease of interest in public transport. More and more people skipped on public transport for the privet vehicles. Along with the decreasing number

of people, the price of ticket went up, which caused even further decrease in the interest in transport services organised by the city. It led to a decision of funding public transport service in this city to a full extent, even though this decision was connected with the increase in public transport maintenance. It is not a solution that changed overnight public transport into free of charge without a change in its form - the number of liens funded by the city of Żory authority was limited but their routes were changed too. The free public transport can be used by everyone, who is in the administrative territory of the city of Żory. It does not concern only the residents it is free for everyone who is in this city occasionally.

Figure 1. A bus serving the free public transport in Żory [10]

The introduction of the free of charge public transport in many cities resulted in different effects. It depended on different factors such as the previous price levels, level of participation in the costs, condition of the fleet, city size, density of the transport network and many more. Some of the indicated examples accomplished more positive effects than it had been predicted, whereas others did not even meet the established minimum. This shows how complicated issue it is and that every city needs to be examined individually, or at least compared to the city that decided to implement the free of charge public transport and that can be considered similar in several aspects. |for instance, it is hard to imaging so large metropolis as London to implement free of charge transport service, because in this city the incomed form the tickets cover 85% of the network operating costs, thus a sudden abolition of the fees would create an enormous budget hole. At the same time, if the ticket prices before the implementation are already very low, one cannot expect that the complete abolition of charges will bring significant changes. Nonetheless, a well thought out and prepared implementation of free of charge transport may bring a very positive effects.

4 BENEFITS AND COSTS THAT MAY RESULT FROM THE IMPLEMENTATION OF THE FREE OF CHARGE PUBLIC TRANSPORT IN A MODEL CITY

As mentioned before, nowadays the funding of the larger public transport networks is necessary. The most significant benefit of the free public transport offer to both the passengers and other road users is definitely a relief of road congestion.

The city of Tychy, as it was selected as the model, can be characterised by high standards in terms of applied solutions regarding the road infrastructure. A significant number of intersections in the form of roundabouts ensures the smooth traffic flow, however there are some traffic congestions that can be observed in the city, especially during the morning rush hours. At that time, people residing in the near communes arrive to the city for work. The free public transport would have to be dedicated, to a large extent, to the residents of those communes. Most of them are located in the area of the transport network organised by the Tychy transportation authority, however it needs to be noticed that this network is less dense outside the limits of the city itself. In order to solve that issue, there is no need to extend this network with new bus stops, but it will be sufficient to create a changing points on edges of the network, so that the interested people have no problem to leave their vehicles in a safe place and travel to the city by public transport.

From obvious reasons the limited congestion would significantly improve the safety in Tychy. Free transport that cause people to switch to collective means of transport would contribute to lowering the road traffic, and that would consequently result in safer roads (less cars equal less accidents). The protection of every life is an economy in many ways.

An often recurring issue regarding the pointlessness of introducing the free public transport is the claim that people require not only the lowest price, but also reliability and speed of journey. But it can be said that by reducing the road traffic the average speed of vehicles will also increase, so a new timetable can be drawn up. It should be noticed that not only the time of journey will be reduced, but also the frequency of running of public transport can also be increased, without incurring additional costs for the fleet and employees. This happens because the buses that completes its route quicker, can sooner be ready for its next course, including the driver. The increase in the efficiency of using the fleet and significant reduction in the time of journey of public transport is a great advertisement for the city, which starts to be perceived as modern.

Another element that ensures the self-funding of the free public transport is the marketing aspect. Nowadays, in the conventional means of public transport the advertising spaces are very expensive. In case of a significant increase in the users of the buses and increase in demand for the advertising space in the fleet is also suspected, hence the current price for those space can be raised.

The last arguments indicate the possibilities of financing that are created by implementation of the free public transport. They refer to the creation of the indirect possibilities for increase in the profitable financial balance of the city. The saving are connected with lack of distribution costs incurred by the city, as well as with an option of limiting the job positions as there is no need to validate tickets. The operators of the ticket offices and devices selling the tickets also becomes unnecessary.

A very large benefit of the implementation of free of charge public transport is comparatively small, for spending related to transport, required investment expenditures. Additionally, the system does not require any infrastructural investments. There is no need for immediate disassembling the ticket systems. Due to that fact there is a possibility of introducing such solution initially only for a trial period. After such time, the tendencies in passengers' behaviour can be precisely studied.

Introduction of the free public transport means not only the positive social aspects, but also the costs that such solution generates. The most common argument against the implementation of the free public transport is the need for funding it entirely form the local government budget, therefore form the money of all local taxpayers. As far as the people who regularly use public transport should not be shocked by this, one cannot ignore the discontent of the people who get to work by privet vehicles and are not willing to change that habit after the free public transport has been implemented. There might also be complaints regarding the squandering of money that otherwise would be used for other purposes such as improvement of the quality of roads.

Another issue is the possible crowding of the fleet vehicles. Should public transport be used by people who so far used the privet vehicles, other people that previously used public transport might now switch to privet vehicles due to the comfort of the journey.

Thus, the planned effect would quickly go back to state from before the implementation of the pioneer idea. In order to ensure the current journey comfort and not to lose the previous clients of public transport in favour of getting new ones, it might be necessary to purchase an expensive fleet, which would significantly increase the costs of introducing the free public transport.

There would also be some issues concerning a new determination of optimal routes and timetables. People wanting to use the possibilities of the new free public transport might indicate the lack of sufficient courses, despite the fact that the transport is funded to the same extent from their money.

5 SURVEY OF RESIDENTS

The purpose of the study was to determine whether the implementation of the free of charge public transport in the cities is economically justified, that is to analyse the benefits and costs that would have to be incurred if decided to abolish the fees for public transport and to collect the opinions of the current public transport passengers and the drivers moving around the roads of the city of Tychy, and to verify the current condition of the fleet. This would allow to for the meaningful conclusions to be drawn and to foresee, to some extent, what changed could happen in the city of Tychy after this innovative system has been introduced.

Opinions of the people traveling in the city allowed to answer the following questions:
– Is public transport in Tychy working in a satisfactory manner, if not what are the most significant disadvantages?
– What is the main determinant that decides on which means of transport are chosen as a main form of travelling by the residents of Tychy?

A short survey was also conducted among the residents of Żory regarding the free of charge public transport functioning. It was assumed that reaction of the Tychy residents might be somewhat similar due to the proximity of those two cities and performing similar functions.

The research method that was adopted to conduct the studies was the interview method, and the used research tool was a survey. The questions asked to the residents of Tychy and Żory were the same (those questions concerned respectively in Tychy - possibility of using discounts and opinion on the effects of the hypothetical implementation of the free public transport, in Żory - possibility of traveling free of charges and opinion on what changes the implementation of the free public transport happened in Żory).

The main places of conducting the surveys were public transport stops and car parks. Such places were not selected randomly as one of the main purpose was to obtain opinions of both the drivers and people traveling in public transport.

The research group included 100 people that are residents of Tychy (or traveling to that city on a daily basis), and respectively 50 people from Żory.

The group of surveyed people in Tychy that agreed to conduct the questionnaire is quite uneven as the majority of them were women. Men are 34% of the surveyed people and women are 66%. In Żory the proportions were only slightly different. In this city a slightly larger part of the surveyed people were men who volunteered to conduct the questionnaire in number of 19, which constitutes 38% of the trial in this city, respectively there were 31 surveyed women, which constitutes 62% of the trial.

The total proportion does not reflect, however, the proportion of women to men in relation to individual groups, and therefore separately for drivers and separately for traveling only by public transport. Among all people surveyed in Tychy, 70 were drivers, and 30 people have no driver's license. In Żory there were almost 10% more of people who have driver's licence.

The division due to having the driver's licence does not expresses in a real manner the actual state, because a part of people that have the driver's licence often choose to travel by public transport anyway. The study was conducted in such a way to learn to a similar degree the opinions of both the people who use the privet vehicles, and public transport, that is why in Tychy 50 surveys were conducted on the car parks and 50 on the bus stops (in Żory, respectively 25 for each place). After summing up the results of surveys conducted in Tychy it turned out that in the surveyed group, the number of people who choose the privet means of transport is slightly higher (Figure 2).

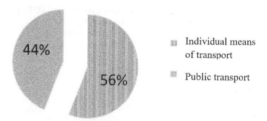

Figure 2. Percentage of people using the particular means of transport in Tychy
Source [own study based on [12]]

The results in Żory turned out to be only slightly different. In this city the people who use the privet means of transport on a daily basis were also in the majority, although the difference was slightly more significant. Even though the majority of the residents of Żory declared using public transport at least from time to time, those who do it regularly are in the minority. It is probably because of the low level of using public transport before the implementation of the free of charge solution.

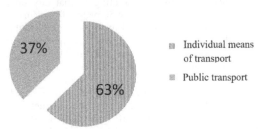

Figure 3. Percentage of people using the particular means of transport in Żory
Source [own study based on [12]]

Urban Transport Management (MZK) in Tychy is a well-functioning company. This confirms the conviction that the most important thing is proving the appropriate network frequency and density. Without them the price regulation would not make changed to the people's transportation habits in a significant manner.

Table 1. The age structure of respondents in Tychy. Source [own study]

	People most commonly choosing means of private transport	People most commonly choosing public transport	TOTAL
<15	0	1	1
16-25	13	6	24
26-40	17	9	21
41-60	18	11	15
>60	8	17	40
TOTAL	56	44	100

Table 2 The age structure of respondents in Żory. Source [own study]

	People most commonly choosing means of private transport	People most commonly choosing public transport	TOTAL
<15	0	2	2
16-25	1	5	6
26-40	11	8	19
41-60	8	9	17
>60	2	4	6
TOTAL	22	28	50

By analysing the age of the respondents, one may notice that the group of people using public transport as a main way of commuting was dominated by the elderly, and the group that uses the privet means of transport is made up of people with age between 26 and 60. When it comes to Żory, the situation is slightly different.

6 PRESENTATION OF RESEARCH RESULTS

The survey was conducted to learn about the predispositions of particular social groups towards choosing means of transport. The researchers wanted to know the main factors that decide on the manner of commuting around the city and how particular social groups would react to the implementation of the free of charge public transport in Tychy. The survey was also conducted in Żory and analysed separately in order to obtain the outlook on the impressions of residents of the cities where the free public transport already functions. The researchers mainly wanted to study the difference in the assessment of the price and comfort between the two cities. They asked about the frequency using public transport, as well as about the assessment of comfort, speed, frequency of public transport courses and its price.

The first question concerned the frequency of using public transport.

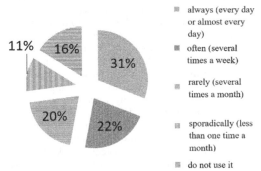

Figure 4. The frequency of using public transport
Source [own study]

A very large group of people, nearly a half, declared that they use public transport almost always, or very often, when they commute. Relatively large is also the group of people that never uses public transport. As a reason for that, they often present having their own car (75% of all people, who declared not using public transport). The remaining answers included reasons such as maladjustment of the timetable or no bus stop near the place of residence (people from outside the city of Tychy commuting to work). It can be assumed that among the people who stated that they do not choose public transport, some of them do not have an alternative for commuting to their work place by other means than a privet car.

The correctness of that thesis is confirmed by the results obtained in the following question. The respondents were asked if they commute on a daily basis to areas not covered by the MZK Tychy service. Affirmative answer was given by 21% of the respondents, whereof over 3/4 turned out to be at the same time people declaring that they most often commute using means of privet transport. This indicates that the last group would not have the full possibility to stop the privet vehicle commutes in favour of using only the MZK Tychy fleet. This is because the network does not cover their whole journey. This group consists of 17% of all respondents. In order for the implementation of the free of charge public transport had an effect also on this group, it would be necessary to create appropriate change centres, allowing people form outside of the areas covered with public transport to easily get to the areas where such network functions and carry on with their journeys without incurring costs, and at the same time reduce the number of vehicles congesting the city. Such solution requires the increase in the number of parking spaces in such locations, however, creating car parks on the outskirts of the city, in locations allowing for changes to public transport would allow for gradual reduction of car parks in the city centre.

In Żory a slightly higher number of people declared that they commute by using their privet vehicles on a daily basis, however, those results differed only slightly. It needs to be noticed that the group that never uses public transport turned out to be much smaller. It might indicate, on the one hand that the people who everyday commute to work to outside of the city, do not use the privet cars after work and take public transport for the inner city journeys. It is the positive side of implementation of the free of charge public transport. On the other hand, such survey results might indicate that the short distances, for instance between the offices and other public places in the city centre, previously travelled by walking, now are travelled by means of public transport.

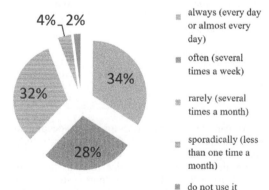

Figure 5. The frequency of using public transport in Żory
Source [own study]

- always (every day or almost every day)
- often (several times a week)
- rarely (several times a month)
- sporadically (less than one time a month)
- do not use it

Fig. 6 Percentage of people traveling on a daily basis to the area not covered by the MZK Tychy service
Source [own study]

- YES
- NO

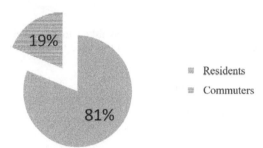

Figure 7. Share of people commuting to the city from the outside of areas covered by the MZK Tychy service in the group not using public transport
Source [own study]

- Residents
- Commuters

The study also showed that the majority of the respondents, who at least several times a week use public transport take advantage of various types of discounts. These were people declaring that most often the purpose of their journey is school, doctor's appointment or social purposes. People who pay the full price for tickets usually selected the answer "work" regarding the purpose of their journey. Most commonly these were people who do not have the driver's licence, thus it can be assumed that they choose public transport as they have no other alternative. When it comes to people not using public transport or using it rarely, the majority of them declared that they are not entitled to any ticket discounts, and their main purpose of everyday journeys with their privet vehicle is said to be work. It is another proof that the price is a factor that has a significant impact on the decision regarding everyday commute. It is definitely the argument in support of the implementation of the free of charge public transport. Among the passengers of public transport the majority of people entitled to travel with the ticket price discount is on the one hand a result of them having no alternative (elderly people, children, people with no driver's licence), and on the other hand, this shows the effectiveness of the price regulations, due to which people carefully analyse the choice of means of everyday transport and often decide to select the option that is less expensive and, at the same time, more ecological and relieving for the traffic congestion.

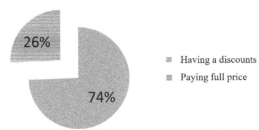

26%

74%

⬚ Having a discounts

⬚ Paying full price

Figure 8. Percentage of people using the discounts for public transport tickets in Tychy
Source [own study]

The respondents were also asked to give their opinion on aspects connected to public transport, in particular:
– price
– level of information
– reliability (punctuality)
– frequency of running
– speed of journey
– comfort of journey
– safety

Table 3. The assessment of public transport parameters by people using it at least a few times a week in Tychy. Source [own study]

Assessed element	Average
Price	3.0
Level of information	4.1
Reliability	4.3
Frequency of running	3.8
Speed of journey	3.6
Comfort of journey	3.7
Safety	4.1
Average result:	3.8

It needs to be noticed that the answers were very different depending on whether the respondent declared frequent use of public transport or often use of means of privet transport. For that reason there were made two tables showing the results and assessment of the indicated parameters, separately for people using public transport at least a few times a week, and people using public transport rarely or not using it at all.

Among people who often use the services of MZK Tychy the individual parameters are characterised by a high average result. People are in general satisfied with functioning of the network. It can be seen that the lower score is given to the ticket prices. However, the group of people who are regularly using public transport is not the target group, which the implementation of the free of charge public transport is directed at. It is obvious that if people uses regularly some solution, they

want it prices to be lower. However, the main purpose of the implementation of the free of charge public transport is to reduce the road traffic, congestion in the city and allow for the spatial development. The remaining evaluated elements were assessed much higher.

Table 4. The assessment of public transport parameters by people using it at least a few times a week in Żory. Source [own study]

Assessed element	Average
Price	5.0
Level of information	3.7
Reliability	3.4
Frequency of running	3.2
Speed of journey	3.2
Comfort of journey	3.4
Safety	4.0
Average result:	3.7

Among the residents of Żory using public transport the majority of the answers were only slightly different. For the obvious reason the price was the parameter, which scored 5 on a scale of 1-5. The comfort of journey received an average score of 3.4, which is the score lower only for about 0.3 than for public transport system in Tychy. One can conclude, therefore, that the introduction of free public transport in Żory has not resulted in drastic decline in passenger comfort. However, it can be seen that public transport operating in Żory, despite the "price" parameter raising the average score, is assessed as worse than the one operating in Tychy. Apart from the price level, all other parameters received lower scores.

Table 5. The assessment of public transport parameters by people who use it rarely in Tychy. Source [own study].

Assessed element	Average
Price	2.9
Level of information	3.8
Reliability	3.3
Frequency of running	3.2
Speed of journey	3.3
Comfort of journey	3.1
Safety	4.3
Average result:	3.4

As can be seen the assessments of people not using public transport every day differ significantly, and the general assessment was much lower. This group also considered the price level as the worst part of using the collective transport system. In this case, however, the reason for that is the reluctance caused by the small difference in the journey costs in using public transport compared to using the privet vehicle, commonly considered as more comfortable and more reliable means of transport. Other reason is the fact that the percentage of people not entitled to any discounts for public transport among the people that do not use it was much higher. When comparing the results it can be seen that the regular car drivers assess much lower the remaining elements that were evaluated. The only element that was assessed higher by them than the regular collective transport users was the safety of public transport. This indicated the size of the disproportions present in the studied groups. The reason for this may be considered, among other things, the low awareness of the current state of affairs among the people choosing to commute by public transport. It needs to be noted that MZK Tychy is a thriving company, which in the past few years developed significantly, and the offered connections are characterised by a high level of comfort. The service network is very dense and has a relatively high frequency of courses. It is possible that if people not using the MZK Tychy services had better awareness, many of them would decide to use it and give up traveling by privet vehicles, even if the free public transport idea is not implemented. However, the implementation of the free public transport may turn out to be an effective tool that draws attention and encourages to use the collective transport. Such people would learn the current reality and maybe they would not give up the ecological solution even after the transport fares are reintroduced, in case the free public transport in Tychy is not there to stay. A very low assessment of the "price" element among the drivers that use their privet vehicles on a daily basis additionally increases the chance actual implementation of the free of charge public transport, due to the fact that it is a flexible tool that allows to control the demand for public transport, which for obvious reasons should be increased when the free public transport is introduced.

Table 6. The assessment of public transport parameters by people who use it from time to in Żory. Source [own study]

Assessed element	Average
Price	3.9³
Level of information	3.6
Reliability	3.4
Frequency of running	2.9
Speed of journey	3.4
Comfort of journey	3.2
Safety	4.2
Average result:	3.5

³ one can add here the respondent's lack of information

Despite a small group of people using public transport in Żory rarely or not at all, their assessment clearly shows the main reason for giving up on using the collective transport. The frequency of running obtains the lowest score. If the respondents had been asked about the network coverage, similar results would definitely be obtained. Also, the "price" element scored low in that group, probably because

that group consists of people who travel to work outside of the areas operated by the free public transport service in Żory.

7 CONCLUSIONS AND INTERPRETATION OF RESULTS

The conducted studies show how different social groups have different approaches to public transport. For obvious reasons, the groups which use various types of discounts are more satisfied with the level of prices. MZK Tychy is highly assessed by users of public transport. One can say that the offered network connections, and frequency of courses, as well as the offered capacity are prepared and characterised by a high level of comfort and reliability. This is important, because people switching from private cars should be provided with the best possible version of public transport, so that they would not be put off and realised that it really is a branch competing at all levels with privet vehicles. The city of Tychy offers much wider network of connection than the city of Żory. For this reason, the possible introduction of free public transport in this city would definitely have even more positive effects, however, it would require a much larger effort. An interesting idea could be the introduction of a pilot program providing, for example, one free ride per hour on each of the offered routes. This way, the residents of Tychy would be given a possibility of free mobility and those that cannot afford to pay the fees could travel fully legally. Such a solution would not require a large capital investment and would help reducing the social exclusion of the poorest.

8 CONCLUSION

The analysis of the survey results, and the review of the effects that free public transport brought in various places in Europe, shows that the idea of free transportation is not certainly a meaningless investment. However, there are a number of factors, which have a greater or lesser impact on the success of its introduction. Omission of the assessment of these factors before implementation of the system can lead to complete failure and to not achieving any of the desired effects. However, when used in the right conditions at the right time, it can bring good results and be a kind of symbol of modernity of regions providing free transport.

It should be noted that the mere abolition of fees for traveling around the city is not enough to reduce the transport congestion and encourage drivers to change their transport habits. It is necessary to prepare adequate transport network, so that it is ready to accept additional streams of passengers without compromising passenger comfort. Such situation could in fact discourage people, who decided to try to take advantage of the fast collective transport, which could result in irretrievable loss of those people as passengers of public transport.

Ideally, the introduction of free public transport should go hand in hand with other actions aimed at encouraging and perhaps even in some sense forcing people to change their transportation habits. An example would be the simultaneous introduction of congestion fees in the areas of the city with the free public transport. This is the simplest use of the popular method of "carrot and stick", where public transport, for which you do not have to pay, is a kind of reward for not traveling around the city by private transport. However, the fees that have to be additionally paid when people still do not decide to change the method of transport are the punishment. It is the bipolar action which greatly increases the possibility of reducing the transport congestion. Moreover, it is likely that people who still use the means of privet transport will also be happy in spite of having to pay fees, and this is due to the fact that time travel will be much shorter than normal, and finding parking places will be much simpler. People the least prone to price regulations are usually the ones for whom price does not play the most important role, so the wealthier people. For them, the time saved with the introduction of this dual plan to reduce transport congestion in the city is often more valuable.

Long-term introduction of free public transport can become, however, fatal to the budgets of authorities organising the transport in cities. Therefore various option should be considered various before implementing the system into effect. It seems that the best solution would be to interpret the needs of specific social groups, and careful adapt the transport according to their needs.

This model of implementation of free transport is obviously not universal and relates mainly to small and medium-sized cities, and areas operated by collective transport network of a particular city. It would rather not work in large cities, where the cost of public transport are much higher, or in those where the revenue from sold tickets cover a large part of the costs generated by public transport, due to the need to involve too large cash outlays. One cannot also count on the positive effect of such action in cities where public transport is already overcrowded or filled on a level close to the maximum, because the real effect could be counterproductive. In addition to the new passengers wanting to try out public transport, it could be left by the previous users due to overcrowding of the fleet. As a result, the number of cars on the road, and thus the level of congestion could be even larger.

REFERENCES

[1] Börjesson, M. et al., Factors driving public support for road congestion reduction policies: Congestion charging, free public transport and more roads in Stockholm, Helsinki and Lyon, Transportation Research Part A: Policy and Practice, Volume 78, 2015, https://ideas.repec.org/a/eee/transa/v78y2015icp452-462.html [dostęp z dnia 08.05.2016]

[2] van Goeverden, C. et al., Subsidies in public transport, European Transport ,32, 2006 https://www.openstarts.units.it/dspace/bitstream/10077/5892/1/vanGoeverden_et_al_ET32.pdf [dostęp z dnia 08.05.2016]

[3] Fearnley N., Free fares policies: Impact on public transport mode s and other transport policy goals, International Journal of Transportation Vol.1, No.1 (2013) http://www.sersc.org/journals/IJT/vol1_no1/5.pdf [dostęp z dnia 08.05.2016]

[4] https://farefreepublictransport.com/city/ [dostęp z dnia 28.12.2016]

[5] http://www.citylab.com/commute/2014/01/largest-free-transit-experiment-world/8231/ [dostęp z dnia 08.05.2016]

[6] http://www.tallinn.ee/eng/freepublictransport/ [dostęp z dnia 08.05.2016]

[7] Darmowa komunikacja miejska w 39 miastach i powiatach Polski https://bezprawnik.pl/darmowa-komunikacja-miejska-w-39-miastach-i-powiatach-polski/ [dostęp z dnia 08.05.2016]

[8] http://www.rp.pl/artykul/1045816-Europejskie-miasta-eksperymentuja-z-bezplatna-komunikacja-miejska.html#ap-1 [dostęp z dnia 08.05.2016]

[9] http://www.transport-publiczny.pl/watki/bezplatna-komunikacja.html# [dostęp z dnia 08.05.2016]

[10] http://www.tvn24.pl/katowice,51/za-darmo-autobusem-po-zorach-ruszyla-bezplatna-komunikacja-miejska,423938.html [dostęp z dnia 08.05.2016]

[11] http://zory.trasownik.net/ [dostęp z dnia 08.05.2016]

[12] Moll J., Economic analysis of introducing free public transport, B.Sc.Eng. Thesis, UE Katowice, 2016

[13] MZK Tychy: Zakres działalności MZK. mzk.pl. [dostęp z dnia 08.05.2016]

Economic Analysis of A Vessel In Service Equipped with an LNG Fueled Ship Engine

M. Bayraktar & M. Nuran
Dokuz Eylul University, Izmir, Turkey

ABSTRACT: Industries need enhancement in their respective fields along with forward-moving technology. In order to diminish further costs, industry associations are trying to find alternative sources. Alternative means of reducing needs to be cheap, sustainable and environmentally friendly. In marine industry, quite a few latest studies have been carried out about ship designs, marine engines and propeller types to minimize owning costs and supply efficient operations. Predominantly, heavy fuel oil (HFO) and marine diesel oil (MDO) is used in the global ship industry. Considering the new environmental legislations, the emission rate of these types of oil won't meet the requirements put forward by the new regulations in maritime industry. Using liquefied natural gas (LNG) as marine fuel will offer a neat solution source in marine applications. Combustion of LNG emits low emission rate compared to heavy fuel oil and marine diesel oil. LNG takes important place in marine industry from the point of the global emission concern. LNG not only in terms of emissions but also economic aspects is a profitable fuel selection. Unit prices of LNG are in a lower level compared to other fuels. Taking into consideration the growing natural gas reserves, natural gas is a fuel that has an important position in the future. In this study, economic analysis of ship equipped with LNG fueled engine has been carried out. The results gained through this study reveal that LNG-fueled engine operations cost less than diesel-fueled engine operations.

1 INTRODUCTION

1.1 *Liquefied Natural Gas General Properties*

Natural gas has a remarkably significant position in meeting world energy needs. Natural gas is extracted from coastal and open sea reserves and its main component is methane. Natural gas can be delivered to consumers in the form of LPG and CNG (Woodyard, 2004).

Liquefied natural gas is obtained by cooling the natural gas between -159 ° C and -162 ° C. (Bahadori,2014).

1.2 *LNG Liquefaction Process*

The process of natural gas liquefaction is a complex process and it is very similar to that used in gas production fields, during the liquefaction process, harmful components of the natural gas are removed and the long hydrocarbon chains are lysed. In this case, LNG fuel becomes a clean burning fuel. (Marine Technology Liquefied Natural Gas (MTLNG), 2014).

During the process of liquefaction, natural gas, consisting of a major portion of the methane, is cooled below the boiling point. As a result of condensation of the hydrocarbons, in the compound of methane, water, carbon dioxide, oxygen, and some amounts of sulfur compounds are reduced or completely eliminated. (Bahadori, 2014).

1.3 *Use of LNG as a Fuel in The Ship Propulsion System*

Considering the increasing environmental awareness of people and the new regulations introduced in the maritime sector, LNG will become a prominent fuel in the maritime sector (Society for Gas as a Marine Fuel (SGMF), 2014).

Dual fuel engines are seen as a reliable technology in passenger ships, tugboats and other marine vessels. As a result of the work done, besides the dual fuel main engines, the main engines which are directly gas-fed were also produced (Wuersig, 2014).

The number of vessels to be transported by LNG fuel by the end of 2018 is shown in Figure 1.

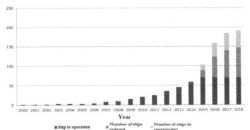

Figure 1. Number of ships shipped with LNG fuel (Andersen, 2015)

Small-scale logistics networks need to be expanded for LNG fuel supplying. LNG-fueled ships initial investment costs are higher than diesel engines. In other respects, the use of LNG as fuel in ships has a notable superiority for ships in the future. The use of LNG as fuel in ships provides a significant advantage for ships in the future. At the top of them is a clean burning situation which will meet the requirements of all future emission regulations (McGhill, Bemley and Winther, 2013).

There is a growing global focus to reduce emissions from ships, according to the US Energy Information Agency (EIA).When the 2014 International Energy Estimate is taken as a reference, the consumption of the world's oil and other liquids are expected to grow 38% between 2010 and 2040. This increasing consumption growth estimates that greenhouse gas emissions will increase by 45% by 2040. Temperature and emission increasing in the long term are well above international targets. As a result of global climate studies, in order to prevent increasing emissions, experts are turning to alternative sources for clean burning and less emissions. In this context, liquefied natural gas is an important alternative fuel source in terms of cost and environmental awareness (International Gas Union (IGU), 2015).

In terms of low cost per ton and creating a high reliability sense, LNG comes to purchasable position as marine fuel. From an economic point of view, the growth of plants and equipments in the LNG sector, there is an important contribution to the development of companies (Bahadori, 2014).

Considering chemical and combustion characteristics of LNG, the use of LNG as fuel provides a significant reduction in NOx, SOx and CO2 emissions. LNG has many emission advantages when compared to heavy fuels (Lowell, Wang and Lutsey, 2013).

Liquefied natural gas is quite clean compared to diesel fuel and heavy fuels used as fuel in the maritime propulsion system, namely release fewer emissions to environment. The emissions from diesel fuel as compared with the LNG, SOx and NOx values are approximately %80 less and CO2 is

less about %15-25 (Center for Climate and Energy Solutions (C2ES), 2013).

Pollution from ships is controlled by the IMO (International Maritime Organization) through the MARPOL (International Convention for the Prevention of Pollution from Ships). MARPOL, during the operation of vessels or in the event of any accident, is the main international convention covering the pollution of the marine environment. MARPOL Annex VI entered into force on May 19, 2005. MARPOL Annex VI "Prevention of Air Pollution from Ships" specifies the limits of NOx, SOx and particulate matter emitted from ships. Moreover, MARPOL Annex VI prohibits intentional emissions of substances that deplete the ozone layer (International Maritime Organization (IMO), 2016).

The global sulfur limit of fuels is now set at 3.5% but this level will be reduced to %0.5% by 2020. In addition, the sulfur limit of fuels in emission control areas has been reduced from 1% to 0.1% (IMO, 2016).

Table 1. SOx and pm emissions

Outside an ECA established to limit SOx and pm emissions	Inside an ECA established to limit SOx and pm emissions
4.50% m/m prior to 1 January 2012	1.50% m/m prior to 1 July 2010
3.50% m/m on and after 1 January 2012	1.00% m/m on and after 1 July 2010
0.50% m/m on and after 1 January 2020*	0.10% m/m on and after 1 January 2015

The use of LNG as fuel, meets existing emission limits without using any emission reduction technology (Bengtsson, Andersson and Fridell, 2011).

1.4 LNG Tanks Settlement Rules for Vessels

According to ABS and DNV class rules, LNG tanks are B / 5m or 11.5m away from the board and must additionally B / 15m or 2m must not be above the bottom (Parsons,O'Hern and Denomy, 2012).

LNG tanks can be placed above or below the deck of the ship. Placing the tanks on the ship deck is cheaper and has a less complicated structure. Placing the tanks below the ship deck requires more details. The tanks should be positioned sufficiently far from the other compartments, have ventilation system, contain explosion-proof means and gas detector system. Tank compartments must have ventilation system, contain explosion-proof means and gas detector system. Additionally, if the tanks are to be positioned above the deck, this zone must be minimum rinsing area to LNG tanks. Moreover, the area where the dynamic movement of the ship is least, must be selected (Boylston, 2011).

In the event of an accident, ships are affected at a minimum level from any accidents so that these rules are specified for LNG Fueled vessels (Mohn, 2012).

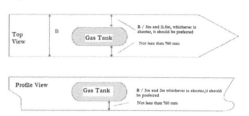

Figure 2. Locations of tanks at underneath ship decks, according to DNV rules (Det Norske Veritas (DNV), 2016)

Locations of tanks at underneath ship decks, according to DNV rules, should meet the following standards at Figure 2.

1.5 *LNG Bunkering Process*

LNG fuelled ship bunkering from the LNG terminal, has to do some operational applications. These applications include particular communication, monitoring, Emergency shutdown system mechanism, prevention of cryogenic conditions, inerting and cleaning of the system, fire extinguishing, electrical insulation of system for proper operations (ABS (American Bureau of Shipping),2015).

A very detailed and progressive system is used in the land transfer process and all connection preparations must be made before the ship arrives for bunkering. Moreover, at the beginning and end of bunkering washing process is applied(Ryste, 2012).

Communication and monitoring are very important in bunkering. Persons should be in contact with each other before the hoses are connected and before the hose connections are disconnected. Any negative situation that may occur in the system can interfere with the monitoring system. There is no established standard for this communication and system monitoring (ABS,2015).

One of the biggest problems in LNG-operated vessels is the bunkering part because the number of stations to supply LNG is quite small and limited (McGhill, Bemley, and Winther, 2013).

1.6 *Dual Fuel Engines*

In this Dual Fuel Engine, natural gas, diesel fuel or heavy fuels are used to operate the system. One of the most prominent and reliable features of the dual-fuel engine system, during operation of the engine with diesel fuel or heavy fuels, without experiencing any disruptions, switch from diesel fuel or heavy fuels to LNG is easily provided. In the same way, during operation of the engine with LNG fuel, switch to diesel or heavy fuels also smoothly is supplied. When this switch is made, the diesel and heavy fuels gradually changes to LNG fuel. As an example, as soon as any gas interruption occurs in the system, regardless of the load of system, the main engine is automatically activated and continues to operate the system with liquid fuel (Wartsila, 2015).

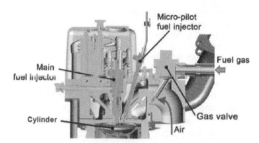

Figure 3. Cross-Section of four-stroke dual-fuel engine (Yanmar,2015)

In this dual fuel system, sufficient amount of air gas mixture enters the system in the first stage. Secondly, this mixture entering the system is compressed by the cylinders and when cylinder reaches the upper top point, the ignition is triggered by the pilot excitation system. In the final stage, the exhaust phenomenon occurs and the system returns to the beginning and repeats the same process. NOx emissions are very low since only a small amount of diesel fuel is used in the pilot excitation (Wartsila, 2015).

2 LITERATURE REVIEW

In the literature review, the reports published by a number of leading industry organizations and academic articles have been examined. A few of the published articles are mentioned in the table below and also each article topics are shown in there.

The study states that the main component of liquefied natural gas is 90% methane and 10% ethane, and specifies the physical and chemical properties of methane. The advantages of using liquefied natural gas as fuel in terms of emission have been pointed out. In the application section, the hazards that will arise from liquefied natural gas are stated. An appropriate model was created by using ETA (event tree analysis) method for hazardous effect. In this model, the hazards that may arise from liquefied natural gas stored under different pressures are indicated. In conclusion, the risks of liquefied natural gas occur in case of release of LNG. If LNG release at above atmospheric pressure, its hazards effect is more significant than LNG kept under atmospheric pressure (Vandebroek and Berghmans, 2012).

Table 2. Passenger ship general features

	Vandebroek and Berghmans (2012)	Herdzik (2012,2013)	Thomson et al. (2015)	Semolinos et al. (2013)	Adamchak et al. (2013)	Kumar et al. (2011)	Brynolf et al. (2014)
Liqufied natural gas price			X	X	X	X	X
Liqufied natural gas emisson rate	X	X	X	X	X	X	X
Storage and transportion of LNG			X	X		X	X
Liqufied natural gas fueled engine		X	X				
Natural gas reserves			X				
Natural gas liquefaction process							
Chemical properties of LNG	X					X	X
Comparison of Other Fuels		X		X		X	X
LNG Safety	X	X		X			
IMO Regulations on Emission		X		X	X		
LNG Bunkering Terminal		X		X	X		

In this study, the comparison of LNG in terms of energy and density with other fuels are expressed and their advantages in terms of additional emissions have been highlighted. Emission control areas are specified and IMO Legislations and regulations are expressed. LNG-operated dual fuel or single fuel machine types are indicated and the installation scheme of the ship shipped with LNG is stated. LNG terminals in Europe have been specified and a list of vessels shipped with LNG has been shown. Measures to be taken in case of using LNG as fuel have been mentioned. In the other article, the LNG conversion system was explained and information was given about the Bit Vikings ship on which conversion system application was made. It has been stated that the use of LNG will increase if the number of LNG bunkering terminals reaches a sufficient amount (Herdzik,2012,2013).

This study refers to new studies on maritime industry and this work covered the new regulations in the maritime sector, environmental perception, emissions from marine vehicles. Moreover, the processes of extraction, liquefaction and storage of natural gas have been expressed; also information is given about the characteristics and routes of the LNG fueled vessels in America and Norway. In terms of environmental, social, technology, economic price and maritime demand, LNG conversion process advantages are described in detail. The comparison of natural gas with other fuels in terms of price and consumption of natural gas according to the regions is shown. Maritime transportation adoption of LNG fuels has been pointed out that it will provide a significant benefit especially in energy (Thomson , 2015).

In this study, the fuels used in the maritime industry are expressed and the new regulations of IMO are pointed out. It has been indicated which types of fuels can be used to avoid exceeding emission limits in ECA (emission control areas). MDO LSHFO and LNG are leading these fuels and the advantages and disadvantages of these fuels have been expressed in this work. Main advantages of LNG fueled vessels has low emission rate and low maintenance cost but cargo space is restricted due to LNG tanks and also the most important concern about this issue is the safety part of the LNG. Information is given about LNG market potential, transportation, storage and bunkering and it is stated how LNG is delivered to the consumer from the production stage. As a result, LNG has become an essential fuel when new restrictions on marine activity are taken into consideration (Semolinos, 2013)

This article also mentions emission limitations in maritime industry. In addition to other works, it is stated that which areas will be included in emission control areas in the future. Considering new regions and regulations, significant increases are expected in the number of LNG fueled vessels. (Adamchak, 2013).

In this study, the general properties of LNG are mentioned and in addition to other studies, information about LPG and CNG is given and their differences are expressed and Comparison of physical and chemical properties of LNG with diesel, gasoline and LPG are mentioned. The use of LNG in transportation is mentioned and the consumption of LNG according to the countries is point out. In addition to transportation, LNG is also used for electricity generation and as in other articles, the advantages of emission are particularly highlighted (Kumar,2011).

In this study, the fuels used in the maritime industry in the future were examined from four different perspectives. These are technical aspects, Economic Aspects, Environmental aspects and other aspects. In maritime sector, all stages of fuels from production to consumption are mentioned in detail and the effects of these fuels on global warming are compared (Brynolf,2014).

3 METHODOLOGY

3.1 Conversion System

Conversion system is applied to a ship which is a passenger ship serving on a constant route. The passenger ship used in the project has 39m length-over-all, 11.6m breadth and the length of water line is 38 m. The service speed of the passenger ship is 18 knots and classified as a DNV + 1A1R4 passenger ship.

Figure 4. Project passenger ship

There are two main machines due to the catamaran construction of the project passenger ship, 1973.98 HP is obtained from these engines. Propeller structure of this passenger ship is controllable-pitch propeller. In addition to these systems, there are bow thrusters on board in order to ease the maneuvering of the ship.

Table 3. Passenger ship general features

Ship Type	Passenger Ship
Gross Tonnage	486 (tons)
Net Tonnage	146 (tons)
Number of Engines	2 (units)
Total Engine Power	1973.98 HP

With the improvements made on the existing main engine, used main engine in the project ship is manufactured so that it has begun to be used in the maritime industry and also it meets IMO Tier II regulation.

Table 4. Main engine catalog values

Main Engine Catalog Values Power		
Hp	Speed (r.p.m.)	Fuel consumption (g/kwh)
900	1800	198
1000	1800	197
1100	1900	200
1200	1950	201

According to the different power values of this main machine, fuel consumption values are stated.

3.2 Engine Load and Fuel Consumption of the Project Passenger Ship During Operation

The values of the main engine load and the fuel consumption of the passenger ship were followed during operation. Obtained values were recorded and these values are given in Table 3. Considering the current velocity, sea wave conditions and wind, there are several minutes of differences round trip between cruising.

Table 5. Project passenger ship main engine load during operation

Departure Time	Arrival Engine Load	Time	Engine Load
09:00	40%	09:20	40%
09:02	34%	09:22	36%
09:04	35%	09:24	34%
09:06	35%	09:26	34%
09:08	35%	09:28	35%
09:10	38%	09:30	35%
09:12	37%	09:32	35%
09:13	35%	09:34	34%
09:14	38%	09:36	35%
09:15	40%	09:38	38%
09:16	40%	09:39	40%

The passenger ship completes its return cruising in 19 minutes, although it completes its first cruising in 16 minutes.

There are two fuel tanks at port and starboard of passenger ship and each tank volume is 400 liters. When the data of the tanks are received from the passenger ship during the operation, port side tank has 260 liters fuels and starboard tank has 210 liters fuels.

Table 6: Project passenger ship fuel consumption values

Diesel Fuel Levels in Tanks	Port Fuel Tank(lt)	Starboard Fuel Tank(lt)
Initial Case	260lt	210lt
First Operation	250lt	187lt
Second Operation	232lt	161lt

The port fuel tank with 260 liters of diesel came to 250 liters level at the end of the first operation; also the fuel tank level on the starboard side came 210 liters to 187 liters. At the end of the operation, a total of 33 liters of diesel fuel was used. The project passenger ship makes 13 operations in a day. In this time, the total fuel consumption observed is approximately 1010 liters.

New dual fuel main engine will be used in our project passenger ship and catalog data of it is as follows.

Table 7. Catalog data of Dual fuel main engine.

Fuel consumption		Gas mode	Diesel mode
Fuel gas consumption at 75% load	kJ/kWh	8326	-
Fuel gas consumption at 50% load	kJ/kWh	8862	-
Fuel oil consumption at 75% load	g/kWh	4,50	196
Fuel oil consumption 50% load	g/kWh	5,40	207

3.3 Tanks Used in Project Passenger Ship

Tanks used in project passenger ship are required for LNG fueled vessels and they are designed according to ship design codes. LNG fuel on the project passenger ship is supplied from such tanks installed in the ship. In these types of tanks, the working pressure is at least 10 bar and tank volume reaches up to 700 m³. The greatest feature of the type C tanks should be a leak proof structure; in addition, there isn't need for a secondary barrier for the tanks. The outer shell of this tank must be resistant to low temperatures and it is typically made of stainless steel.

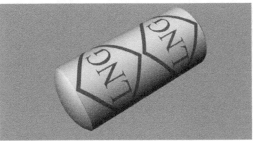

Figure 5. 20m3 LNG tank of project passenger ship

Most of the C type tanks are produced for gas fueled ships and for LNG carrier ships and vacuum isolation is used in these types of tanks. In project passenger ship, two pieces 20m³ LNG C type tanks will be used.

3.4 *Dual Fuel Engine Installation*

In the new system, the existing main engines will be replaced with the dual fuel main engines. In addition to this system, two LNG tanks will be installed to support main engines. Gas valve unit will be installed to deliver the gas from the LNG tank to the main machine as specified.

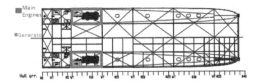

Figure 6. Project passenger ship main engine layout plan

Table 8. Costs of the equipments used in the system

Equipment used in the project passenger ship	Cost(€)
Dual Fuel Main Engine I	650.000,0 €
Dual Fuel Main Engine II	650.000,0 €
Cost to prepare for operation I	20.000,0 €
Cost to prepare for operation II	20.000,0 €
Bridge Control System	7.000,0 €
Bridge Control System	7.000,0 €
LNG Tank (20 m3)	25.000,0 €
LNG Tank (20 m3)	25.000,0 €

Considering the above inputs, our total cost is calculated as approximately 1.5M €.When comparing the old system with the new system, the system meets the initial investment cost in approximately 10 years. Considering the costs of the two systems, maintenance and repair cost of LNG fueled vessel is less than diesel fueled vessel. Similar to this work, according to the data obtained from the shipyard, the installation cost of vessel with 68 m³ tanks and dual fuel main engine system is 1,25M € roughly.Additionally, establishment of this system on the vessel covers a certain period of time so that ship will not be able to provide service during establishment.

4 RESULTS

Considering literature review and case study, evaluating the system in terms of cost, it is stated that liquefied natural gas has more advantage than diesel fuels and heavy fuels.

To begin with, liquefied natural gas has low unit costs and natural gas production increases continuously. Moreover, LNG fueled vessel has clean burning. In other words, it emits less to the environment and also meets new regulations of maritime authority.

LNG-operated engines are less noise than diesel engines and provides more favorable working environment for employees and passengers.

LNG as a fuel is safer than diesel fuel, in the event of a leak, the gas evaporates directly because LNG only burns contacted with air at a concentration of 5-15% when it is evaporated into the environment. However, in the event spillage of diesel fuels or heavy fuels, they have the characteristic of combustion and explosion. In addition, in case of spillage, diesel and heavy fuels pollute soil and water.

Liquefied natural gas plants have inadequate network around the world, this situation leads to the fact that the use of LNG remains in restricted areas and this fuel does not become widespread. LNG fuel cannot be used on ocean-going vessels because there isn't any bunkering terminal for these type of vessels.

Initial investment costs of LNG fueled vessels are higher than diesel fueled main engines so that the ship-owners are not very interested in these types of vessels.

LNG fueled vessel and bunkering terminal has to perform some operational applications during bunkering process. As mentioned in the literature review operational applications include special communications and monitoring, cryogenic material precautions, emergency shutdown, inerting and purging, firefighting and electrical isolation. Owing to these applications respond to failures as soon as possible or avoiding any possible mistakes.

Although LNG fuel tanks can be stored above and below decks, non-storability in small scale bottom tanks is a disadvantage for liquefied natural gas.

As a result, the use of LNG as fuel in ships provides a significant advantage for ships in the future. LNG can be obtained at low prices and clean burning conditions bring this fuel to an important position. Due to these characteristics, the number of LNG-operated vessels has been increasing day by day in the world.

In this working, the passenger ship that was operated with diesel fuel was converted to the system operated with LNG fuel and the achievements of vessels transported with LNG fuel in terms of operators are shown in this project.

Applications reveal briefly that, the initial investment cost of such a system is estimated at approximately € 1.5M. In case the project passenger ship is operated with the new system, the system will be able to meet initial investment costs in about 10 years considering the provided financial gain. This period seems to be over in the first stage but developing technology, increasing reserves and environmental awareness will shorten this period.

REFERENCES

Adamchak, F., and Adede, A. (2014). LNG as marine fuel. 25 June 2016, http://www.gastechnology.org/Training/Documents/LNG17-proceedings/7-1-Frederick_Adamchak.pdf.

Andersen, H.J. (2015). LNG basics and status of LNG fuelled ships. 5 June 2016, http://www.caribbeanshipping.org/images/Documents/LNG_Basics_and_Satus_of_LNG_Fuelled_Ships__Jan_Hagen_Anderson.pdf.

American Bureau of Shipping (2015). LNG bunkering: technical and operational advisory. 30 April 2016, http://ww2.eagle.org/content/dam/eagle/publications/2015/LNG_Bunkering%20A dvisory.pdf.

Bahadori , A. (2014). Natural gas processing technology and engineering design (1.Edition). Waltham: Gulf Professional.

Bengtsson, S. Andersson, K. and Fridell, E. (2011). A comparative life cycle assessment of marine fuels liquefied natural gas and three other fossil fuels. Journal of Engineering for the Maritime Environment, 225 (2), 97-110.

Boylston, J.W. (2011). LNG as a fuel for vessels some design notes. 15 April 2016, http://leg.wa.gov/JTC/Documents/Studies/LNG/LNGFuelDesignNotes_060911.pdf.

Brynolf, S., Fridell, E., and Andersson, K. (2014). Environmental assessment of marine fuels: liquefied natural gas, liquefied biogas, methanol and bio-methanol. Journal of cleaner production, 74, 86-95.

Center for Climate and Energy Solutions (2013). Leveraging natural gas to reduce greenhouse gas emissions. 14 May 2016, https://www.c2es.org/docUploads/leveraging-natural-gas-reduce-ghgemissions.pdf.

Det Norske Veritas (2016). Rules and standards. 30 May 2016,https://rules.dnvgl.com/ServiceDocuments/dnvgl/#!/home#!%2Fhome.

Herdzik, J. (2012). Aspects of using LNG as a marine fuel. Journal of KONES, 19, 201-209.

Herdzik, J. (2013). Consequences of using LNG as a marine fuel. Journal of KONES, 20(2), 159-166.

International Gas Union (2015). World LNG report-2015 edition. 3 December 2016. http://www.igu.org/sites/default/files/node-page field_file/IGUWorld%20LNG%20Report2015%20Edition.pdf.

International Maritime Organization (2016a). International convention for the prevention of pollution from ships. 25 April 2016, http://www.imo.org/en/About/Conventions/ListOfConventions/Pages/International-Convention-for-the-Prevention-of-Pollution-from-Ships-(MARPOL).aspx.

International Maritime Organization (2016b). Status of conventions. 12 May 2016, http://www.imo.org/en/About/Conventions/StatusOfConventions/Documents/List%20of%20instruments%20%20as%20at%2031%20May%202016.pdf.

International Maritime Organization (2016c). Nitrogen oxides (NOx) – regulation 13.25 May 2016,http://www.imo.org/en/OurWork/Environment/PollutionPrevention/AirPollution/Pages/Nitrogen-oxides-(NOx)-%E2%80%93-Regulation-13.aspx.

International Maritime Organization (2016d). Prevention of air pollution from ships.25 May 2016, http://www.imo.org/en/OurWork/Environment/PollutionPrevention/AirPollution/Pages/Air-Pollution.aspx.

International Maritime Organization (2016e). Sulphur oxides (SOx) – regulation 14. 25 May 2016, http://www.imo.org/en/OurWork/Environment/PollutionPrevention/AirPollution/ Pages/Sulphur-oxides-(SOx)-%E2%80%93-Regulation-14.aspx.

Kumar, S., Kwon, H. T., Choi, K. H., Lim, W., Cho, J. H., Tak, K., and Moon, I. (2011). LNG: An eco-friendly cryogenic fuel for sustainable development. Applied Energy, 88(12), 4264-4273.

Lowell,D., Wang, H. and Lutsey N. (2013).Assessment of the fuel cycle impact of liquefied natural gas as used in international shipping. 20 May 2016, http://www.theicct.org/sites/default/files/publications/ICCT whitepaper_MarineLNG_130513.pdf.

McGhill, R.,Bemley, W. and Winther K. (2013). Alternative fuels for marine applications. 29 October 2016, http://www.ieaamf.org/app/webroot/files/file/Annex%20Reports/AMF_Annex_41.pdf.

Mohn, H. (2012). LNG as a ship fuel: operational experience and safety barriers. 20 June 2016, http://cbss.idynamic.lv/component/option,com_attachments/id,1446/lang,en/task,download.

MTLNG (2014). Liquefied natural gas general knowledge. 27 May 2016, http://www.goLNG.eu/files/Main/news_presentations/rostock/Liquefied%20Natural%20Gas%20General%20Knowledge.pdf.

Parsons, M. G., O'Hern, P. J., and Denomy, S. J. (2012). The potential conversion of the U.S. great lakes steam bulk carriers to liquefied natural gas propulsion-initial report. Journal of Ship Production and Design, 28(3), 97-111.

Semolinos, P., Olsen, G., and Giacosa, A. (2013). LNG as marine fuel: challenges to be overcome. 28 October 2016, http://www.gastechnology.org/training/documents/LNG17-proceedings/7-2-pablo_semolinos.pdf.

Society for Gas as a Marine Fuel (2014). Gas as a marine fuel an introductory guide. 20 April 2016, http://www.sgmf.info/media/8087/SGMF-Gas-as-a-Marine-Fuel- An-introductory-Guide.pdf.

Thomson, H., Corbett, J. J., & Winebrake, J. J. (2015). Natural gas as a marine fuel. Energy Policy, 87, 153-167.

Vandebroek, L., and Berghmans, J. (2012). Safety Aspects of the use of LNG for Marine Propulsion. Procedia engineering, 45, 21-26.

Wartsila (2015). Dual-Fuel engines – Wärtsilä 20df, 34df, 46df And 50df. 25 May 2016, http://cdn.wartsila.com/docs/default-source/product-files/engines/dfengine/brochure-o-e-df-engines 2015.pdf?sfvrsn=6.

Wartsila (2015). Wartsila 20DF product guide. 29 July 2016, http://www.wartsila.com/docs/default-source/product files/engines/dfengine/20df-product-guide.pdf?sfvrsn=0.

Woodyard, D. (2004). Pounder's marine diesel engines and gas turbines.Butterworth-Heinemann.

Wuersig G.M. (2014). LNG as ship fuel the future – today. 5 May 2016, https://www.dnvgl.com/Images/LNG_report_2015 01_web_tcm8-13833.pdf.

Yanmar (2015). Dual-Fuel Marine Engine (Highly Reliable Environmentally Friendly Engine), 25 October 2016, https://www.yanmar.com/global/technology/technical_review/2015/0727_2.html.

Mathematical Models, Methods and Algorithms

Safety Analysis of a New and Innovative Transhipping Concept: a Comparison of two Bayesian Network Models

L. Clarke, G. Macfarlane, I. Penesis, J. Duffy & S. Matsubara
Australian Maritime College, University of Tasmania, Launceston, Tasmania, Australia

R. Ballantyne
Sea Transport Corporation, Runaway Bay, Queensland, Australia

ABSTRACT: The floating harbour transhipper (FHT) is a new and innovative concept designed to improve the efficiency of bulk commodity export. As with all new and untested maritime systems, a rigorous safety analysis is essential prior to the operation's launch. Sophisticated tools, such as Bayesian networks, are increasingly being adopted for safety analysis in the maritime industry. Bayesian networks have predominately been used to replace fault trees due to their flexibility and improved handling of uncertainty. However, when applied to complex socio-technical systems Bayesian networks, as currently used, inherit many of the same shortcomings as traditional risk-based methodologies. These limitations are not indicative of the methods potential. This paper compares the traditional approach for developing Bayesian networks with an approach based on a foundation of resilience engineering. Both models were applied to an aspect of manoeuvring during the FHT's operation. The resilience-based approach successfully developed a representative network. The two approaches led to vastly different models, both in terms of structure and parameters. However, the descriptive approach of resilience-based model produced more proactive safety recommendations and, unlike the traditional model, serves as a knowledge base for future investigations.

1 INTRODUCTION

Systematic safety analysis was developed in the 20th century to prevent increasingly complicated technological systems from becoming unmanageable. The analysis of systems through decomposition brought much needed structure and manageability. The componentisation also brought economic benefits as it was much more efficient and cost effective to replace a single, malfunctioning component than to redesign an entire system.

The componentisation that allowed systems to remain manageable simultaneously acted as a driver for their development. Technological systems have been purposely designed so that components can be analysed independently. Effectively analysing systems through decomposition is not self-evident but is instead the result of purposeful design.

The general approach to advancing technological systems has been to increase their functionality by adding more components and improve their reliability by identifying and replacing weak links.

It is hard to argue with the success of this find-cause, fix-cause approach which led to relatively stable technological systems by the latter half on the twentieth century and made accidents caused by technological malfunctions increasingly rare.

However, the stabilisation of technology did not eliminate accidents or incidents entirely and during the 1980's the maritime industry, along with many others, suffered a series of high profile casualties (Anderson, 2003). The subsequent hunt for causes concluded that 8 out of 10 maritime accidents were caused by human error (Baker & Seah, 2004). The solution was a series of new regulations which included; better standards of crew training, improved organisational structures and more comprehensive procedures, all of which were designed to 'improve the human' as the weak component.

Humans, however, are rarely incorporated into socio-technical systems to act as an independently functioning component. In general, humans are excellent reasoners under uncertainty and can adjust their work to meet variable conditions (Knill & Pouget, 2004). Because of this, humans often serve as a guard against the unknown, providing improved system outcomes across a broader range of uncertain operating conditions. Incorporating humans into systems for this purpose violates the assumptions which validated componentisation.

This realisation was partly responsible for driving the development of resilience engineering. When adopting a resilience-based approach to safety analysis, the goal is not to identify and improve components which may fail, but to improve a system's ability the adapt to variable conditions (Hollnagel, Woods, & Leveson, 2007). Differences in system outcomes are not explained as components acting unexpectedly, but are instead viewed as a natural consequence of uncertain operating conditions. The ability for humans to vary their performance to meet these conditions is the reason things usually go right, as well as the reason things occasionally go wrong (Hollnagel, 2014).

In addition to freeing safety analysis from the unwarranted assumption of componentisation, there are two additional benefits to adopting an approach based on resilience engineering:

1 An emphasis is placed on developing an understanding of how socio-technical systems truly function, particularly concerning the work undertaken by humans.

2 The resilience of a system can always be improved without the prerequisite of new risk or hazard identification. This improvement comes not only from system modifications but also from developing a more complete understanding of how the system functions.

The differences between traditional and resilience safety analyses are explored through application of each approach to a novel transhipping concept, the floating harbour transhipper. The analyses do not aim to replicate a comprehensive safety or risk assessment but instead to illustrate the differences in the processes and outcomes.

2 THE FLOATING HARBOUR TRANSHIPPER

The floating harbour transhipper (FHT) is a new concept for exporting bulk commodities like iron ore or coal (Macfarlane, Johnson, Clarke, Ballantyne, & McTaggart, 2015; Macfarlane, Matsubara, Clarke, Johnson, & Ballantyne, 2015). Transhipping is commonly used to load ocean-going vessels in locations where deep water ports are not viable due to geographic, political or economic constraints. Transhipping typically involves ferrying small quantities of bulk product from a shore-based stockpile to a bulk carrier anchored offshore. The ferrying vessel is typically of shallow draft to allow access to the stockpile. The ferrying vessel is typically a barge towed by tugs or specially designed, self-propelled vessel. The product is transferred from the ferrying vessel to the bulk carrier using reclaiming equipment integral to the ocean-going vessel, the ferrying vessel or a purposely designed tertiary vessel.

Figure 1. A bulk carrier (right, red) being loaded alongside the floating harbour transhipper (left, grey), while a feeder vessel (small, near) is unloaded in the FHT's well dock. The sheltered well dock affords a significant increase in the allowable seastate for feeder vessel cargo operations.

The operable window of current transhipping methods is limited to small seastates. In larger seas the relative motions between the vessels create difficulties, both with mooring and cargo transfer. The transfer of cargo in an open environment also raises environmental issues such as dust emission and product spillage.

The FHT offers an alternative transhipping system which overcomes many of the limitations common to current operations. The FHT operates as a large storage and offloading vessel for bulk commodities. The FHT is semi-permanently moored offshore to a single point mooring. Self-propelled, ferrying vessels transport relatively small quantities of product to the anchored FHT where they enter a sheltered, aft well dock. The feeder vessels are subsequently unloaded using the FHT's reclaiming equipment. The transfer of product in a sheltered environment affords a significant increase in the operable sea state.

When required, a bulk carrier is loaded alongside the FHT either from the FHT's stockpile or directly from a feeder vessel moored in the FHT's well dock. Although the FHT is unpowered, the design includes stern thrusters that allow it to keep station during the mooring operation for the bulk carrier. Figure 1 shows the bulk carrier (right, red) being loaded alongside the FHT (left, grey) while the feeder vessel (small, near) is unloading her cargo in the FHT's aft well dock.

The concept has many benefits over traditional transhipping operations, including:

- A substantial increase in loading rate from less than 5000t/h to potentially 8000-10,000t/h.
- A significant increase in the allowable seastate from less than 1.5 metres to more than 4 metres. (Macfarlane, Ballantyne, Ballantyne, & Lilienthal, 2012).
- The elimination of environmental issues such as dust and product spillage.
- A reduced requirement for large negative pressure sheds and other shore-based infrastructure.

The FHT could revolutionise the way bulk commodities are exported and lead to dramatic improvements in supply chain efficiency along with better environment outcomes. There are many similarities between the FHT concept and existing operations. For example, rafting the bulk carrier alongside the FHT is similar to current ship-to-ship transfer operations. However, it is beyond the scope of normal operations to manoeuvre a large bulk carrier, in ballast condition, alongside a floating structure in potentially unsheltered water. This paper uses this rafting operation to illustrate the traditional and resilience-based approaches to safety analysis. Both approaches utilise Bayesian networks as a quantitative safety analysis tool.

3 BAYESIAN NETWORKS

Bayesian networks belong to a family of probabilistic graph models that integrate graph theory and probability theory. A Bayesian network consists of a directed acyclic graph (DAG) and a joint probability distribution over a set of random variables (Jensen, 1996). Bayesian networks are used to represent the state of knowledge in an uncertain domain and have been applied across an extremely diverse range of fields to fulfil a variety of purposes.

The DAG consists of nodes and arcs. A node represents a random variable which is defined by the possible states the variable can take. The states of a node are typically discrete but can also represent continuous outcome states. The directed arcs, which connect the nodes, give information about dependencies between the variables. The dependency specified by an arc can, but does not have to, represent a causal relationship (Pearl, 2009) and although adopting a causal DAG structure will produce a more efficient network (Pearl, 2014), information can pass through the network in any direction.

Encoded in the structure of the DAG are the conditional independencies that allow the joint probability distribution to be condensed. These independency assumptions are governed by the rules of d-separation and allow the joint probability distribution to be reduced to the product of the conditional probability distributions of each node (X_i) given the set of its direct predecessors (pa_i) as given by Equation 1. This assumption of conditional independence usually provides a substantial decrease in the number of parameters required to specify the joint probability distribution (Neapolitan, 2012).

$$P(X_1,...,X_m) = \prod P(X_i|pa_i) \qquad (1)$$

The application of Bayesian networks to risk assessment and safety analysis has increased dramatically over the past decade (Weber, Medina-Oliva, Simon, & Iung, 2012). Bayesian networks are commonly applied as direct replacements to more traditional methods such as fault trees and bow-tie analysis due to their increased flexibility when developing a model's structure and their improved treatment of uncertainty (Bobbio, Portinale, Minichino, & Ciancamerla, 2001). As a direct replacement for traditional models, Bayesian networks have predominately been used to model a finite list of adverse events rather than attempting to represent the state of knowledge in an uncertain domain.

The process of building a Bayesian network can be separated into two parts: structural learning and parameter learning.

The structural learning involves shaping the DAG to suitably represent the domain being modelled. The structure of a Bayesian network is typically learnt through expert judgement or by utilising available data through automated algorithms. Automatically learning Bayesian network structures is an important research topic in artificial intelligence and machine learning (Neapolitan, 2004). However, the majority of Bayesian networks developed for safety analysis use expert judgement to learn the DAG structure (Celeux, Corset, Lannoy, & Ricard, 2006).

Parameter learning is the assigning of conditional probability distributions for each node giving their direct predecessors. As with structural learning, parameter learning can be expert or data driven and there are many algorithms available for automatically learning parameters (Neapolitan, 2004). For a typical safety analysis application, a combination of data and expert judgement is often used.

This paper contrasts the traditional application of Bayesian networks in safety analysis with a resilience-based approach.

4 BAYESIAN NETWORK 1: A TRADITIONAL APPROACH

To date, Bayesian networks have been applied in safety analysis to identify and mitigate faults. Often Bayesian networks form direct extensions of traditional risk-based models such as fault trees. This approach is illustrated in the following subsections with an application to aspects of the FHT's rafting operation.

4.1 Step 1: Identify unfavourable events and their causes

A typical risk assessment or safety analysis begins with identifying potentially unfavourable events such as accidents or incidents as well as their potential causes, the process is known as failure analysis.

There are two commonly adopted approaches to failure analysis. A 'top-down' approach begins with the identification of a set of unfavourable outcomes and proceeds by tracing the events back to a list of 'root causes'. Fault trees are an example of a top-down approach to failure analysis. Conversely, a bottom-up approach starts with a single failure mode or hazard and then explores the potential consequences arising from the event, failure modes and effects analysis (FMEA) is an example of a bottom-up approach. It is common to use a combination of both approaches.

This analysis focuses on two technical malfunctions; a loss of propulsion, and a loss of steering. A combination of top-down and bottom-up approaches were used to deduce some of the potential causes and consequences of each event.

As is common for this type of analysis, the causes were identified using a combination of expert judgement and existing data.

4.2 Step 2: Learn the structure of the Bayesian network

The structure of the Bayesian network was learnt directly from the failure analysis undertaken in Step 1. Often a fault tree or similar graphical representation is mapped directly to a Bayesian network. The nodes of the network are the failure events and their root causes identified in Step 1. The states of the nodes are typically bimodal truth values indicating the occurrence of each failure event.

The arcs indicate cause-effect relationships and the network flows from the root causes to final consequences or 'top events'. The structure of the network is dependent on the failure analysis approach used, if the network has been directly converted from a fault tree the structure will typically converge to a single, final set of consequences.

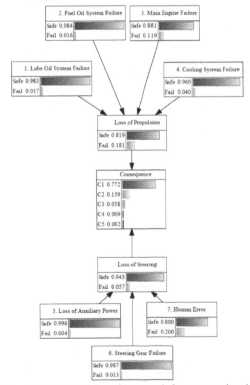

Figure 2. An updated Bayesian network for manoeuvring the Bulk carrier alongside the FHT, developed using a traditional approach to risk analysis and safety management

A node representing potential consequences resulting from the occurrence of the two malfunctions was also used in the analysis. This node includes a finite set of system outcome states, ordered by severity, the states included: Safe operation, C1; Aborted Operation, C2; Near Miss, C3; Minor Collision, C4; and Major Collision, C5; The structure of the network can be seen in Figure 2.

4.3 Step 3: Learn the parameters of the Bayesian network

With the exception of human error, the prior probability of each root cause was determined based on estimated failure probabilities (Cross & Ballesio, 2003). Human error was estimated from values used in similar maritime safety assessments. The failure rate of each root cause is given in Table 1 as an expected occurrence per year.

These probabilities are not Bayesian in nature. Instead, they are frequentist statistics taken from long range failure rates and accident data.

Table 1. The prior probabilities of root causes for each identified scenario

I.D.	Root Cause	Associated Scenario*	Failure Rate**
1	Lube oil system failure	LP	0.017
2	Fuel oil system failure	LP	0.016
3	Main engine failure	LP	0.119
4	Cooling system failure	LP	0.040
5	Loss of auxiliary power	LS	0.004
6	Steering gear failure	LS	0.013
7	Human error	LS	0.200

* LP: Loss of propulsion, LS: Loss of steering
** Expected occurrence per year (Cross & Ballesio, 2003)

The conditional probabilities were evaluated using an 'or-gate'. Other combination schemes, such as 'noisy or' gates or the use of expert judgment are also occasionally utilised.

The combination of root probabilities and conditional probabilities completely define the joint probability distribution of the network. Probabilities are updated either using an exact method or an approximate algorithm. Due to the network's structural simplicity an exact method was used. Figure 2 depicts the Bayesian network developed using the traditional approach and shows the updated probabilities for each event.

4.4 Step 4: Mitigate risks by addressing the root causes or preventing their effect

Improving system safety is achieved by lowering the probability of one or more failure modes or by mitigating the failure modes ability to affect a negative consequence.

Bayesian networks allow information to be passed in any direction. This allows the likelihood of

each root cause to be determined given a 'top event' has occurred. This analysis provides a type of sensitivity assessment that allows root causes to be prioritised. For example, given a major collision has occurred, the likelihood of each root cause having occurred can be estimated as shown in Table 2.

The results indicate that failure of the main engine is the most likely root cause of a major collision followed by human error.

Table 2. The estimated likelihood of each root cause given that a major collision has occurred

I.D.	Root Cause	Associated Scenario*	Failure Rate**
1	Lube oil system failure	LP	0.075
2	Fuel oil system failure	LP	0.070
3	Main engine failure	LP	0.527
4	Cooling system failure	LP	0.022
5	Loss of auxiliary power	LS	0.022
6	Steering gear failure	LS	0.067
7	Human error	LS	0.332

* LP: Loss of propulsion, LS: Loss of steering

Reducing the prior probabilities of these failure modes is a priority for improving system safety. Countermeasures typically include, increasing the reliability of the main engine or improving the standards of crew training.

An alternative approach to improving system safety is to implement barriers which inhibit a root cause's ability to affect an unfavourable consequence. An example of a safety barrier for this system would be to use tugs to assist the manoeuvring operation. While the tugs do not directly lower the probability of any root cause directly, they reduce the likelihood that these failure modes will lead to a collision.

5 BAYESIAN NETWORK 2: A RESILIENCE ENGINEERING APPROACH

The authors are unaware of any Resilience-based Bayesian network application in safety analysis. However, the resilience engineering approach is closely linked to modelling an uncertain domain, and networks with similarities to this approach are common in other fields. As with the traditional model, the resilience-based Bayesian network is focussed on manoeuvring the bulk carrier alongside the FHT. The methodology and results are detailed in the following subsections.

5.1 Step 1: Identify the system functions and background variables

Rather than modelling events, the resilience-based Bayesian network characterises the functions which must be completed for the system to fulfil its intended purpose. Modelling a system based on what it has to do is a natural starting point for an untested concept. This is because knowledge of what the system must do is often known, even at the earliest stages of design, and is often the driving force for the system's development. The way components should behave and the interactions between them is rarely understood early in the design phase.

This model focusses on a single function, 'manoeuvring the bulk carrier alongside the FHT'. Further subdivision of this task may be appropriate to enable analysis at a higher degree of resolution but further subdivision has not been undertaken in this paper.

5.2 *Step 2: Describe how each function is typically undertaken*

Many variables affect vessel manoeuvring in a variety of different ways. The resilience-based approach aims to develop a detailed description of the operation before any explanatory efforts are made. The description of each function is a qualitative depiction of the work normally undertaken during the activity. In addition each function was characterised based on six aspects: Control, Output, Resources, Preconditions, Input and Time.

These aspects were taken from the qualitative, functional resonance analysis method (Hollnagel, 2012), which is closely linked to resilience engineering.

This qualitative description and characteristics serve as the basis for the development of the Bayesian network. When new information becomes available these descriptions are updated prior to updating the model.

5.3 *Step 3: Develop the structure of the Bayesian network.*

The structure of the network was developed from the descriptions in Step 2. Learning the structure from the descriptions is not straightforward compared to the traditional approach.

The nodes of the resilience-based Bayesian network represent one of three sources of variability.
1 Function nodes are the outputs of other functions within the system which either directly or indirectly affect the manoeuvring operation. e.g. 'ready tug assistance'.
2 Background nodes are external and inherent sources of variability affecting the manoeuvring operation e.g. 'environmental conditions'.
3 Output nodes represent variability in manoeuvring outcomes. The output variability has been assessed in terms of timing and precision.

The nodes incorporated in the resilience-based Bayesian network are outlined in Table 3.

The structure is a fluid representation of the state of knowledge about the system and can be changed at any time considering new information.

Function nodes can affect manoeuvring outputs either directly, e.g. reducing variability in manoeuvring precision by having tug assistance; or indirectly by altering the background variables, e.g. reducing the availability of resources and crew competence by providing inadequate crew briefings.

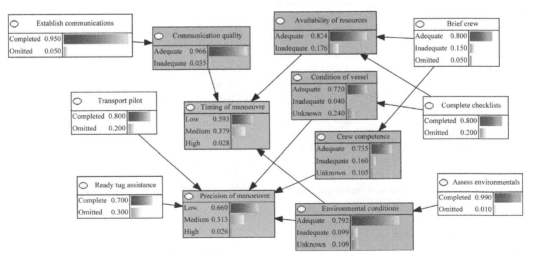

Figure 3. The resilience-based Bayesian network showing the elements of manoeuvring outputs (blue), other affecting functions (white) and background variables (grey)

Table 3 The nodes modelled in the resilience-based Bayesian networks for the manoeuvring operation

I.D.	Variable	Type
1	Communication quality	External variable
2	Availability of resources	External variable
3	Condition of vessel	External variable
4	Crew competence	External variable
5	Environmental conditions	External variable
6	Establish secure communications	Other function
7	Transport pilot to ocean-going vessel	Other function
8	Prepare tug assistance	Other function
9	Brief crew on operation	Other function
10	Complete checklists	Other function
11	Assess environmental conditions	Other function
12	Timing of the manoeuvre	Function output
13	Precision of the manoeuvre	Function output

5.4 *Step 4: Learn the parameters of the network*

The states of function nodes and background nodes are given as qualitative assessments of their level of adequacy. The states of the manoeuvring output nodes are qualitative assessments of variability. The set of conditional probabilities were developed individually without or-gates or other logical operators. This is because, unlike the treatment of failure modes, the functions cannot be assumed to act independently or according to fixed laws and their effects may combine or resonate leading to unforeseen outcomes.

The actual probabilities are not hard frequentist values. Instead the numbers provide concise, representations of qualitative relationships. These relationships are derived from the descriptions and characterisers detailed in Step 2. Figure 3 shows the resilience-based Bayesian network for the manoeuvring operation.

5.5 *Step 5: Develop strategies to improve the safety of the system*

The goal of a resilience-based safety analysis is to improve the system's ability to adapt to variable operating conditions. In this simplified example the goal is to try to dampen the output variability of the manoeuvring function. Detailed suggestions for improving system resilience can be found in Hollnagel, Woods, and Leveson (2007).

An approach to improving safety is to identify the conditions where variability from many sources may combine and lead to unexpectedly large outcome variability. There is a small subset of conditions where the output variability of the manoeuvring function is especially high. Modelling the system as a Bayesian network allows these conditions to be investigated with more concision than using a purely qualitative approach.

Without assumptions of causal independence, the resilience-based approach enables the full effect of counter measures to be assessed across the entire system. Simply developing a better understanding of the system's operation can also lead to less outcome variability as the probabilities are themselves representations of knowledge and uncertainty.

6 DISCUSSION

Although both methods are focussed on a similar operation and utilise the same quantitative technique, their methodology and handling of information vary considerably between the two approaches.

The traditional approach identifies and mitigates unfavourable events. While this approach offers simplicity it also limits expressiveness solely to the events being modelled. The resilience-based approach however, develops a description of functions the system must perform to achieve its goal. The primary purpose of the resilience-based model is to store information about an operational domain rather than to identify and mitigate risks. The emphasis on understanding systematic variability means the resulting model is not just representative of a finite subset of identified malfunctions but of the entire spectrum of system outcomes.

Traditional safety assessments typically remain static until new risks are identified, the resilience-based Bayesian network is in a better position to continually improve the system's safety. This is because the purpose of analysis is to continually understand the system in more detail. Additionally, the development of a knowledge base is not only useful for safety analysis but can also be used for training purposes or to improve system performance and efficiency. However, the approach is less convenient than traditional methods as a significant time investment is necessary to procure and analyse the required information.

The treatment of information varies between the two approaches. Naturally, the traditional Bayesian network uses information relating to things malfunctioning. The demand for this type of information means the majority of data collected focusses on the occurrence of adverse outcomes. Failure rates and accident statistics are examples of this type of information.

The purpose of the resilience-based model is to develop an understanding of how the system achieves its goals and the variability it must overcome to do so. Therefore, the information and data used must concern itself with how every day work is undertaken and how this varies on a situational basis. This type of information is not easy to obtain and, at present, rarely exists in a formal format, often the sole source of how work is undertaken is the experience of operators.

Countermeasures also vary between the two Bayesian network applications. The traditional approach seeks to improve safety by eliminating things going wrong, whereas the resilience-based approach seeks to improve safety by maximising the resilience of the system. Under the traditional approach the full effect of countermeasures cannot be understood as the assumption of causal independence restricts the ability to predict the effect across the whole system. Under a resilience-based approach decisions and countermeasures are investigated across the entire system.

Improving the rigor for developing quantitative, resilience-based safety methodologies would encourage more practitioners to adopt the approach, perhaps in conjunction with traditional risk-based methodologies.

7 CONCLUSIONS

The traditional method for conducting safety analysis is approaching maturity and has proved to be an extremely effective scheme for the advancement of complicated technological systems. However, an approach based on eliminating the causes of accidents cannot ensure an absence of adverse outcomes in complex socio-technical systems. Resilience engineering is a new concept of safety engineering that aims to explore system safety by first developing an understanding of the how the system adapts to variability to achieve its goals.

Bayesian networks were first developed as a knowledge modelling tool for uncertain domains. To date, a traditional risk-based approach has dominated the application of Bayesian networks in safety analyses. However, as demonstrated in this paper, Bayesian networks can also model socio-technical systems from alternative perspectives that incorporate the latest thinking on human factors.

ACKNOWLEDGEMENTS

This work was supported by the Australian Research Council (Grant no. LP130100962).

REFERENCES

Anderson, P. (2003). Cracking the code: the relevance of the ISM code and its impact on shipping practices. London: Nautical Institute.

Baker, C. C., & Seah, A. K. (2004). Maritime accidents and human performance: the statistical trail. Paper presented at the MarTech Conference, Singapore.

Bobbio, A., Portinale, L., Minichino, M., & Ciancamerla, E. (2001). Improving the analysis of dependable systems by mapping fault trees into Bayesian networks. Reliability Engineering & System Safety, 71(3): 249-260.

Celeux, G., Corset, F., Lannoy, A., & Ricard, B. (2006). Designing a Bayesian network for preventive maintenance from expert opinions in a rapid and reliable way. Reliability Engineering & System Safety, 91(7): 849-856.

Cross, R. B., & Ballesio, J. E. (2003). A quantitative risk assessment model for oil tankers. Transactions-Society of Naval Architects and Marine Engineers, 111: 608-623.

Hollnagel, E. (2012). FRAM: the functional resonance analysis method: modelling complex socio-technical systems. Aldershot: Ashgate Publishing, Ltd.

Hollnagel, E. (2014). Safety-I and safety–II: the past and future of safety management. Farnham: Ashgate Publishing Ltd.

Hollnagel, E., Woods, D. D., & Leveson, N. (2007). Resilience engineering: concepts and precepts. Farnham: Ashgate Publishing Ltd.

Jensen, F. V. (1996). An introduction to Bayesian networks (Vol. 210). London: UCL press.

Knill, D. C., & Pouget, A. (2004). The Bayesian brain: the role of uncertainty in neural coding and computation. Trends in neurosciences, 27(12): 712-719.

Macfarlane, G. J., Ballantyne, R., Ballantyne, S., & Lilienthal, T. (2012). An experimental study on the relative motions between a floating harbour transhipper and a feeder vessel in regular waves. Trans RINA, 154.

Macfarlane, G. J., Johnson, N. T., Clarke, L. J., Ballantyne, R. J., & McTaggart, K. A. (2015). The Floating Harbour Transhipper: New-Generation Transhipment of Bulk Ore Products. Paper presented at the ASME 2015 34th International Conference on Ocean, Offshore and Arctic Engineering.

Macfarlane, G. J., Matsubara, S., Clarke, L. J., Johnson, N., & Ballantyne, R. J. (2015). Transhipment of bulk ore products using a floating harbour transhipper. Paper presented at the Australasian Coasts & Ports Conference 2015: 22nd Australasian Coastal and Ocean Engineering Conference and the 15th Australasian Port and Harbour Conference.

Neapolitan, R. E. (2004). Learning bayesian networks (Vol. 38). Upper Saddle River, NJ: Pearson Prentice Hall.

Neapolitan, R. E. (2012). Probabilistic reasoning in expert systems: theory and algorithms: CreateSpace Independent Publishing Platform.

Pearl, J. (2009). Causality. Cambridge: Cambridge university press.

Pearl, J. (2014). Probabilistic reasoning in intelligent systems: networks of plausible inference. New York: Morgan Kaufmann.

Weber, P., Medina-Oliva, G., Simon, C., & Iung, B. (2012). Overview on Bayesian networks applications for dependability, risk analysis and maintenance areas. Engineering Applications of Artificial Intelligence, 25(4): 671-682.

Extensions of the Cayley-Hamilton Theorem to Transfer Matrices of Linear Systems

T. Kaczorek

Bialystok University of Technology, Bialystok, Poland

ABSTRACT: The classical Cayley-Hamilton theorem is extended to the square and nonsquare transfer matrices of linear systems. The theorems are proposed for rational and polynomial matrices.

1 INTRODUCTION

The classical Cayley-Hamilton theorem (Gantmacher 1959, Kaczorek 1992, 1998b) says that every square matrix satisfies its own characteristic equation. The Cayley-Hamilton theorem has been extended to rectangular matrices (Kaczorek 1995a, 1995c), block matrices (Kaczorek 1995a, 1998a), pairs of block matrices (Kaczorek 1998a) and standard and singular two-dimensional linear (2-D) systems (Kaczorek 1995b, 1994).

The Cayley-Hamilton theorem and its generalizations have been used in control systems, electrical circuits, systems with delays, singular systems, 2-D linear systems, etc. (Chang & Chan 1992, Kaczorek 2006, 2007a, b, c, 1992, 2012, Kaczorek & Przyborowski 2007, Lancaster 1969, Lewis 1982, 1986, Mertizios & Christodoulous, 1986, Smart & Barnett 1989, Theodoru 1989, Victoria 1982).

In Kaczorek 2005b the Cayley-Hamilton theorem has been extended to n-dimensional (n-D) real polynomial matrices. An extension of the Cayley-Hamilton theorem for continuous-time linear systems with delays has been given in Kaczorek 2005a.

The Cayley-Hamilton theorem has been extended to the fractional continuous-time and discrete-time linear systems in Kaczorek 2016a and to fractional descriptor systems in Kaczorek, T. 2016b.

In this paper the Cayley-Hamilton theorem will be extended to the transfer matrices of linear systems.

The paper is organized as follows. In section 2 the classical Cayley-Hamilton theorem is extended to square transfer matrices of linear systems. Extensions of the Cayley-Hamilton theorem for nonsquare transfer matrices are given in subsection 3.1 for $m > p$ and in subsection 3.2 for $m < p$, where m is the number of inputs and p is the number of outputs of the system. Concluding remarks are given in section 4.

2 CAYLEY-HAMILTON THEOREM FOR SQUARE TRANSFER MATRICES

Theorem 1. (Kaczorek 1998b) (Cayley-Hamilton) Let $A \in \mathfrak{R}^{n \times n}$ and

$$\det[I_n s - A] = s^n + a_{n-1} s^{n-1} + ... + a_1 s + a_0 \qquad (1)$$

be its characteristic polynomial.
Then

$$A^n + a_{n-1} A^{n-1} + ... + a_1 A + a_0 I_n = 0 \qquad (2)$$

Consider the linear continuous-time linear system

$$\dot{x}(t) = Ax(t) + Bu(t), \qquad (3a)$$

$$y(t) = Cx(t) + Du(t), \qquad (3b)$$

where $x(t) \in \mathfrak{R}^n$, $u(t) \in \mathfrak{R}^m$, $y(t) \in \mathfrak{R}^p$ are the state, input and output vectors and $A \in \mathfrak{R}^{n \times n}$, $B \in \mathfrak{R}^{n \times m}$, $C \in \mathfrak{R}^{p \times n}$, $D \in \mathfrak{R}^{p \times m}$.

The transfer matrix of the system (3) is given by

$$T(s) = C[I_n s - A]^{-1} B + D \in \mathfrak{R}^{p \times m}(s), \qquad (4)$$

where $\mathfrak{R}^{p \times m}(s)$ is the set of $p \times m$ rational matrices with real coefficients. The transfer matrix (4) can be written in the form

$$T(s) = \frac{P(s)}{d(s)}, \qquad (5)$$

where $P(s) \in \mathfrak{R}^{p \times m}[s]$ is the set of $p \times m$ polynomial matrices with real coefficients and $d(s)$ is the minimal degree denominator of all entries of the matrix $T(s)$.

If $p = m$ then we may calculate the polynomial

$$\det[I_m \lambda - T(s)] = \lambda^m + \bar{a}_{m-1}(s)\lambda^{m-1} + ... + \bar{a}_1(s)\lambda + \bar{a}_0(s), \tag{6}$$

where $\bar{a}_k(s)$, $k = 0,1,...,m-1$ are rational functions of s.

Substituting $\lambda = T(s)$ into (6) by Cayley-Hamilton theorem we obtain

$$[T(s)]^m + \bar{a}_{m-1}(s)[T(s)]^{m-1} + ... + \bar{a}_1(s)[T(s)] + \bar{a}_0(s)I_m = 0 \tag{7}$$

and after substitution (5) and multiplication by $[d(s)]^m$

$$[P(s)]^m + d(s)\bar{a}_{m-1}(s)[P(s)]^{m-1} + ... + d^{m-1}(s)\bar{a}_1(s)P(s) \\ + d^m(s)\bar{a}_0(s)I_m = 0 \tag{8}$$

Therefore, the following theorem has been proved.

Theorem 2. If $p = m$ and the transfer matrix has the form (5) then it satisfies the equation (8).

Example 1. Consider the transfer matrix

$$T(s) = \begin{bmatrix} \dfrac{1}{s} & \dfrac{1}{s+1} \\ \dfrac{2}{s+2} & \dfrac{1}{s(s+1)} \end{bmatrix} \tag{9}$$

which can be written in the form

$$T(s) = \frac{1}{s(s+1)(s+2)} , \tag{10a}$$

$$= \begin{bmatrix} (s+1)(s+2) & s(s+2) \\ 2s(s+1) & s+2 \end{bmatrix} = \frac{P(s)}{d(s)}$$

where

$$P(s) = \begin{bmatrix} (s+1)(s+2) & s(s+2) \\ 2s(s+1) & s+2 \end{bmatrix}, \tag{10b}$$

$$d(s) = s(s+1)(s+2). \tag{10c}$$

In this case the polynomial (6) has the form

$$\det[I_2 \lambda - T(s)] = \begin{vmatrix} \lambda - \dfrac{1}{s} & -\dfrac{1}{s+1} \\ -\dfrac{2}{s+2} & \lambda - \dfrac{1}{s(s+1)} \end{vmatrix}, \tag{11a}$$

$$= \lambda^2 + \bar{a}_1(s)\lambda + \bar{a}_0(s)$$

where

$$\bar{a}_1(s) = -\frac{s+2}{s(s+1)}, \quad \bar{a}_0(s) = \frac{-2s^2 + s + 2}{s^2(s+1)(s+2)}. \tag{11b}$$

Substitution of (10b), (10c) and (11b) into (8) for $m = 2$ yields

$$[P(s)]^2 + d(s)\bar{a}_1(s)P(s) + d^2(s)\bar{a}_0(s)I_2$$

$$= \begin{bmatrix} (s+1)(s+2) & s(s+2) \\ 2s(s+1) & s+2 \end{bmatrix}^2$$

$$- (s+2)^2 \begin{bmatrix} (s+1)(s+2) & s(s+2) \\ 2s(s+1) & s+2 \end{bmatrix}$$

$$+ (s+1)(s+2)(-2s^2 + s + 2)\begin{bmatrix} 1 & 0 \\ 0 & 1 \end{bmatrix} = \begin{bmatrix} 0 & 0 \\ 0 & 0 \end{bmatrix}. \tag{12}$$

Therefore, the transfer matrix (9) satisfies the equation (12).

If

$$\bar{a}_0(s) = \det T(s) \neq 0 \tag{13}$$

then premultiplying (7) by $T^{-1}(s)$ we obtain

$$[T(s)]^{m-1} + \bar{a}_{m-1}(s)[T(s)]^{m-2} + ... \\ + \bar{a}_1(s)I_m + \bar{a}_0(s)T^{-1}(s) = 0 \tag{14}$$

and

$$T^{-1}(s) = -\frac{\bar{a}_1(s)}{\bar{a}_0(s)}I_m - \frac{\bar{a}_2(s)}{\bar{a}_0(s)}T(s) - ... \\ - \frac{\bar{a}_{m-1}(s)}{\bar{a}_0(s)}[T(s)]^{m-2} - \frac{1}{\bar{a}_0(s)}[T(s)]^{m-1} \tag{15}$$

Knowing the coefficients $\bar{a}_k(s)$, $k = 0,1,...,m-1$ and $[T(s)]^j$, $j = 0,1,...,m-1$ and using (15) we may compute the inverse matrix of $T(s)$.

Example 2. (Continuation of Example 1) The transfer matrix (9) is nonsingular and using (15) and (11b) we obtain

$$T^{-1}(s) = \begin{bmatrix} \dfrac{1}{s} & \dfrac{1}{s+1} \\ \dfrac{2}{s+2} & \dfrac{1}{s(s+1)} \end{bmatrix}^{-1} = -\frac{\bar{a}_1(s)}{\bar{a}_0(s)}I_2 - \frac{1}{\bar{a}_0(s)}T(s)$$

$$= \frac{s(s+2)^2}{-2s^2 + s + 1}\begin{bmatrix} 1 & 0 \\ 0 & 1 \end{bmatrix}$$

$$- \frac{s^2(s+1)(s+2)}{-2s^2 + s + 1}\begin{bmatrix} \dfrac{1}{s} & \dfrac{1}{s+1} \\ \dfrac{2}{s+2} & \dfrac{1}{s(s+1)} \end{bmatrix} \tag{16}$$

$$= \frac{1}{-2s^2 + s + 2}\begin{bmatrix} s(s+1) & -s^2(s+2) \\ -2s^2(s+1) & s(s+1)(s+2) \end{bmatrix}.$$

Theorem 3. If the characteristic polynomial of the nonsingular transfer matrix $T(s)$ has the form (6) then

$$\bar{a}_0(s)[T(s)]^{-m} + \bar{a}_1(s)[T(s)]^{1-m} + ... \\ + \bar{a}_{m-1}(s)[T(s)]^{-1} + I_n = 0 \tag{17}$$

Proof. Postmultiplying (7) by $[T(s)]^{-m}$ we obtain (17) since $[T(s)]^k[T(s)]^{-k} = I_m$ for $k = 1,...,m$. $\square$

Example 3. (Continuation of Examples 1 and 2) The transfer matrix (9) is nonsingular and its inverse is given by (16), Using (11b), (16), (17) we obtain

$$\bar{a}_0(s)[T(s)]^{-2} + \bar{a}_1(s)[T(s)]^{-1} + I_2 =$$

$$-\frac{-2s^2+s+2}{s^2(s+1)(s+2)}\begin{bmatrix} \dfrac{s(s+1)}{-2s^2+s+2} & \dfrac{-s^2(s+2)}{-2s^2+s+2} \\ \dfrac{-2s^2(s+1)}{-2s^2+s+2} & \dfrac{s(s+1)(s+2)}{-2s^2+s+2} \end{bmatrix}^2$$

$$-\frac{s+2}{s(s+1)}\begin{bmatrix} \dfrac{s(s+1)}{-2s^2+s+2} & \dfrac{-s^2(s+2)}{-2s^2+s+2} \\ \dfrac{-2s^2(s+1)}{-2s^2+s+2} & \dfrac{s(s+1)(s+2)}{-2s^2+s+2} \end{bmatrix} + \begin{bmatrix} 1 & 0 \\ 0 & 1 \end{bmatrix}$$

$$= \begin{bmatrix} 0 & 0 \\ 0 & 0 \end{bmatrix}. \tag{18}$$

3 CAYLEY-HAMILTON THEOREM FOR NONSQUARE TRANSFER MATRICES

The following two cases for $m > p$ and $m < p$ will be addressed.

3.1 Case 1, m > p

The transfer matrix $T(s) \in \Re^{p \times m}(s)$ can be written as

$$T(s) = [T_1(s) \quad T_2(s)] \in \Re^{p \times m}(s), \tag{19a}$$

where

$$T_1(s) \in \Re^{p \times p}(s), \quad T_2(s) \in \Re^{p \times (m-p)}(s). \tag{19b}$$

Theorem 4. Let

$$\det[I_p \lambda - T_1(s)] = \sum_{k=0}^{p} \hat{a}_k(s)\lambda^k, \quad \hat{a}_p(s) - 1. \tag{20}$$

The matrices (19b) satisfy the equation

$$\sum_{k=0}^{p} \hat{a}_{p-k}(s)[T_1^{m-k}(s) \quad T_1^{m-k-1}(s)T_2(s)] = 0, \quad \hat{a}_p(s) = 1. \tag{21}$$

Proof. By induction it is easy to prove that

$$\begin{bmatrix} T_1(s) & T_2(s) \\ 0 & 0 \end{bmatrix}^k = \begin{bmatrix} T_1^k(s) & T_1^{k-1}(s)T_2(s) \\ 0 & 0 \end{bmatrix} \in \Re^{m \times m}(s),$$

$$k = 1,2,\dots \tag{22}$$

From (19a) we have

$$\det\begin{bmatrix} I_p\lambda - T_1(s) & -T_2(s) \\ 0 & I_{m-p}\lambda \end{bmatrix} = \lambda^{m-p}\det[I_p - T_1(s)]$$

$$= \sum_{k=0}^{p} \hat{a}_{p-k}(s)\lambda^{m-k}, \tag{23}$$

$$\hat{a}_p(s) = 1.$$

Using (6) we obtain

$$\sum_{k=0}^{p} \hat{a}_{p-k}(s)\begin{bmatrix} T_1^{m-k}(s) & T_1^{m-k-1}(s)T_2(s) \\ 0 & 0 \end{bmatrix} = 0. \tag{24}$$

Taking into account only first p rows of (24) we obtain (21). □

Let $\hat{d}(s)$ be the minimal common denominator of (19). Then the transfer matrix can be written in the form

$$T(s) = \begin{bmatrix} \dfrac{\hat{P}_1(s)}{\hat{d}(s)} & \dfrac{\hat{P}_2(s)}{\hat{d}(s)} \end{bmatrix}, \tag{25}$$

where $\hat{P}_1(s) \in \Re^{p \times p}[s]$, $\hat{P}_2(s) \in \Re^{p \times (m-p)}[s]$.

Substitution of (25) into (21) yields

$$\sum_{k=0}^{p} \hat{a}_{p-k}(s)\begin{bmatrix} \dfrac{\hat{P}_1^{m-k}(s)}{\hat{d}^{m-k}(s)} & \dfrac{\hat{P}_1^{m-k-1}(s)}{\hat{d}^{m-k-1}(s)}\dfrac{\hat{P}_2(s)}{\hat{d}(s)} \end{bmatrix} = 0, \quad \hat{a}_p(s) = 1. \tag{26}$$

Multiplying (26) by $\hat{d}^m(s)$ we obtain

$$\sum_{k=0}^{p} \hat{a}_{p-k}(s)[\hat{P}_1^{m-k}(s)\hat{d}^k(s) \quad \hat{P}_1^{m-k-1}(s)\hat{P}_2(s)\hat{d}^k(s)] = 0. \tag{27}$$

Therefore, the following theorem has been proved.

Theorem 5. If the characteristic polynomial of the transfer matrix $T_1(s)$ has the form (20) then the matrix (25) satisfies the equation (27).

Example 4. Let

$$T(s) = \begin{bmatrix} \dfrac{1}{s} & \dfrac{1}{s+1} & \dfrac{s+1}{s(s+2)} \\ \dfrac{2}{s+2} & \dfrac{1}{s(s+1)} & \dfrac{s^2}{(s+1)(s+2)} \end{bmatrix}$$

$$= [T_1(s) \quad T_2(s)] = \begin{bmatrix} \dfrac{\hat{P}_1(s)}{\hat{d}(s)} & \dfrac{\hat{P}_2(s)}{\hat{d}(s)} \end{bmatrix}, \tag{28a}$$

where $m = 3$, $p = 2$ and

$$T_1(s) = \begin{bmatrix} \dfrac{1}{s} & \dfrac{1}{s+1} \\ \dfrac{2}{s+2} & \dfrac{1}{s(s+1)} \end{bmatrix}, \quad T_2(s) = \begin{bmatrix} \dfrac{s+1}{s(s+2)} \\ \dfrac{s^2}{(s+1)(s+2)} \end{bmatrix},$$

$$\hat{d}(s) = s(s+1)(s+2), \tag{28b}$$

$$\hat{P}_1(s) = \begin{bmatrix} (s+1)(s+2) & s(s+2) \\ 2s(s+1) & s+2 \end{bmatrix}, \quad \hat{P}_2(s) = \begin{bmatrix} (s+1)^2 \\ s^3 \end{bmatrix}.$$

Note that $T_1(s) = T(s)$ given by (9) and $\hat{d}(s) = d(s)$ given by (10c) and

$$\det[I_2\lambda - T_1(s)] = \lambda^2 + \bar{a}_1(s)\lambda + \bar{a}_0(s), \tag{29}$$

where $\bar{a}_1(s)$ and $\bar{a}_0(s)$ are defined by (11b).

Using (21) and (29) we obtain

$$\sum_{k=0}^{2}\hat{a}_{2-k}(s)[T_1^{3-k}(s) \quad T_1^{2-k}(s)T_2(s)]$$

$$=[T_1^3(s) \quad T_1^2(s)T_2(s)]+\bar{a}_1(s)[T_1^2(s) \quad T_1(s)T_2(s)]+\bar{a}_0(s)[T_1(s) \quad T_2(s)]$$

$$=\left\{\begin{bmatrix} 1 & 1 \\ s & s+1 \\ 2 & 1 \\ s+2 & s(s+1) \end{bmatrix}^3 \quad \begin{bmatrix} 1 & 1 \\ s & s+1 \\ 2 & 1 \\ s+2 & s(s+1) \end{bmatrix}^2 \begin{bmatrix} s+1 \\ s(s+2) \\ s^2 \\ (s+1)(s+2) \end{bmatrix}\right\}$$

$$\frac{s+2}{s(s+1)}\left\{\begin{bmatrix} 1 & 1 \\ s & s+1 \\ 2 & 1 \\ s+2 & s(s+1) \end{bmatrix}^2 \quad \begin{bmatrix} 1 & 1 \\ s & s+1 \\ 2 & 1 \\ s+2 & s(s+1) \end{bmatrix}\begin{bmatrix} s+1 \\ s(s+2) \\ s^2 \\ (s+1)(s+2) \end{bmatrix}\right\}$$

$$+\frac{-2s^2+s+2}{s^2(s+1)(s+2)}\left\{\begin{bmatrix} 1 & 1 \\ s & s+1 \\ 2 & 1 \\ s+2 & s(s+1) \end{bmatrix} \quad \begin{bmatrix} s+1 \\ s(s+2) \\ s^2 \\ (s+1)(s+2) \end{bmatrix}\right\}=\begin{bmatrix} 0 & 0 & 0 \\ 0 & 0 & 0 \end{bmatrix} \quad (30)$$

Therefore, the transfer matrix (28) satisfies its characteristic equation (30). Similarly, substituting the matrices $\hat{P}_1(s)$, $\hat{P}_2(s)$ defined by (28b) to (27) for $p=2$ and $m=3$ we obtain

$$\sum_{k=0}^{2}\hat{a}_{2-k}(s)[\hat{P}_1^{3-k}(s)\hat{d}^k(s) \quad \hat{P}_1^{2-k}(s)\hat{P}_2(s)\hat{d}^k(s)]$$

$$=\left\{\begin{bmatrix} (s+1)(s+2) & s(s+2) \\ 2s(s+1) & s+2 \end{bmatrix}^3 \quad \begin{bmatrix} (s+1)(s+2) & s(s+2) \\ 2s(s+1) & s+2 \end{bmatrix}^2\begin{bmatrix} (s+1)^2 \\ s^3 \end{bmatrix}\right\}$$

$$\frac{s+2}{s(s+1)}\left\{\begin{bmatrix} (s+1)(s+2) & s(s+2) \\ 2s(s+1) & s+2 \end{bmatrix}^2 s(s+1)(s+2)\right.$$

$$\begin{bmatrix} (s+1)(s+2) & s(s+2) \\ 2s(s+1) & s+2 \end{bmatrix}\begin{bmatrix} (s+1)^2 \\ s^3 \end{bmatrix}s(s+1)(s+2)\right\}$$

$$+\frac{-2s^2+s+2}{s^2(s+1)(s+2)}\left\{\begin{bmatrix} (s+1)(s+2) & s(s+2) \\ 2s(s+1) & s+2 \end{bmatrix}[s(s+1)(s+2)]^2\right.$$

$$\begin{bmatrix} (s+1)^2 \\ s^3 \end{bmatrix}[s(s+1)(s+2)]^2\right\}=\begin{bmatrix} 0 & 0 & 0 \\ 0 & 0 & 0 \end{bmatrix} \quad (31)$$

Thus, the transfer matrix (28) satisfies also its characteristic equation (31).

3.2 Case 2, $p>m$

In this case the transfer matrix $T(s)\in\Re^{p\times m}(s)$ can be written in the form

$$T(s)=\begin{bmatrix} T_1(s) \\ T_2(s) \end{bmatrix}\in\Re^{p\times m}(s), \quad (32a)$$

where

$$T_1(s)\in\Re^{m\times m}(s), \quad T_2(s)\in\Re^{(p-m)\times m}(s). \quad (32b)$$

Theorem 6. Let

$$\det[I_m\lambda-T_1(s)]=\sum_{k=0}^{m}\tilde{a}_k(s)\lambda^k, \quad \tilde{a}_m(s)=1. \quad (33)$$

The matrices (32) satisfy the equation

$$\sum_{k=0}^{m}\tilde{a}_{m-k}(s)\begin{bmatrix} T_1^{p-k}(s) \\ T_2(s)T_1^{p-k-1}(s) \end{bmatrix}=0, \quad \tilde{a}_m(s)=1. \quad (34)$$

Proof. By induction it is easy to show that

$$\begin{bmatrix} T_1(s) & 0 \\ T_2(s) & 0 \end{bmatrix}^k=\begin{bmatrix} T_1^k(s) & 0 \\ T_2(s)T_1^{k-1}(s) & 0 \end{bmatrix} \text{ for } k=1,2,\ldots. \quad (35)$$

From (32) we have

$$\det\begin{bmatrix} I_m\lambda-T_1(s) & 0 \\ -T_2(s) & I_{p-m}\lambda \end{bmatrix}$$

$$=\lambda^{p-m}\det[I_m\lambda-T_1(s)]=\sum_{k=0}^{m}\tilde{a}_{m-k}(s)\lambda^{p-k}, \quad (36)$$

$$\tilde{a}_m(s)=1.$$

Using (6) we obtain

$$\sum_{k=0}^{m}\tilde{a}_{m-k}(s)\begin{bmatrix} T_1^{p-k}(s) & 0 \\ T_2(s)T_1^{p-k-1}(s) & 0 \end{bmatrix}=0. \quad (37)$$

Taking into account only the first m columns of (37) we obtain (34). $\square$

Let $\tilde{d}(s)$ be the minimal common denominator of the matrix (32). Then the matrix (32) can be written in the form

$$T(s)=\begin{bmatrix} \dfrac{\tilde{P}_1(s)}{\tilde{d}(s)} \\[2mm] \dfrac{\tilde{P}_2(s)}{\tilde{d}(s)} \end{bmatrix}, \quad (38a)$$

where

$$\tilde{P}_1(s)\in\Re^{m\times m}[s], \quad \tilde{P}_2(s)\in\Re^{(p-m)\times m}[s]. \quad (38b)$$

Substitution of (38) into (34) yields

$$\sum_{k=0}^{m}\tilde{a}_{m-k}(s)\begin{bmatrix} \left[\dfrac{\tilde{P}_1(s)}{\tilde{d}(s)}\right]^{p-k} \\[3mm] \left[\dfrac{\tilde{P}_2(s)}{\tilde{d}(s)}\right]\left[\dfrac{\tilde{P}_1(s)}{\tilde{d}(s)}\right]^{p-k-1} \end{bmatrix}=0. \quad (39)$$

Multiplying (39) by $\tilde{d}^p(s)$ we obtain

$$\sum_{k=0}^{m}\tilde{a}_{m-k}(s)\begin{bmatrix} \tilde{P}_1^{p-k}(s)\tilde{d}^k(s) \\ \tilde{P}_2(s)\tilde{P}_1^{p-k-1}(s)\tilde{d}^k(s) \end{bmatrix}=0. \quad (40)$$

Therefore, the following theorem has been proved.

Theorem 7. If the characteristic polynomial of the transfer matrix $T_1(s)$ has the form (33) then the matrix (38) satisfies the equation (40).

Similar results can be obtained for transfer matrices of discrete-time linear systems. The considerations can be easily extended to fractional continuous-time and discrete-time linear systems and also to impulse response matrices and to unit impulse response matrices of linear systems.

4 CONCLUDING REMARKS

The classical Cayley-Hamilton theorem has been extended to square and nonsquare transfer matrices of linear systems. The theorem has been extended to rational and polynomial matrices (Theorem 2) and also to inverse matrices (Theorem 3). Next, the theorems have been extended to nonsquare rational (Theorems 4 and 6) and polynomial matrices (Theorems 5 and 7). The theorems have been illustrated by examples of matrices of transfer functions of linear systems. The presented theorems can be easily extended to transfer matrices of discrete-time linear systems and to continuous-time and discrete-time fractional linear systems.

ACKNOWLEDGMENT

This work was supported under work number S/WE/1/16.

REFERENCES

Chang, F.R. & Chan, C.M. 1992. The generalized Cayley-Hamilton theorem for standard pencils, System and Control Lett., 18(3): 179-182.

Gantmacher, F.R. 1959. The Theory of Matrices. Chelsea Pub. Comp., London.

Kaczorek, T. 1995a. An extension of the Cayley-Hamilton theorem for non-square block matrices and computation of the left and right inverses of matrices, Bull. Pol. Acad. Sci. Techn. 43(1): 49-56.

Kaczorek, T. 1995b. An extension of the Cayley-Hamilton theorem for singular 2D linear systems with non-square matrices, Bull. Pol. Acad. Sci. Techn. 43(1): 39-48.

Kaczorek, T. 1998a. An Extension of the Cayley-Hamilton theorem for a standard pair of block matrices, Int. J. Appl. Math. Comput. Sci. 8(3): 511-516.

Kaczorek, T. 2006. An Extension of the Cayley-Hamilton theorem for nonlinear time-varying systems, Int J. Appl. Math. Comput. Sci. 16(1): 141-145.

Kaczorek, T. 2016a. Cayley-Hamilton theorem for fractional linear systems, Proc. RRNR, 20 – 21 Sept.

Kaczorek, T. 2016b. Extensions of the Cayley-Hamilton theorem to fractional descriptor linear systems, Proc. Conf. MMAR, Międzydroje, 29 Aug. – 1 Sept.

Kaczorek, T. 2005a. Extension of the Cayley-Hamilton theorem for continuous-time systems with delays, Int. J. Appl. Math. Comput. Sci. 15(2):231-234.

Kaczorek, T. 1994. Extensions of Cayley-Hamilton theorem for 2D continuous-discrete linear systems, Int. J. Appl. Math. Comput. Sci. 4(4): 507-515.

Kaczorek, T. 1995c. Generalization of the Cayley-Hamilton theorem for non-square matrices, Proc. Inter. Conf. Fundamentals of Electrotechnics and Circuit Theory XVIII-SPETO, 77-83.

Kaczorek, T. 2005b. Generalizations of Cayley-Hamilton theorem for n-D polynomial matrices, IEEE Trans. Autom. Contr. 50(5): 671-674.

Kaczorek, T. 2007a. Generalizations of the Cayley-Hamilton theorem with applications, Archives of Electrical Engineering. LVI(1): 3-41,.

Kaczorek, T. 1992. Linear Control Systems, vol. I, II, Research Studies Press.

Kaczorek, T. 2007b. Positive 2D hybrid linear systems, Bull. Pol. Acad. Sci. Techn. 55(4): 351-358.

Kaczorek, T. 2007c. Positive discrete-time linear Lyapunov systems, Journal of Automation, Mobile Robotics and Intelligent Systems 2(2): 13-19.

Kaczorek, T. 2012. Selected Problems of Fractional Systems Theory. Springer-Verlag, Berlin.

Kaczorek, T. 1998b. Vectors and Matrices in Automation and Electrotechnics, WNT, Warsaw, (in Polish).

Kaczorek, T. & Przyborowski, P. 2007. Positive continuous-time linear time-varying Lyapunov systems, Proc. of XVI Intern. Conf. on Systems Science, 140-149, 4-6 September, Wrocław-Poland.

Lancaster, P. 1969. Theory of Matrices, Acad. Press.

Lewis, F.L. 1982. Cayley-Hamilton theorem and Fadeev's method for the matrix pencil [sE-A], Proc. 22nd IEEE Conf. Decision Control, 1282-1288,.

Lewis, F.L. 1986. Further remarks on the Cayley-Hamilton theorem and Fadeev's method for the matrix pencil [sE-A], IEEE Trans. Automat. Control, 31: 869-870.

Mertizios, B.G. & Christodoulous, M.A. 1986. On the generalized Cayley-Hamilton theorem, IEEE Trans. Automat. Control, 31: 156-157.

Smart, N.M. & Barnett, S. 1989. The algebra of matrices in n-dimensional systems, Math. Control Inform. 6: 121-133.

Theodoru, N.J. 1989. M-dimensional Cayley-Hamilton theorem, IEEE Trans. Automat. Control, 34(5): 563-565.

Victoria, J. 1982. A block Cayley-Hamilton theorem, Bull. Math. Sco. Sci. Math. Roum, 26(1): 93-97.

Proceedings of 12th International Conference on Marine Navigation and Safety of Sea Transportation, TransNav 2017
21-23 June 2017, Gdynia, Poland

A Method of Assessing the Safety of Technical Systems of the Ship

W. Nowakowski, Z. Łukasik & J. Wojciechowski
University of Technology and Humanities in Radom, Radom, Poland

ABSTRACT: Ships as means of transport are crucial elements which have a large impact on the safety of maritime transport. Therefore, the safety analysis can be found as an important element of process safety management. The paper presents systematic approach to modelling the ship's safety, in which hazard identification and risk assessment are performed for every subsystem of the ship. In this case, methods considering safety requirements (applying in safety engineering) have been used. They allow engineers to take into account all safety aspects during development, modification and maintenance of technical objects. These methods also include hazard analysis and risk analysis. Hazard analysis identifies hazard and its cause-effect relationship. On the other hand, risk analysis determines frequency and effects of dangerous events. The paper discusses procedures allowing for validation of safety for subsystems ship: Performance Level (PL) and Safety Integrity Level (SIL).

1 INTRODUCTION

One of the most essential elements of the sea safety is technical safety of the ship, which is understood as the lack of unacceptable risk (Gerigk 2005, Celik et al. 2010). This means that the risk related to the work of technical systems of the ship needs to be on an acceptable level (Ruud & Skjetne 2014, Pomeroy 2014). Which is why, there is a need to take into account risk analysis when designing, building and maintaining technical systems of the ship. Risk should be understood as a possibility of occurrence of an unwanted event and related to it losses. In methods of risk analysis probability is often used, which is a consequence of a random character of unwanted event's occurrence (Kolowrocki & Soszynska 2008a,b). The unwanted event is an event (malfunction, damage, failure, human error), which occurrence causes hazard. In each technical object, at a different time, new unwanted events may arise and they might trigger a chain of secondary events and transition of the hazard state to losses called accidents or catastrophes. The size of losses usually concerns life and human health, material losses i.e. hazard to the ship and its freight and ecological losses – hazard for the sea environment. In the process of safety analysis of technical systems one can distinguish: hazard analysis, which includes identification of dangerous situations and their causes and risk analysis, which concerns identification of frequency and outcome of hazardous events. One of the most often used risk analysis methods include: check-list, primary hazard analysis, hazard and operability studies (HAZOP), structure « What if? » (SWIFT). Among the methods of technical risk analysis one might differentiate: scenario analysis, FMEA (Failure Modes and Effects Analysis), FMECA (Failure Modes, Effects and Criticality Analysis), ETA (Event Tree Analysis), FTA (Fault Tree Analysis), Markov analysis, Monte Carlo simulation, Bayesian statistics and Bayes Nets, layer protection analysis (LOPA) and decision tree (Nowakowski et al. 2008, Łukasik et al. 2017, Łukasik et al. 2014, Ahvenjarvi 2001, Nicholson 2013).

2 RISK ASSESSMENT OF TECHNICAL SYSTEMS OF THE SHIP

In the process of risk assessment one can distinguish a few basic stages. The first stage is the identification of a technical object. The aim of this stage is a deep cognition of the study object, its work conditions, way of operation and the like. In our case it is about distinguishing all technical systems of the ship. For example there might be the following systems (Dziula et al. 2008, Aichhorn et al. 2012, Baldauf et al. 2011):

1 Navigation systems:
   - route planning system (ECDIS),
   - route monitoring system (GPS, ECDIS, log, echo sounder, sonar),
   - anti-collision and detecting system (ARPA, AIS),
   - radio communication system (GMDSS),
   - voyage data recording system (VDR, S-VDR).
2 Propulsion and steering system:
   - main engine and auxiliary systems,
   - automation and alarm systems,
   - steering gear and system,
   - power supply systems.
3 Cargo compartments and loading system
4 Hull and outfit
5 Safety system
6 Mooring and anchoring system

Next, hazard identification is conducted, during which hazardous events are identified, the events might occur while technical object's exploitation. In this stage a thorough description of potential events is prepared, also the event's causes, outcome and possible safety systems are distinguished. On the basis of the gathered information one can estimate risk. All three stages are included in risk analysis. After risk estimation, risk evaluation should be conducted and a decision regarding the acceptability or unacceptability of the risk should be made. If it is a lack of acceptability of the risk, it is required to take up actions aimed at minimizing it, the actions are the safety function, and then running the whole procedure again (Fig. 1).

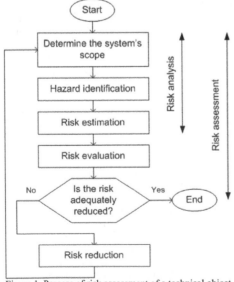

Figure 1. Process of risk assessment of a technical object.

The level of the technical system's safety can be described with one of two parameters (Piggin 2006):
- Performance Level (PL) – it can be used for electric, mechanical, pneumatic and hydraulic solutions serving the improvement of safety,
- Safety Integrity Level (SIL) – can only be used for the assessment of electric, electronic and programmable solutions serving the improvement of safety.

2.1 Performance Level (PL)

PL has five levels: a, b, c, d, e, where the PL is the highest safety level (Tab. 1) (Bornemann et al. 2015a,b).

Table 1. Descriptions of the PL levels.

Performance level (PL)	Probability of a dangerous failure per hour (PFH$_d$) 1/h
a	$\geq 10^{-5}$ to $< 10^{-4}$
b	$\geq 3 \times 10^{-6}$ to $< 10^{-5}$
c	$\geq 10^{-6}$ to $< 3 \times 10^{-6}$
d	$\geq 10^{-7}$ to $< 10^{-6}$
e	$\geq 10^{-8}$ to $< 10^{-7}$

Risk assessment of a technical system should be started with the identification of the risk's sources. Next, for each source once should estimate the risk based on the following factors:

S - severity of injury:
- S_1 slight (normally reversible injury),
- S_2 serious (normally irreversible injury or death).

F - frequency and/or exposure time to the hazard:
- F_1 seldom to less often and/or the exposure time is short,
- F_2 frequent to continuous and/or the exposure time is long.

P - possibility of avoiding the hazard or limiting the harm:
- P_1 possible under specific conditions,
- P_2 scarcely possible.

It allows to know whether it is necessary to minimize the risk or if safety is sufficient. If a situation requires reducing risk, for each safety function it is necessary to determine PL_r (Performance level required). For this purpose one uses the risk graph presented in Figure 2.

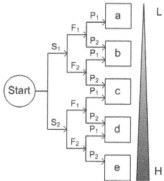

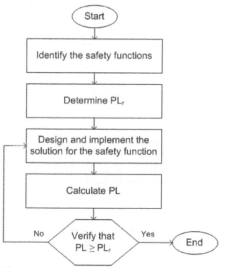

Figure 2. Estimation of the required PL.

The next step is designing and implementing a technical solution of safety function, after which its PL can be calculated. At the same time one must ensure that the PL is as high as PL_r (Fig. 3).

Figure 3. Procedure for verification of the safety functions.

When calculating the PL the easiest way is to divide a system into subsystems, for example: entry, logic block, exit. If one knows the PL of each subsystem (e.g. when they are defined by manufacturers), then the PL of the system results from the lowest PL for the subsystem and the number of the subsystems (Tab. 2).

If the subsystem's PL is unknown it must be calculated on one's own on the basis of:
– hardware architectures (*categories*),
– mean time to dangerous failure (*MTTFd*),
– diagnostic coverage (*DC*),
– common cause failures (*CCF*).

Table 2. Determining the system's PL on the basis of the subsystem's PL.

Lowest PL	Number of lowest PL	Entire system PL
a	> 3	not allowed
	≤ 3	a
b	> 2	a
	≤ 2	b
c	> 3	b
	≤ 3	c
d	> 3	c
	≤ 3	d
e	> 3	d
	≤ 3	e

One can distinguish the following categories: B, 1, 2, 3, 4, where the three first categories concern single channel systems. A difference between category 1 and 2 is that category 1 refers to subsystems built from reliable elements including tested and tried safety rules (well-tried components and well-tried safety principles shall be used), whereas category 2 includes subsystems with fault detection (safety functions shall be checked at suitable intervals). Categories 3 and 4 concern subsystems with redundancy and diagnostics (redundant architectures with diagnostics). In category 3 one assumes that single fault does not lead to the loss of the safety function, and in category 4 that the single fault is detected at or before the next demand upon the safety function.

Value of the average time between hazardous damages can be established when including a type of subsystem's elements. For mechanical and hydraulic parts, this parameter can be determined on the basis of a norm. Where it comes to the pneumatic, mechanical and electro mechanic components, the $MTTF_d$ can be ascertained on the basis of the formula:

$$MTTF_d = \frac{B_{10d}}{0,1 \cdot n_{op}} \tag{1}$$

$$n_{op} = \frac{d_{op} \cdot h_{op} \cdot 3600 s / h}{t_{cycle}} \tag{2}$$

where B_{10d} = average number of operating cycles achieved before 10% of the samples fail to the dangerous condition; n_{op} = number of activity cycles per years; d_{op} = operation days per years [d/y]; h_{op} = operation hours per day [h/d]; t_{cycle} = mean time between two activity cycles [s/cycle].

Where it comes to the electronic elements, the $MTTF_d$ can be determined on the basis of the following formula:

$$\frac{1}{MTTF_d} = \sum_{i=1}^{N} \frac{1}{MTTF_{di}} \tag{3}$$

where N = number of elements.

Next, for the appointed parameter one can establish one of three ranges: (low, medium, high) (Tab. 3).

Table 3. Various methods of calculating or estimating the MTTF$_d$.

MTTF$_d$	
Denotation of each channel	Range of each channel
Low	3 years $\leq$ MTTF$_d$ < 10 years
Medium	10 years $\leq$ MTTF$_d$ < 30 years
High	30 years $\leq$ MTTF$_d$ < 100 years

Diagnostic coverage (*DC*) for the subsystem is determined by a ratio between failure rate of detected dangerous failures and the failure rate of total dangerous failures:

$$DC = \frac{\sum \lambda_{DD}}{\sum \lambda_{DD} + \sum \lambda_{DU}} \quad (3)$$

where λ_{DD} = detected dangerous failure rate; λ_{DU} = undetected dangerous failure rate.

Where it comes to the whole system, the following formula can be applied:

$$DC_{avg} = \frac{\dfrac{DC_1}{MTTF_{d1}} + \dfrac{DC_2}{MTTF_{d2}} + ... + \dfrac{DC_n}{MTTF_{dn}}}{\dfrac{1}{MTTF_{d1}} + \dfrac{1}{MTTF_{d2}} + ... + \dfrac{1}{MTTF_{dn}}} \quad (4)$$

In practice, the diagnostic coverage can be estimated on the basis of the norm, where DC_{avg} takes value in ranges shown in Table 4.

Table 4. Ranges of the diagnostic coverage DC$_{avg}$.

Evaluation	Diagnostic coverage (DC)
none	DC < 60%
low	60% $\leq$ DC < 90%
medium	90% $\leq$ DC < 99%
high	99% $\leq$ DC

The *CCF* is a point-based scale with the maximum number of 100 points. Checking the *CCF* consists in using the right protective means and summing up points applied to them. A requirement is fulfilled when the minimum number of points is 65.

After assessing all four parameters, mentioned above, one can start defining the subsystem's PL using Figure 4.

In order to assess the PL of the whole system one can use Table 2 one more time. If the resulted PL from Table 2 is higher or equal than the PL$_r$ required for the safety function, then one can state that the safety system fulfils the requirement of resistance to devices.

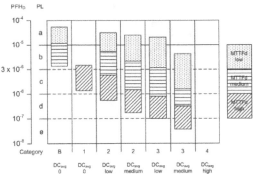

Figure 4. Relations between: categories, DC$_{avg}$, MTTF$_d$ and the subsystem's PL.

2.2 *Safety Integrity Level (SIL)*

In order to assess SIL one needs to determine (Bornemann et al. 2015a,b):
- *Se* - severity of the potential harm,
- probability of the harm occurring:
  - *Fr* - frequency and duration of exposure,
  - *Pr* - probability of a dangerous event occurring,
  - *Av* - probability of avoiding or limiting the harm.

Parameter *Se* concerns severity of injuries or damage to health. Classification is presented in Table 5.

Table 5. Severity of the potential harm.

Severity of the potential harm	Se
Irreversible: death, loss of an eye or an arm	4
Irreversible: shattered limb, loss of a finger	3
Reversible: requires the attention of a medical practitioner	2
Reversible: requires first aid	1

Frequency and duration of exposure *Fr* is linked to the need to access the hazardous zone, which is presented in Table 6.

Table 6. Frequency and duration of exposure.

Frequency and duration of exposure	Fr
$\leq$ 1 hour	5
> 1 hour to $\leq$ 1 day	5
> 1 day to $\leq$ 2 weeks	4
> 2 weeks to $\leq$ 1 year	3
> 1 year	2

When estimating the *Pr* parameter (Tab. 7) two factors need to be taken into account:
- the predictability of the dangerous components of the machine in its various operating modes (normal, maintenance, troubleshooting),
- behaviour of the persons interacting with the machine, such as stress, fatigue, inexperience, etc.

Table 7. Probability of a dangerous event occurring.

Probability of a dangerous event occurring	Se
Very high	5
Probable	4
Possible	3
Almost impossible	2
Negligible	1

The last of the parameters allows for a possibility to avoid or limit damage and is connected with the system's construction (Tab. 8).

Table 8. Probability of avoiding or limiting the harm.

Probability of avoiding or limiting the harm	Av
Impossible	5
Almost impossible	3
Probable	1

The sum of the parameters: *Fr*, *Pr* and *Av* determines the class of probability *Cl*. Estimation of the SIL is made with the help of the Table 9.

Table 9. Estimation of the SIL.

Se	Class Cl				
	3-4	5-7	8-10	11-13	14-15
4	SIL2	SIL2	SIL2	SIL3	SIL3
3	-	-	SIL1	SIL2	SIL3
2	-	-	-	SIL1	SIL2
1	-	-	-	-	SIL1

If the analyzed system was built from subsystems with the following ascribed parameters:
- SILCL - SIL Claim Limit,
- PFH$_d$ - Probability of Dangerous Failure per Hour

then the SIL of the whole system is determined with the series circuits of all subsystems taken into account:

$$PFH_d = PFH_{d1} + PFH_{d2} + ... + PFH_{dn} + PT \qquad (5)$$

where *PTE* = probability of dangerous transmission error.

Table 10 presents SIL levels depending on the aggregate value of the PFH$_d$.

Table 10. Checking the required SIL.

Safety Integrity Level (SIL)	Probability of a dangerous failure per hour (PFH$_d$)
1	$\geq 10^{-6}$ to $< 10^{-5}$
2	$\geq 10^{-7}$ to $< 10^{-6}$
3	$\geq 10^{-8}$ to $< 10^{-7}$

Next, one need to check if the SIL of the whole system is not lower than the target SIL determined with the help of Table 8. If this condition is not fulfilled, then one has to make construction alterations and repeat the whole procedure.

If one of the subsystem's SIL is unknown, it is necessary to define it. In order to do that subsystem logical architectures need to be determined, there are four types of them (A, B, C, D). The subsystem logical architectures with accompanying formulae are shown in Figures 5 through 8.

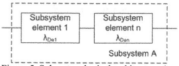

Figure 5. Subsystem logical architecture A.

$$\lambda_{DssA} = \lambda_{De1} + ... + \lambda_{Den}$$
$$PFH_{DssA} = \lambda_{DssA} \times 1h \qquad (6)$$

λ_D is the dangerous failure rate. λ_{DssA} is the dangerous failure rate of subsystem A. It is the sum of the failure rates of the individual elements: e_1, e_2, ..., e_n. The probability of dangerous failure is multiplied by 1 hour to create the probability of failure within one hour. λ can be determined for each of the components of the subsystem using the following formulas:
- $\lambda = 1/MTTF$ (electronic component)
- $\lambda = 0,1*C/B10$ (electromechanical components)

Figure 6 shows a single fault tolerant system without a diagnostic function.

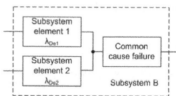

Figure 6. Subsystem logical architecture B.

$$\lambda_{DssB} = (1-\beta)^2 \cdot \lambda_{De1} \cdot \lambda_{De2} \cdot T_1 + \beta \cdot (\lambda_{De1} + \lambda_{De2})/2$$
$$PFH_{DssB} = \lambda_{DssB} \times 1h \qquad (7)$$

where β = susceptibility to common cause failures; T_1 = proof test interval or lifetime (whichever is smaller).

Figure 7 shows the functional representation of a zero fault tolerant system with a diagnostic function. Diagnostic coverage is used to decrease the probability of dangerous hardware failures. The diagnostic tests are performed automatically.

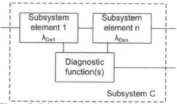

Figure 7. Subsystem logical architecture C.

$$\lambda_{DssC} = \lambda_{De1}(1 - DC_1) + ... + \lambda_{Den}(1 - DC_n)$$
$$PFH_{DssC} = \lambda_{DssC} \times 1h \qquad (8)$$

where DC = diagnostic coverage for each of the subsystem elements.

The last example of a subsystem architecture is shown in Figure 8. This subsystem is single fault tolerant and includes a diagnostic function.

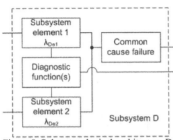

Figure 8. Subsystem logical architecture D.

If the subsystem elements are identical in each channel, the following formula is used:

$$\lambda_{DssD} = (1 - \beta)^2 \cdot \begin{Bmatrix} \left[\lambda_{De}^2 \cdot 2 \cdot DC\right] \cdot T_2 / 2 + \\ + \left[\lambda_{De}^2 \cdot (1 - DC)\right] \cdot T_1 \end{Bmatrix} + \beta \cdot \lambda_{De} \qquad (8)$$

$$PFH_{DssD} = \lambda_{DssD} \times 1h$$

where T_2 = diagnostic test interval.

3 CONCLUSIONS

Sea safety, such as life, property and sea environment safety is becoming a subject of greater attention. That is why it is important to develop methods of assessing the level of the sea safety and, at the same time, be able to choose possibilities and efficient ways of influencing this level. One of the most essential elements of the sea safety is the technical safety of the ship. If one treats the ship as a complex technical object then, for risk assessment one can use methods used in safety engineering. It applies to all stages of the ship's life cycle, including designing, building and maintaining technical systems of the ship. The authors of this article have presented a concept of assessing risk of the technical systems of the ship with the use of methods concerning technical realization of machines. In detail, two concepts of assuring safety through technical subsystems of the ship have been described, these concepts are the Performance Level (PL) and the Safety Integrity Level (SIL).

REFERENCES

Ahvenjarvi, S. 2001. Failure mode and effect analysis of automation systems of ships, 4th International Conference on Marine Technology, Szczecin, Poland, 2001, Marine Technology IV, Marine and Maritime, Volume 2, pp. 401-407

Aichhorn, K., Bedoya, C.C., Berglez, P. et al. 2012. Maritime Volumetric Navigation System, Proceedings of the 25th International Technical Meeting of the Satellite Division of the Institute of Navigation (ION GNSS 2012), pp. 3651-3657

Baldauf, M., Benedict, K., Fischer, S. et al. 2011. Collision avoidance systems in air and maritime traffic, Proceedings of the Institution of Mechanical Engineers, Journal of Risk and Reliability, Volume 225, Issue O3, pp.333-343

Bornemann, A., Froese, Y., Landi, L. et al. 2015a. Probabilities in safety of machinery-Part 1: Risk profiling and farmer matrix, Safety and Reliability: Methodology and Applications, CRC Press, pp. 1933-1942

Bornemann, A., Froese, Y., Landi, L. et al. 2015b. Probabilities in safety of machinery-Part 2: Theoretical and practical design, Safety and Reliability: Methodology and Applications, CRC Press, pp. 1943-1950

Celik, M., Lavasani, S.M. & Wang, J. 2010. A risk-based modelling approach to enhance shipping accident investigation, Safety Science, Volume 48, Issue 1, pp.18-27

Dziula, P., Jurdziński, M. ,Kołowrocki, K. & Soszyńska J. 2008. Modelowanie bezpieczeństwa morskich systemów i procesów transportowych, Prace Wydziału Nawigacyjnego Akademii Morskiej w Gdyni, z. 22, str. 17-26, Wydawnictwo Akademii Morskiej w Gdyni

Gerigk, M. 2005. Challenges of modern assessment of safety of ships in critical conditions. Options for preliminary design, 11th International Congress of the International Maritime Association of the Mediterranean (IMAM 2005), Lisbon, Portugal, 2005, Maritime Transportation and Exploitation of Ocean and Coastal Resources, Volume 1: Vessels for Maritime Transportation, pp. 1529-1536

Kolowrocki, K. & Soszynska, J. 2008a. Safety Analysis of Ship Technical Systems Related to Variable Operation Conditions, International Conference on Industrial Engineering and Engineering Management, Singapore, 2008, IEEM, pp. 1324-1329

Kolowrocki, K. & Soszynska, J. 2008b. A General Model of Ship Technical Systems Safety, International Conference on Industrial Engineering and Engineering Management, Singapore, 2008, IEEM, pp. 1346-1350

Łukasik, Z., Ciszewski, T., Młynczak, J., Nowakowski, W. & Wojciechowski J. 2017. Assessment of the safety of microprocessor-based semi-automatic block signalling system, Springer-Verlag, Series: Advances in Intelligent Systems and Computing, pp. 137-144

Łukasik, Z., Nowakowski, W. & Kuśmińska-Fijałkowska, A. 2014. Zarządzanie bezpieczeństwem infrastruktury krytycznej, Logistyka 4/2014, str. 758-763

Nicholson, M., 2013. Computer Safety For Modern Bridge Systems, Journal of Navigation, Volume 66, Issue 5, pp. 789-797

Nowakowski, W., Łukasik, Z. & Bojarczak P. 2016. Technical safety in the process of globalization, Proceedings of the 16th International Scientific Conference Globalization and Its Socio-Economic Consequences, Rajecke Teplice, Slovakia, 2016, Part IV, pp. 1571-1578

Piggin, R. 2006. What's happening with machine safety standards and networks?, Assembly Automation, Volume 26, Issue 2, pp. 104-110

Pomeroy, V. 2014. On future ship safety - people, complexity and systems , Journal of Marine Engineering and Technology, Volume 13, Issue 2, pp. 50-61

Ruud, S. & Skjetne, R., 2014. Verification and Examination Management of Complex Systems, Modeling Identification and Control, Volume 35, Issue 4, pp. 333-346

Proceedings of 12th International Conference on Marine Navigation and Safety of Sea Transportation, TransNav 2017
21-23 June 2017, Gdynia, Poland

Reliability Assessment of Vessel's Main Engine by Combining Markov Analysis Integrated with Time Dependent Failures

M. Anantharaman, F. Khan, V. Garaniya & B. Lewarn
Australian Maritime College (Utas), Tasmania, Australia

ABSTRACT: Safe operation of a merchant vessel is dependent on the reliability of the vessel's main propulsion engine. Overall reliability of the main propulsion engine is interdependent on the reliability of a number of subsystems including lubricating oil system, fuel oil system, cooling water system and scavenge air system. The reliability of various components of certain system such as gear pumps in a fuel oil system or filters in a lubricating oil system, which exhibit constant failure rate (random failure) independent of their history of operation, therefore could be analysed using Markov modelling. Other vital component such as turbochargers exhibits time dependent failure rate (wearing out). The wearing out failure rate (increasing failure rates) can be analysed using Weibull analysis. This paper presents integration of Markov model (for constant failure components) and Weibull failure model (for wearing out components) to estimate the reliability of the main propulsion engine. This integrated model will provide more realistic and practical analysis. It will serve as a useful tool to estimate the reliability of the vessel's main propulsion engine and make efficient and effective maintenance decisions.

1 INTRODUCTION

The reliability of a vessel's main propulsion engine is dependent on a number of essential sub systems, including fuel oil system, lubricating oil system, cooling water system and scavenge air system. Each of this subsystem has its own individual system components, the reliability of them would dictate the reliability of the corresponding subsystem, (EPSMA 2005), (Mollenhauer & Tschöke 2010, Matika & Manovic,2011).To determine the reliability of the various system components one need to look at the failure pattern depicted by these components. Previous studies have shown that most of the system components in commercial vessels, propelled by large two-stroke engine will fall in the second and third phase of the bath tub curve (shown in Fig 1), which is a constant failure rate followed by an increasing failure rate.

The reliability of various components of other systems such as gear pumps in a fuel oil system or filters in a lubricating oil system, exhibits constant failure rate (random failure) independent of their history of operation, therefore they could be analysed using Markov modelling. Other major components such as turbochargers exhibits time dependent failure rate. The wearing out failure rate can be analysed using Weibull analysis. Reliability of the fuel oil and lubricating oil system are computed using the Markov modelling, since they exhibit constant failure rate. The reliability of the sacvenge air system s determined using Weibul analysis, because they exhibit an increasing failure rate. Since the systems are in series, the reliability of the main propulsion engine would be equal to the product of the system reliability.This integrated model will provide more realistic and practical analysis. It will serve as a useful tool to estimate the reliability and make efficient and effective maintenance decisions.

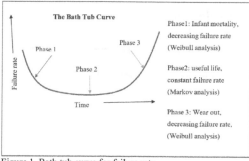

Figure 1. Bath tub curve for failure rate

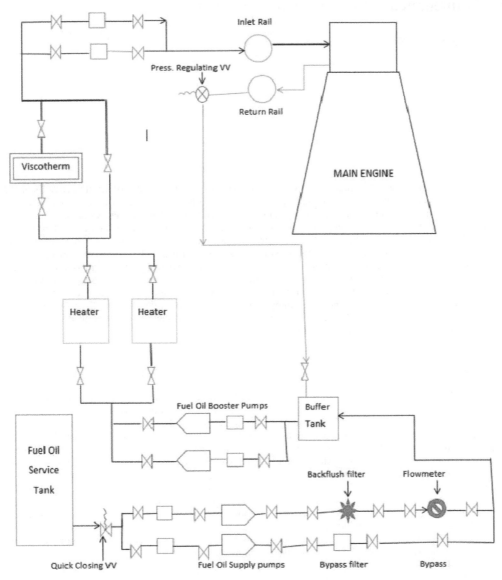

Figure 2. Main Engine fuel oil system

Figure 3. RBD for Main Engine Fuel Oil System

2 RELIABILITY OF FUEL OIL SYSTEM

Presently all large two stroke marine diesel engines, burn residual fuel oil the viscosity of which ranges between 380 to 700 cSt at 50 degs C. These fuel need to be purified, filtered and heated to a temperature as high as 150 degs C, to obtain a viscosity of 10-12 cSt at the inlet to the main engine fuel injection pump, for effective and efficient burning of the fuel. A fuel oil circuit for a large engine is shown in Fig. 2. This comprises of a service tank where purified fuel oil is stored at a temperature of 90 degs C. The tank is provided with a quick closing valve, which can be operated remotely in case of any emergencies, to cut off fuel to the main engine. The oil from the tank is drawn by means of a supply pump, which is of the gear type, where a suction filter is provided at the downside of the pump to filter out any sediments. The supply pressure of the pump varies between 5 to 7 bars, The pump discharges the fuel oil to a discharge filter which is of the automatic back flush type and has a fine mesh where filtration of the fuel oil can be done up to a sediment size of 25 microns. In case of any problem with the automatic back flush filter, it is possible to isolate and bypass this automatic back flush filter by means of a bypass filter, until the automatic back flush filter is recommissioned. The fuel is then led to a flow meter, where the fuel consumption is monitored. In case of any malfunction of the flowmeter it is possible to bypass this flowmeter, by means of a bypass valve. The next component in the system is a buffer tank, where the fuel mixes with fuel returned from the return rail of the engine. A booster pump on the upstream of the buffer tank boosts the system pressure up to 10 bars and discharges it to steam heaters, where the fuel is heated to the requisite temperature. The temperature is controlled by a viscometer, located at the upstream of the heaters. Normally one heater is in use, the other being a standby. It is to be noted that the fuel supply pumps and booster pumps are identical. Normally one pump is in use, the other being on standby, (Liberacki).The fuel oil at the correct viscosity between 10 to 12 cSt is supplied to the inlet rail, from where it is led to the individual fuel injection pump. The fuel injection pump, one for each cylinder, supplies the fuel at high pressures up to 600 bars, to fuel injectors on the cylinders. The return oil from the injectors comes to a return rail and led back to the buffer tank. A pressure regulator valve on the return line is provided, to maintain a back pressure of 8 to 9 bars.

Fig. 3 shows a reliability block diagram (RBD) for a main engine fuel oil sysytem. QC represents the Quick Closing valve, FS represents the Fuel Supply pumps, FL is the Discharge filters, FM is the Flowmeter, BT is the Buffer tank, BP represents the Booster pumps, HT represents the steam heater and VIS the Viscotherm. The next step is the analysis of evaluating the reliability of the main engine fuel oil sysytem, by using Markov analysis(Gowid, Dixon & Ghani 2014).

The following points are taken into consideration.
1 Each block represents the maximum number of components in order to simplify the diqgram.
2 The function of each block is easily identified
3 Blocks are mutually independent in that failure of one should not affect the probability of failure of another.(Anantharaman 2013; Xu 2008), (Bhattacharjya & Deleris 2012).

2.1 *Reliability of the Quick Closing Valve*

The quick closing valve is the main tank outlet valve, which can be operated remotely in case of an emergency. If we assume a constant failure rate λ,(PCAG 2012), then the reliability of this component may be expressed as

$$R_{QC}(t) = e^{-\lambda t},$$

where the mean time to failure MTTF $= 1/\lambda$.

2.2 *Reliability of the Fuel Oil Supply pump FS*

The fuel oil supply pumps FS are of the gear type and identical in design and construction. The state diagram for the pumps is shown in Fig. 4 The reliability function is an exponential function of time t and the failure rate λ expressed as number of failures per running hours, (Bhattacharjya & Deleris 2012).

Figure 4. Markov Model analysis for the fuel oil supply pump FS

Table 1. State of Fuel oil supply pump

State	Pump 1	Pump 2
1	Operating	Standby
2	Failed	Operating
3	Failed	Failed

From Table 1 it is clear that there are 3 states. In this case the two fuel oil supply pumps are identical units, (Liberacki, 2015), one of which is on line and the other standby.The reliability of the two identical systems is derived as,

$$R_s(t) = e^{-\lambda t} \sum_{i=0}^{1} \frac{(\lambda t)^i}{i!}.$$

In this case $R_s(t) = e^{-\lambda t}(1 + \lambda t)$ and MTTF (Mean time to failure) $= 2/\lambda$

Markov analysis is used to compute the reliability of the other componennts in the fuel oil system. The reliability of main engine fuel oil system will be given by

$$R_{F.O.}(t) = R_{QC}(t) R_{FS}(t) R_{FL}(t) R_{FM}(t) R_{BT}(t) \\ R_{BP}(t) R_{HT}(t) R_{VIS}(t) \quad (1)$$

3 RELIABILITY OF LUBRICATING OIL SYSTEM

The next step in the analysis of evaluating the reliability of the main engine lubricating oil sysytem is shown below:

The following five (5) cases s are analysed:
1 Failure of suction strainer S
2 Failre of pumps P
3 Failure of discahrge filter F
4 Failure of Temperature Control Valve TCV
5 Failure of cooler CLR

Each block represents the maximum number of components in order to simplify the diqgram.The function of each block is easily identified. The analysis and assumptions are the same as for the fuel oil system.

M.E. Lube Oil Pump Strainer	M.E. Lube Oil Pumps	M.E. Lube Oil Pump discharge filters	M.E. Lube Oil TCV	M.E. Lube Oil Cooler
S	P	F		CLR

Figure 5. Detailed RBD for M.E. Lube Oil system, with all system components

3.1 State diagram for the Main Engine Lube Oil Strainer S

The first component suction strainer S is a basket type strainer, located before the lubricating oil pumps, (Khonsari & Booser 2008).This is a duplex type of filter with a change over cock for isolation of filters. One of the filters is in use, the second one being a standby. Clogging of the strainer can result in pump's inability to draw suction from the sump,

which may sound a low pressure alarm. This provides time for changing over to the standby strainer. Failure of this change over will result in pump's inability to supply lubricating oil to the engine, finally resulting in an engine failure, (Cicek & Celik 2013).These filters will be identical as shown in Fig. 6. The state diagram for the filters is shown in Fig.7. The reliability function is an exponential function of time t and the failure rate λ expressed as number of failures per running hours , (Navy 1994).

Figure 6. Lube oil suction strainers for the Main Engine Lube oil system

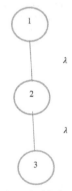

Figure 7. Markov Model analysis for the M.E. Lube oil Strainer S

Table 2. State of Lube oIl strainer S

State	Strainer 1	Strainer 2
1	Clean	CIlean
2	Clooged	Clean
3	Clogged (Failed)	Clogged(Failed)

As shown in Table 2, there are 3 states. In this case the two main engine lube oil pump strainers are identical standby units, one of which is on line and the other standby.The reliability of the two identical systems isderived as,

$$R_s(t) = e^{-\lambda t} \sum_{i=0}^{1} \frac{(\lambda t)^i}{i!}.$$

In this case $R_s(t) = e^{-\lambda t}(1 + \lambda t)$ and MTTF (Mean time to failure) $= 2/\lambda$

3.2 Reliability of the Main Engine Lube Oil System

The state diagrams for all other components of the system are analysed on the same lines, as done for the suction strainer S. Markov analysis(Smith 2011; Troyer 2006), carried out to determine the reliability of the system components . Finally the relaiility of the lubricating oil system is determined.

$$R_{L.O.}(t) = R_s(t) R_p(t) R_F(t) R_{TCV}(t) R_{CLR}(t) \qquad (2)$$

where
$R_p(t)$ is the reliability of the Pumps
$R_F(t)$ is the reliability of the Filter
$R_{TCV}(t)$ is the reliability of the temperature control valve
$R_{CLR}(t)$ is the reliability of the cooler.

4 RELIABILITY OF SCAVENGE AIR SYSTEM

Reliability of a scavenge air system for a large propulsion engine consists mainly of an exhaust gas turbocharger, (Takashi and Susumu, 1994), .The heat energy of the exhaust gas drives the exhaust gas turbine coupled to a rotary air compressor, which draws air from the engine room. The compressor compresses the air and then is cooled in an air cooler before being sent to the engine cylinder. One such turbocharger is shown in Fig.8. In short the turbocharger and the cooler form the main elements of the scavenge air system, failure of any one of the component could lead to failure of the main engine, as shown in the fault tree, (Zhu 2011), diagram in Fig.9.

Failure of either the Turbocharger, (ATSB,2006), or Air Cooler would result in failure of the Main Engine Scavenge system, (Conglin Dong 2013, Laskowski 2015).

Figure 8. Turbocharger for a large two stroke engine at test bed in QMD, Qungdao, China.

4.1 Fault tree for Main Engine failure

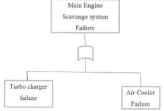

Figure 9. Fault tree for a Main Engine Scavenge system

4.2 RBD for Scavenge air system

A reliability block diagram for the scavenge air system is shown n Fig.10.

$$R_{Scaveneg\ air} = R_{Turbocharger} * R_{Air\ Cooler}$$

Figure 10. RBD for Main Engine Scavenge system

The turbocharger and air cooler are in series, hence the reliability of the scavenge air system could be computed. These two components form a very robust part of the scavenge air system. Depending upon the engine capacity there could be one or more turbochargers or air coolers fitted to the main propulsion engine. This arrangement has more to do with the engine capacity and not based on a redundancy factor.

4.3 Reliability of the Turbocharger

The turbocharger assembly consists of air filter, blower casing, turbine casing, rotor and bearings, (Schieman, 1992). Modern turbochargers are manufactured with sleeve type bearings which have very high operating life ranging up to 50,000 running hours, (MAN Diesels & Turbo). mHence while determining the reliability of the turbochargerit is necessary to look into the phase 3 of the bath tub curve where the end of life wear out could be considered, rather than the phase 1 or phase 2 of the bath tub curve. In the phase 3 the reliability of the Turbocharger may be computed using Weibull distribution, (Dhillon, 2002).The Reliability of the Turbocharger could be expressed as a function of time t.

$$R_{Turbo}(t) = e^{\left(-\frac{t}{\theta}\right)^{\beta}}$$

and the hazard rate function will be given by

$$\lambda_{Turbo}(t) = \frac{\beta}{\theta}\left(-\frac{t}{\theta}\right)^{\beta-1}$$

where θ is the scale parameter that influence both the mean and the spread or dispersion of the distribution and is the characteristic life and has units to those of time t, in this case hours, $\theta > 0$. β is referred as the shape parameter and $\beta > 0$. The Weibull hazard rate function can be increasing or decreasing depending on the value of β. If $\beta = 1$, $\lambda_{Turbo}(t)$ is constant and equal to $1/\theta$, the distribution being identical to the exponential.

4.4 Reliability of the Air cooler

The air cooler plays a very vital role in the scavenging system. The high temperature air discharged by the turbocharger needs to be cooled before sending it to the engine cylinders. These air coolers are generally sea water cooled, the sea water being passed through bronze alloy tubes, by means of a two pass cooling arrangement, to provide effective cooling of the charge air. The air flow will be one pass through the aluminium fins which are soldered to the brass alloy tubes, to avoid excessive pressure drop.The failure rate of the air cooler depends on its age factor, calling for consideration of the phase 3 of the bath tub curve. On similar lines to detecting reliability of the turbocharger, the reliability of the air cooler may be computed using Weibull distribution, (Kriya, 2002). The reliability of the air cooler could be expressed as a function of time t.

$$R_{Airclr}(t) = e^{\left(-\frac{t}{\theta}\right)^{\beta}}$$

and the hazard rate function will be given by

$$\lambda_{Airclr}(t) = \frac{\beta}{\theta}\left(-\frac{t}{\theta}\right)^{\beta-1}$$

where θ is the scale parameter that influence both the mean and the spread or dispersion of the distribution and is the characteristic life and has units to those of time t, in this case hours, $\theta > 0$. β is referred as the shape parameter and $\beta > 0$.

4.5 Reliability of the Scavenge air system

The turbocharger and air cooler being in a serial configuration, both needs to function for the scavenge air system to function. Both the components are critical and if either one of them fails the scavenge system will fail. The combined Weibull system reliability can be computed as below

$$R_{scavenege\ air} = \prod_{i=1,2} e^{-(t/\theta_i)^{\beta_i}} \qquad (3)$$

where i=1 is the Turbocharger and i= 2 is the Air cooler.

5 RELIABILITY OF THE MAIN PROPULSION ENGINE

Having determined the Reliability of the Fuel oil system by Markov analysis, Lubricating Oil system by Markov analysis which are both modelled using constant failure rate principle, and also determined the Reliability of Scavenge air system as a time dependent failure model,the Reliability of the Main propulsion engine is positioned to be determined as follows:

$$R_{MainEngine} = \prod R_{iI=1,2,3,} \qquad (4)$$

i=1 is the fuel oil system from Equation1 (Markov modelling)
i=2 is the lubricating oil system from Equation2 and Markov modelling
i=3 is the scavenge air system from Equation3 (Weibull modelling)

5.1 Improving Reliability

Reliability of the main engine can be improved by improving the individual system reliability as seen in the above Equation 4 above, For instance in the case of the scavenge air system, a modern high performance turbocharger will improve the reliability of the turbocharger.This cost for improvement of reliability need to be assessed and the cost benefit for the incremental reliability to be determined. If the original value of reliability R_O at cost x is improved to Reliability R_I at cost y, then the incremntal reliability for the differential cost $\frac{R_I - R_O}{y-x}$ should be compared with the base relaibility to cost ratio which in this case is $\frac{R_O}{x}$.

For cost benefit $\frac{R_I - R_O}{y-x} > \frac{R_O}{x}$.

This could be a feasible proposition for some components, but not for all components. An appropriate maintenance program to strike the right balance between reliability required and the cost penalty likely to be incurred, could be designed. Attention is drwan to the air cooler in the scaveneg air system as an example. All modern air coolers manufactured by major engine manufactureres have a cleaning in place system incorporated for maintenance of air coolers, which involves no dismantling of the air cooler whilst carrying out maintenance. Accordingly the maintenance intervals for air coolers could be shorter, at the same time increasing the maintenance intervals of turbochargers and stil provide a more effcient and reliable main engine.

6 CONCLUSION

In this paper we have looked at methods of determining the reliability of three subsystems of a vessel's main engine which includes the fuel oil system, lubricating oi system and the scavenge air system. The fuel oil and lubricating oil sytem was modeleld by Markov analsis and the scaveneg air system was analysed using Weilbull distribution which is time dependent failure model. An attempt has been made to make reliability assessment of vessel's main engine by combining Markov analysis integrated with time dependent failures.We have also discussed the increamental reliability to increamental cost ratio for the main enginewhich should always be greater than the original reliability to original cost ratio, for cost benefits in the long run. Finally we looked at some examples of effectively altering the maintenance intervals of certain system components, whwerby the overall reliability of the system could be improved.

ACKNOWLDAGEMENTS

Authors thanfully acknladge the support provided National Centre for Port and Shipping (NCPS), Australian Maritime College, University of Tasmania to enable this study.

REFERENCES

Anantharaman, M 2013, 'Using Reliability Block Diagrams And Fault Tree Circuits, To Develop A Condition Based Maintenance Model For A Vessel's Main Propulsion System And Related Subsystems', *TransNav, the International Journal On Marine Navigation And Safety Of Sea Transportation*, Vol. 7, No. 3.

Australian Transport Safety Bureau, 2006, 'Independent investigation into the equipment failure on board the Australian registered bulk carrier Golaith in Bass Strait'.

Bhattacharjya, D & Deleris, La 2012, 'From Reliability Block Diagrams To Fault Tree Circuits', *Decision Analysis*, Vol. 9, No. 2, Pp. 128-137.

Brandowski, A 2009, 'Estimation Of The Probability Of Propulsion Loss By A Sea Going Ship Based On Expert Opinion.', *Polish Maritime Research(1)*, Vol. 16, Pp. 73-77.

Cicek, K & Celik, M 2013, 'Application Of Failure Modes And Effects Analysis To Main Engine Crankcase Explosion Failure On-Board Ship', *Safety Science*, Vol. 51, No. 1, Pp. 6-10.

Conglin Dong, Cy, Zhenglin Liu, Xinping Yan 2013, 'Marine Propulsion Sysytem Reliability Research Based On Fault Tree Analysis', *Advanced Shipping And Ocean Engineering*, Vol. 2, No. 1, P. 7.

Dhillon,S 2002, 'Engineering Maintenance, A Modern approach', CRC press, Washington, D.C.

EPSMA 2005, 'Guidelines To Understanding Reliability Prediction', No. 24 June 2005, P. 29.

Gowid, S, Dixon, R & Ghani, S 2014, 'Optimization Of Reliability And Maintenance Of Liquefaction System On Flng Terminals Using Markov Modelling', *International Journal Of Quality & Reliability Management*, Vol. 31, No. 3, Pp. 293-310.

Khonsari, Mm & Booser, Er 2008, *Applied Tribology: Bearing Design And Lubrication*, Vol. 12, John Wiley & Sons.

Kiran 2002, 'Statistical Study on Reliability of Ship Equipment and Safety Management–Reliability Estimation for Failures on Main Engine System by Ship Reliability Database System– ', Bulletin of the JIME, Vol. 29, No.2

Laskowski,R,2015, ' Fault tree Analysis as a tool for modelling the marine main engine reliability structure', Scientific Journals of the Maritime University of Szczecin, 2015, 411(113), PP 71-77.

Liberacki, R 'Influence Of Redundancy And Ship Machinery Crew Manning On Reliability Of Lubricating Oil System For The Mc-Type Diesel Engine', In Gdansk University Of Technology Ul. Narutowicza 11/12, 80-950 Gdańsk, Poland.

MAN Diesel and Turbo, 'TCA Turbocharger the benchmark'.

Matika,D & Manovic,L 2011,' Reliability of a Light High Speed Marine Diesel Engine ', Original Scientific Paper in BRODOGRADNJA, 62(2011), PP 28-36.

Mollenhauer, K & Tschöke, H 2010, 'Handbook Of Diesel Engines', Handbook Of Diesel Engines, Edited By K. Mollenhauer And H. Tschöke. Berlin: Springer, 2010., Vol. 1.

Schieman, John 1992-1996, 'ABB Turbocharging Operating Turbochargers', collection of articles,published in Turbo magazinem 1992-1996.

Smith, Dj 2011, *Reliability, Maintainability And Risk 8e: Practical Methods For Engineers Including Reliability Centred Maintenance And Safety-Related Systems*, Elsevier.

Troyer, D 2006, 'Reliability Ngineering Principles For Plant Engineers'.

Takashi,A & Susumu, Y 1994,'Investigation in Damages of Turbochargers for Marine Appplication :-Analysis of Database for Ship Reliability in Japan', Marine Engineering 29(2), PP 215 218, The Japan Institute of Marine Engineering.

Navy, Us 1994, *Handbook Of Reliability Prediction Procedures For Mechanical Equipment*Pcag 2012, 'Failute Rate And Event Data For Use Within Risk Assesments.'

Xu, H 2008, 'Drbd: Dynamic Reliability Block Diagrams For System Reliability Modelling', University Of Massachusetts Dartmouth.

Zeszyty, N 2012, 'An analysis of cause and effect relations in diagnostic relations of marine Diesel engine turbochargers', 2012, Scientific Journals, Maritime University of Szczecin 31(103) pp. 5–13.

Zhu, Jf 2011, 'Fault Tree Analysis Of Centrifugal Compressor', *Advanced Materials And Computer Science, Pts 1-3*, Vol. 474-476, Pp. 1587-1590.

Method of Vehicle Routing Problem with Fuzzy Demands

T. Neumann
Gdynia Maritime University, Gdynia, Poland

ABSTRACT: The routing planning one of the classic problems in discrete mathematics concepts. Its application have various practical uses ranging from the transportation, civil engineering and other applications. The resolution of this paper is to find a solution for route planning in a transport networks, where the description of tracks, factor of safety and travel time are ambiguous.

1 INTRODUCTION

The goal of every routing algorithmic program is to straight bargain from origin to destinations maximizing meshwork accomplishment while minimizing charge. In this way, the universal question of regulate an ideal passing algorithmic rule can be established as a multi objective optimization problem in a non-stationary stochastic surrounding. Additional constraints are determined by the fundamental meshwork switching and transmission technology.

Distribution system device is essential to insure that the growth need of electricity is compensate by the distributors. Planning begin at purchaser even, arrangement system openly joined to purchaser any failing in the system would soften the customers. Therefore individual delineation of the arrangement system is very weighty for cohesion of influence. Distribution System Planning (DSP) surround ideal quotation of encourager course, count of encourager, substation adjust and situation. In this work quotation of ideal encourager course is hold by the proposed system. Several optimization techniques have been accomplish to explain the proposition of encourager course. In the past mathematical advance were appropriate such as branch and boundary process for the optimisation of arrangement system, blended number scheme devote to the classification system proposition was found practicable, explain the ideal encourager routing using dynamic programming and geographical information systems GIS facilities, which is powerful. Another implement to realize the optimisation goal is ant colony system algorithm (ACS). This methodology is meta-heuristic in quality and is very manageable, strong in minimising the vestment expense. The subjection in the charge during the planning of arrangement system, meeting the constraints is hold by branch interchange process. The causativeness of Genetic Algorithm is versed in the intriguing of the dispersion system by reducing the disruption period. Simulated annealing is also proved to be practicable in contrivance of the arrangement meshwork. In this course the minimum charge explanation is hold by steepest degradation approximate, further the hold release is moderate by feigned annealing. This process is faster, attracting less waste time.

In this paper Dijkstra algorithmic program is the explanation generalship for the ideal encourager passing in the planning of the radiated arrangement system. Dijkstra algorithmic program is shortest passing algorithmic rule that contemplate the judgment of the minimum charge (variance) from an fountain to a intention through some copulative graph, used in intriguing the distribution meshwork. Even in the enlargement of the feeders at least charge the intend algorithmic program is found practicable. The effectiveness of the algorithmic rule is proved in influence system replacement. The imperfect division is detached by the converse algorithmic program and the furnish is return in the system. The Dijkstra's method is posterior because it confide more on the numeral of arcs than nodes. The proposed algorithmic program works on directed weighted graph and the margin should be non-negative. The optimum passing are hold, further to minimise the amount charge, common and voltage excellence are required. For this lading melt analysis is unfold. (Jha and Vidyasagar, 2013)

Routing problems in networks are the problem in the context of sequencing and in recent times, they have to receive progressive note. Congruous issues usually take places in the zones of transportation and communications. A schedule problem engages identifying a route from the one point to the other because there are many of optional tracks in miscellaneous halting place of the passage. The cost, time, safety or cost of travel are different for each routes. Theoretically, the method comprises determining the cost of all prospective tracks and the find with minimal expense. In fact, however, the amount of such options are too large to be tested one after another. A traveling salesman problem is a routing problem associated with preferably strong restrictions. Different routing problem emerges when it can to go from one point to another point or a few points, and choose the best track with the at the lowest estimate length, period or cost of many options to reach the desired point. Such acyclic route network problem easily can be solved by job sequencing. A network is defined as a series of points or nodes that are interconnected by links. One way to go from one node to another is called a path. The problem of sequencing may have put some restrictions on it, such as time for each job on each machine, the availability of resources (people, equipment, materials and space), etc. in sequencing problem, the efficiency with respect to a minimum be measured costs, maximize profits, and the elapsed time is minimized. The graph image and the example of costs of borders are given in the figure 1. In this hypothetical idea the tract network is illustrated by a graph. Presented graph is given with an ordered pair $G: = (V, E)$ comprising a set V of vertices or nodes together with a set E of edges (paths), which connect two nodes. The task is to reach the N1 node from N3 node in the graph at smallest cost.(Neumann, 2016a)

The analysis of the navigational calculations indicates the use of different calculating algorithms based on the same geometric model as well as various geometric models of the Earth (spheroid, sphere, Euclidean plane, triaxial ellipsoid, etc.) in the algorithm used without defining the boundary conditions (criteria) for their use and change today in modelling locally and globally. (Weintrit and Kopacz, 2012)

The International Maritime Organization (IMO) in 2013 has reviewed the preliminary list of potential e-Navigation solutions and prioritized five potential main solutions, presented the finalized Cost Benefit and Risk Analysis, considered the further development of the detailed ship and shore architecture, giving an example of a technical infrastructure to support seamless information/data exchange in e-Navigation. (Weintrit, 2013)

The solutions have served as the basis for the creation of Risk Control Options that were believed to be tangible and manageable in terms of quantifying the risk reducing effect and the related costs. The Risk Control Options listed below demonstrate cost-effectiveness according to the International Maritime Organization Formal Safety Assessment criteria:

- RCO1: Integration of navigation information and equipment including improved software quality assurance,
- RCO2: Bridge alert management,
- RCO3: Standardised mode(s) for navigation equipment,
- RCO4: Automated and standardised ship-shore reporting,
- RCO5: Improved reliability and resilience of PNT systems,
- RCO6: Improved shore-based services,
- RCO7: Bridge and workstation layout standardisation.

Facility location problems investigate where to physically locate a set of facilities (resources, stations, etc.) so as to minimize the cost of satisfying some set of demands (customers) subject to some set of constraints. Location decisions are integral to a particular system's ability to satisfy its demands in an efficient manner. In addition, because these decisions can have lasting impacts, facility location decisions will also affect the system's flexibility to meet these demands as they evolve over time.

Facility location models are used in a wide variety of applications. These include, but are not limited to, locating warehouses within a supply chain to minimize the average time to market, locating hazardous material sites to minimize exposure to the public, locating railroad stations to minimize the variability of delivery schedules, locating automatic teller machines to best serve the bank's customers, locating a coastal search and rescue station to minimize the maximum response time to maritime accidents, and locating a observatory stations to cover monitored area. These six problems fall under the realm of facility location research, yet they all have different objective functions. Indeed, facility location models can differ in their objective function, the distance metric applied, the number and size of the facilities to locate, and several other decision indices. Depending on the specific application, inclusion and consideration of these various indices in the problem formulation will lead to very different location models.(Neumann, 2015)

2 PROPOSED ALGORITHMS

In this section author explore some canonical methods and concepts whose relevancy in conveyance standard is very worldwide. It intend an inadequate critique of their applicability and suitableness in the framework. This disputation does not overspread all ideas that have been intimate in the sector. Some further ideas will also be contemplate as influential refute to the animadversion on these classics. (Neumann, 2017)

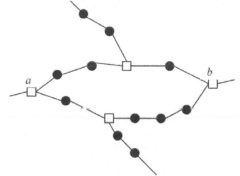

Figure 1. Face with class four; the white regularity delineate the nodes

The straight explanation technique is basically supported on penetrating the optimal footway for a swelling among all the practicable paths. Starting from a substation there may be many practicable divergent paths to gain a node. Then the minimum charge route among all the divergent paths for feeding a especial host will be the optimal route for the node. For this Dijkstra Algorithm is employment. This algorithmic program solves as a shortest-route proposition that contemplate the judgment of the minimum charge (variance) from an source to a destiny through some connection diagram. After course of nodes, weight melt is refer to find the common to compute the power detriment and stoppage damage. Henceforth the optimum passing is formed by the intend algorithm.

2.1 Classical algorithms

There are several algorithms similar Dijkstra's or Bellman algorithmic program which is a individual source- single destiny shortest passage algorithmic program, Bellman-Ford algorithmic program endeavor to unfold individual origin shortest route algorithmic program with negative weights, A* search algorithm solves separate pair shortest path problems using heuristics, Floyd Warshall algorithm and Johnson''s algorithm find all-pairs shortest passage. For the employment, Dijkstra Algorithm is found compatible. Dijkstra's Algorithm is a graph explore algorithm usage for support the shortest route from a fixed protuberance to all the other nodes in the meshwork the criteria for the explore, here is the distance between two nodes. The divergent arrangement system is always a directed route, where influence is directed from a substation to the lading nodes. A directed diagram with logical

pair D= (V, A), where V is a set of vertices or nodes and A is a prepare of arcs, addressed edges or arcs. Dijkstra algorithmic program being individual origin shortest passage is apt for divergent arrangement system which has individual origin, which is the substation from where several paths or feeders are radiated out to the load nodes at incontrovertible disagreement, which is the weight for the algorithmic program. For the Dijkstras Algorithm to employment it should be directed- weighted graph and the margin should be non-negative. Each Each arc(i, j) has a positive charge cij combined with it. If there is no arc between nodes i and swelling k then the alienation between them is boundless. This algorithm appoint every node a tag: constant or momentary. Initially all the protuberance except the origin node is appurtenance as momentary tag. The nodes which are not associated with the origin are granted eternity momentary tag. Once it is determined that the especial node pertain to minimal path its tag becomes constant.

Bellman-Ford's algorithmic program can also be employment with directed graphs with negative weights that have no negative cycles. The Bellman-Ford algorithm tolerate an effectual calculation of array paths and travel times by computation them over extending wave fronts. By segmentation the complete proposition of support the shortest paths from an origin, into smaller problems, we hold an algorithmic program that is more computationally effectual. The process originate a complete travel time and array extent to all characteristic in track, so there are no problems with darkness girdle. A counterpart implementation of the algorithmic rule should demonstrate very beneficial for three dimensional applications.

1 Start with the origin node: the stem of the tree; which is the substation in this performance.
2 Assign a charged of 0 to this node and companion it the first constant node.
3 Examine each neighbour node of the node that was the last constant node.
4 Assign a aggregated charge to each node and require it provisional.
5 Among the list of provisional nodes
 – Find the node with the smallest aggregated charge and stamp it as constant. A constant node will not be curbed ever again, its cost testimony now is conclusive.
 – If a node can be reached from more than one direction, choice the direction with the shortest aggregated cost.
6 Repeat steps 3 to 5 until every node becomes durable.

In this section there is presented a generic algorithm to compute single-source shortest distances for semirings covered by our framework. The algorithm is generic in two senses: it works with any semiring that falls within this framework and with any choice of the queue discipline. A generic algorithm for solving single-source shortest-distance problems. This algorithm is based on a generalization of the classical relaxation technique. As seen earlier, a straightforward extension of the relaxation technique would lead to an algorithm that would not work with non-idempotent semirings. To deal properly with multiplicities in the case of non-idempotent semirings, we keep track of the changes to the tentative shortest distance from s to q after the last extraction of q from the queue. In Figure 2 the pseudocode of the algorithm is presented. (Mohri, 2002)

GENERIC-SINGLE-SOURCE-SHORTEST-DISTANCE (G, s)

```
1   for  i ← 1 to |Q|
2       do  d[i] ← r[i] ← 0̄
3   d[s] ← r[s] ← 1̄
4   S ← {s}
5   while  S ≠ ∅
6       do  q ← head(S)
7           DEQUEUE(S)
8           r' ← r[q]
9           r[q] ← 0̄
10          for  each e ∈ E[q]
11              do if  d[n[e]] ≠ d[n[e]] ⊕ (r' ⊗ w[e])
12                  then  d[n[e]] ← d[n[e]] ⊕ (r' ⊗ w[e])
13                        r[n[e]] ← r[n[e]] ⊕ (r' ⊗ w[e])
14                  if  n[e] ∉ S
15                      then  ENQUEUE(S, n[e])
16  d[s] ← 1̄
```

Figure 2. Pseudocode of a generic algorithm for solving single-source shortest-distance problems

2.2 Mathematical theory of evidence and fuzzy values

So far it looked at the transportation planning problem as a static problem. Of course this is in fact not the case. Uncertainty can through events such as errors in the communication between automated guided vehicles and the system maintained stay reservations, break-down of a mobile unit (engine failure) or failures are caused (for example due to traffic accidents) in the transport network. Uncertainty can also be caused by a change in the transport requests. For example, does the arrival of a new transport request a current plan unworkable.(Boominathan and Kanchan, 2014)

Uncertainty and especially incidents can be dealt with proactive or active. Proactive methods try to create robust plans, while reactive methods of incidents actually recover they occur. A typical proactive approach is to insert limp in plans, so that, for example, delays have no consequences and new demands can be easily inserted. If nothing unexpected happens these plans take much longer than necessary. (Zutt et al., 2010)

When the assessment of the situation undergoes solely a subjective expert rating, the results are only to be obtained in form of linguistic variables. Theories presented show (Zadeh, 1975) possibility of transforming such values into figures with use of the fuzzy sets theory, a concept created by L.A. Zadeh in the sixties of the 20th century and developed ever since (mainly by its author), which increasingly intercedes in various economic issues. According to Zadeh, the aforementioned theory has not been sufficiently employed for the purpose of detection analysis of marine units. A more extensive use of possibilities offered by the fuzzy sets theory appears as a necessity for rational construction of new maritime traffic monitoring systems. (Neumann, 2011)

The mathematical theory of evidence deals with function combining information contained in two sets of assignments, subjective expert ratings. This process may be interpreted as a knowledge update. Combining sets results in forming of new subsets of possible hypotheses with new values characterising probability of specific options occurrence. The aforementioned process may continue as long as provided with new propositions. This function is known as Dempster's rule of combination.

A fuzzy nature can be attributed to events which may be interpreted in fuzzy manner, for instance, inaccurate evaluations of precisely specified distances to any point. Subjective evaluations in categories: near, far, very far may be expressed with fuzzy sets defined by expert opinions. Such understanding of fuzzy events is natural and common. Introduction of events described by fuzzy sets moderates the manner in which the results of processing are used, expands the versatility of such approach, as well as changes the mode of perceiving the overall combining procedure. Deduction of specific events involved in the process of combining pales into insignificance, as obtaining information on related hypotheses is of greater interest. Combining evidence of fuzzy values brings new quality into knowledge acquisition due to the usage of combination results as a data base capable of answering various questions. After combing many fuzzy distances, the results allow to set the support level for the veracity of statement claiming a distance between a vessel and a barrier is very close, safe or yet another. Other possibilities of the mathematical theory of evidence in problems of navigation can be found in (Filipowicz, 2010).

2.3 Possibility theory

To allow decision making based on fuzzy numbers there is a need for a system that will allow ranking of fuzzy numbers. There are several such systems however most of them consider only one point of view on the problem (Dubois and Prade, 1983). The complete set of ranking indices in the framework of possibility theory was proposed in (Dubois and Prade, 1983). This ranking system uses possibility and necessity measures to determine relation of two fuzzy numbers. (Caha and Dvorsky, 2015)

Utilization of possibility theory allows also semantically describe fuzzy numbers as possibility distributions (Zadeh, 1975). This semantic than help us explaining what such fuzzy numbers mean. The values with membership value 1 are believed to be absolutely possible or unsurprising, thus they should cove the most likely result. With decreasing degree of membership the possibility of obtaining given result decreases and the surprise rises. When membership value reaches 0 then such result is impossible (or almost impossible at some cases) and the surprise that such result would present is maximal. Such semantics helps with practical explanation what the results truly mean. (Neumann, 2016b)

To asses position of fuzzy number $\tilde{X}$ to the fuzzy number $\tilde{Y}$ four indices are needed (Dubois and Prade, 1983). Two of them define possibility and necessity that $\tilde{X}$ is at least equal or greater than $\tilde{Y}$:

$$\Pi_{\tilde{X}}\left(\left[\tilde{Y},\infty\right)\right) = \sup_x \min\left(\mu_{\tilde{X}}(x), \sup_{y \le x}\mu_{\tilde{Y}}(y)\right) \qquad (2)$$

$$N_{\tilde{X}}\left(\left[\tilde{Y},\infty\right)\right) = \inf_x \max\left(1 - \mu_{\tilde{X}}(x), \sup_{y \le x}\mu_{\tilde{Y}}(y)\right) \qquad (3)$$

The other two determine if $\tilde{X}$ is strictly greater than $\tilde{Y}$:

$$\Pi_{\tilde{X}}\left(\left[\tilde{Y},\infty\right)\right) = \sup_x \min\left(\mu_{\tilde{X}}(x), \inf_{y \ge x} 1 - \mu_{\tilde{Y}}(y)\right) \qquad (4)$$

$$N_{\tilde{X}}\left(\left[\tilde{Y},\infty\right)\right) = \inf_x \max\left(1 - \mu_{\tilde{X}}(x), \inf_{y \ge x} 1 - \mu_{\tilde{Y}}(y)\right) \qquad (5)$$

Together these indices allow comparison of any two fuzzy numbers, based on pairwise comparison any set of fuzzy numbers can be sorted.

For both set of indices there are four situations of the combinations of possibility and necessity that can be outcome of the calculation. In this paragraph both relations – at least equal or greater, and strictly greater – are referred as relation, because the descriptions are valid for both pairs of indices. The first situation is when $\Pi_{\tilde{X}}\left(\left[\tilde{Y},\infty\right)\right) = N_{\tilde{X}}\left(\left[\tilde{Y},\infty\right)\right) = 0$, which means that $\tilde{X}$ is definitely does not fulfil the given relation to $\tilde{Y}$. Then there is opposite situation in which $\tilde{X}$ completely satisfy the relation. The other two relations contains some uncertainty, because they indicate certain results but they cannot provide them absolutely. The first of those is situation when $\Pi_{\tilde{X}}\left(\left[\tilde{Y},\infty\right)\right) > 0$ and $N_{\tilde{X}}\left(\left[\tilde{Y},\infty\right)\right) = 0$. This means that there is possibility that X might satisfy the relation, but it is not necessary. Obviously that means that the indicators are not strong. The last possible combination of values is $\Pi_{\tilde{X}}\left(\left[\tilde{Y},\infty\right)\right) = 1$ and $N_{\tilde{X}}\left(\left[\tilde{Y},\infty\right)\right) > 0$. In such case again it the relation is not satisfied absolutely but the indicators are much stronger than in previous case (Caha and Dvorsky, 2015).

3 CONCLUSIONS

In the contrivance and sketch shape of dispersion, the scope to procure ideal encourager passing and to diminish the absolute annual expense, which hold the controlling restoration, energy detriment and undelivered power charge is accomplish in this work. The proposed algorithmic program, Dijkstra Algorithm is proved to be effectual for support the minimum charge passing from the substation to the request side. The computational effectiveness and expedition of Backward and Forward load melt in arrangement system is relatively useful comparison to the canonical methods. Due to the plainness in the load flow, it is extensively usage. From the standard proceed on 25 nodes system, it is decide that the intend algorithmic program is effectual for hold the best encourager passing and shorten the computational period. Hence the usage of Dijkstra Algorithm can be appropriate for arrangement system contrivance.

REFERENCES

Boominathan, P., and Kanchan, A. (2014). Routing Planning As An Application Of Graph Theory. International Journal of Scientific & Technology Research 3, 61–66.

Caha, J., and Dvorsky, J. (2015). Optimal path problem with possibilistic weights. In Geoinformatics for Intelligent Transportation, (Springer International Publishing), pp. 39–50.

Dubois, D., and Prade, H. (1983). Ranking Fuzzy Numbers in the Setting of Possibility Theory. Information Sciences 30, 183–224.

Filipowicz, W. (2010). Fuzzy Reasoning Algorithms for Position Fixing. Pomiary Automatyka Kontrola 12, 1491–1494.

Jha, P., and Vidyasagar, S. (2013). Dijkstra Algorithm for Feeder Routing of Radial Distribution System. IOSR Journal of Engineering 3, 01–06.

Mohri, M. (2002). Semiring Frameworks and Algorithms for Shortest-Distance Problems. Journal of Automata, Languages and Combinatorics 7, 321–350.

Neumann, T. (2011). A Simulation Environment for Modelling and Analysis of the Distribution of Shore Observatory Stations - Preliminary Results. In Transport Systems and Processes - Marine Navigation and Safety of Sea Transportation, A. Weintrit, and T. Neumann, eds. (London, UK: CRC Press/Balkema), pp. 171–176.

Neumann, T. (2015). Good choice of transit vessel route using Dempster-Shafer Theory. (Omsk: IEEE), pp. 1–4.

Neumann, T. (2016a). The Shortest Path Problem with Uncertain Information in Transport Networks. In Challenge of Transport Telematics, J. Mikulski, ed. (Katowice-Ustroń: Springer International Publishing), pp. 475–486.

Neumann, T. (2016b). Routing Planning As An Application Of Graph Theory with Fuzzy Logic. TransNav, the International Journal on Marine Navigation and Safety of Sea Transportation 10, 661–664.

Neumann, T. (2017). Automotive and Telematics Transportation Systems. (Astana, Kazakhstan: IEEE), pp. 1–4.

Weintrit, A. (2013). Technical Infrastructure to Support Seamless Information Exchange in e-Navigation. In

Activities of Transport Telematics, J. Mikulski, ed. (Katowice-Ustroń: Springer International Publishing), pp. 188–189.

Weintrit, A., and Kopacz, P. (2012). Computational Algorithms Implemented in Marine Navigation Electronic Systems. In Telematics in the Transport Environment, J. Mikulski, ed. (Katowice-Ustroń: Springer International Publishing), pp. 148–158.

Zadeh, L.A. (1975). The concept of a linguistic variable and its application to approximate reasoning. Information Sciences 8, 199–249.

Zutt, J., van Gemund, A., de Weerdt, M., and Witteveen, C. (2010). Dealing with Uncertainty in Operational Transport Planning. Intelligent Infrastructures, Intelligent Systems, Control and Automation: Science and Engineering 42.

Fishery

Logistical Approach to a Fishing-Industrial Complex Functioning

S.S. Moyseenko & L.E. Meyler
Baltic Fishing Fleet State Academy of the Kaliningrad State Technical University, Kaliningrad, Russia

ABSTRACT: The paper describes a concept of the logistical approach to a fishing-industrial complex functioning and it is grounded the necessity of its improving. The concept is determined by the specifics of such basic processes as: fishery, production of fish-products directly on fishing vessels, the transport service of the fishing fleet, ensuring the safety of navigation and fishery, processing, storage of raw materials and finished products, their delivery and selling on the domestic and foreign markets. The main tasks, goal and features of management of logistics processes in the fishing-industrial complexes are formulated. It is considered the structural elements of the fishing industry which have both internal logistic communication and with the external environment: netting manufacturers, shipbuilding production, product markets, etc. The paper considers a scheme of the logistic chain of a fishing-technological cycle of a fishing vessel and a method of a choice of the optimal strategy of fishing vessels relocation under conditions of uncertainty.

1 CONCEPT OF THE LOGISTICAL APPROACH TO FUNCTIONING THE FISHING-INDUSTRIAL COMPLEX

1.1 *The statement of the problem*

There are many areas of the logistics such as transportation, warehousing, commercial, information, purchasing, manufacturing, distribution logistics, etc. Each area or a kind of the logistics has its own features which are determined by the specifics of activity and the nature of the processes that implement these activities. The logistical approach to a fishing-industrial complex functioning can be formulated as the "Fishing-industrial logistics". Generally a fishing-industrial complex can be considered as a complicated system. Such elements of its structure as fishing companies; fishing and transport fleet; fishing gear factories; fish processing enterprises; supply companies, scientific and educational institutions can be elements of its structure.

Only the structural elements of a fishing-industrial complex having both internal logistics links and links to the external environment are considered in the paper.

1.2 *What is the fishing-industrial logistics?*

Nettings manufacturers, shipbuilding production, product markets, etc. are understood here as the external environment.

The commercial fishing should be considered as an actually basic process that includes catching fish and other bio-resources of the World Ocean. In its turn, this basic process is determined by the specificity of such processes as: designing and manufacturing fishing gear, the construction of fishing vessels, the organization of fishing expeditions, searching fish schools, the fish catching proper, manufacturing fish products directly on vessels during fishing, the transport service of the fishing fleet, a delivery of raw materials to the port, cargo handling at the port, storage, manufacturing fish products in the coastal enterprises, its subsequent delivery to the sales organizations and the actual product sales, etc.

The whole set of the above processes can be defined as the fishing-industrial logistics. Also the fishing-industrial logistics is an integrated formation using methods and knowledge of various kinds of the logistics.

1.3 The main task, goal and features of the fishing-industrial logistics

The primary task and the function of the fishing-industrial logistics is the rational organization of interaction of all these processes.

The goal of the fishing logistics - finding optimal/rational decisions in the fishing industry is determined by the known rules of logistics taking into account the specific character of the activity.

The set of above mentioned processes forms in total the space-time process model. It is necessary to solve many problems related to the areas of warehousing, production, transport, commercial, information, distribution and other kinds of logistics to optimize this process model. Thus, the fishing-industrial logistics is an integrated system formation where methods and knowledge of various kinds of logistics are used. All components of the system function as a unit in order to achieve certain goals.

The main features of this system are:
- integrity, that is the feature focused on the implementation of the objective function as a whole by the logistic system but not its separate elements, i.e. supplying the market with fish products;
- structuredness, that is characterized with the presence of definite links and relations between elements of the system and their distribution on hierarchy levels, i.e. beginning from a fishing vessel to the management of the fishing industry;
- adaptability, that is defined as the ability of the system to implement effectively the given functions within the definite range of changed conditions of catching, production, transportation, warehousing, marketing, etc.

The logistic chain of the fishing-industrial logistics is a set of elements that are linearly ordered on technological, material, informational and financial processes. Logistic chains in the fishing industry may include a lot of fishing companies with their fleet and fish processing enterprises, transport means using for transportation of raw materials or fish products to ports and to consumers, etc.

In particular, the logistic chain of a fishing-technological cycle of a fishing vessel consists of such processes as: searching for schools of fish, catching (trawling, casting a seine net, setting gill nets, etc.), processing the catch (production of finished products and/or semi-finished products, preparation of raw materials for loading on a mother ship or a transport refrigerating ship), warehousing, etc. The generalized scheme of such supply chain is shown in Figure 1.

Level of physical processes

searching for objects of catching	catching	processing the catch	warehousing	loading the production on the transport ship

Level of management processes

management of fish catching	management of fish production	management of a vessel unloading

Figure 1. Generalized logistic chain of the technological cycle of a fishing vessel.

2 NECESSITY OF IMPROVING THE LOGISTICAL APPROACH TO FUNCTIONING THE FISHING-INDUSTRIAL COMPLEX

2.1 Areas of using logistic research tools in fishery

It is known that application of the logistic approach to management in many spheres of human activity gives positive results (Hameri & Palsson 2003, Jensen et al. 2009, Islam & Habib 2013). The use of logistics at the macro level allows to reduce inventory levels by approximately 30% and time to transfer products from the manufacturer to the customer by 25-45%. The establishment of logistic systems in economy gives reducing the overall cost of production at 12-35%, including the reduction of transportation costs up to 20% in the developed countries of Europe (Papavassiliou 2008, Gagalyuk et al. 2010).

There was the experience of using logistic research tools in fishery, such as the systems theory, operations research (linear and dynamic programming, scheduling theory, inventory management theory, game theory, graph theory), simulation methods, etc. carried out by Russian scientists (Teplitskiy & Sheinis 1975, Moiseenko 1988, Nesterenko 2012).

Results of the research in the field of optimization of fishery and fish production, organization of the fish searching and fishing fleet allocation, its transport servicing at the fishing grounds have shown that the effectiveness of these processes, ceteris paribus, will increase by 15-25%. It can be achieved due to the optimal planning and management of the fleet and processing enterprises (Moiseenko 1988, 2009, Moyseenko & Meyler 2011, Jang 2010).

Nevertheless, it can be stated that there is no a common logistic approach to the fishing-industrial complex operation so far, mostly at the regional level. There are problems to be solved for the development of logistics in the fishing industry, in particular: the imperfect system of fishing logistics standardization; low levels of the logistic information system and the service with the added value; a lack of participants of logistic processes training (Moyseenko 2009, Wang 2012).

However, the use of the mathematical methods only without corresponding organizational measures cannot give the desired effect. For example, plans for products manufacturing and its transportation calculated using optimization logistic models will fail very soon if the concrete project of these plans implementation and an effective system of operational management and traffic control were not developed at the same time. Thus, the rational combination of organizational and analytical components of the logistic is a main condition for improving the efficiency of a fishing-industrial complex.

Many companies working in the fishing industry use the supply chain management in order to improve an organizational efficiency of the company, the best use of resources and their competitiveness enhance. Mathematical models of supply chains in the fishing industry are used to achieve this purpose. Analyzing the recent publications (Hameri & Palsson 2003, Papavassiliou 2008, Jensen et al. 2009, Asche & Smith 2010, Jang 2010, Gagalyuk et al. 2010, Moiseenko et al. 2013), it can be concluded that regional problems of supply chain management and logistics of interaction between companies, suppliers, customers and consumers are solved in most cases in the field of fishery.

2.2 Logistical approach to management in fishing - industrial systems

It is known (Mirotin 2005, Bowersox et al. 2009) that the logistical approach to the management is primarily a systematic approach to designing transport, technological and logistical systems on the principles of an equilibration between all the elements/subsystems, integrity and optimality. Management decisions should be aimed for achieving the objectives of the system which is achieved mainly by optimizing the organization of all processes.

The efficiency of the system depends on the interrelated functioning of all its elements, interaction with the external environment, minimizing costs, the level of operational and strategic management. Thus, the efficiency of the fishing-industrial system is ensured, ceteris paribus, through the introduction of the logistical approach to the management of fishing, transport and service processes.

Logistics of the fishing industry acts as a set of techniques and methods of the management of catching bio-resources of the ocean and optimization of material - physical flows (seafood products) in

time and space. It is necessary to consider the most effective methodological approaches to the formation of fishing-industrial logistical systems.

The practical implementation of the logistical approach to the management of the fishing-industrial system is based on the concept of the common costs management and the theory of compromises. It means that costs in each stage of the product distribution are considered together, i.e. carried out simultaneously. Reducing costs in one area can affect on the costs of other areas. It is required to find such a compromise of costs distribution when the total costs will be minimal. The costs on the logistics in the fishing industry can be divided into two groups:
1 the costs generated in the area of bio-resources catching and seafood production;
2 the costs in the sphere of circulation.

The introduction of the logistical approach to the management in the fishing industry is primarily affected favorably on the sphere of circulation (transportation and storage of products). According to expert estimations (Mirotin 2005), the application of logistics allows to reduce by 30-50% the inventory level and to shorten by 25-45% time of production movement. Thus, it allows to reallocate costs for the operation of goods circulation in order to achieve the maximum economic benefit.

The logistic approach to the management of fishing and transport processes have been reported by authors earlier, in particular, as the example of planning the route of the goods delivery according to the chosen efficiency criterion (minimizing the length of the route) (Moiseenko et al. 2011). Other examples illustrating the effectiveness of the logistical approach to accepting management decisions was presented as the choice of a fishing ground under different usefulness criteria and selection of an optimal solution for vessels relocation (Moyseenko & Meyler 2011a), optimizing the transport service and risk assessment of fishing vessels at fishing grounds (Moyseenko & Meyler 2011b, 2015, Moyseenko et al. 2015).

2.3 Current state of the fishing industry in Russia

As to the current state in the fishing industry in Russia, it should be noted that the main problem is to improve the efficiency and to develop this sector of the economy. There is a high degree of depreciation of fixed assets (fishing fleet, fish processing enterprises and the infrastructure) as well as the relatively low level of organization and management of the commercial fishery. Russia has now many small fishing companies, but the system of cooperation both among them and with authorities of the fishing industry is poorly developed.

The current financial and economic crisis in Russia has led to the decline in the number of operations in the field of the long term fish searching and forecasting fishing grounds in the World Ocean, designing new fishing gear. The same can be said about the marketing of fishing products, organization of the transport service and ensuring the safety of the fleet at the fishing grounds, etc. Also, there is a slow renewal of the fishing fleet, technical and technological modernization of fishing ports. These reasons cause stagnation in the fishing industry of the country. However, as the analysis of the situation shows, there are real opportunities to improve the efficiency of the industry due to the development of logistical methods and their application in the practice of management. An analysis of the described above logistic problems in the fishing industry shows that it is appropriate to put into practice the concept of the regional fishing-industrial logistic system. Such system should include subsystems of searching, catching, production, transportation, cargo handling, financial-economic and legal supporting, marketing and monitoring, organization and planning, quality control and safety, training, etc. Various logistics systems specializing in deep-water and coastal fisheries, their scientific supporting and designing, etc. can be developed within the framework of the fishing-industrial regional logistic system. These systems are designed to ensure a balance of activities in the field of the fishing industry. They are closely interrelated and aimed on improving the efficiency of management processes of the fishery and increasing the efficiency of the fishing industry as a whole.

It should be noted that the solution of many mentioned above problems is stipulated by the State Programme of the Russian Federation "Development of the fishing-industrial complex". Its realization was planned for the period 2013 – 2020.

3 CHOOSING THE OPTIMAL STRATEGY OF THE FISHING VESSELS RELOCATION UNDER CONDITIONS OF UNCERTAINTY

3.1 Formulation of the problem

In practice decisions are quite often made under conditions of the limited information, i.e. in conditions of uncertainty. The main feature of the problem is that the decision maker doesn't know the real state of the environment which can be in one of the states A_1, ..., A_i. Depending on the degree of awareness the decision maker can only express subjective assumptions concerning the state of the environment based sometimes on the past experience. To develop a reasonable decision under conditions of uncertainty the decision maker can using the logistical approach, in particular, criteria

of optimality (Wald, Hurwitz, Savage) known from the theory of operations research (Ackoff & Sasieni 1971).

Let us illustrate the process of decision making under conditions of uncertainty on a concrete practical example. The Head of the fishing company (the decision maker) plans to send fishing vessels in a poorly studied fishing ground. But, according to the output of the long term fish searching it is possible to catch fish schools with purse seines.

Depending on the state of the environment A_i the numbers of successful castings of net Z_k are equal to 0, 5, 10, 15, 20, 30 units. The amount of the daily norm of the catch is equal to $u = 10$ t.

The decision maker can accept some of the "coefficients of optimism" in the range $\alpha = 0.1 \div 0.9$ concerning to real conditions of the environment. These coefficients are determined by expert assessments. The possible number of the fleet is equal to $X_i = 1, 5, 10, 15, 20, 30$ vessels.

It is necessary to find a decision: how many vessels are most reasonable to send to the fishing ground?

According to the given task the possible states of the environment A_i are identified with the number of successful castings of net Z_k and the possible strategies of the decision maker are the amount of vessels X_i sent to the fishing ground.

3.2 Decision making

Let us assume that one possible outcome $P(A_i/Z_k) = \sigma_{ik}$ corresponds to each of the state of the environment.

$$\sigma_{ik} = \begin{cases} 1, \text{if } i = k \\ 0, \text{if } i \neq k \end{cases}$$

In this case the mathematical model of decision-making is determined by a variety of strategies $X = \{X_i\}$, a variety of environmental states $A_i = \{Z_k\}$ and the initial matrix (Table 1).

Elements of the matrix are a usefulness of the result (the value of the catch) Z_k when the strategy (a number of vessels) X_i is used.

$$G_{ik} = f(Z_k \, X_i); \quad k = 1,...,L; \quad i = 1,...,m$$

The set $\{P(A_i)\}$ is unknown.

Table 1. Decision (initial) matrix.

X_i	A_i			
	Z_1	Z_2	...	Z_K
X_1	q_{11}	q_{12}	...	q_{1K}
...	...	...	...	...
X_m	q_{m_1}	q_{m_2}	...	q_{mK}

Let us suppose that the fishing vessel may make no more than one casting of net in a day. Then the elements of the matrix are determined from expressions:

$$G_{ik} = \begin{cases} Z_k u, & \text{if } x_i \le Z_k \\ Z_k u - x_i u, & \text{if } x_i > Z_k \end{cases} \tag{1}$$

$$G_{ik} = u x_i \tag{2}$$

The restriction (2) supposes that the total catch of all vessels exceeds the planned catch when a vessel makes no more than one casting per a day. Using expressions (1) and (2) it is possible to calculate elements of the matrix of catches per a day of fishing (Table 2).

Negative values in the matrix of daily catches correspond to underfishing, i. e. the departure of the actual catch from the planning one. Recurring elements of the matrix in one line are caused to the restriction (2).

Table 2. Elements of the catches matrix (t/day)

X_i	Z_k					
	0	5	10	15	20	30
1	-10	10	10	10	10	10
5	-50	50	50	50	50	50
10	-100	-50	100	100	100	100
15	-150	-100	-50	150	150	150
20	-200	-150	-100	-50	200	200
30	-300	-250	-200	-150	-100	300

The decision maker has to choose an optimality criterion depending on the awareness of the state of the environment, the concrete objectives and specificity of fishing. It is certainly the most important stage in the decision-making process. Finding the optimal solution of the problem is shown below.

3.2.1 Using the Wald criterion

The Wald criterion gives a guaranteed gain under the worst state of the environment. The optimal strategy is determined by the decision rule:

$$\begin{array}{c} \max \min \\ X_i \ Z_k \end{array} f(X_i, Z_k) \tag{3}$$

The optimal strategy $X_{opt} = 1$ vessel is found from the initial matrix (Table 1) according to the Wald criterion. Thus, if the risk is unacceptable, the decision maker will send a vessel to a poorly studied fishing ground. The main purpose of the vessel activity will be further investigation of the raw material base at the fishing ground.

3.2.2 Using the Hurwitz criterion

The Hurwitz criterion is based on the assumption that the environment can be in the most unfavorable state with the probability $(1 - \alpha)$ or in the most favorable state with the probability α, where α is the coefficient of optimism. The decisive rule for finding the optimal strategy according to this criterion has the form:

$$X_{opt} = \max \left[\alpha \max_k G_{ik} + (1-\alpha) \min_k G_{ik} \right] \tag{4}$$

To realize the Hurwitz criterion the catches matrix (Table 3) is calculated for the different coefficients of optimism α using the initial matrix (Table 1) and the decisive rule (4).

$$H = \|h_{i\alpha}\|,$$
$$h_{i\alpha} = \left[\alpha \max G_{ik} + (1-\alpha) \min G_{ik} \right] \tag{5}$$

Table 4 shows the optimum number of vessels depending on the coefficient of optimism calculated from expression (4) and the matrix (Table 2).

Table 3. Catches matrix for the different coefficients of optimism

X_i	α			
	0.3	0.4	0.6	0.7
1	-4	-2	2	4
5	-20	-10	10	20
10	-40	-20	20	40
15	-60	-30	30	60
20	-80	-40	40	80
25	-100	-50	50	100
30	-120	-60	60	120

Table 4. Optimum number of vessels for different coefficients of optimism

α	0.3	0.4	0.6	0.7
X_{opt}	1	1	30	30

Let us suppose that the control castings of the vessel at the poorly studied fishing ground as well as results of the hydrological reserves allow to adopt the coefficients of optimism $\alpha = 0.6 \div 0.7$. In this case, the optimal decision is to send 30 vessels to the fishing ground.

3.2.3 Using the Savage criterion

The Savage minimax regret criterion assumes that the state of the environment is worst differed from the expected one. The decisive rule has of the form:

$$\max_{X_i} \min_{Z_k} f_r(X_i, Z_k) \tag{6}$$

The regrets function:

$$f_r(X_i, Z_k) = f(X_i, Z_k) - \max_{X_i} f(X_i, Z_k) \tag{7}$$

The regrets matrix (Table 5) is calculated from the initial matrix (Table 1) using the formula (7)

Table 5. Regrets matrix

X_i	Z_k					
	0	5	10	15	20	30
1	0	-40	-90	-140	-190	-290
5	-40	0	-50	-100	-150	-250
10	-90	-100	0	-50	-100	-200
15	-140	-150	-150	0	-50	-150
20	-190	-200	-200	-200	0	-100
25	-240	-250	-250	-250	-250	-50
30	-290	-300	-300	-300	-300	0

Using the regrets matrix and the decisive rule (6) it is possible to find:

$$\max_{X_i} \min G_{ik}(r) =$$
$$= \max_{X_i}\{-290,-250,-200,-150,-200,-250,-300\} = -150$$

$X_{opt} = 15$ vessels.

Let us suppose that there are positive results of the control castings at the concerned fishing ground. Based on the need to develop new fishing grounds the decision maker considers a possibility to permit a certain risk. In this case, according the Savage criterion it is advisable to send 15 vessels to the fishing ground.

4 CONCLUSIONS

The paper gives the concept of the logistical approach to functioning the fishing- industrial complex and definition of the "Fishing-industrial logistics". The basic logistical objects of the complex, relationships between them, the main tasks of functioning the complex and logistical management of catching and production processes, transport and storage of raw materials are defined.

The necessity of improving the efficiency of the fishing-industrial complex on the basis of the systematic logistical approach, in particular, taking into account the current state of the fishing industry in Russia.

Methods of decision-making of the fishing vessels relocation under conditions of the limited information, i.e. conditions of uncertainty are proposed as an example of the logistical approach.

REFERENCES

Ackoff, R. & Sasieni, M. 1971. *Fundamentals of Operation Research*. Moscow: Mir (in Russian)

Asche, F. & Smith, M. 2010. Trade and Fisheries: Key issues for the World Trade Organization. *World Trade Organization Economic Research and Statistics Division Staff Working Paper ERSD-2010-03*.

Bowersox, D. Closs, D. & Cooper M. 2009. *Supply chain logistics management*. New-York: McGraw-Hill.

Gagalyuk, T, Hanf, J & Herzlieb, C. 2010. Managing supply chains successfully: An empirical testing of success of supply chain networks in the German fish sector: *Acta Agriculturae Scandinavica, Section C - Food Economics. Special Issue: Food Industry and Food Chains in a Challenging World*. 7(2-4): 139-150.

Hameri, A. & Palsson, J. 2003. Supply chain management in the fishing industry: the case of Iceland. Int. *J. Logistics Res. Appl: A Lead. J. Supply Chain Man*. 6(3): 137-149.

Islam, S.B & Habib, M.M. 2013. Supply chain management in fishing industry: A case study. *Int. J. Supply Chain Manag*. 2(2): 40-49.

Jang, Y. 2010. Problems and countermeasures of logistics in the marine fisheries industry. *Asian Agric. Res*. 2(6): 34-36.

Jensen, T., Nielsen, J., Larsen. E. et al. 2009. The fishing industry - toward supply chain modelling. *Kongens Lyngby (Denmark): Technical University of Denmark, Department of Management Engineering*.

Mirotin, L. *Transport logistics* 2005. Moscow: Examen.

Moiseenko, S. 1988. *Methods of the optimal management of the fishery*. Kaliningrad: BFFSA Publ. House. (in Russian)

Moyseenko, S. & Meyler, L. 2011a. Optimal management of fleet relocation at deep-sea fishing grounds. In: B. Lachynski (ed.) *Green ships, Eco shipping, Clean seas; Proc. 12-th Annual General Assembly of the Int. Assoc. of Maritime Universities (IAMU) AGA12. 12-14 June 2011*, Gdynia, Poland.

Moyseenko, S. & Meyler, L. 2011b. *Safety of marine cargo transportation*. Kaliningrad: BFFSA Publ. House (in Russian).

Moiseenko, S., Meyler, L. & Semenkov V. 2011. Organization of fishing fleet transport service at ocean fishing grounds. In: T. Blecker, W. Kersten & C. Jahn (eds). *Maritime Logistics and International Supply Chain Management*. Proc; 2011, 08-09 September; Hamburg, Germany: 193-203.

Moiseenko, S., Meyler, L.& Fursa S. 2013. Methods and models to optimize functioning of transport and industrial cluster in the Kaliningrad region. In: A. Weintrit &T. Neumann (eds), *STCW, Maritime Education and Training, Human Resources and Crew Manning, Maritime Policy, Logistics and Economic Matters - Marine Navigation and Safety of Sea Transportation*, CRC Press/ Balkema, Taylor & Francis Group, London: 225 – 232.

Moyseenko, S. & Meyler, L. 2015. Optimization of the transport service of fishing vessels at ocean fishing grounds. In: A. Weintrit & T. Neumann (eds), *Safety of Marine Transport - Marine navigation and safety of sea transportation.*, CRC Press/ Balkema, Taylor & Francis Group, London: 293 -296.

Moyseenko, S. 2009. *Management of the fleet operation*. Kaliningrad: BFFSA Publ. House (in Russian)

Moyseenko, S., Meyler, L. & Bondarev, V. 2015. Risk assessment for fishing vessels at fishing grounds. *TransNav, the International Journal on Marine Navigation and Safety on Sea Transportation*. 9(3), September 2015: 351 – 355.

Nesterenko, N. 2012. Current state of logistic management of material flows in the fishing industry of the Kamchatka krai. *Rus. Entrepren*. 9(207): 152-156. (in Russian)

Papavassiliou, N. 2008. Fishery products distribution and logistics in Greece. *J. Food Prod. Mark*. 15(1): 38-63.

Teplitskiy, V & Sheinis, L. 1975. *Optimization of planning in the fishing industry*. Moscow: Food Industry. (in Russian)

Wang, Y. 2012. Logistics coordination management mechanisms. *Adv. Eng. Forum*. 6-7: 778-782.

The Effects of Burnout Level on Job Satisfaction: an Application on Fishermen

M. Buber & A.C. Töz
Dokuz Eylul University, Izmir, Turkey

ABSTRACT: Production in fishery industry makes substantial contribution to employment and so it plays an important role in Turkish economy. Obviously, fishing profession is so difficult due to such as prevalent risk of unemployment, low income, dangerous job conditions, mental stress, heavy weather conditions, self-reported stress levels and so on. The aim of this study is to investigate the effects of burnout level on job satisfaction of fishermen and evaluate the burnout and job satisfaction levels of fishermen. A questionnaire which is about burnout and job satisfaction has been adapted for fisheries sector and conducted by means of face to face interviews engaged in Bay of Izmir. The purpose of the conduction is to reveal the effects of burnout level on job satisfaction among 41 fishermen in five sub-regions of Aegean sea. The data obtained from the surveys were transferred to the IBM SPSS (Statistical Package for Social Science) V-22 program, and significance tests of the difference between the means of the groups and necessary descriptive statistics (frequency) and multi-dimensional linear regression method were performed to reach the aims of the study.

1 INTRODUCTION

There is no better expressed description anywhere than by the International Maritime Health Association when it says *"It has been established that seafaring is one of the most physically demanding professions in one of the most dangerous work environments: the sea."* (IMHA, 2012; Iversen, 2012; Andruskiene et al., 2016).

Many of the factors such as long working hours, sleep disturbance and disturbed sleep rhythm, harsh working conditions such as noise, ship motions, mental stress, bad weather condition associated with fatigue (Hovdanum et.al, 2014). Even when things are going right, mental and physical fatigue are unavoidable and frequently lead to injury (Bhondve et al., 2013).Thus, fatigue and reduced performance give rise to environmental damage, mental health problems, reduced life span, job dissatisfaction and also burnout (Smith et al., 2006).

Despite the growing awareness of the negative effects of fatigue in the transportation sector, burnout and job satisfaction at sea particularly fishery, has been much less studied than other industries (Hovdanum et.al, 2014). This study has been carried out to investigate the effects of burnout level on job satisfaction of fishermen and evaluate the burnout and job satisfaction levels of fishermen.

2 LITERATURE REVIEW

2.1 *Job satisfaction of seafarers*

According to Spector (1997:2), job satisfaction is the extent to which people like (satisfaction) or dislike (dissatisfaction) their job. Job dissatisfaction brings about lack of loyalty, increased absenteeism, increased number of accidents, and lead to negative consequences and so on.

Social scientists have developed various scales to measure job satisfaction, and some have become more popular than others. The studies involving job satisfaction of fishermen have especially been based on Maslow's hierarchy of needs (Bavinck and Monnereau, 2007; Bavinck et al., 2012), including three basic categories: basic need, social need, self actualization.

There are few job satisfaction studies about fisheries sector. Pollnac and Poggie (1988) were the first to emphasize on job satisfaction in fisheries sector. Their research had a questionnaire consisting of 22 items referred to Maslow's hierarchy of needs. Then they added two questions. The first question is about to whether fishermen would still go into fishing if he or she had his/her life to live over again; the second question is about to whether or not he or she would advise a young person to go into fishing. Pollnac et al. (1975) found that characteristics of fishermen's job (e.g., uncertainty of catch and periodicity of income) were especially affected by their attitudes toward financial planning and perception of the future. Bavinck (2012) who explored job satisfaction among shrimp trawl fisheries of Chennai, India focused on general satisfaction of fishermen. However over three-fifths of fishermen said they would be willing to change and crew members on shrimp trawlers expressed dissatisfaction with state of the stock and performance of management. A study done by Pollnac et al. (2001) randomly had chosen sample of Philippines fishermen. Their findings confirmed the hypothesis that fishers like their occupation and only a minority would change to another occupation. It is a significant factor that they are satisfied with job because of not having a boss, but beauty of the sea, daily income, easy source of food, tradition of their family job.

In summary, previous research has revealed that job satisfaction is a significant factor related with variety of social and economic variables. In other words, higher job satisfaction level correlated with positive social and economic variables, Lower job satisfaction with negative impacts of them.

2.2 Burnout of seafarers

Burnout is defined to become exhausted; loss of creativity; loss of commitment to work; disaffection from clients, co-workers, job or agency; a response to the chronic stress of making it to the top (Cordes and Dougherty, 1993). Another explanation of burnout is a syndrome of emotional exhaustion and cynicism that occurs frequently among individuals who do "people-work" of some kind. As their emotional resources are consumed, workers feel exhausted and dissatisfied with their accomplishment of their job (Maslach and Jackson, 1981).

The most commonly accepted burnout inventory has three components. Firstly, *"emotional exhaustion"* is characterized by lack of energy and feel emotional resources depleted. Second of all, *"depersonalization"* is expressed by the treatment of clients as objects rather than people. The last component is *"diminished personal accomplishment"* is characterized by a tendency to evaluate oneself negatively (Maslach and Jackson, 1981; Cordes and Dougherty, 1993). Higher scores in the dimensions of emotional exhaustion and depersonalization and scores lower in the personal accomplishment dimension are associated with higher levels of burnout.

There are a few researches on burnout syndrome for seafarers. Oldenburg et.al (2013) investigated the respective indicator about burnout syndrome for seafarers. Study revealed that the seafarers who stayed on ships for a long time far away from their familiar social contacts ashore are limited. Therefore, it is assumed that seafarers are at high risk of experiencing burnout. Sliskovic and Penezic (2015) studied work related stressor in seafaring population. Work related stressor of seafaring included typical sources of work stress lead to such as insomnia, anxiety, anger, loss of control, heart problem and also burnout.

Considering the studies reviewed, Burnout and job dissatisfaction bring about negative influences on health and wellbeing.

3 MATERIALS AND METHOD

3.1 Aim of the study

Fishing is a dangerous profession (Allen et al., 2010; Sliskovic and Penezic, 2015). Obviously, fishing profession is as difficult due to such handicaps as unemployment, low income, dangerous job conditions, mental stress among fishermen, harsh weather conditions, psychological sense of mastery, self-reported stress levels and so on. The research questions assessed in this article are as follows; (1) Is there a relationship between aspects of burnout and perceptions of job satisfaction? (2) the extent to which changes in aspects of burnout and job satisfaction among fishermen according to demographic characteristic of fishermen (3) the extent to how many percent of the effect of burnout level on job satisfaction is affected by the regression analysis. The aim of this study is to investigate the effects of burnout level on job satisfaction of fishermen. A questionnaire adapted for fisheries sector has been conducted by means of face to face interviews through the fishermen in Bay of Izmir.

3.2 Data collection and research sample

The study was carried out in Bay of Izmir by coastal fisheries. It focuses on five landing sites in Cesme, Urla, Karaburun, Guzelbahce, Foca. The main sources of making living in these areas are small scale fisheries which include traditional - small-scale fishing, coastal drift nets, coastal shores, barracudas, coastal lengthening nets, coastal purse-seines, fisheries with lagoons.

In order to find the correlation between the burnout and job satisfaction, the research design adopted for fishermen. In conducting the survey, we used the "dockside intercept" method: we went to the fishing docks to find people willing to complete an interview that lasts from thirty minutes, including filling out our 2-page questionnaire. Interviews were conducted by 41 fishermen.

The questionnaire form that was used to collect the data was divided into four parts. The first part consisted of open-ended and close-ended personal information and demographic questions. The second part consisted of 22 questions from the MBI developed by Maslach and Jackson (1981). Likert scale measuring tool was used for the MBI measurement. Respondents answer the question of how often they feel in a particular way on a 1–5 scale, where 1 indicates "never," and 5 means "every time". The scale consisted of 22 questions, each of which had five points and was composed of three dimensions: *"emotional exhaustion, depersonalization and personal accomplishment."* Third part of questionnaire is about job satisfaction which is evaluated by Pollnac (2010- 2011). This part consisted of 9 indicators which has 3 dimensions: *"basic need"* components: actual earning, predictability of earning, safety. *"Social or psychological"* need components: time away from home, physical fatigue of the job, healthfulness. *"Self actualization"* needs components: adventure of the job, challenge of the job, opportunity to be own master. Job satisfaction scale is ranging from 1 to 5 from "very dissatisfied" to "very satisfied". Additional questions which were previously used by Pollnac and other investigators (Would you advise a young person to go into fishing? and would you still fishermen if you had your life to live over?) is carried out and extra question were asked (Could you indicate the factors that affect you negative in relation to your occupation?). Responses to these questions were answered either yes or no.

3.3 Statistical Analysis and Results

IBM SPSS (Statistical Package for Social Science) V-22 program was used for statistical analysis of fishermen demographic data. Total 41 fishermen whose details are shown in Table 2 were involved in study.

There are 14 people who are involved in seine fishery, 10 hand-line fisheries, 7 people purse seine fishery and 3 long liners. 87.8% of the study group was males.

Of the total 41 fishermen, 70.7% were married, nearly 80% had at least one child and 28% fishermen had only primary education.85.4% of fishermen continued family profession who is engaged in fishing.

But the main problem is that the inability to make fishery profession of the young population puts the continuity of the family fishery in danger. According to the bulk of workforce, 87.8% belonged to more than minimum wage of the country. In terms of experience of fishing, 73.2% of fishermen had job experience of more than 10 years.

As expected, captains of fishing boat tend to be older than crew, with more experiences and earning more money. A significant point is that crew members are more educated than captains.

Table 2. Demographic structure of fishermen.

Variable	n	%
Gender		
Male	36	87.8
Female	5	12.2
Family Status		
Single	12	29.3
Married	29	70.7
Having a Child		
Yes	31	75.6
No	10	24.4
Age		
18 and younger	1	2.4
19-25	1	2.4
26-35	9	22
36-45	13	31.7
46 and older	17	41.5
Education		
Primary	5	12.2
Middle school	8	19.5
High school	19	46.3
Faculty	7	17.1
Postgraduate	2	4.9
Income		
0-1500TL	5	12.2
1501-3000TL	19	46.3
3001-4500TL	8	19.5
4501-6000TL	7	17.1
6001 +	2	4.9
Have a family member as a fishermen		
Yes	35	85.4
No	6	14.6
Rank		
Open sea skipper	4	9.8
Skipper	18	43.9
Deck boy	19	46.3
Competency		
Open sea skipper	4	9.8
Skipper	7	17.1
Deck boy	9	22
Yacht captain	2	4.9
Restricted Watch keeping officer	4	9.8
Amateur Sailor	8	19.5
Ordinary seaman	5	12.2
Able seaman	2	4.9
Total sea experience		
0-10	11	26.8
11-20	8	19.5
21-30	11	26.8
31-40	6	14.6
41 +	5	12.2
Hunting type		
Seine fishery	14	34.1
Hand line fishery	10	24.4
Purse seine fishery	7	17.1
Other seine fishery	7	17.1
Long line fishery	3	7.3
Landing site		
Çeşme	3	7.3
Foça	5	12.2
Güzelbahçe	11	26.8
Urla	17	41.5
Karaburun	5	12.2

According to regulations of seafarers' competency, only 4.9 % had yacht captain competency, 19.5 % had amateur sailor competency,

9.8% were open sea skipper, 4.9% had able seaman competency.

As per additional question, younger crew members were more willing to leave the job because of lower income, low status in society, absence of governmental rights, heavy weather conditions. Older fishermen complained about physical fatigue, sustainability of fish stocks, damage caused by trawl nets and marine pollution.

One of the research findings; the elderly fishermen answered the additional question results that they willingly prefer the profession of the fishermen, they recommended the next generation to the profession of fishermen, and they do not want to change the fishery profession to a different profession.

But, another side of the research findings is that elderly people consistently told that governmental rights are insufficient. They mostly told that landing site facilities were insufficient and also complained about the shortage of bilge and garbage storage tank and unconscious consumption of fishery stocks.

Table 3. Internal consistency and other descriptive findings of burnout.

Dimension	Item	N	Mean	Scale Variance	Std. Deviation
Emotional Exhaustion	9	41	23,073	59,37	7,7051
Depersonalization	5	41	10,731	12,90	3,5918
Personal Accomplishment	8	41	29,658	32,38	5,6903

Note. Emotional Exhaustion: 0-16 = Low, 17-26 = Moderate, >27 = High; Depersonalization: 0-6 = Low, 7-12 = Moderate, >13 = High; Personal Accomplishment: >39 = Low, 32-38 = Moderate, 0-31 = High.

Of the 22 questions, 9 assessed the respondent's level of emotional exhaustion, 8 assessed the respondent's sense of personal accomplishment and 5 questions referred to depersonalization. Higher scores in the dimensions of emotional exhaustion and depersonalization and scores lower in the personal accomplishment dimension are associated with higher levels of burnout.

This questionnaire has been validated in Turkish language and achieved the following alpha values for the scales. According to factor analysis of questionnaire, cronbach's alphas for Maslach Burnout Inventory dimensions as 0.409 emotional exhaustion, 0.562 depersonalization, 0.786 reduced personal accomplishment. Other part of the questionnaire is about job satisfaction. Cronbach's alphas of 0.52 for social need dimension, 0.573 for basic need dimension and 0.745 for self-actualization. The total consistency analysis of the items composing the MBI subscales presented a Cronbach's alpha of 0.608 (0.60< α<0.80 is reliable).

According to through the data obtained from the fishermen, when the burnout score averages were evaluated according to the interpretation table of burnout scale scores. It was seen that the sample has moderate level emotional exhaustion and depersonalization and high level of personal accomplishment. This indicates that the sample has moderate level in terms of emotional exhaustion and depersonalization dimensions and low level of burnout in terms of personal accomplishment dimension. This indicates that the sample did not feel physically and emotionally tired and being exhausted and they were not so much affected by work stress, and relationships with people in their environment were regular and humane. However, the personal accomplishment level is high, in other words, the low level of exhaustion experienced in the personal accomplishment dimension; is a sign that fishermen are satisfied with their own experience.

Table 4. Internal consistency and other descriptive findings of job satisfaction.

Dimension	Item	N	Mean	Scale Variance	Std. Deviation
Basic Need	3	41	8,9756	5,974	2,4443
Social-Pyschological	3	41	9,2683	5,651	2,3772
Self-Actualization	3	41	10,609	7,494	2,7375

Note: The three components were measured by summing responses to questions belonging to the specific components, thus resulting in scales ranging from 3 to 15, reflecting fishermen's levels of job satisfaction with each component.

The results of the job satisfaction surveys per category (see Table 4) show that the category self actualization scored highest, followed by social-psychological needs and basic needs. Mean values for all scores, fall above the midpoint of 3 indicating general satisfaction.

The basic needs relates to fishers' satisfaction to measure job safety, the actual earning and predictability of earning. In fact, the question on the respondents' satisfaction with 'job safety' is the highest scoring item of all. The scoring for the first of these three items is something not very difficult to understand taking into account their wages they earn in Turkey standards are above the minimum wage. Concerning the scoring for the "Level of Earning" this is also something easy to understand in a population of small-scale fishermen, with lower and unstable incomes as fishers depend on the size of the catches because of the ban of hunting season and heavy weather. But, fishermen are satisfied about their earning seeing that they can provide a family subsistence.

Social-psychology needs category could be related to the fact all respondents are small-scale fishers who lives in a coastal community where they take part in regular life. Respondents are satisfied with the items of "healthfulness of their job" and "the time spent away from home". The results about being away from home refer to their adventures of fishermen.

Self-Actualization needs also have satisfied scores. The score of self actualization shows fishers feel proud of their job. This relates in specific to the high score on 'worth of the Job'. In fact, the question on the respondents' satisfaction with 'adventure of the job' is the highest scoring item of all. It is shown that time spent at sea and their adventurous feeling cause them to be satisfied.

Regression analysis examines the dependence of one variable on another or more variables. Linear regression analysis which is widely used in social sciences has been used in order to be able to reveal the effect levels of perceived burnout dimension variables on job satisfaction.

For this purpose, multi-dimensional linear regression analysis was used in this part of the research. The method used to explain the cause-effect relationships between two and more independent variables affecting a variable with a linear model and to determine the effect levels of these independent variables is called multiple linear regression analysis.

Table 5. Multi-dimensional regression analysis results.

Model	R	R^2	Model Summary Adjusted R^2	Std. Error of the Estimate	Sig. F Change
1	.495ᵃ	.245	.184	5.82737	.014

a. Predictors: (Constant). Total dimensions of burnout

In order to determine the effect of fisherman burnout levels on job satisfaction level, Multiple regression analysis method was applied. As a result of the regression analysis according to the Table 5, a meaningful result (p<0.05 F: 5.83) was obtained according to the effect on the independent variable (emotional exhaustion, depersonalization, personal accomplishment dimensions) of dependent variable (job satisfaction).

R^2 is the measure of explained variance in the dependent value and indicates how much the dependent variable changes are explained by the independent variables. As a result of the regression analysis, R^2 value was 0.245. The independent variables in this research model are the percentage of comments on the dependent variable. According to this result, dependent variable indicates that 24.5% of "fisherman's job satisfaction" levels are explained by the burnout variables.

As Table coefficient of regression analysis results shows, it is significant that the p value of 1 of 3 independent variables is smaller than 0.05. The personal accomplishment sub-dimension (p <0.05 B: 0.300) has a direct effect on the level of job satisfaction of fishermen. One unit increase in personal accomplishment is increased job satisfaction with 0, 34 on the scale of job satisfaction and there is a positive and meaningful relationship between each other.

4 CONCLUSION AND DISCUSSION

Fishing is one of the oldest living resources that meet people's needs for food and livelihoods. In this profession, men and women have worked together. Today, participation in official records and working life is an unbalanced situation in favor of men in terms of gender. One of the limitations of our study is the low number of female fishermen.

In this study, it has been found that 41 fishermen has moderate level emotional exhaustion and depersonalization dimensions level and high level of personal accomplishment. As a result revealed that fishermen are exhausted but satisfied with their personal experience and success. To reduce emotional exhaustion on board, it is recommended, to improve leadership, communication and work-team management to evaluate family social contacts or to diminish stress level.

The results of the job satisfaction level show that fishermen are satisfied with their job. But an interesting point is that the responses to some additional questions reveal that younger person complains about lower income, heavy weather condition and the lack of governmental right. Younger fishermen are more tend to quit the job and they do not want advice the next generations to do this job.

For the purpose of the study according to multi-dimensional regression analysis, According to this result, dependent variable indicates that 24.5% of "fisherman's job satisfaction" levels are explained by the burnout variables. However, there are not significant relationship between emotional exhaustion, depersonalization dimensions and job satisfaction. Personal accomplishment dimension is related with positive effect on job satisfaction. This dimension affected 30% meaningful and positive effect on job satisfaction.

In conclusion, it is observed that there are numbers of factors associated with experiencing burnout and job satisfaction. To explain the relationship such as personality, mental disorder, coping styles or else. But a wide range of fishermen working at different ports should be reached and the causal relationships of the occupational burnout levels of fishermen should be investigated in more detail and close to the truth. It is expected that the result of the study provides certain outcomes and guidelines for related organizations dealing with fishing operations as well as certain proposals for effective and efficient coordination among the relevant institutions.

Table 6. Coefficient of regression analysis results

Coefficient of Regression Analysis Results	Unstandardized Coefficients		Standardized Coefficients	t	Sig.
	B	Std. Error	Beta		
1 α	27.203	5.885		4.622	.000
Emotional Exhaustion	-.125	.144	-.149	-.868	.391
Depersonalization	-.517	.307	-.288	-1.680	.101
Personal Accomplishment	.340	.166	.300	2.041	.048

Note: Dependent variable: Job Satisfaction Independent Variable : Subdimensions of Burnout

Since conduction of face to face interview in our study, a limited quantity of fishermen was reached in short period of time. Further studies will lead to different outcomes of burnout studies on more samples. This study, which is the preliminary stage of the master thesis, is going to increase the number of samples in order to draw a general framework in the future as limited sampling in limited time. At the end of the thesis, we are planning to get more samples in consequences of representatives of population. Other limitation of research is that the definition of what is satisfaction and burnout may vary from respondent to respondent. Thus, it may be subject to reporting bias.

REFERENCES

Allen,P., Wellens, B. & Smith, A. 2010. Fatigue in British Fishermen. *International Maritime Health Association*.61(3): 154-158.

Andruskiene, J.,Barseviciene, S. & Varoneckas, G. 2016. Poor Sleep, Anxiety, Depression and Other Occupational Health Risks in Seafaring Population. *TransNav, the International Journal on Marine Navigation and Safety of Sea Transportation*.10(1): 19-26.

Bavinck, M., & Monnereau, I. 2007. Assessing the social costs of capture fisheries: An exploratory study. *Social Science Information*. 46(1): 135–152.

Bavinck, M. 2012. Job Satisfaction in the Shrimp Trawl Fisheries of Chennai, India. *Social Indicators Research*.109 (1): 53-66.

Bhondve, A., Kasbe, A., Mahajan, H. & Sharma, B. 2013. Assesment of Job Satisfaction Among Fishermen in Southern East Coastal Area of Mumbai, India. *International Jjournal of General Medicine and Pharmacy*. 2 (2): 65-74.

Cordes, C.L. & Dougherty, T.W. 1993. A Review and An Integration of Research on Job Burnout. *Academy of Management Review*. 18(4): 621-656.

Hovdanum, A.S., Jensen,O.C., Petursdottir, G. & Holmen, I.M. A Review of Fatigue in Fishermen: A Complicated and Underproritised Area of Research. *International Maritime Health Association. 65(3):*166-172

IMHA. International Maritime Health Association. Newsletter, January 2012: 14.

Iversen, R.T.B. 2012. The Mental Health of Seafarers. *International Maritime Health Association*. 63(2), 78-89.

Maslach, C. & Jackson S.E. 1981. The Measurement of Experienced Burnout. *Journal of Occupational Behaviour*. 2: 99-113.

Oldenburg, M., Jensen, H.J. & Wegner R. 2013. Burnout Syndrome in Seafarers in the Merchant Marine Service. *International Archives of Occupational and Environmental Health*.86(4): 407-416.

Pollnac, R. B., Gersuny, C., & Poggie, J. J. 1975. Economic Gratification Patterns of Fishermen and Millworkers in New England. *Human Organization*. 34(1): 1–7.

Pollnac, R.B. & Poggie, J.J.1988. The Structure of Job Satisfaction Among New England Fishermen and its Application to Fisheries Management Policy. *American Anthropologist*. 90: 888–901.

Pollnac, R.B., Pomeroy,R.S. & Harkes, I.H.T. 2001. Fishery Policy and Job Satisfaction in Three Southeast Asian Fisheries. *Ocean & Coastal Management*. 44:531-544.

Pollnac, R. B. 2010. Using multivariate analysis to select indicators for surveys in fishing communities. *Presented at the NOAA Indicators Workshop*, Miami, FL, May 4–6.

Pollnac, R. B. 2011. Using multivariate analysis to select indicators for large scale surveys. *Presented at the NOAA Well-Being Indicator Workshop*, Charleston, SC, March 8–9.

Sliskovic, A. & Penezic, Z. 2015. Occupational Stressors, Risks and Health in The Seafaring Population. Review of Psychology. 22 (1-2): 29-39.

Smith, A., Allen, P. & Wadsworth, E. 2006. Seafarer Fatigue: The Cardiff Research Programme. *Cardiff, Seafarers International Research Centre*.

Spector, P.E. 1997. Job Satisfaction. Application, Assessment, Causes, and Consequences. Thousand Oaks, CA: Sage Publications.

Legal Aspects

Legal Issues Concerning the UN Convention on the Conditions for Registration of Ships (1986)

E. Xhelilaj, K. Lapa & L. Danaj
University of Vlore "Ismail Qemali", Vlore, Albania

ABSTRACT: UN Convention on the Condition for Registration of Ships (1986) reflects differing aims and interests, and has as its salient feature coastal state' responsibility for ships registered in that state. As a result, a requirement for registration is a genuine link between a particular territory and ship. In the Convention obligations are enforced on coastal states to implement legislation for the shipping industry to make sure that ship within the jurisdiction of each member state obligate genuine link with that state and to safeguard states adversely impacted by the control of the state of registration. Nevertheless, this instrument reflects many legal issues as well as political disagreements which have brought a situation where many states have rejected to ratify the Convention and therefore, blocked its entry into force. In this respect, this paper's purpose is to identify the relevant legal issues which made impossible the ratification of this Convention. The authors' opinion is that the genuine link and dual registration issue as well as other legal concerns have been the main causes of disagreement between states which subsequently prevented the Convention to entry into force.

1 INTRODUCTION

On recommendation of the Trade and Development Board, the UN General Assembly decided in 1982 to convene a plenipotentiary conference to consider the adoption of an international agreement concerning the conditions under which vessels should be accepted on national shipping registers. After several efforts expressed in meetings and international conferences, the UN Convention on the Conditions for Registration of Ships (UNCROS) was finally adopted in 1986. The purpose of the Convention is ensuring or strengthening the genuine link between a state and ships flying its flag, and to exercise effectively its jurisdiction and control over such ships with regard to identification and accountability of ship-owners and operators as well as with regard to administrative, technical, economic and social matters. Considered an important international legal instrument which promotes relevant legislative standards regarding safe and secure shipping on uncontaminated maritime environment, the Convention *per se* has reflected several legislative and political issues representing thus a complex and unpleasant situation within the maritime industry.

Consequently, a further analysis and discussion of the main issues of this convention which have mirrored legal deficiencies may assist the international maritime community as well as maritime organizations to have a better and comprehensive understanding of its legislative problems, particularly on the genuine link requirement, aiming thus at a possible entry into force or a potential revision of the Convention with the purpose of making efficient its legal standards. In light of these considerations, this paper will analyze the legal issues concerning UNCROS (1986) by discussing first its main objectives, then analyzing the legitimization of its *status quo*, continuing subsequently with the issue of dual registration vs. encumbrances as well as considering the genuine link matter in the context of the possible failure of this important convention.

2 MAIN OBJECTIVES OF THE CONVENTION

The Convention (1986) mirrors as its fundamental feature state's responsibility for ships registered in that state and as a consequence, a prerequisite for ship registration is considered a genuine link between sovereign territory and ship flying its flag. According to the Convention there is recognition that a ship may be registered not only because of the connection through ownership but also that of bareboat-charter One of the main objectives of the Convention is expressed to be that of ensuring or, as the case may be, strengthening the genuine link between a State and ships flying its flags (Article 1). In addition, the Convention introduces the notion of *economic link,* providing for the participation by nationals of the flag State in the ownership, manning and management of ships (Coles, 2002).

UNCROS is considered a public law convention in the sense that responsibilities are imposed on member states to adopt a framework for the shipping industry to ensure that ships within the *control* of each state have genuine link with that state and to protect states adversely affected by the control of the state of registration. The Convention sets out the basic structure for linking registration in a state with ownership or nationality and the right to fly the flag of the state of nationality. Under its provisions, each member state is required to have a national maritime administration supervising and coordinating the administration of shipping and the implementation of international rules concerning shipping (Article 5.1). The Convention introduces for the first time in international treaty law a distinction between the Flag State, meaning the a state whose flag a ship flies and is entitled to fly, and the State of registration, meaning the State in whose register of ships a ship has been entered (Article 2).

The Convention applies to all ship used for international trade and for the first time provided detailed provisions regulating bareboat chartering (Article 11-12). Its provisions contain measures to protect the interests of labor supplying countries and to minimize adverse economic effects that might occur within developing countries consequent upon its implementation (Article 14-15). Article 1 reaffirms the broad requirement of a genuine link, and Article 4 the same, including that ships have nationality of the State whose flag are entitled to fly and that ships should fly under the flag of one state.

One significant innovation found in Article 2 and 5 is the requirement that each flag state have a national maritime administration. The Convention allows flexibility for flag states to meet national manning or ownership tests for their ships (Article 7-9). Moreover, the Convention offers a solution when the developed states accepted a proposal of the developing states in Article 7 to allow a country to opt either for a provision on manning or on ownership, or both, thus allowing the open registries states to move toward complete control in gradual manner (Sohn & Noyes, 2004).

Article 7 allows s states an option; either they may comply with the ownership requirement in Article 8, or the manning requirement in Article 9; they may however comply with both. However it seems that the wording of the Article 8 concerning ownership and Article 9 regarding manning contain some indeterminate language stipulating that national laws should have provisions for participation of that State or its national as owners of ship registered in the state, and for the level of participation of the crew and officers that should be sufficient to permit the flag state to exercise effectively its jurisdiction and control over ships flying its flag (Sohn & Noyes, 2004).

Article 10 sets out the role of the flag state in respect of the management of ship, owning companies and ships on its national register. Before entering a ship in its register, the State of registration shall ensure that the ship owning company and/or a subsidiary ship owning company is established or has its principal place of business within its territory in accordance with its national laws and regulations. However, where these circumstances do not exist the requirement may be satisfied by the appointment of a representative or another management person who should be national of the flag state or domicile therein (Coles, 2002).

3 LEGITIMIZATION OF THE *STATUS QUO*

During 1950-60 pressure from a number of close registry states as well as several traditional maritime powers was exerted on the international community with the sole purpose to restrict open registry practice through strict national requirements or the adoption of international maritime standards. This legal approach however appeared unsuccessful at the Geneva Conference, where the obvious legal issue of infringement on state sovereignty and the accepted right to grant nationality came to the forefront. Consequently, the *status quo* prevailed: registration was synonymous with nationality and the requirements of genuine link or economic link were of no consequence as were considerations of the beneficial and true ownership (Odeke, 1998).

Be that as it may, the Convention of 1986 laid down important new standards related to the conditions for registration of ships, including the controversial genuine link requirement. Accordingly, it gives the impression that the *status quo* issue was finally solved by the adoption of the innovative legal norms reflected in the main provisions of the Convention. Nevertheless, the wording of many provisions found in the new convention appear to laid down legal aspects which *per se* legitimized further on the *status quo* already present in maritime industry for many decades. Thus, although expressed in mandatory terms, the articles relating to ownership, manning and management leave so much of their detail implementation and interpretation to the flag state that their effect may largely be negated. Viewed in this regard, the provisions can be seen as littler more than statements of principle.

Accordingly, Article 8 leaves the decision as to the level of national participation of the ownership of a vessel to the flag State. The only requirement is that the relevant laws must be sufficient to permit the flag state to effectively exercise its jurisdiction and control over ships flying its flag.

Another legal issue which legitimizes the *status quo* of the main provisions found in the Convention, creating thus further problems in terms of ambiguity and confusion, is the content of Article 9. With regard to manning matter laid down in this provision, the determination of what is "satisfactory" level of crewing by nationals is again left to the flag State, taken into consideration the: 1) the availability of qualified seafarers with the state of registration, and, 2) the sound and economically viable operation of its ship. Presumably, therefore, if the flag state determines, for instance, that the high level of its national wages precludes the economically viable operation of its ship, the Convention would not prevent the whole of the complement of the officers and crew being non-nationals (Coles 2002). Likewise, the procedures in Article 10 allowing for the appointment of a representative or manager as an alternative to national management, largely negates the object of the provision; ship-owners seeking to register their ships under open registration system would have little difficulty in locating a nominees in the flag State (Ready, 1998).

4 DUAL REGISTRATION VS.ENCUMBRANCES

The UNCROS (1986) presents the first international legal instrument which makes the distinction between Flag State and State of Registration, but it fails to address the question of how mortgages and liens are affected by dual registration. In this respect, the convention fails to address the matter under which law the mortgagees will be subject to in a case of a dispute, primary registration State legislation, or the Flag State legislation. Cyprus, Panama Vanuatu and Philippines have different approaches on this issue. For instance, Cyprus parallel registration provisions provide that in a case a vessel temporarily entered in the Cyprus Ship Register on the basis of bareboat charter, the registration in the original registry should be suspended, save as regards transfer of ownership and the creation and registration of mortgages or other encumbrances on the ship (Ready, 1998, p.46).

In addition, the Convention was the first international legal instrument which expressly approved the basic concept of bareboat-charter, which is defined as a contract for the lease of a ship for a stipulated period of time, by virtue of which the lessee has complete possession and control of the ship, including the right to appoint the master and crew of the ship, for the duration of the lease. The provisions enabling bareboat registration are contained in Articles 11 and 12. The provisions presupposes not merely the grant of the right to fly the flag of the State where the vessel is bare boated-in, but a grant of registration; accordingly, particulars of any mortgages or other similar charges upon the ship should be recorded (Article 11 (2) i). This wording brings ambiguity and lead to confusion since in many civil law countries like Spain, due to the fact that national law in such countries does not allow the de-registration of the vessel's mortgages and liens from the Commercial Registry.

In the provisions of the Article 11 (1), (2) h, is stipulated that the ship might be entered in the register of the Flagging-in State either in the name of the owner, or, where nationals laws and regulations so provide, the bareboat charterer. In any event, the name address, and nationality of the bareboat charterer must be recorded. These provision might lead to a certain legal issues in the international law, within which is not allowed for a ship to claim two nationalities, otherwise it might be classifies as a stateless ship and be seized by different jurisdictions. In the case *United States v. Gonzalez,* (1987) the vessel was judged to be stateless, apparently because it sailed under authority of two nations and made false claim of nationality and flag.

The same conclusion was found by the court in the case *United States v. Matute* (1985), underlining the ruling that the ship was assimilated to one without nationality because she was sailing under hybrid Colombian/Venezuelan flag. Another issue is reflected in the Article 11 (5). Thus, in the case of a ship bareboat-chartered in, a State should assure itself that the right to fly the flag of the former State is suspended. As mentioned in the above analysis again nothing is mention about de-registering the ship from the former State Registry, which probably led to the conclusions that the ship might be registered in two different States.

On the other hand, according to Article 12 (4) a state should ensure that a ship bareboat-chartered in and flying its flag will be subject to its full control and possession. The charterer according to the Convention, Article 12(3), will be considered the owner, but the Convention does not provide for ownership rights in the chartered ship other than those stipulated in the contract (Ready, 1998). Another legal deficiency of the Convention is that is ambiguous regarding the harmonization of national laws and practices to minimize opportunities for fraud and to protect the rights of ship mortgages to the greatest extend possible in view of the still evolving system of bareboat-charter (Odeke, 1998).

5 GENUINE LINK REQUIREMENT

The introduction of genuine link had occurred at a period of time when open registries were beginning to become very widespread preference for ship-owners. The genuine link concept was seen from maritime powers and close registry states as a mean against open registries since introduced strict legal requirement pertaining to ownership and manning of ships. Accordingly, the implementation of this strategy gave rise to the argument that since open registries countries were not in position to exercise the genuine link element, effective jurisdiction and control over vessels under their flag, no genuine link could be established, hence the practice could be described antithetical to international maritime law. Nevertheless, the exact meaning of effective jurisdiction and control concept in this respect according to many scholars and maritime law experts is considered far from clear and quite ambiguous (Pamborides 1999).

The concept of genuine link, as it emerged from the *Nottebohm Case,* was quickly adopted by International Law Commission which included the concept in its 1955 draft Geneva Convention. The genuine link requirement was an attempted solution by International Law to avoid the abuse or misuse of the flag and requires a legal link between the registry and the vessel where there is no economic link (Odeke, 1998). However, the concept was not included in the final draft since was considered not to be practicable. The 1958 Geneva Convention was faced with two different principles regarding the nationality of ships; one was the general principle of international law which allowed each state to fix the conditions according to which a ship could fly its flag, as discussed in *Muscat Dhows case,* and the other to adopt the newly emerged genuine link principle established by *Nottebohm Case* (Pamborides, 1999). Nevertheless, the conference incorporated both principles in Article 5.

By 1958 this principle applied to the nationality of merchant vessels was enshrined in the international treaty law. However, so ambiguous this concept remains, that the genuine link has had little apparent effect in stemming the flow of ships to open registries. In the 1958 Geneva Convention is not stipulated or described what is meant by genuine link in terms of preconditions for the grant of nationality; nor is there any sanction indicated in the case of the absence of genuine link, whatsoever that expression may mean. Rather the link seems to arise *ex post facto,* being expressed in terms of jurisdiction and control exercised by the flag State after the ship has been registered in that State. The socio-economic and political aims behind the introduction of the requirement of connecting factors are, however, thereby defeated. The concept of genuine link was dealt a further blow in 1960 when

ISJ was requested to deliver an advisory opinion to MSC of IMO (Ready, 1998 p.12).

Despite the global debate for genuine link both as to what it means and as to how it should be implemented, and despite the fact that it created ambiguity and uncertainty instead of clarifying the issue of ship registration the UNCLOS (1982) incorporated in its article 91 which seems to be reproduction of Article 5 in the Geneva Convention. The removal of the phrase in particular to the genuine link in the text of Geneva Convention and that the requirement for the existence of genuine as it appears in Article 91 of UNCLOS (1982) is not linked in any way with the requirements of Article 94 entitled Duties of the Flag State indicates that the phrase genuine link is subject to even more liberal interpretation than those witnesses under Geneva Convention. Nevertheless, the lack of clarity and vagueness introduced by Geneva Convention was still preserved in UNCLOS 1982 (Pamborides, 1999). Be that as it may, many in the shipping industry as well as scholars take the view that in both the Geneva Convention (1958) and UN Convention on the Law of the Sea (1982) the genuine link concept is applied to merchant vessels probably as a challenge to the practice of flags of convenience or open registries (Odeke, 1998).

With regard to international customary law, Churchill and Lowe (1991) argue that genuine link requirement laid down in the aforementioned conventions does not represent customary international law, contrary to the statement included in the preamble, which describes the provisions of the UNCROS (1986) as being generally declaratory of established principles of international law. The LOS Tribunal must apply the LOS Convention and other rules of international law not incompatible with this Convention (Art 293, Annex vi, Art 23). But when international law leaves to each state to fix the condition for granting nationality to ships and for registering ships, the Tribunal must take account of national laws. This legal concept is applied by the court to the ruling of the judgment *St. Vincent and the Grenade v. Guinea* as well as in the legal case *Belize v. France* (2001).

The debate about the meaning of genuine link concept was to continue with further implications in the comprehensive global debate for the abolishment of the international system of open registries, led to the circumstances wherein UNCTAD was somehow obligated to convene an important international conference in order to address this essential matter and attempt to promote a long-lasting solution to the problem (Pamborides, 1999).

The *ad hoc* Intergovernmental Working Group established under the auspices of UNCTAD, on the economic consequences of the existence or lack of genuine link between vessel and flag of registry concluded in its Report that the following elements are normally relevant when establishing whether e genuine link exists between a vessel and its country of registry, that is, the merchant fleet contributes to the national economy of the country, revenues and expenditures of shipping, as well as purchases and sales vessels, are treated in the national balance-of-payments of the vessel, the employments of nationals on vessel and, the beneficial ownership of the vessel (Ready, 1998 p.18-19).

6 REASONS BEHIND THE POSSIBLE FAILURE

The objective of UNCROS was to abolish the open registries, therefore since the beginning this was probably a failed objective. Moreover, the Convention circumvented most controversial aspects of ship registration, although attempted to provide the first blueprint on conditions for providing nationality to ships (Pamborides, 1999). Under the Convention, the state of bareboat-charter must notify the state of original registration when the bareboat-charter terminates, but this procedure raises many questions. First, if the charterer disputes the owner's right to terminate the charter, there might be considerable difficulty in determining where such vessel is properly registered, particularly if the bareboat-charter state has a deletion procedure. There should be a balance between public and private law in this regard. When a vessel exchanges a flag during bareboat-charter-out, national laws are currently not adequate to deal with the problem of where the mortgages covering that vessel should be recorded whether in the underlying registry or bareboat-charter registry (Odeke, 1998).

In the Articles 11 and 12 is underlying the granting of the right to fly the flag of the country where the ship is bareboat-charter-in without closing the register in the original state of registry (suspended). Panama and Germany are two countries that engage in this practice. Although, in the instance, the vessel is not flying two flags at the same time, however, as a bareboat-charter it remains a *Parallel* or *Dual Registry*. In fact it was the Convention which codified *parallel* or *Dual Registry* practice of bareboat charterers. It was the finance, commodity, merchant and ship owning interests that lobbied for this at the Conference (Sohn & Noyes, 2004). To achieve the goal of compliance and for the purpose of applying the requirements of this Convention in the case of a ship so bareboat chartered-in the charterer will be considered to be the owner. This Convention, however, does not have the effect of providing for any ownership rights in the chartered ship other than those stipulated in the particular bareboat charter contract.

The State where the bareboat chartered-in ship is registered shall ensure that the former flag State is notified of the deletion of the registration of the bareboat chartered ship. First is not clear from the first sentence of paragraph 3 whether the charterer shall be considered the owner, but it is assumed that the charterer is considered to be the owner of the ship, so bareboat charter-in (this leads to a legal issue because the register of ship is *prima facie* title of ownership). Second it is not clear from paragraph 5 when the state where the ship is bareboat chartered–in must notify the former flag state of the registration of the bareboat chartered vessel.

The later omission may be serious as a state may neglect for some considerable time to inform the former flag state, hence creating uncertainty as to what flag the vessel is entitled to fly. This can create a classical case of a ship flying two flags concurrently and be penalized under the article 6 of Geneva Convention and article 92 of UNCLOSS 1982. In the event which is by no means certain, that the Convention does ultimately enter into force, the question arises whether, given its greater explicitness in seeking to address the problems of genuine link, one can expect to see the abandonment of open registries or flags of convenience and a return of the traditional maritime nations. The answer would seem to be no (Coles, 2002, p 12).

7 CONLUSIONS

Consequently, over four decades after genuine link concept was first put forward as a means of undermining the growth of the open registries, and notwithstanding the efforts of on the part of international community, little progress has been made in establishing this norm as an effective principle of international maritime law. The same difficulty has encountered even the Convention *per se* which reflects the genuine link requirement in its provisions, that is, UNCROS (1986).

The Convention as discussed in this analysis reflects many legal issues as well as has been in the focus of criticism by many coastal states in terms of political settlements. The most prominent issue has been the intolerance that many traditional maritime powers, which implement close registries, have shown towards the wording and essence of the genuine link requirement, an important provision leaning in favor of open registry system. In support of this argument, open registration system has been criticized by close registry countries for implementing lax safety policies resulting therefore in substandard shipping as well as security and safety issues.

Moreover, other legal issues such as bareboat-charter requirements and its ambiguous wording regarding the encumbrances, the possible failure of dual registration concept and problems which its function has created, as well as the potential *status quo* of the main legal elements and some of the important provisions of the Convention which somehow has been legitimized during the last decades, has made very difficult the ratification and entry into force of this crucial legal instrument.

Accordingly, the opinion of these authors is that UNCROS should undergo essential legal amendments, mainly aiming to improve the concept of genuine link and bareboat-charter registration in order to make it more acceptable for coastal states which continue to refuse the ratification of the Convention; taken into consideration of course the interests of open registry countries. One efficient proposal regarding this issue might probably be discussing and possibly introducing into the Convention the legal concept of international registry, a *sui generis* notion, which interrelate both close and open registry legal elements and at the same time maintaining appropriate safety and security levels.

REFERENCES

Coles, R. M. E (2002). *Ship Registration Law and Practice*, First Edition, Informa Law - Routledge, London.

Churchill, R.R & Lowe, A (1999). *Law of the Sea*, 3rd ed, Manchester University Press, London.

Grand Prince Cases *Belize v. France*, Case No. 8 (ITLOS, 2001), http://www.itlos.org/start2_en.html.

M/V Saiga (no 2) Case, *St. Vincent and the Grenade v. Guinea* (ITLOS).

Odeke, A (1998). *Bareboat Charter ship Registration*, Kluwer Law International, The Hague/London/Boston.

Ready N.P. (1998). Ship *Registration*. 3rd. ed. London, LLP Reference Publishing.

Sohn, L. B & Noyes, J. E (2004), *Cases and Materials on the Law of the Sea, Transnational Publishers, University of California, USA*.

Sturmey, S.G. (1983). *The Open* Registry *Controversy and the Development Issue*, 8 vols. Bremen: Institute of Shipping Economics

Pamborides, G.P (1999). *International Shipping Law: Legislation and Enforcement*, Kluwer Law International, The Hague/London/Boston, Ant. N. Publishers, Athens.

UN (1986). United Nations Convention on the Conditions for Registration of Ships, 1986.

UN (1985). *Report of the United Nations Conference on Conditions for Registration of Ships on its Third Part*, Doc. TD/RS/CONF/19/Add.1

UN (1982). United Nations Convention on the Law of the Sea, 1982, UN Publications, London.

United States v. Gonzalez (810 F.2d 1538, 1541-42, 11th Cir. 1987).

United States v. Matute, (767 F.2d 1511, 1512-13 (11th Cir. 1985)

United Nations Convention on the High Seas 1958.

Development of Competencies when Taking into Consideration Nowadays Challenges Inside Organizations

A. Duse, C. Varsami, R. Hanzu-Pazara & P. Arsenie
Constanta Maritime University, Constanta, Romania

L. Hanzu-Pazara
Ovidius University, Constanta, Romania

ABSTRACT: The economic and social changes bring new professional competences requirements. The satisfaction of these requirements is one of the main objectives of the education system. Starting here it is necessary to redefine the education process, to become focused on competences and comply with the organization's needs. This paper intends to present some aspects of this redefinition of activity areas where competences are the most important and the main activities are based on human capabilities. Competences have been all the time treated as important components of all human activities. Along the time, with every society evolution, there have been researches on the purpose of particular competences in specific activity sectors. Today, the competences become more complex due to organizations exigencies. The necessary competences' study has to start from the present stage of requirement and to see which will be the future tendencies in different activities in order to be able to focus on these. The methodology used in this case has been based on the study of two different activity sectors, medicine and maritime transport, considering the changes which take part during the last period, both technological and conceptual and also, the impact of these on human performance. The human performance evaluation has been considered a reflection of competences acquired. Using performance, the analysis has resulted in the effect of competences in the specified activity field. The main objective of this study was the quantification of personal competences in activity fields where human life is very important, and how can be these improved according to organizations changing process and needs. The study is addressed to the educational sector, where competences are developed for the first time in any activity field. Results are important for academics, in order to remodel competences according with the actual organizations requests, and for practitioners, as applicants and beneficiaries of these competences.

1 INTRODUCTION

All economic and social changes are bringing new requirements for professional competences. Satisfying these requirements is one of the main objectives of the education system. This is why the education and training process have to be redefined, to become focused on competences and to comply with organization's needs.

Change, innovation, quality, competency, expertise and creativity represent the background of organizational changes. During the last decades numerous concepts and theories have been developed to face the challenges of complex societies and to fundamentally improve organizational structures. In this context, the organizations needed to gradually introduce new organizational concepts and to use new technologies for improving activities, exposing themselves to

multitude of changes in order to meet the growing challenges posed by other competitors on markets. To manage the changes is today one of the major challenges, affecting not only organizations, but individuals also, who, in a way have to stay in line with organization and to pass through the process of change, or, on the other way, are in charge of enabling change by implementing and coordinating processes of change and innovation within the organizational structures. [1]

These changes that affect all activity sectors have generated a need for competencies to face them. This aspect leads to a prominent role of competency development in enhancing the success of employees and organizations and promote the introduction of competency development as a central part of the organizations human resource practices. [2] Unfortunately, the biggest interest of the economical entities in competency development was not fully translated into the academic world, creating a gap between theory and practice.

At the same time, the changes suffered by the organizations and the entire social and economic environment, have altered the concept of career and contributed to the development of new models for career management. [3] Central to the notion of so-called 'new-career' is that organizations can no longer guarantee employees' career success by providing lifetime employment. Instead, employees need to create their own career by pursuing lifetime employability, acquiring or creating work through the optimal use of one's competencies. [4] As such, these new career patterns make it increasingly important for employees to continuously invest in the development of their competencies. For organizations, focusing on the continuous development of their employees' competencies is also necessary, since it gives them the opportunity to stand out to their competitors. The management literature increasingly acknowledges the importance of competency development in enhancing the competitiveness and performance of an organization. As such, competency development becomes a crucial strategic management tool in today's work environment, in particular where competencies are vital for a safe work environment, like maritime transportation and health care systems.

The growing importance of competence orientation in academic education resides in the increasing complexity which people face in modern activities. In the actual globalized world, organizations are confronted with a fierce economic competition and volatile markets. The uncertainty of the environmental context leads to a high dependency of organizations on the capability of their employees to learn and acquire new skills and competences in order to adapt to the changing external situations and work requirements. [5]

Organization competitiveness on market is well illustrated in the maritime transportation sector, where, after the crisis years only those companies which invest in new technologies and development of new competencies for their employees succeeded. This trend generates the necessity for the training system to review the offer and to comply with these requirements. The increased requirements for graduates' job profiles on actual labour markets are pushing the maritime academic system to improve the training concepts and to reconfigure the curricula structures from knowledge transmission to competence-oriented learning outcomes and to try hard to reduce the gap between the theoretical competencies acquired during study and the practical ones required on board ship.

Competitiveness in the health care system is particularly met in the private sector and the medical staff competency and knowledge is the main advantage in front of competitors. Also, the medical system has to comply with the cultural and human interactivity competency requirements. This way, the education system needs to understand the factors that are forcing the health care system to become more competent and to modify the perspective of future medical staff into recognizing and understanding the cultural and clinical dynamics when interacting with patients.

2 IMPROVEMENT OF COMPETENCIES IN THE HEALTH CARE SYSTEM

The development of competence is thus relying on a high level of individual activity. However, they play an important role in the educational scenarios as the ultimate objective of professional development. From this point of view, competence was defined as a prerequisite to master specific challenges in an actual field of activity. It was assumed that individuals can improve and gain new competences through learning and experience and give a first hint on how competences can be developed through learning. Learning, which takes place, and experience, which is gained in authentic situations, is seen as the basis for a process of individual or collective competence acquisition. In one of his papers, Weinert states that learning is a necessary condition for the acquisition of prerequisites that enable a successful mastery of complex tasks, which is one description for competence. [6]

An important aspect for the role of learning in competence development is the unstable character of the learning process. Learning is sparked and initiated through a state of irritation, which is caused by action that takes place in an unstable, non-routinized, and complex context. In this unfamiliar and complex context, the effect of individual or collective action is not predictable, as any experience on the effect of action is lacking. Challenges under such uncertain conditions lead to a labialization of the existing value system, the learners have to learn through concrete experience about the effects of their actions in a new and complex context. When the action has been completed, the gained experience and knowledge is incorporated into the existing value system and thereby modifies existing attitudes of the learner. Thus, developing competences requires authentic challenges in an uncertain context.

The health care system is assuming the development of competences in a complex context. To cope with complexity, individual actors have to acquire and to integrate new knowledge, to apply this knowledge within a specific action, and to assess and to value the results of the action. In this case, competences are acquired in confrontation with the immediate environment.

Competence in health care describes the ability to provide care to patients with diverse values, beliefs and behaviours, including tailoring health care delivery to meet patients' social, cultural and economic needs. An important aspect of the health care systems is the development of the cultural competence to their members. Cultural competence produces numerous benefits for the organization, patients and society. Organizations that are culturally competent have improved health outcomes, increased respect and mutual understanding from patients, and increased participation from the local community. Even more, this kind of organizations may have lower costs and fewer care disparities. [7]

For a harmonious development of professional competences in the health care system it is necessary to respect and apply a number of principles coming from the special relation between doctor and patient. It is well-known that patients are receiving the best care when they cooperate properly with doctors. This partnership implies abilities of human interaction and good communication for the doctors, which facilitate the closeness between practitioner and patient. Studies on the subject suggest that the majority of patients have an excellent and trusting relationship with their doctors. Developing abilities like respect and understanding make possible the existence of such a relationship. The feedback and support provided by patients are vital for the medical personnel to offer help, to manage medical situations effectively and to continue improving standards of practice.

"Good doctors" should listen carefully the patient's explanation of why he or she has come to see them, and take into account the patient's previous and current health and illness, as well as the lifestyle and generic lifestyle views. This ability needs a good capacity of understanding the real problem from explanations given by a non-specialized person. Also, another important ability is to connect the present health state with previous conditions and other important issues that can offer a complete image about the patient's health status.

Medical personnel must listen to the patients and respond to their questions and concerns. They should give information and advice in a way that patients are able to understand. A helpful indication is to stimulate patients to explain anything that they do not understand, including any technical words or jargon.

Health care personnel must develop competence of being able to take prompt action if they think that patient's safety, dignity or comfort is compromised. This is why, doctors, nurses or any other health personnel should always be able to analyse and understand if anything that the patient felt can put him in a dangerous situation. The medical personnel has to be capable to report when things go wrong for the patient, for example accidents and mistakes, or when patients get serious side effects from medication.

Another important ability in health care is to be able to review the work and results and compare their own performance with other doctors' or other teams'. This review of activity will help to identify any areas of practice possible to be improved. This action involves the capacity to manage information from all patients and to evaluate them both in an independent and combined system.

The medical personnel have to treat all their patients fairly and without discrimination. It is necessary at all times to be able to improve offered services. They must obey the rules and the law and act honestly in any patient related matters. They must be honest and open with patients if something has gone wrong with the patient's care or treatment. They have to be able to explain what happened and apologize when appropriate.

There are no boundaries in the medical scientific world anymore. Medical information is travelling beyond the speed of light and patients are aware of current protocols and new strategies of treatment worldwide. A fair medical service has to be able to offer all possibilities available in different parts of the world, or at least to be able to gather accurate and easy understanding information about medical possibilities.

For the medical education system, training of the next generation of doctors is a key part of the process. It was established that medical students cannot learn all they need to know from books and will have to participate directly to the medical act. This participation comes to help them to develop practical competences in working with the patient and applying the medical treatment.

In an international world with a constant migration of people of different race, ethnicity and culture, also facing a current exchange of background and habits, doctors should be prepared to understand and manage medical conditions and treatments that are forcing knowledge beyond a school of medicine. The language skills, the way of approaching an issue and managing cultural mixtures may represent a challenge for the practitioners.

Also, a good medical service is based on a right understanding of the community and the role played within the community. In order to become culturally competent, a health care organization has to follow some steps, like: analysing data and micro targeting surveys to improve service for the community, communicating survey findings to determine prioritizations, educating personnel and aligning programming and resources to meet community needs.

The medical personnel needs to understand the factors that are forcing the health care system to become culturally competent. Becoming culturally competent involves developing and acquiring the skills needed to identify and assist patients from diverse cultures. With the necessary competences and skills, the medical personnel can quickly identify the services required by a patient, thereby increasing positive health outcomes.

An effective educational program for cultural competence of the future medical personnel correlates with a lasting awareness and understanding. Although there are several approaches to educate this kind of personnel, a successful educational program should include: cultural assessment, multiple training methods, on-going education and measurement and tracking.

3 CHALLENGES IN COMPETENCE DEVELOPMENT FOR THE MARITIME ACADEMICS

In the last years many of the maritime academies and universities have changed their education approaches due to new regulations implemented or due to evolution of the technology and needs for training in accordance with the new required competences. Technologies used today for development of competences in maritime transportation field are various and include simulators, computerized programs and all types of virtual learning.

Today, computerized technology is part of life for many people and becomes indispensable for many activities, computers being part of the educational processes and leverage for competence acquiring. It seems that the latest technologies have power to change and upgrade the concepts regarding the way of training and how competences and skills can be reached in different activity fields. In 1999 Hensen explained that simulator-based training courses were introduced primarily to train the skills of passage planning and the importance of human interaction on board ships. [8] In 2002, other two authors have observed that many types of simulators used for competency development tended to emphasise a physically realistic environment in which specialized exercises occur, although on the PC-based simulators for training some tasks are increasingly widespread [9] and the use of simulation in providing solutions to the problems of risk and crisis management and the optimal use of crew resources have a long established pedigree in maritime education and training. [10]

Being regulatory, the only mandatory requirements for development of skills and competences in the maritime activity field are those of the International Maritime Organizations, expressed by the Seafarer's Training, Certification and Watchkeeping Convention and Code, where the minimum standards of competences for those persons with responsibilities in the safety of maritime operations are specified. Non-regulatory, are the requirements for competences stated by those companies and organizations with economical interest in the maritime transport. The education sector has to comply both with regulatory and non-regulatory requirements for competences and skills in order to facilitate a proper career development for future employees.

In the field of maritime education, introduction of technologies for training and skills development have been done for years, before beginning the process of changing from theory-based to competence-based education. Perhaps, the most important aspect of this revolution in training was marked by the changes of graduate's opinion about their future career and what really supposes this. All these challenges, like curricula structure, the use of the latest technologies and training based on competences, lead to the change of the older concepts about how training must be done, what materials and methodologies are used and produced changes on trainers' concepts and visions about their position and meaning in the education process.

Time is necessary in order to quantify the impact of these challenges on graduate's competency, and a number of generations must graduate and apply their knowledge and skills in daily activities, sufficient to receive feedbacks about their level of competence. In the context of actual changes made in the training methodology, focusing on competences' development, is too early to have a complete image of the impact of these on graduates and especially on their duties on board ships, as a place where the assessment process is made by people without contact with the education system. [11]

Beside the technical skills, another important competence field in the maritime industry is human interaction. In the last period of time, due to the increase of multicultural crews, it has become necessary to develop competences for the future employees on human relationships and cultural interactions. This approach is based on consideration that human society is based on cooperation and interaction. Without these characteristics, there will not be possible to stimulate social development and wellbeing. By bringing this concept on board vessel, the ship can be considered as the society and seafarer - the individual who needs to add value to the wellbeing of the entire society. This desideratum can be applied when people involved have developed the abilities to manage human interactions and interpersonal conflicts.

The complexity of most of the on board operations places them beyond the control of one person, and the only efficient way to tackle process improvement is to work in group, as a team. Working together has various advantages, like: a greater variety of complex issues that can be finalized by pooling expertise and resources, problems are exposed to a greater diversity of knowledge, skill and expertize, the approach boosts moral and ownership through participative decision making and, recommendations that are more likely to be implemented than if they come from an individual. [12]

4 RESULTS

In order to evaluate the importance of competences development during the education process and the impact of these on the actual challenges of the maritime industry, interviews have been realized and discussions with involved entities like students, graduates, employers and on board ship evaluators have been developed. Most of the subjects discussed were related to the changes brought by the new approach and how these covered the remarked lacks of knowledge and skills considered during previous studies. Also, increasing practical periods and using simulated applications were discussed leading to an optimum level of competence and we asked about their opinion and suggested ideas for future approaches on these subjects.

Comparison has been done with results gathered from questionnaires targeting maritime medical attendance personnel, that were focused on medical activity, the level of stress accumulated via language barrier, cultural differentials, social position of patients, time necessary for making a proper anamnesis and for establishing a trusting relationship with the practitioner. Results are offering more than 70% openness to training into intercultural relationships and need for improving social skills via targeted training, 14% considered sufficient actual educational instruction, and the rest were indifferent or not willing to respond.

After analysing the answers of personnel involved in the maritime field, resulted that more than half of the interviewers consider most useful the new approach and saw the competence development as the main target of every education process, twenty per cent considered normal to combine practical skills with theoretical background in order to strengthen the competences resulted, and other fifteen per cent expressed that the system based on competence development is not well fitted to an engineering-based education process. The rest of the interviewers have no direct considerations on the subject or didn't express any opinion about, but considering normal to redefine the training process according to the present requirements of the sector labour market.

For the study of competence developed related to human interaction, different working configurations and various scenarios were used, to allow the evaluation of the abilities to interact during normal or emergency situations. Groups of students from Maritime University divided into students with experience in working on board ship's, familiar with the multi-nations crews, and students without this experience, that never met a multicultural environment, and students from the Faculty of Medicine were invited for the study.

The study allows seeing how the future employees in a multicultural organization understand the group working principles and known how to work in a diverse cultural environment. There have been some difficulties in adopting the necessary attitude during applications according with the position occupied in the scenario. After the completion of applications it has been observed a wide opening of the participants to the group working concept, an acceptance of its principles and an increased receptivity to new challenges for the group.

Also, it has been observed a changing of attitude toward group working, most of the participants who had some reservations in the beginning, have started to interact more with the others and to improve their involvement in group activities.

5 CONCLUSIONS

The authors' opinion is that students are open for changes which involve an increase of their capabilities and skills, which will facilitate their access on labour market and will offer them the ability to use the latest technologies. Anyway, the actual students are less attracted to the classical education methodologies. Graduates, also, appreciate these new orientations, especially after finding the major use in their daily duties, where it is necessary to have properly developed skills and to be competent in the use of modern technology.

REFERENCES

[1] GIZ (2012). Managing Change and innovation: A Challenge for Modern Organizations. Beitrage aus der Praxis der beruflichen Bildung (15).
[2] De Vos, A., De Hauw, S., Willemse, I. (2011), Competency Development in Organizations: Building an Integrative Model Through a Qualitative Study. Vlerick Leuven Gent Working Paper Series 01/2011.
[3] Arthur, M.B., Inkson,K., Pringle, J.K. (1999). The New Careers: Individual Action and Economic Change. Sage, London.
[4] Van der Heijde, C.M., Van der Heijden, B.I.J.M. (2006). A Competence-Based and Multidimensional Operationalization and Measurement of Employability. Human Resources Management (45), pp. 449-476.
[5] Ehlers, U.D. (2013). Open Learning Cultures: A Guide to Quality, Evaluation, and Assessment for Future Learning. Springer. 978-3-642-38173-7.

[6] Weinert, F.E. (1999). Definition and Selection of Competencies: Concepts of Competence. Max Planck Institute for Psychological Research, Munich, Germany.

[7] Health Research & Educational Trust. (2013). Becoming a Culturally Competent Health Care Organization. Chicago, IL, Accessed at www.hpoe.org

[8] Hansen, H. (1999). Ship Bridge Simulators: A Project Handbook. Nautical Institute. London.

[9] Peterson, A.M.M. (2002). Simulation and Modeling in Innovation. Proceedings of the 29th Annual General Meeting of the International Marine Simulator Forum.

[10] Barnett, M.L., Gatfield, D., Habberley, J. (2002). Shipboard Crisis Management: A Case Study. Proceedings International Conference Human Factors in Ship Design and Operarion, pp. 131-145.

[11] Blagovest, B., Hanzu-Pazara, R., Nistor, C. (2009). Strategic Human Resources Management in the Maritime Knowledge Based Organization. Proceedings of 15th International Conference The Knowledge-Based Organization, Vol. 2, pp.35-37.

[12] Department of Trade and Industry (2012). "People Development and Teamwork". Accessed at www.dti.gov.uk

Investigation of the Piracy Causes: an Quantitative Research

Ü. Özdemir
Mersin University, Mersin, Turkey

M.S. Cetin
Karadeniz Technical University, Trabzon, Turkey

ABSTRACT: Seafarers were faced with many negative factors in this profession. Piracy is one of the factors of this negative contributing cause for seafarer. In this study causes of piracy occurred in the seas, and measures to be taken were discussed in order to reduce or prevent such adverse events. Besides, factors have been identified that lead to piracy occurred in seas and have tried to present the alternative solutions can be applied on this issue. However, causes of piracy can be extremely complex. Therefore, scientific methods should be used to evaluate the causes and determine measures need to be taken. Severity of the reasons that led to the piracy and relationships with each other identified by using fuzzy AHP (Analytic Hierarchy Process) method.

1 INTRODUCTION

Maritime transport is open to many national and international effects when compared to other sectors that this makes running of the system more important. Typical running of this system is possible with global security of world's seas (Gugercin, 2007; Berg, 2010; Bulut, 2013; Drobetz et al., 2014; Özdemir, 2016). Privateering incidents are shown as one of the factors, which threaten the security of the sea (Vanek et al., 2013; European Union, 2014; Mazzarella et al., 2014; Wang et al., 2014; Vespe et al., 2015). While economic losses are experienced as a result of experienced privateering incidents together with the failure of maritime trade, many seamen are kidnapped, wounded and killed in consequence of the attacks aimed at the crew (Alvarez et al., 2015; Vespe et al., 2015).

Concepts of privateering and piracy are evaluated in different categories. While privateering is called as groups that are under a flag or are supported by national forces and attack with a certain plan, system and organizational activities; piracy is defined as a person or organization that performs petty theft, kidnapping, robbery etc. only locally without belonging to any political organization and being subject to any union (Germond, 2015). Although all privateering actions in the seas were prohibited with the Congress of Paris in 1856, they are still experienced in some strategic points. Also, it is known that as from 2013, budget around 7 billion US dollars was set to protect maritime trade against privateering incidents in the world on a global scale and this amount is in higher rates in our day (Germond, 2015).

It is explained that these illegal activities, which occurred in risky regions, have many reasons. Increase in uncontrolled and weak regions as a result of the growth of maritime trade activities, constraints of an economic bottleneck over the public of a region (unemployment, lower salaries, higher food prices etc.), inadequate security measures of littoral and port states and political splits can be shown as several factors, which cause privateering and piracy activities to occur (Axel, 2013; Pristom et al., 2016; Özdemir and Güneroğlu, 2017). As these and similar reasons weren't presented and due to scarcity of studies related to this subject, it was thought that a study like this should be conducted. From this point of view, with the study, the factors, which cause privateering and piracy activities in world's waters, and solution offers aimed at eliminating this problem were identified with the help of scientific methods. When the reasons of these experienced actions are analyzed, it is seen that these factors have interrelated, various causes and effects. In solving these problems, it is accepted that looking for a solution with fuzzy multi-criteria decision making methods, which include the most common methods of decision making, making a selection among alternatives with the evaluation of decision criteria, grouping and listing the alternatives, is a right method. The use of fuzzy multi-criteria decision making methods in the solutions of this problem and similar problems makes obtaining more concrete cluster of alternatives and reaching more realistic solutions possible (Jahanshahloo et al., 2006; Yang and Hung, 2007; Awad et al., 2013; Ozdemir and Guneroglu, 2015; Guneroglu et al., 2016; Özdemir and Güneroğlu, 2017). In this study, criteria which lead to privateering incidents, was identified with the use of fuzzy multi-criteria decision making methods. Then appropriate methodology was proposed for the solution of the problem. By listing established criteria between each other fuzzy AHP which may offer solution to the problem within the framework of these criteria, are benefited. Within the direction of obtained results, a practical and applicable model was proposed by confirming new alternative options and new hypotheses in order to help the authority while organizing many international conventions, agreements and regulations about maritime safety and security for maritime industry.

2 METHODOLOGY

Privateering and piracy activities, which threatened world maritime sector, were considered in this study. Variables, which caused these illegal activities to occur and affected decision making period, were established with comprehensive and long-term literature reviews, evaluation of security reports, interviews made with authorized sector employees and academicians, reviewing conventions and agreements related to maritime safety and reviewing accessible reports related to privateering incidents. When the factors, which cause these illegal activities to occur, are analyzed, it is seen that there are many interrelated, convergent and complicated systems, which require decision making within the structure of existing problem (Dabrowski and Villiers, 2015; Pristom et al., 2016; Özdemir and Güneroğlu, 2017). Fuzzy multi-criteria decision making methods are used in various fields in the literature in order to reach ideal and applicable results in problems, which are hard to model and for which views of competent people are required (Jahan and Edwards, 2015; Wang et al., 2016). Because of these reasons, in the applications, which are carried out within the framework of the study, fuzzy AHP for weighting criteria for the evaluation and listing of alternatives were benefited. The methods, which are used during the implementation phase, are summarized below:

2.1 Fuzzy AHP

The Fuzzy AHP technique by Buckley (1985) can be summarized as follow (Buckley, 1985; Özdemir and Güneroğlu, 2017). The first step of the technique contains defining the main criteria that potentially affecting the problem under investigation by experts and decision makers. Afterwards, in order to convert the expert evaluations to the fuzzy numbers, a predefined linguistic scale is used. Then, expert evaluations were received as pairwise comparisons that converted to the fuzzy numbers with in matrix form as shown in Eq.1:

$$\tilde{A}^k = \begin{bmatrix} 1 & \tilde{A}_{12} & \tilde{A}_{1m} \\ \tilde{A}_{21} & 1 & m \\ \cdot & \cdot & \cdot \\ \cdot & \cdots & \cdot \\ \cdot & \cdot & \cdot \\ \tilde{A}_{m1} & \tilde{A}_{m2} & 1 \end{bmatrix} \quad (1)$$

where " $\tilde{A}^k$ " is the response matrix by each expert. Linguistic scale and corresponding fuzzy numbers used in the Buckley's technique were given in Table 1 (Ting et al., 2004, Xu and Yager, 2008).

Then all data received as result of experts' evaluations is compiled by using weighted mean formula as given in Eq.2:

$$\tilde{A}_{xy} = \frac{Z_1 A_{xy}^1 + Z_2 A_{xy}^2 + \ldots + Z_k A_{xy}^k}{Z_1 + Z_2 + \ldots + Z_k} \quad (2)$$

In Eq.2 " $\tilde{A}_{mn}$ ", is the joined comparison value of the critieria "x" and "y"; " Z_k " is the weighted value of "k." expert; " A_{xy}^k " is the comparison value of "k." expert evaluations corresponding to "x" and "y" criteria. The decision matrix formed by weighted means of all experts' scores can be shown in following matrix form (Eq.3)

$$\tilde{A} = \begin{bmatrix} 1 & \tilde{A}_{12} & \cdots & \tilde{A}_{1m} \\ \tilde{A}_{21} & 1 & \cdots & \tilde{A}_{2m} \\ \vdots & & \cdots & \\ \vdots & \vdots & \cdots & \vdots \\ \tilde{A}_{m1} & \tilde{A}_{m2} & \cdots & 1 \end{bmatrix} \quad (3)$$

After obtaining the decision matrix the weight of each criterion can be calculated in two steps, the first step is computing the geometric mean of each row in decision matrix as shown in Eq.4.

$$\breve{b}_i = \left(\widetilde{a_{i1}} \otimes \widetilde{a_{i2}} \otimes \ldots \otimes \widetilde{a_{in}} \right)^{1/n} \quad (4)$$

where "n" stands for total number of criteria, "$\widetilde{a_{in}}$" is the fuzzy comparison value between two criteria "i." and "n." and " b_i " is the fuzzy geotmetric mean of the all compared criteria. As a second step, a fuzzy weight of each criterion is calculated by applying Eq.5.

$$\widetilde{w}_i = \widetilde{b}_i \otimes \left(\widetilde{b}_1 + \widetilde{b}_2 + \ldots + \widetilde{r}_n \right)^{-1} \tag{5}$$

where, "$\widetilde{w}_i$" is the fuzzy weight of criterion "i.". The remaining part of the Buckley's Fuzzy AHP technique requires convertion of the fuzzy numbers to corresponding absolute values and calculation of relative weights among all criteria.

Where,"B" is referred to triangular fuzzy number, defuzzification of "B" can be applied using Eq.6.

$$\breve{B} = \frac{\widetilde{b}_1 + \widetilde{b}_2 + \widetilde{b}_3}{3} \tag{6}$$

For better evaluation of the obtained values, normalization is applied on the main criteria as it is in Eq.7.

$$\left(w_i^R \right)^N = \frac{w_i^N}{\sum_{i=1}^n w_i^N} \tag{7}$$

where "$\left(w_i^R \right)$N"is referred to normalized weight of each criteria and "n" is the total number of the criteria. Similarly, normalization of sub-criteria is also performed for each elements of the matrix.

Finally, relative fuzzy and absolute weights of main and sub-criteria are computed by multiplying main and related each sub-criteria in the matrix as shown in Eq.8 and Eq.9.

$$\left(\widetilde{w}_i^R \right)^{SN} = \left(\widetilde{w} \right)^N \otimes \left(\widetilde{w}_i \right)^{SN} \tag{8}$$

where, "$\left(\widetilde{w}_i^R \right)$SN" is the fuzzy relative weight of "i." sub-criterion, "$\left(\widetilde{w} \right)$N" is the fuzzy weight of the related main criterion and "$\left(\widetilde{w}_i \right)$SN" is the fuzzy weight of the same sub-criterion.

$$\left(w_i^R \right)^{SN} = \left(w^R \right)^N \times \left(w_i^R \right)^{SN} \tag{9}$$

In Eq.9, "$\left(w_i^R \right)$SN" is the normalized relative absolute weight of the "i." sub-criterion, "$\left(w^R \right)$N" is the normalized absolute weight of the related main criterion and "$\left(w_i^R \right)$SN" is the normalized absolute weight of the same sub-criterion.

3 CASE STUDY

With an aim of obtaining data sets by establishing criteria within the framework of research model; sector employees, academicians, analysts and ship's crew were interviewed, safety reports were evaluated and criteria were established by considering the agreements and conventions published for maritime security. As a result of the evaluations made, it was decided that 5 main criteria shown in Table 1, would be studied.

Table 1. Criteria determined for the study.

C1	Regional or local political vacuum and impotencies
C2	Inadequate protection and lack of inspection in risky regions
C3	Ideological and political reasons
C4	Inadequate legislation and regulations related to safety measures
C5	Economical insufficiencies

Maritime management is comprised of 8 people (Oceangoing master/3, sector employees/2, academician/3) and these people have experiences about ship's crew and international maritime law. Questionnaire forms of compiled criteria shown in Table 1, questionnaires were applied to the participants with an aim of obtaining opinions of decision makers. Paired comparison matrixes of each expert related to all the criteria were obtained in the form of verbal statements as a result of the evaluation of all the criteria. Due to the fact that all the questionnaire data, which were collected from the experts, were in verbal forms, they need to be converted to triangular fuzzy numbers in accordance with fuzzy number equivalents of linguistic scale, which was specified before (Ting et al., 2004, Xu and Yager, 2008). The values in Table 2 were used in these conversions.

Table 2. Linguistic terms used for Buckley's Fuzzy AHP (Chen et al., 2005; Kafalı et al., 2014; Özdemir, 2015, Özdemir and Güneroğlu, 2016; Özdemir and Güneroğlu, 2017).

Linguistic Terms	Fuzzy Number
Slightly more important (Row)	(1, 3, 5)
Strongly more important (Row)	(3, 5, 7)
Highly more important (Row)	(5, 7, 9)
Absolutely more important(Row)	(7, 7, 9)
Equally important	(1, 1, 3)
Absolutely more important (Column)	(0.111, 0.111, 0.143)
Highly more important (Column)	(0.111, 0.143, 0.200)
Strongly more important (Column)	(0.143, 0.200, 0.333)
Slightly more important (Column)	(0.200, 0.333, 1.000)

Table 3. Aggregated fuzzy decision matrix for the criteria

	C1	C2	C3	C4	C5
C1	1,000	0,254	0,650	0,251	1,210
	1,000	1,127	1,000	0,254	2,214
	1,000	1,325	1,447	0,621	3,221
C2	0,341	1,000	0,244	0,366	1,324
	0,600	1,000	1,010	0,215	1,276
	1,417	1,000	1,303	1,250	4,663
C3	0,525	0.241	1,000	0.333	1.552
	1,000	0.650	1,000	0.654	2.180
	1.334	1.530	1,000	1.000	3,256
C4	1.245	2.330	1.550	1,000	1.320
	1.355	1,024	1.270	1,000	2.480
	3.156	2.200	3.440	1,000	4.544
C5	0.452	0.415	0.270	0,250	1,000
	0.333	0.227	0.280	0,410	1,000
	1,000	1.000	0.333	0,325	1,000

Incorporated fuzzy decision matrixes of the data, which were obtained as a result of paired comparison of main criteria by using formula 10 and 11, were calculated as in Table 3.

$$\tilde{C}_{ij} = (1/N) \otimes (\tilde{c}_{ij}^1 \oplus \tilde{c}_{ij}^2 \oplus \ldots \oplus \tilde{c}_{ij}^N) \qquad (10)$$

$$\tilde{D}_{ij} = (1/N) \otimes (\tilde{d}_{ij}^1 \oplus \tilde{d}_{ij}^2 \oplus \ldots \oplus \tilde{d}_{ij}^N) \qquad (11)$$

The step after calculating incorporated fuzzy decision matrixes calculations was the calculation of criterion weights according to Buckley approach and it is carried out in the second step. As a first step, geometric average of each line of incorporated fuzzy decision matrixes is calculated. This process is expressed in formula 4. In the second step, geometric average of matrix is calculated and then its fuzzy weight value is calculated with the help of formula 5. This process was applied for all main and sub-criteria. Weighted fuzzy decision matrix, which was calculated for main criteria, was shown in Table 4.

Table 4. Weighted fuzzy decision matrix for criteria

C1	0.025	0.203	0.387
C2	0.030	0.321	0.205
C3	0.070	0.152	0.241
C4	0.204	0.227	0.761
C5	0.048	0.018	0.327

Table 5. Defuzzified criteria weights

C1	0.067	0.131	0.648
C2	0.035	0.210	0.400
C3	0.040	0.126	0.347
C4	0.174	0.278	0.498
C5	0.098	0.214	0.145

Table 6. Normalized criteria weights

Criteria	Normalization	Percentage (%)	Order
C1	0.1368	13,68	5
C2	0.2168	21,68	2
C3	0.1768	17,68	4
C4	0.2138	21,38	3
C5	0.2558	25,58	1
Total	1.000	100	

According to Buckley approach, the next step is conversion of fuzzy values into absolute values. According to this, defuzzification and normalization processes are carried out. Formula 6 was used for this calculation. Formula 7 was benefited in order to evaluate absolute weights in a better way. The results, in which calculated defuzzification process was included for the criteria, were shown in Table 5 and normalization results were shown in Table 6.

4 RESULTS

According to the results of the study, the reasons that caused privateering activities emerging in the seas are specified as follows in order of priorities: Economical insufficiencies. (C5), inadequate protection and lack of inspection in risky regions (C2), inadequate legislation and regulations related to safety measures (C4), ideological and political reasons (C3), regional or local political vacuum and impotencies (C1). According to the results of the study, criteria of economical insufficiencies (C5), which is ranked in the first place, and pressure of global financial crisis upon the countries reveal that main reasons of privateering activities were experienced due to economic reasons.

When the previous studies about privateering are considered, it is thought that this study will significantly contribute to the literature and future studies due to the lacking number of quantitative studies about privateering incidents, which are accepted as a major problem for maritime industry. Also, with the model approach used in the study, decision maker was provided a solution, which is reliable, flexible and easy to use. Adaptation process of this method to the problem and obtained results show that the method is simple, intelligible and useful enough for maritime sector and also it shows that it could be converted into a flexible structure in order to use it in a full range of maritime industry decision making problems easily. Therefore, it is thought that applicable solutions could be proposed in the future for different problems in maritime sector by using various multi-criteria decision making methods.

REFERENCES

Güğerçin, G. (2007). *Deniz Taşıma İşletmelerinde Maliyet Yapısının İncelenmesi ve Yük Taşımacılığı Uygulaması.* Zonguldak Karaelmas University. Department of Business Administration, MsC. Thesis, Zonguldak, Turkey (in Turkish).

Berg, J. (2011). You're Gonna Need a Bigger Boat: Somali Piracy and the Erosion of Customary Piracy Suppression. *New England Law Review*, 44, 343–385.

Bulut, E. (2013). *Economic Analysis on Ship Investment and Management Strategy of Dry Bulk Shipping.* Kobe University. Doctor of Philosophy in Maritime Science and Technology, PhD. Thesis, Kobe.

Drobetz, A., Merikas, A., Tsionas M.G., and Merika A. (2014). Tsionas Corporate social responsibility disclosure: the case of international shipping Transport. Res. Part E: Logistics *and Transport. Review*, 71, 18–44.

Özdemir, Ü. and Güneroğlu, A. (2016). Cargo Type Selection Procedure Using Fuzzy AHP and Fuzzy TOPSIS Techniques: The Case of Dry Bulk Cargo Ships. International *Journal of Shipping and Transport Logistics* (in progress).

Vanek, O., Jakob, M., Hrstka, O. and Pechoucek, M. (2013). Agent-based model of maritime traffic in piracy-affected waters. *Transportation Research* , 36, 157–176.

European Union, (2014). *European Union maritime security strategy*. Council of the European Union Doc. 11205/14, Brussels: European Union.

Mazzarella, F., Vespe, M., Damalas, D. and Osio, G. (2014). *Discovering vessel activities at sea using AIS data: mapping of fishing footprints*. Paper presented at the17th international conference on information fusion (FUSION), Salamanca, 7-10 July 2014. New Jersey: IEEE Explore.

Wang, Z.J., Songa, Y., Lei, L. (2016). Combination of unreliable evidence sources in intuitionistic fuzzy MCDM framework. *Knowledge-Based Systems*, 97, 24–39.

Vespe, M., Greidanus, H. and Alvarez, A. M. (2015). The declining impact of piracy on maritime transport in the India Ocean: Statistical analysis of 5-year vessel tracking data. *Marine Policy*, 59, 9–15.

Alvarez M., Gammieri, V., Alessandrini, A., Ziemba. L., Mazzarella, F., Delaney, C. (2015). PMAR viewer user manual. *JRC Technical Report*, Luxembourg: Publications Office of the European Union, Luxembourg.

Germond, B. (2015). The geopolitical dimension of maritime security. *Marine Policy*, 54, 137–142.

Axel, K. (2013). The moral economy of Somali piracy organized criminal business or subsistence activity? *Global Policy*, 4(1), 94–100.

Pristrom, S., Yang, Z., Wang, J. and Yan, X. (2016). A novel flexible model for piracy and robbery assessment of Merchant ship operations. *Reliability Engineering and System Safety*,155, 196–211.

Özdemir, Ü. and Güneroğlu, A. (2017). Quantitative Analysis of the World Sea Piracy by Fuzzy AHP and Fuzzy TOPSIS Methodologies. *International Journal of Shipping and Transport Logistics*, in progress.

Jahanshahloo, G.R., Hosseinzadeh, L.F. and Izadikhah, M. (2006). Extension of the TOPSIS Method for Decision Making Problems with Fuzzy Data. *Applied Mathematics and Computation*, 181, 1544-1551.

Yang, T. and Hung, C. C. (2007). Multiple-Attribute Decision Making Methods for Plant Layout Design Problem. *Robotics and Computer-Integrated Manufacturing*, 23(1), 126-137.

Awad, M.R., Nazmy, T. and Ismael, I.A. (2013). Integrating Approach for Multi Criteria Decision Making (Case Study: Ranking for Bulk Carrier Shipbuilding Region. *International Journal of Scientific & Technology Research*, 2(10), 77-86.

Özdemir, Ü. and Güneroğlu, A. (2015). Strategic Approach Model for Investigating the Cause of Maritime Accidents. *Scientific Journal on Traffic and Transportation Research*, 27(2), 113-123.

Güneroğlu, N., Özdemir, Ü. and Güneroğlu, A. (2016). Decisions on Quality Assurance Criteria of Recreational Beaches. *Proceedings of the Institution of Civil Engineers Municipal Engineer*, 169(4), 233-242.

Dabrowski, J. J., Villiers, J.P. (2015). Maritime piracy situation modelling with dynamic Bayesian networks. *Information Fusion*, 23, 116–30.

Jahan, A. and Edwards, L.K. (2015). A state-of-the-art survey on the influence of normalization techniques in ranking: Improving the materials selection process in engineering design. *Materials & Design* (1980-2015), 65, 335–342.

Wang, Z.J., Songa, Y., Lei, L. (2016). Combination of unreliable evidence sources in intuitionistic fuzzy MCDM framework. *Knowledge-Based Systems*, 97, 24–39.

Buckley, J.J. (1985). Fuzzy Hierarchical Analysis. *Fuzzy Sets and Systems*, 17, 233- 247.

Ting, Y. H., Shih, T. L. and Gwo-Hshiung, T. (2004). Fuzzy Mcdm Approach for Planning and Design Tenders Selection in Public Office Buildings. *International Journal of Project Management*, 22, 573-584.

Xu, Z. and Yager, R. R. (2008). Dynamic Intuitionistic Fuzzy Multi-Attribute Decision Making. *International Journal of Approximate Reasoning*, 48(1), 246–262.

Chen, C.T., Torng, C.L., Huang, Sue-Fn. (2005). A Fuzzy Approach for Supplier Evaluation and Selection in Supply Chain Management. *International Journal of Production Economies*, 38, 1-13.

Kafalı, M., Özkök, M. and Çebi, S. (2014). Evaluation of Pipe Cutting Technologies in Shipbuilding. *Brodogradnja*, 65(2), 33-48.

Özdemir, Ü., 2016. Çok Kriterli Karar Verme Yöntemleri Kullanılarak Gemiler için Uygun Yük seçiminin Analizi, PhD. Thesis, Karadeniz Technical University, Trabzon, Octtober (in Turkish).

The Lusitania Tragedy: Crime or Conspiracy?

E. Doyle
Cork, Ireland

ABSTRACT: The British passenger liner RMS Lusitania, built for the transatlantic trade between Europe and North America, was one of the largest such vessels of her time. For much of the first year of conflict during the Great War her owners and operator, Cunard Line, employed the vessel on scheduled passenger services between Liverpool and New York. On 7 May 1915, while steaming off the south coast of Ireland on the homeward passage towards Liverpool, Lusitania was torpedoed by the German submarine, U-20. Though only one torpedo was fired in the attack, two separate and distinct explosions followed, and the liner sank in 18 minutes with the loss of 1201 lives. Amongst the dead were 128 citizens of the then neutral USA. The sinking and its collateral effects became the focus of a major propaganda campaign by Great Britain, and was ultimately an influential factor in drawing America into the war as a British ally.

Favourable propaganda for the British side was contingent on media acceptance that Lusitania was unarmed and carried no contraband cargo. German justification for the attack was the conviction that the ship was carrying substantial consignments of war munitions thus making her a legitimate military target. This paper reviews aspects of the incident in light of the original manifest for the fateful voyage.

1 INTRODUCTION

1.1 *The British Shell Crisis of 1915*

During the first year of conflict in the Great War the belligerents found themselves locked in a resource-sapping stalemate, bogged down in trench-warfare with no immediate end in sight. The early euphoria, when both sides anticipated victory by Christmas 1914, was well faded and by early 1915 the high expenditure of munitions due to the unforeseen reliance on artillery exchanges had led to 'shell shortages' on both sides. Germany and the Central Powers had appeared more successful in ramping up production to replenish depleted stocks, but the 'Shell Crisis' (or 'Shell Scandal') of 1915 led to a political crisis in Britain from which Lloyd George would ultimately rise to prominence. It has been argued that the strategic plans of the belligerents on the Western Front caused an over-reliance on high-explosive shrapnel shells to counter infantry formations, especially in the open, and the unanticipated high rate of fire over a long period gave rise to seriously depleted stocks of shells.

1.2 *The War at Sea*

In the maritime sphere, exclusively naval engagements were infrequent while the dominant hostile actions in the North Sea and the Western Approaches to the British Isles were focused on blockade and disruption of trade, with each side seeking to deny or degrade strategic commerce to the ports and harbours of its opponent. In this regard, the British blockade of the North Sea was notably effective in eliminating German commerce.

1.3 *Blockade Law*

Historically, the rules of blockade were based on customary international law, but codified to a limited extent by the Paris Declaration of 1856:
− privateering was abolished;
− neutral goods, except contraband, were not subject to capture;
− a neutral flag protected a belligerent's goods, except for contraband;
− and a blockade to be legal must be effective.

A more ambitious attempt at codifying the law of war at sea was developed in 1909 by the Declaration of London, but Britain could not ratify the new accord because of opposition in the House of Lords. However, several maritime powers acknowledged the partial progress as comprising a new statement of customary law, and at the outset of the war Germany declared her intention to abide by its terms. The United States sought to persuade Britain to do likewise, but without success.

1.4 The British Blockade

The implementation of the British blockade, within days of the outbreak of war, coincided with the announcement that the entire North Sea was considered a war zone and subject to mining. These and other measures, especially the stretched definition of 'contraband' as including all food and historically exempt goods, violated the traditional laws of blockade. Enforcement of lawful blockade required a belligerent to effect 'close blockade' of its enemy's ports and coastal waters only; the blockade should not be so extensive as to directly affect neutral ports. Britain had, in effect, adopted a contentious strategy of 'distant blockade'. But it must also be recognized that Britain had the added strategic objective of enticing/provoking the German High Seas Fleet into a classic fleet action, an engagement from which she might reasonably feel confident of victory.

1.5 The German War Zone

For her part, Germany declared a war zone around the British Isles on 4 February 1915. She announced that all enemy merchant ships in that area, armed and unarmed, were subject to destruction, possibly even without warning. Equally, this blockade measure was also a clear breach of the law of armed conflict at sea, which required that merchant ships be warned, and their crews and passengers be allowed to abandon ship, before they were sunk. Germany countered that Britain's violations, indeed abandonment, of the 'cruiser rules' left her with no choice.

Figure 1. First Lord of the Admiralty, Winston Churchill

1.6 The Cruiser Rules

The rules of war allowed a belligerent warship, including a U-boat, to challenge any merchant ship to accept a 'visit and search' party. In the most civilized encounter the merchantman would stop to facilitate boarding when signaled to so do. Failure to respond to such a signal invited a shot across the bows, and any continued obstinacy would free the warship to fire 'for effect' at or into the merchant ship. If the boarded ship proved to be neutral and not carrying contraband she would be allowed resume her voyage. Enemy ships, on the other hand, could be seized as a prize of war, or destroyed, but only after the safety of crew and passengers was assured, ie, they were allowed take to the lifeboats.

1.7 Transition to Unrestricted Submarine Warfare

At the outbreak of war, German U-boats generally attempted to abide by the cruiser rules. Submarines were much more vulnerable on the surface but there was a certain financial logic to sinking enemy shipping, where possible, using the single deck gun rather than firing a very expensive torpedo (a scarce asset, in any case). Very soon after hostilities started the British Admiralty (Winston Churchill was the political head, or First Lord, at this time) sought to exploit the U-boat's surface vulnerability by instructing merchant ships to resist a submarine's right to effect 'visit and search', and to escape if possible. In particular, they advised "...if a submarine comes up suddenly close ahead of you with obvious hostile intention, steer straight for her at your utmost speed..."

The cruiser rules had become history. And taken with Churchill's other infamous instructions to British merchantmen, to fly the colours of neutral states and especially the United States, it is not difficult to understand Germany's reluctance thereafter to have its U-boat fleet abide by rules of war that its enemy had seemingly abandoned without compunction. The world was about to grapple with the implications of 'unrestricted submarine warfare'.

2 HUNTER AND PREY

2.1 Status of LUSITANIA on the North Atlantic

At 240m length overall and displacing 38000 tons, LUSITANIA was the largest ship in the world when launched in 1906. She and her sister ship, the slightly larger and longer MAURETANIA, both entered service on the North Atlantic the following year. They were fast ships, regaining for Britain the notional Blue Riband for the fastest Atlantic passage. The record, initially taken by LUSITANIA, was exchanged several times with her marginally faster sister, until September 1909 when *MAURETANIA* set a record at fractionally over 26 knots, which was to stand for the next 20 years.

Figure 2. RMS Lusitania, in wartime 'black funnel' livery.

2.2 *Conception as Auxiliary Cruiser*

These two ships had an interesting financial pedigree at a time when the British liner trade was enduring a period of crisis. The owners, Cunard Line, did not have the resources to build new ships and there was an unfolding prospect that American financial strength and stiff German competition would squeeze British tonnage out of the lucrative and prestigious Atlantic liner business. Cunard saw opportunity in the crisis and persuaded the British Government to advance a 20-year loan of £2.6 million to finance two ships. The deal was further sweetened with receipt of an annual government subvention of £75,000 for each ship, and a mail contract worth £68,000. All Cunard had to do in return was to have the ships built to Admiralty specifications to facilitate their use as auxiliary cruisers in times of war. Thus were *LUSITANIA* and *MAURETANIA* conceived.

2.3 *Wartime Deployment*

At the outbreak of war the British government agreed to the withdrawal of *MAURETANIA*, and her 'half-sister' *AQUITANIA* from liner service and to lay-up in Liverpool – they would later serve as troop transports and hospital ships. *LUSITANIA* was allowed to continue in commercial passenger service, though, in the interests of fuel economy, six of her twenty-five boilers were shut down (No. 4 Boiler Room), reducing her service speed to 21 knots.

2.4 *Alleged Armament*

It would appear that the very high fuel consumption of these ships (800/900 tons of coal per day) in normal pre-war liner service was a significant factor in the Admiralty's decision to forego their requisition for conversion to armed merchant cruisers (AMC's). It is almost certainly the case that the construction of these ships, funded and designated to fill that wartime role (AMC), incorporated suitably strengthened deck structures to facilitate the rapid mounting of appropriate armament in times of crisis. But there is no evidence that any such guns were ever fitted on *LUSITANIA*.

Some publications have suggested that the ship had up to twelve 6" guns concealed in the forward shelter deck and so configured that they could be wheeled out of hiding and positioned on their mounting rings in 20 minutes. Leaving aside the "minor" difficulty of handling such a heavy weapon and its mounting into position at sea, how was it to be so secured as to safely withstand the considerable recoil when fired. The very notion is ludicrous beyond comprehension.

2.5 *U-20*

One of the newer submarines, U-20 was built in Danzig (Gdansk) and commissioned on 5 August 1913. Her length was 64.2m, submerged displacement 827 tons and surface speed 15.5 knots. She had four 50cm torpedo tubes (two bow and two stern) and carried six torpedoes, and a single 88mm deck gun. Her commander was Kapitänleutnant Walther Schwieger, and he had started his patrol on 30 April, sailing northabout thence southwards to the west of Ireland until he sighted Fastnet Rock on 5 May.

Figure 3. Kapitänleutnant Walther Schwieger, of U-20.

2.6 *Room 40 tracks U-20*

With U-20 now on station, Schwieger began a vigilant patrol along the south coast of Ireland. Undoubtedly, his patrol orders included intelligence that all British merchant traffic had been re-routed to the south of Ireland and, in particular, he could expect *LUSITANIA,* steaming eastward, through this area about 7/8 May. If German intelligence was good, British SIGINT was better; the Admiralty's Room 40 was able to intercept U-20's situation report on that evening of 5 May – she had sunk the schooner *Earl of Lathom* by gunfire at 1730, and later reported her position as 51.32N 8.22W (Old Head of Kinsale area). At 2130 that same evening she was reported 12M south of Daunt's Rock (outer approaches to Cork Hbr and Queenstown). Only a fool or a knave could fail to appreciate the magnitude of the threat that U-20 now posed to *LUSITANIA*.

2.7 More Victims

The next day, 6 May, saw U-20 continue her patrol eastward towards St George's Channel, at the southern extremity of the Irish Sea. By early afternoon she had sunk two British merchant ships, *Candidate* and *Centurion*, before leaving the area to head westward overnight. Once again Room 40 was tracking her activity.

3 THE ATTACK

3.1 Prudent Decisions

As *LUSITANIA* approached the Irish coast on the morning of 7 May, her master, Captain William Turner was clearly cognizant of the heightened submarine risk attending this final stage of the voyage. Thus far, he had maintained the full service speed of 21 knots, but he now reduced speed to 18 knots so as to avoid arriving at Liverpool Bar before daybreak on 8 May; *LUSITANIA* could then proceed directly into port and thus avoid loitering about in the port approaches (becoming an easy U-boat target) by arriving too early. Turner took other sensible precautions: he had ordered all life-boats swung out and all bulkhead doors to be closed except those required open in order to work the ship; portholes were also closed (some survivors would later dispute this, but it is quite probable that some portholes above main-deck level were open); the lookout positions were doubled and the strength of the bridge watch was increased. Very high steam pressure was ordered, so as to give the vessel all possible speed if the bridge so required.

3.2 Old Head of Kinsale

Shortly after 0800, when speed had been reduced to 18 knots, *LUSITANIA* encountered fog necessitating a further speed reduction to 15 knots. Speed was restored to 18 knots when the fog lifted, a little before noon. Though land was in sight at this time, nothing positive could be identified so Turner maintained his course, S87E. At 1240 he felt confident enough to alter course to N67E, taking the ship closer to shore in the expectation of establishing a positive landfall. His judgment was correct, and

when the Old Head of Kinsale was identified he resume his former course (S87°E) at 1340. *LUSITANIA* now maintained this course so that the ship's position could be accurately fixed by the 'four-point bearing' method, a form of 'running fix' that would see the ship maintain her course until the Old Head of Kinsale came directly abeam.

3.3 U-20 in Fortuitous Attack Position

Meanwhile, U-20 was prowling westward and on the lookout for more prey, not too far from the position of her first 'success' (*Earl of Lathom*) two days previously. At 1320 (recorded as 1420 in U-20's war diary, but all relevant times in this paper are referenced to GMT.) the lookouts reported distant smoke on the starboard bow. Schwieger, also on the conning tower, identified a four-funnel ship, about 12-14 miles distant and possibly heading for Queenstown. He altered course to intercept and submerged for a possible attack. As U-20's officers figured out the probable identity of the target and worked on the attack solution, Schwieger must have been amazed to see his target alter course almost directly towards him (Turner's 1340 alteration to S87°E, although U-20 logs this event at 1350), presenting a near perfect attack position.

3.4 Single Torpedo Strike

Schwieger fires a single bow torpedo at 1410, into *LUSITANIA's* starboard side. He has reckoned his firing solution on a range of 700m, intersection angle of 90° and a target speed of 22 knots. His over-estimate of *LUSITANIA's* speed, by 4 knots, means that the torpedo will strike the target further forward than his midship aiming point. The precise point of the torpedo impact has never been established definitively. But the weight of circumstantial evidence points relentlessly towards a location between the bridge and foremast, three metres below the waterline, into or adjacent to the cargo hold.

3.5 The Second Explosion

What happens next is not in dispute; a second, much larger explosion follows that of the torpedo, almost immediately. Through his periscope, Schwieger, notes a level of destruction well in excess of his expectations for a single torpedo hit. His log notes: *"The ship stops immediately and heels over to starboard quickly, immersing simultaneously at the bow. It appears as if the ship is going to capsize very shortly."* LUSITANIA does not capsize; her increasing list appears to be checked at some point and it seems as if the vessel is being righted to a more upright condition. But she continues to sink by the bow and then resumes heeling to starboard again. The great liner sinks in 18 minutes, taking 1201 souls with her.

3.6 Fabrication of the 'Second' Torpedo

The Formal Investigation (chaired by Lord Mersey) into the loss of LUSITANIA convened in London the following month. Mersey duly reports, that on board LUSITANIA, the first warning of the torpedo attack comes from the lookout, high in the crows nest. The warning is acknowledged by the Second Officer who shouts a warning: *"There is a torpedo coming, sir"*. Captain Turner is on the port side of the bridge and, looking to starboard, he sees *"...a streak of foam in the wake of a torpedo travelling towards his ship."* Mersey continues: *"Immediately afterwards the "Lusitania" was struck on the starboard side somewhere between the third and fourth funnels. The blow broke number 5 life-boat into splinters. A second torpedo was fired immediately afterwards, which also struck the ship on the starboard side. The two torpedoes struck the ship almost simultaneously."* Leaving aside the unambiguous 'one torpedo' evidence of the U-20 war diary (which would not come to light for many years), Mersey's findings challenge reality: no competent submariner would fire two torpedoes simultaneously, risking a double miss as easily as a double hit; he would be much more likely to fire a consecutive salvo, timed to achieve a realistic 'spread' at the moving target. Secondly, torpedo damage so far aft (*"...between the third and fourth funnels."*) could not possibly cause the ship to trim by the head and submerge the bow so rapidly. A damage location this far aft, abaft the ship's tipping centre, would have had quite the opposite effect, in that significant flooding here would have caused the ship to trim by the stern.

3.7 Nasty Consignments?

It seems quite clear from all the Mersey evidence that the 'invention' of a second torpedo hit was an Admiralty attempt to explain the very real second explosion. Equally, and very obviously, this conspiracy attempted to locate the torpedo strike as far away from the cargo hold as possible. But there is no longer any doubt: one single torpedo was fired and struck *LUSITANIA* well forward, at or near the cargo space, and a massive secondary explosion followed almost immediately. Over the intervening years, many theories have been advanced to explain the cause of the second explosion. However, it is impossible to review the *LUSITANIA* literature and evidence and not conclude that her last cargo must have comprised some very nasty consignments, details of which the British authorities were not then, and are still not, keen to reveal. In this regard, it is instructive to note Captain Turner's appearance at a Coroner's Inquest at Kinsale the day after the sinking.

3.8 The Kinsale Inquest

When the local coroner learned that some of the victims had been landed at Kinsale he immediately convened a jury inquest for 8 May and subpoenaed Turner and some other survivors who had been landed at Queenstown. This development caused evident consternation at the Admiralty who issued instructions to the local Crown Agent in Cork to stop the proceedings; under no circumstances, was Captain Turner to give evidence. But the Admiralty agent was too late because the coroner, possibly smelling a conspiratorial rat, had acted with extraordinary eagerness and held his court that very afternoon. In giving his evidence, Turner explained that he was constrained by wartime secrecy in what he could say, but he did testify that only one torpedo had been fired.

Six weeks later, at the Mersey enquiry, the one torpedo reality had become a two torpedo fabrication. It is self evident that the Admiralty seemed determined to create an 'alternative fact' by presenting a second torpedo as the explanation for the second explosion.

4 THE CARGO

4.1 LUSITANIA's Perceived Immunity

From the earliest days of the war, *LUSITANIA's* passenger liner status was seen to confer a condition of near invulnerability upon the vessel; in the first instance, her speed would minimize the risk of a successful submarine attack, and secondly, her clearly advertised employment in an exclusively civilian role presumed immunity from hostile interference. An innocent civilian role implied that the ship must not carry contraband, but this definition had been so degraded by the British side, at the outset, that in reversed circumstances, almost all of *LUSITANIA's* cargo would have been judged contraband by a British warship.

4.2 Non-Explosive War Material

Given the myth of semi-invulnerability attached to *LUSITANIA*, and the frequency and regularity of her transatlantic service, and the government subvention to which the ship was bound, it should be no surprise that the British authorities would demand absolute preference on eastbound cargo space for whatever war material it needed. The finer points of blockade law had already been ignored and, in Winston Churchill, the Admiralty had a First Lord who blatantly advocated 'alternative' rules of war that endangered neutrals, in the process. The American press had carried frequent reports about munitions being shipped to Britain and France and highlighted the risks this trade posed to US neutrality. The public concern about the use of passenger ships from US ports carrying munitions to Europe had been somewhat assuaged by the contentious assurance from port authorities that passenger ships could only load 'non-explosive' war material. And .303 rifle ammunition was deemed to fall into this category. For a belligerent power, under political and military pressure to solve a growing 'shells crisis', the temptation to use *LUSITANIA's* cargo space must have been irresistible.

4.3 *The Cargo and Manifest(s)*

4.3.1 Within days of the tragedy *LUSITANIA's* two-page manifest was released to the New York newspapers. It was a condensed version of the real thing, but the same document would be submitted to the Mersey inquiry, and for many decades afterwards was considered the only record of *LUSITANIA's* last cargo. The pejorative label 'sanitized' has been attached to it, with some justification.

4.3.2 The actual, "Supplementary", manifest runs to twenty-five manuscript pages and was filed on 5 May, four days after the ship sailed from New York. This document was 'supplemental' to a one-page "Shippers Manifest" that was filed with customs on 30 April in order to effect outward clearance for the ship. It is hardly surprising that the Shippers Manifest is devoid of any entry relating to munitions or other goods of a military character.

4.3.3 The combined (Shippers and Supplementary) manifest was retained by the Collector of Customs for New York, until the incumbent, Harry Durning, sent it to President Roosevelt in 1940. He had it locked away and it has remained in the Roosevelt Papers ever since, but circulation was released several years ago and it is now downloadable on Lusitania internet sites.

4.3.4 The sanitized manifest includes many suspect items, but also, two obvious entries that beg further explanation:
– *Ammunition, cases 1271*
– *Cartridges and ammunition, cases 4200*
    It is clear from the original 'combined' manifest that the latter item above refers to seven entries on Sheet 18, totaling 4200 *"cases cartridges"*. This was a consignment of .303 rifle ammunition, being shipped by Remington. Such small calibre ammunition was considered relatively safe in the context of loading onto a passenger ship, and in *LUSITANIA's* case there was no violation of

American law in so doing. Port of New York practice saw small arms ammunition as "Non-Explosive". By way of comparison, modern Dangerous Goods rules would most likely label such ammunition as Class 1.3 (minor blast/projection hazard).

4.3.5 The sanitized manifest, relating to the entry *"Ammunition, cases 1271"*, is an altogether different matter. In this case, the only possible match with the original manifest appears on Sheet 2, as follows:
– *1250 cases Shrapnel*
– *12 cases Fuses*
– *6 cases Fuses*
    All three lots were shipped by Bethlehem Steel and consigned to the Royal Arsenal, Woolwich. The related shipping note for this consignment is more illuminating, wherein Bethlehem Steel describes the shrapnel as: '1,248 cases of three-inch calibre shrapnel shells, filled; four shells to each case'. There can be no ambiguity here; these were not containers of raw shrapnel material for later processing in the manufacture of live shrapnel rounds. They were the real thing, live rounds being shipped to Britain for ultimate use in the QF 13 Pounder Field Gun. And when packed as described, 'four shells to each case', they were packed in precisely the same manner as artillery units expected to receive them. 'QF' signifies quick firing ammunition which means that gun batteries must be supplied with *fixed* rounds, ie, the filled shrapnel projectile is fitted or *fixed* to the cordite filled shell case, as shown:

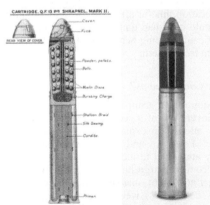

Figure 4. Example and cutaway of QF 13pdr shrapnel round. Image courtesy IWM ©

4.4 *Make-up of a 13pdr Shrapnel Round*

4.4.1 The 13 pounder shrapnel projectile, weighing 12.5 pounds without its fuze, was filled with 234 lead musket balls and a small bursting charge in the base cavity of the shell. Under battlefield conditions the bursting charge would be initiated by the timer/percussion fuze fitted to the nose of the projectile. But these were only fitted in the rear lines just before the complete round was sent forward to the gun batteries for use. And those rounds would then be marked 'Fuzed'. In general transport, the fuzes were stored separately and the nose of the shell was fitted with a protective plug or transit cap instead. The 18 cases of fuzes, part of this consignment of munitions, were stowed in LUSITANIA's aft magazine, and were discovered by a diving team, in 1982.

4.4.2 The complete round weighed 17 pounds, including the brass shell case containing the propellant charge of just over 1.25 pounds of cordite. Under no circumstances could such rounds be considered "Non-Explosive", as the manifest implied and as Cunard and the Admiralty later alleged. Applying the modern Dangerous Goods classification we would expect to find such munitions in Class 1.1 (a substance or article with a mass explosion hazard).

4.5 *LUSITANIA's Explosive Cargo*

4.5.1 *LUSITANIA* was laden with 5000 live shrapnel rounds (1250 cases), stowed in the lower cargo hold, having an explosive mass of 6250 pounds of cordite and supplemented by the additional effect of 5000 bursting charges. A torpedo strike in this location must have had the most adverse consequences; how could a mass explosion of 5000 artillery rounds not have been triggered, together with any other explosive material stowed in that area. If the British Admiralty were minded to expose *LUSITANIA* to such a risk, and it is not possible to construe a more innocent interpretation, why would they not equally have taken the opportunity to load other consignments of munitions under the guise of false descriptions.

4.5.2 Approximately 90 tons of butter, cheese and lard, also stowed in the lower hold, were consigned to the Admiralty Weapons Testing Establishment at Shoeburyness. This strange consignment to such an establishment has never been explained, and although insured, no claim was ever made for its loss. Over the intervening years, many commentators have identified guncotton as the most plausible munition shipped under the guise of butter and cheese. While there is no direct evidence supporting this contention the circumstantial evidence is compelling.

4.5.3 The ship's cargo stowage plan shows "4927 Boxes Cartridges" in the baggage room area, directly above the lower hold. The manifest (original and sanitized) lists *"Cartridges and ammunition, cases 4200"*, giving a discrepancy of 727 boxes of cartridges. There is no mention of any such additional quantity of ammunition on the manifests, suggesting that these 727 boxes of cartridges, identified on the stowage plan, were declared as something quite different on the manifest.

4.5.4 It is interesting to note that reports from the most recent dive on *LUSITANIA* (2012) confirm a significant quantity of undamaged .303 ammunition in the debris field within and close to the wreck. This clearly indicates that the particular consignment of 4200 cases did not explode en masse, as the current wisdom held. No dives have yet detected any signs of shrapnel rounds, exploded or intact.

4.5.5 The diving survey operation of 2012 included the deployment of a manned submersible and ROV, and was conducted in conjunction with a series of real-time explosive trials and computer simulations at the High Explosives Applications Facility (HEAF) at Lawrence Livermore National Laboratory (LLNL), California. The entire dive programme and the HEAF trials were the central feature of a National Geographic video documentary, *Dark Secrets of the Lusitania*. Four possible 'second explosion' scenarios were examined by LLNL:
– Aluminium powder (listed on manifest);
– Coal dust (as in empty bunker space);
– Guncotton (widely suspected munition);
– Boiler explosion, as a computer simulation.

Aluminium powder and coal dust explosions were quickly eliminated as the explosive power generated was considered insufficient to cause major hull rupture. A guncotton explosion would certainly have yielded the necessary power to cause major structural failure. A boiler explosion (simulated) following the torpedo strike was seen as very likely, but would not have had the power to cause catastrophic hull failure – the HEAF team seemed unaware that there were survivors from all three working boiler rooms and no one reported a boiler exploding.

Significantly, a second-explosion scenario involving the 5000 rounds of 13 pounder shrapnel ammunition was not tested.

4.6 *Evidence Ignored by Mersey*

One harshly treated witness at the Mersey inquiry, Joseph Marichal, recalled the sound of the second explosion thus: *"The second explosion might have been due primarily to the explosion of a torpedo, but not to a torpedo alone. The nature of the explosion was similar to the rattling of a machine gun for a short period."* It is not difficult to understand how bursting shrapnel rounds could fit that description perfectly. Mersey chose to disbelieve this evidence.

Another witness likened the secondary sound to that of "shattering glass". And Senior Third Officer John Lewis testified: *"A few seconds afterwards, whether it was an explosion or not I couldn't say, but there was a heavy report and a rumbling noise like a clap of thunder"*. All these survivors believed the source and location of the torpedo strike and the second explosion lay towards the forward part of the ship.

5 CONCLUSIONS

1 On the outbreak of war, Germany declared that her offensive naval operations would be conducted according to the then accepted customary rules of war, and in the early months of hostilities seemed to honour that declaration. Great Britain gave no such undertaking, and it quickly became obvious to her enemy, the Central Powers, that she was in violation of the rules of blockade and had abandoned the cruiser rules. Germany responded in kind. By early 1915 we see both main belligerents fight the war at sea, in open violation of accepted international law. And with the advent of unrestricted submarine warfare, *LUSITANIA* could never be an immune target. She was much more; when used to carry war munitions, *LUSITANIA* the prize target, became an even more vulnerable one.

2 The sustained attempts by the British Admiralty to advance the notion of a second torpedo hit had one sole purpose, namely, to offer an explanation for the second, even greater, explosion within seconds of the torpedo explosion. The clear reality of a single torpedo strike is no longer contentious to impartial researchers. Equally perfidious is the Admiralty construction that the torpedo strike lay somewhere between No. 3 and No. 4 funnels; if this were so, the ship could not possibly have sunk by the bow, still less sink so rapidly.

3 U-20 fired a single torpedo and struck *LUSITANIA* at or near the ship's hold, exactly as her commander, Sweiger, recorded in his war diary. To suggest that the second explosion was anything other than a munitions explosion is delusional and akin to clutching at 'alternative facts'. Coal dust and aluminium powder may well have been a sub-set of the second explosion event, but the fact that there were survivors from all Boiler Rooms seems to negate a boiler explosion as a constituent of that event. However, the mis-declared consignment of 5000 artillery rounds could not have been anything other than a critical constituent of the second explosion.

4 If we are looking for criminality in the *LUSITANIA* saga we find it on both sides.

Conspiracy, on the other hand, resides almost exclusively within the British Admiralty. The specific signals addressed to and responses from *LUSITANIA* (as distinct from the general broadcast signals to all British ships) have never been released, leading some researchers to believe they may have been destroyed. The gaps in the files are indicative of conspiracy or cock-up, or probably both.

5 In Germany, the *LUSITANIA* sinking was hailed as a stunning success, but once the British propaganda machine swung into action, it quickly became a strategic disaster for Germany and a winning PR hand for Britain. Whatever the fallout from the *LUSITANIA* disaster, it was seen to be imperative to the British national interest to preserve the myth of the ship's non-explosive cargo; neutral sympathy was too valuable to squander. *LUSITANIA* did not bring America directly into the war, since it would be another two years before the US position changed. But the sinking, with such loss of American lives became a festering wound in US-German diplomacy which would ultimately have but one outcome.

6 More than a century later, the one enduring tragedy of the *LUSITANIA* affair is surely the continuing failure of the British authorities to finally release all relevant classified archival material.

REFERENCES

Liddell Hart, B.H., *A History of the First World War* (London, Book Club Associates, 1973)

Churchill, Winston, *The World Crisis (Vol II)* (London, Thornton Butterworth, 1923)

Mersey, *Formal Investigation into the Loss of the Steamship "Lusitania", Proceedings and Report of the Court* (London, Board of Trade, 17 July 1915) www.titanicinquiry.org/lusitania/index.php

IMO, *International Maritime Dangerous Goods (IMDG) Code*, Resolution MSC.122(75) (London)

Franklin D. Roosevelt Presidential Archive, *Lusitania Complete Manifest*, http://www.lusitania.net/deadlycargo.htm

The Lusitania Resource http://www.rmslusitania.info/lusitania/

Lusitania Online http://www.lusitania.net/deadlycargo.htm

Imperial War Museums, © IWM (MUN 5805) http://www.iwm.org.uk/collections/item/object/30024005

O'Sullivan, Patrick, *The Sinking of the Lusitania: Unraveling the Mysteries* (Cork, The Collins Press, 2014)

Simpson, Colin, *Lusitania* (London, Penguin Books, 1983)

Beesly, Patrick, *Room 40: British Naval Intelligence 1914-1918* (London, Harcourt, 1983)

National Archives, Kew, UK: http://germannavalwarfare.info/indexbr.html

ADM 137/4152 U-20 operations known to British Intelligence.

ADM 137/4353 Lusitania signals and copies.

ADM 137/3923 U-20 War diary.

Aviation

Current Challenges within Security Systems at International Airports

G. Nowacki & B. Paszukow
Military University of Technology, Warsaw, Poland

ABSTRACT: The paper refers to aviation threat assessment in light of current dimension of terror against international airports, in particular taking into consideration current threats and modus operandi of terrorist's organizations. Within this document highlighted current significant risk for airport environment, as well as possible methods and measures to mitigate thereof.

1 INTRODUCTION

Dangerous The recent attacks in Europe, carried out by Al-Qaeda (Paris, January 2015) and IS inspired or affiliated individuals and groups (Paris in November 2015 or Brussels in March 2016), have increased concerns about possible future terrorist attacks by jihadist individuals and other extremist groups, as well as radicalized units within the territory of EU Member States.

To meet the challenges of increased security regulations and growing number of passengers, and to keep up date on the latest anti terrorism strategies, airports in Europe are looking for tailor made solutions and employing experts capable of performing passenger and baggage screening at a high service level.

The only way to meet these specific demands and challenges facing the aviation industry is to have a focused and dedicated security organization, as well defined and implemented security systems based on means and measures aiming to prevent against acts of unlawful interference.

As the threat of terrorism stands serious scenario in today's aviation industry, deep concerns are being raised about the quality, effectiveness and efficiency of security controls at airports. The same security controls intended to be implemented in order to save human lives and protect airport critical infrastructure facilities (Bammer, 1997).

The EU is currently witnessing an upward trend in the scale, frequency and impact of terrorist attacks in the jurisdictions of Member States. France in particular has recently been hit hard by a series of terrorist attacks, perpetrated both by groups of terrorists and lone actors – so called "lonely wolfs". Islamic State (IS) is actively propagating terrorist acts on EU soil by any means available, increasingly inspiring radicalised individuals to act. IS already had proven to be very effective in both moving people to commit terrorist acts and to set these actions against particular targets.

The success generated by these terrorist attacks, as seen from an IS perspective, will encourage the group to hit even more targets within the territory of European Union. Although France appears to be the primary focus of IS, the organisation has threatened all countries that are part of the US led coalition against IS in Syria and Iraq, including Germany, the UK and other Member States, which have specifically been mentioned as enemies of IS in several video messages. New attacks by both terrorist groups and lone actors are thus to be expected, most probably following the patterns that have been used in earlier attacks. New variations in attack, for example in the use of CBRN weapons, may also be developing.

Taking into consideration mentioned above, airport authorities are looking for solutions, how to improve overall security at airports, without simultaneously jeopardizing passengers experience and still maintaining the throughput indicators at reasonable level. (Benny, 2013)

Therefore, within current aviation security experts' panels became revised concepts of advanced risks assessment concepts, not limited to profiling or behavioural detection elements – all supported with advanced technological solutions, aiming to detect the potential threats in advance.

Aforementioned aims in primary to revise airport's operational security focus, in order to instead search for prohibited articles, rather focus on people potentially representing threat to airport or airlines.

2 RISK BASED SCREENING

Within the field of risk based screening, the key elements are focusing on passenger risk assessment responsibility.

It is assumed that the authority responsible for screening at the point of departure will have primary responsibility for the final determination of risk assessment for an individual passenger by using the agreed National criteria and any additional information collected at the airport.

The relevant government authority for the country of transfer or arrival may also make an initial assessment for security purposes (Haliżak & others, 2004).

Where Government assessment of risk is undertaken, it will be a contributing factor to a number of other criteria, with the final categorization being made at the checkpoint entrance.

The actual level of screening that each category of passenger undergoes might depend on the measures applied by the screening authority (Kostur-Balcerzak, 2015).

Differentiation between known travellers, normal and enhanced may mean different levels of divestiture, different sensitivity of equipment or different levels of randomness for example.

World Wide Efforts for Risk-Based Security

Agency	Initiative
ACIE / AEA	better SECURITY
ACI WORLD	Next Generation Screening TaskForce
European Commission	Smarter aviation security for the EU
UK DfT	Outcome-based Security
IATA	Checkpoint of the Future
ICAO	Next Generation Screening TAG
TSA	TSA Pre✓™

Figure 1. Word Wide Efforts for Risk Based Security (Case Study Fraport, 2013)

Aforementioned deliberations brought again, in more intensive way, concept of risk based screening. Risk based screening considers in primary applying appropriate and proportionate levels of screening towards particular type of persons. Specifically, risk based screening is where system level security is adapted using technology and processes in response to the threat environment and a government's security risk needs.

3 PASSANGER PROFILING

Passenger profiling does not have the luxury of hard evidence, instead it has to make the assumptions before the crime has happened. Passenger profiling basically relies on the behaviour of passengers and other persons within the airport environment. Behavioural experts are highly trained in reading body language to pick up on the signs of potential nervousness or kinds of unusual behaviour (Olszewski, 2011).

This process shall never rely on appearance or the race of an individual. It should only pick up on passengers' behavioural patterns. This method being questioned, stating that terrorists are able to probe the security system to determine which members of their group fall into the low-risk passenger profiling categories. In the future these members could be used to pass unchallenged through security checks.

Convincing the public that the advantages of passenger profiling outweigh the disadvantages is proving to be a challenging task. Unfortunately, the nature of this process means that it is only when things go wrong and people do slip through the net that we will truly understand if it is working or not.

The disadvantages of passenger profiling, including concerns about stereotyping, are preventing the process from gaining widespread use. But surely if it helped towards deterring or better still foiling, a possible terror plot, and then it would be a process worth living with

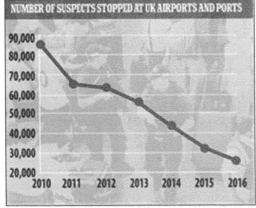

Figure 2. Number of suspects (Anti-terror police, 2016)

The number of people being stopped at airports has dropped 23 per cent It comes after authorities were accused of 'racially profiling' passengers.

Home Office insists other methods are now being used to stop attacks by Richard Spillett for mailonline.

4 NEW SCREENING TECHNOLOGIES

4.1 *Millimetre Wave Imaging*

Perceptions of increased threats from explosives and non-metallic weapons have prompted the investigation of new passenger screening technologies, including chemical, biological and explosive trace detection techniques and imaging methods that can see through clothing. Demands for additional space, utilities, labour costs, and increased operator skills that these technologies could impose on air carriers and airports also will have to be considered within every security system to be created.

For the time being several emerging technologies can detect metallic and non-metallic weapons, explosives, and other contraband material concealed under multiple layers of clothing by creating images that can be examined to discern these materials. No physical contact is necessary to be involved. (Price & Forrest, 2016)

Imaging technologies either scan subjects for natural radiation emitted by the human body (passive imaging) or expose subjects to a specific type of radiation reflected by the body (active imaging). In either case, materials such as metallic weapons or plastic explosives, which emit or reflect radiation differently from the human body, are distinguishable from the background image of the body.

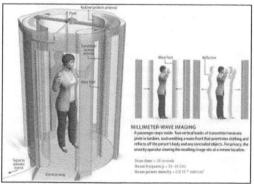

Figure 3. TSA imagine machines explained (w/poll) by Cartoon Peril Saturday, Dec 04, 2010

Images are viewed by an operator trained to identify potential threat objects in these images, sometimes with the assistance of image enhancing software that highlights unusual features or any other structural anomalies.

Although these technologies cannot detect objects concealed inside the body or in skin flaps, they are being considered for airport passenger screening because they would enable air carriers to screen for a wider variety of materials than they can with present screening systems (Siadkowski, 2014).

4.2 *Trace detection technologies*

Trace-detection technologies are based on the direct chemical identification of either particles of explosive material or vapour containing explosive material. Thus, the presence of a threat object or bomb is inferred from the presence of particulate matter or vapour.

The main difference between trace-detection and electromagnetic or imaging is that in trace detection, a sample of the explosive material must be transported to the instrument in concentrations that exceed the detection limit. Trace-detection technologies cannot be used to detect the presence of metallic weapons or any other metallic articles commonly classified as prohibited for transport on board an aircraft.

The two distinct steps in trace detection are sample collection and chemical identification. To identify the presence of explosives, both steps have to function at the same time. The sample-collection phase of the procedure is the main point of contact between the technology and the subjects being screened.

Explosive substances can be transported from the carrier to the detection instrument as vapour or as solid particles. Initial efforts in the development of trace-detection technology were focused on collecting vapour around the person or baggage. However, because many modern explosives do not readily give off vapour at room temperature, the focus has expanded to include detection of particulates of explosive materials on the skin and other surfaces. The methodology and technique of trace sampling procedure stands the key element influencing effective detection of possible threat.

If traces of explosive material are to be detected, they must be concentrated from an air sample (vapour technologies) or dislodged from a substrate (particulate technologies). In vapour detection, large amounts of air must be collected, from which small amounts of the substances of interest must be extracted. In particle detection, pieces of explosive material must be removed from the surface to which they are adhering. Both trace-detection approaches have strengths and weaknesses, depending on the type of explosive material being sought. Vapour technologies are more effective for detecting explosive materials with high vapour pressures, while particulate technologies are more appropriate for explosive materials with low vapour pressure, such as military plastic explosives.

Although using a hand-wand device is a potentially efficient sample-collection technique, it is more labour intensive and more time consuming than collecting samples using an automated portal. The optimum solution may be to attach a hand-wand device to a portal-based trace-detection system as a higher-level surveillance accessory.

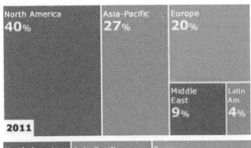

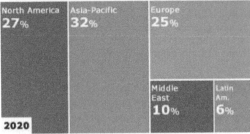

Figure 4. Explosives & Narcotics Trace Detection (ETD): Technologies & Global Market – 2015-2020

Following the January 2015 terror attacks in Paris we forecast that the demand for ETD systems will increase in the European aviation & public transportation security sector, secured facilities, first responders and public venues security.

4.3 Explosive Detection Dog Teams - EDDT

The dog-operation is very cost-efficient, due to a couple of facts. A dog can smell a million times better then a human being and is therefore able to scan 6 pallets or containers in just a few minutes' time. For example, in relation transport of shipments, the disruption of the normal cargo logistic chain is limited to an absolute minimum. In the result thereof the sniffing process is both quick and efficient and does not affect the logistic process itself. Moreover its one of the most flexible method of security control, bearing in minds its mobility characteristics.

With reference EDDT method each explosion-detection team consists of one dog handler together with one or two dogs. The dogs are selected after a number of tests. These tests have to determine their attitude and endurance while searching for explosives, their obedience, their psychological strength and their ability to learn. The dogs must fit not only for the sniffer work itself, but also for the specific training method. The sniffer dogs are trained to recognize a number of different substances that can be used, either on their own, or in combination with each other or other substances, to create explosives.

When the dog is ready with its basic explosive recognition training, the dog is certified by passing a test carried out by the authorized national authorities – usually by Police, Border Guards or any other Law Enforcement Units.

This certification test is committed periodically according to the agreed protocol. The dog trainers have their own kits with different explosive materials that are being used during internal training-sessions; only frequent trainings (announced and unannounced) will keep the teams in good shape.

4.4 CBRN-E detection methods

CBRN are weaponized or non-weaponized Chemical, Biological, Radiological and Nuclear materials – including Explosives - that can cause great harm and pose significant threats in the hands of terrorists. Weaponized materials can be delivered using conventional bombs (e.g., pipe bombs), improved explosive materials (e.g., fuel oil-fertilizer mixture) and enhanced blast weapons (e.g., dirty bombs). Non-weaponized materials are traditionally referred to as Dangerous Goods (DG) or Hazardous Materials (HAZMAT) and can include contaminated food, livestock and crops.

IS and associated extremist groups have a wide variety of potential agents and delivery means to choose from for chemical, biological, radiological, or nuclear (CBRN) attacks. (Sweet, 2008)

End goal is the use of CBRN to cause mass casualties, however, most attacks by the group—and especially by associated extremists—probably will be small scale, incorporating relatively crude delivery means and easily produced or obtained chemicals, toxins, or radiological substances.

CBRN weapons represent significant challenges for aviation security professionals who have over the decades grown used to hijacking and conventional explosives. Airports infrastructure become an ideal target. Current security checkpoints do not effectively screen for chemicals, or bio-weapons, so terrorists can gain ready access to aircraft. Therefore high concentration of people would allow rapid dissemination of pathogens nationally and internationally.

Given the difficulty in obtaining nuclear devices and the extent of damage, an airport would likely be a secondary target compared to a political or financial center.

However, airports authorities could provide terrorists with aerial access to better reach these more attractive targets. More likely is the detonation of a radiological dispersal device (RDD) in an airport. This would cause fatalities and a damage radius comparable to a conventional explosive device, along with variable radiation effects. To face such challenges, detection systems form an increasingly vital part of the overall countermeasures strategy.

In relation to CBRN threats still the manual collecting samples using swabs (similar to explosives detection at checkpoints) is likely to be the best sampling approach. The crucial elements are that detection must be rapidly linked to an isolation/ detoxification response in order to ensure that the detection procedure does not cause involuntary or intended release of toxins.

In relation to disease and risk symptoms detection and identification possible solutions are infrared cameras to assess whether particular individuals are sick. This approach was used extensively and effectively at Asian airports during the SARS epidemic. Detection methods are being investigated that involve measuring variation in output of sweat and other physiological changes that indicate a passenger may be incubating a serious or rare infectious disease (Siadkowski, 2014).

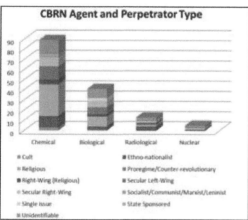

Figure 5. CBRN Events and Agent Type, 1990-2013 (Bale &others, 2005, Ackerman & others, 2014)

The potential for CBRN attacks provides a broad and substantial challenge to aviation security professionals. Although highly unpredictable in their effects, tools such as those presented here are needed for prioritizing threats and responses to assure that our limited resources are focused on the most critical threats.

Strengthening intelligence gathering, focused detection technology deployment, and improved training coupled with rapid response and containment following an incident will form the best short-term strategies for countering a long-established and worrying threat that may increase in its potential to do harm on a mass scale.

5 CONCLUSIONS

The airports are a very demanding environment: seasonal traffic, fluctuating passenger volumes and last minute changes mean there is a lot of flexibility required in order to meet specific needs of airport authorities and their clients – passengers (Olszewski, 2011).

Especially the providing and planning of certified security agents is of paramount importance. An important factor in ensuring a high level of security is both at technological level, taking into consideration implementation of adequate solutions

It is required implementation of specialised security equipment's and quality, training and experience of the security officers themselves. When screening passengers or luggage, security officers are usually always under pressure. Within seconds they have to perform specified tasks designed to determine whether the passenger or luggage entering the restricted area may be a threat or not.

Experience counts in this environment and the job is not for everyone, therefore applied human resources needs to be recruited and trained within staff that will enjoy the challenges and can handle them. Regular scheduled training for all security officers is designed to keep their skills sharp and up to date on changing technology, new regulations and changing threats.

Additional element within every airport security system is to works pro-actively with the industry organizations, governments, airports and airlines on a number of levels to create a high quality sustainable security service.

REFERENCES

Anti-terror police cut back on stopping and searching passengers at airports and ports amid fears of racial profiling - despite severe terror threat, http://www.dailymail.co.uk/news/article-3803676/Anti-terror-police-cut-airport-stop-searches.html, 23 September 2016.

Ackerman Gary, Binder Markus, Iarrocci Emily, A global picture of non-state actors and cbrn, https://warontherocks.com/2014/08/a-global-picture-of-non-state-actors-and-cbrn/, August 13, 2014.

Bale Jeffrey M. and Ackerman Gary, Recommendations on the Development of Methodologies and Attributes for Assessing Terrorist Threats of WMD Terrorism, Center for Nonproliferation Studies, Monterey Institute of International Studies, 2005.

Bammer H., Reshaping the European Air Transport System, „International Forum on Worldwide Liberalisation on the Aviation Industry", 7.04.1997.

Benny Daniel J., General Aviation Security: Aircraft, Hangars, Fixed-Base Operations, Flight Schools, and Airports. Taylor & Francis Group, 2013.

Case Study Fraport AG, Phillip Kriegbaum. Vice President Security and Quality Management Fraport AG, http://www.slideshare.net/RussellPublishing/airport-security-2013-philipp-kriegbaum, 19 September 2013.

Haliżak E., Lizak W., Łukaszuk L., Śliwka E., Terrorism in contemporary world (Terroryzm w świecie współczesnym). Werset, Warszawa-Pieniężno 2004.

Haaretz braking news, haaretz.com/israel-news/1.743291, Sep 20, 2016.

Kostur-Balcerzak, Security passengers service in air transport (Bezpieczeństwo obsługi pasażerów w transporcie lotniczym), Dęblin: Wydawnictwo Wyższej Szkoły Oficerskiej Sił Powietrznych, 2015.

Olszewski P., Considerations of security civil airport (Uwarunkowania bezpieczeństwa cywilnego portu lotniczego), [w:] „Bezpieczeństwo i Administracja: zeszyty naukowe Wydziału Bezpieczeństwa Narodowego Akademii Obrony Narodowe". R. 2, nr 1/2011.

Price Jeffrey, Forrest Jeffrey, Practical Aviation Security: Predicting and Preventing Future Threats. 27th July 2016.

Siadkowski A., Security & protection of civil aviation (Bezpieczeństwo a ochrona lotnictwa cywilnego: akty bezprawnej ingerencji w lotnictwie cywilnym i metoda ich badania), [w:] Bezpieczeństwo Polski: współczesne wyzwania , 2014.

Siadkowski A., Counteratacking of terrorism threats on civil aviation in Poland (Przeciwdziałanie zagrożeniom terrorystycznym w lotnictwie cywilnym w Polsce), Poznań. Wydawnictwo Wyższej Szkoły Bezpieczeństwa, 2011.

Sweet Kathleen, Aviation and Airport Security: Terrorism and Safety Concerns, Second Edition, CRC Press December 23, 2008.

Tackling Chemical, Biological, Radiological And Nuclear Terrorism, https://www.unodc.org/unodc/en/terrorism/news-and-events/nuclear-terrorism.html, 25 March 2014.

Transportation Security Administration, https://www.tsa.gov, 22 December 2016.

Ecological Aspects Associated with an Operation of Aviation Electronic Support Systems

M. Dzunda, D. Cekanova, L. Cobirka, P. Zak & P. Dzurovcin
Technical University of Košice, Kosice, Slovakia

ABSTRACT: The aim of the article is to analyze the impact of electromagnetic radiation of aviation electronic support systems on environmental segments and a human organism. We were looking for effects of electro-magnetic radiation on inhabitants and environment in the vicinity of airport radars. We accomplished measuring and found out the level of radiation harmfulness of electromagnetic radiation sources. In the conclusion we suggest to eliminate the negative impact of electromagnetic radiation on the human organism. At the same time we present the ways how workers, performing their jobs in the vicinity of strong electromagnetic radiation source, can protect themselves in compliance with legislative of the Slovak Republic.

1 INTRODUCTION

Ecology is the science, the branch of biology that studies the relationships between the organisms and the environment and the relationships between living organisms themselves. The definitions of ecology generally express that the subject of studying the ecology is a living matter on different stages of organization, its external environment which consists of open Biosystems [Trnka A., Peterkova P. & Prokop P. 2006]. At the same time, studied objects of ecology may be on the different biological level. Elements of the environment are divided into biotic, abiotic and abiotic-biotic. The biotic elements include living organisms (plants, animals, humans), an abiotic part consists of the inanimate elements (air, water, rocks) and abiotic-biotic component consists of a combination of animate and inanimate components.

Electromagnetic radiation is characterized as a transfer of energy in the form of electromagnetic waves. The individual electromagnetic wave is therefore a locally formed change in the electromagnetic field.

Electromagnetic smog is increasingly being mentioned in respect of electromagnetic radiation to the environment. Electromagnetic smog is produced in a dense tangle of electromagnetic fields, mostly in densely populated areas or near sources of electromagnetic radiation, where several sources simultaneously occur.

Aviation electronic safety technology as a source of electromagnetic radiation is a potential risk [Cekan P., Hovanec M. & Sabo, J. 2016] to the surrounding ecosystem emits continuous radiation waves into the surrounding space.

2 OVERVIEW OF AIR ELECTRONIC SAFETY TECHNOLOGY

Air Traffic Services of the SR are using communication, radio navigation and radar systems to control air traffic. Communication systems are operating at frequencies of 100-150 MHz. The output power of these systems is 5.0 to 20.0 W. The primary surveillance radar is operating at frequencies of 2-4 GHz. Transmitted power of these systems is 14-25 kW. The power of older types of radar is several hundred kW. [Dzunda M., & Csefalvay Z. 2013] Secondary radars operate at frequencies of 1030 MHz and 1090 MHz. Transmission power of a device is about 2 kW.

NDB navigation systems and ADF work on the frequency kHz 200-525. Transmission power of a device is 25-50 W. VOR operates on a frequency MHz 108-112. Transmission power has a 25-100 W. DME measures the distance and works on the frequency 960-1215 MHz. Transmission power of a device is 100 W.

The ILS precision approach:

- Localizer LLZ (device frequencies: 108-112 MHz, transmission power 2W)
- GP Glide path beacon (device frequencies: 328.6 to 335.4 MHz, transmission power 2W)
- VHF marker beacons (frequency 75 MHz, transmission power 3W)

3 MEASUREMENT OF THE ELECTRIC FIELD IN VELKA IDA

Near the village Velka Ida in the place with the coordinates N48° 36'51,9 "E21 ° 08 '47.14" we made measurements of the primary radar electric field (PRL).

The measuring instruments were used as follows:
- Spectrum Analyzer - Advantest R3271A, Serial No .: J001158
- Antenna - EMCO 3115 DOUBLE GUIDE RIDGE HORN, Serial No .: 4974
- Attenuator - 50Ω / 36 db
- Coaxial cable with N connector - 50Ω Coax 3m, type M17 / 75 RG-214
- Generator - 220V / 50Hz: HONDA EM30 PC

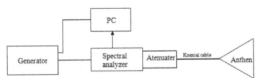

Figure 1. The connection diagram of devices during the measurement

3.1 Measured values

On the site we observed the field strength of six channels, and we reached the following measured values:

Table 1. The values of measurement

Frequency [GHz]	Measured Values [mV]	Intensity [V.m-1]
2,709	214,2	7,98
2,829	807,8	31,37
2,955	69,2	2,81
2,978	4398,0	179,84
3,016	699,5	28,99
3,098	741,0	31,58

The values of the electric field are in the range of 2.81 to 179.84 W / m. The measured values are the peak values of the electric field.

Allowed values of the electric field are assigned in the decree of the Slovak Republic Government No. 329 on the minimum health and safety requirements protecting workers from the risks related to exposure to electromagnetic fields. The action values are expressed in effective values for continuous exposure [Pavolová H. & Tobisová A. 2013]. Allowed values of the exposure to the

electric field for the population are set out in the Decree No. 534 of the Ministry of Health of the Slovak Republic on the details of the requirements for sources of electromagnetic radiation and limits for citizens' exposure to electromagnetic radiation in the environment. The action values are expressed in effective values for continuous exposure.

We evaluated the electric field impact in order to protect workers and citizens in accordance with the Slovak Government regulations No. 329/2007 and the Ministry of Health Decree No. 534/2007.

The value of the intensity of the electromagnetic field for continuous exposure of employees in the frequency range of 2 GHz - 300 GHz is 137 V / m. The specified value for continuous exposure of the population in the same frequency range is 61 V / m.

3.2 Results of measuring

The effective value in the measuring point is equal to the square root of the time average of the square of the field strength E (t) over the period:

$$E_{ef} = Eef = ((1 / T) \int E(t)dt)^{0,5},$$

wherein T = 1/f - is the period of the oscillating parameters, and f is the frequency (s^{-1} = Hz). After substituting into the above equation we can formulate E_{ef} , as follows:

$$E_{ef} = 0{,}707.E_{\text{šp}}$$

After an analysis of how the radar works (PRL) close to Velka Ida, it can be assumed that the pulsed electromagnetic field has a pulse character with a width of 0.9 to 3.1 microseconds and a repetition period of 300-1275 Hz.

The maximum measured value of the electric field in the introduced measuring point is equal to $E_{\text{šp}}$ = 179.84 V/m. The maximum effective value of the electric field at the measuring point is equal to E_{ef} = 127.14 V/m. After comparing the measured effective values of the electric field at all frequencies with the maximum permissible action values for the exposure of employees, we have found out that the measured values do not exceed the maximum permitted levels.

If there is a simultaneous exposure to fields from several sources with different frequencies, the thermal effect applying at frequencies exceeding 100 MHz shall not exceed the action value if the followed inequality is valid:

$$\sum_{100kHz}^{1MHz} \left(E_i / c \right)^2 + \sum_{f<1MHz}^{300GHz} \left(E_i / E_{L,i} \right)^2 \le 1$$

where E_i denotes the electric field with a frequency of "i", $E_{L,i}$ is action value of the electric field for the i-th frequency, c = $610.10^6/f$, V/m. When measuring we found out the presence of an electric field with a frequency greater than 1.0 MHz, so the first member of an inequality on the left side equals zero. By substituting we obtain the inequality of 0.93 <1.0. By calculation, we have confirmed that the heat of an electric field in the measurement does not exceed the action value.

The maximum measured peak value of the electric field of the $E_{\text{šp}}$ = 179.84 V/m does not exceed 32 times the allowable action field intensity equal to E_L = 61 V/m.

The radiant heat exposure to an electric field, shorter than those determined for centring, or a series of short-term exposures acting during the time shorter than designed for centring doesn't exceed the action value if the exposure time " t_i "and the measured levels of fields Ei in the range from 100 kHz to 10 GHz meet the inequality:

$$\sum\left(E_i^2 t_i\right) \le \left(6 E_{L,i}^{\ 2}\right)$$

where t_i is the time of the i-th exposure expressed in minutes. For i-th exposure during the 6 minute interval, we determined the parameters of the transmitted signal PRL. The width of the transmitted pulse PRL is 3,1 µs, repetition period of 1275 Hz. Over a period of 6 minutes t_i will be equal to 0.0237. After substituting into the relationship (4) we have received the inequality 417.40 <22326.0.

The inequality is met. We hereby confirm that the radiant heat exposure to an electric field does not exceed the action values for the field.

In the range of frequencies from 100 kHz to 300 GHz, we assessed the effect of field temperature with regard to identified action levels during the exposure to an electric field and magnetic field of the same frequency or different frequencies (according to the Decree 534/2007 Z.z.). Substituting the parameters into a relationship, we get the inequality:

2.5. 10^{-3}<1

Radiant heat at the point of measurement does not exceed the action value because the above mentioned inequality has been met.

4 CONCLUSION

The measurement results obtained by the analysis demonstrate that the effect of the electric field in the vicinity of Velka Ida, which is in the range of 2.709 to 3.098 GHz, does not exceed the action values for exposure of workers and the citizens. Even though, it is appropriate to be protected from the high-frequency radiation [Cekan P., Korba P. & Sabo J. 2014].

The basic types of protection from non-ionizing electromagnetic radiation:
- distance,
- time,
- shielding of the workplace,
- protective work equipment
   Protection methods:
- Protection by distance
   The workplace close to the electromagnetic radiation should be kept as far away from this resource as possible. The workplace should not miss signalling and warning that the one is close to the electromagnetic field.
- Protection time
   Protection time means the exposure of workers to electromagnetic fields only for a certain period of time, to avoid possible serious consequences of prolonged exposure close to the source field. Likewise, in the case of mobile phones, one of the recommendations to minimize the consequences of the radiation is to shorten the talk time to a minimum.
- Protective shielding of the workplace
   Workplace protection against electromagnetic radiation is performed in a similar manner as the shielding for radiation. Protective sheets, foils and network in this case are installed in walls of a particular job. There are other ways of protecting the space from radiation such as different coatings and spraying with metal particles, window films, curtains and blinds containing metals.
- Personal protective equipment
   Personal protective equipment for workers is used in cases where there is perhaps no other protection of the above. Most often it is a complete suite for the protection of the entire body (suit, helmet, gloves and boots). They may also be independent parts of clothing for individual parts of the human body. They must be made of materials that do not prevent worker's free movement.

REFERENCES

Cekan P., Korba P. & Sabo J. 2014. Human Factor in Aviaion - Models Eliminating Errors. Kaunas Univ Technol Press Conference. Lithuania, pp. 464-467.

Cekan P., Hovanec M. & Sabo, J. 2016. Human factor in conversation between subordinates and managers. Nase More 63 (3), pp. 241-243.

Trnka A., Peterkova P. & Prokop P. 2006. Ekologia. Trnavska univerzita, Trnava, pp. 81, ISBN 800-8082-002-3.

Pavolova H. & Tobisova A. 2013. The model of supplier quaity management in a transport company. Nase More 60 (5-6), pp. 123-126.

Dzunda M., & Csefalvay Z. 2013. Selected methods of ultra-wide radar signal processing. In A. Weintrit (ed): Advances in Marine Navigation - Marine Navigation and Safety of Sea Transportation, CRC Press/ Balkema, Taylor & Francis Group, London, pp. 239-242.

AUTHOR INDEX

Printed and bound by CPI Group (UK) Ltd, Croydon, CR0 4YY

24/10/2024

01778293-0010